CHELATING AGENTS AND METAL CHELATES

CHELATING AGENTS AND METAL CHELATES

EDITED BY

F. P. DWYER

AUSTRALIAN NATIONAL UNIVERSITY, CANBERRA, A.C.T., AUSTRALIA

D. P. MELLOR

UNIVERSITY OF NEW SOUTH WALES, KENSINGTON, N.S.W., AUSTRALIA

ACADEMIC PRESS New York and London 1964

ACADEMIC PRESS INC.
111 Fifth Avenue, New York 3, New York

United Kingdom Edition published by
ACADEMIC PRESS INC. (LONDON) LTD.
Berkeley Square House, London W.1

LIBRARY OF CONGRESS CATALOG CARD NUMBER: 63-16969

PRINTED IN THE UNITED STATES OF AMERICA

To G. J. Burrows
able scientist and inspiring teacher

List of Contributors

Numbers in parentheses refer to the page on which the author's contribution begins.

D. A. Buckingham (237), *Biological Inorganic Chemistry Section, The John Curtin School of Medical Research, Australian National University, Canberra, A.C.T., Australia**

D. P. Craig (51), *Department of Chemistry, University College, London, England*

F. P. Dwyer (335 and 383), *Biological Inorganic Chemistry Section, The John Curtin School of Medical Research, Australian National University, Canberra, A.C.T., Australia‡*

J. E. Falk (441), *Division of Plant Industry, C.S.I.R.O., Canberra, A.C.T., Australia*

F. L. Garvan (283), *Biological Inorganic Chemistry Section, The John Curtin School of Medical Research, Australian National University, Canberra, A.C.T., Australia†*

Harold A. Goodwin (143), *School of Chemistry, The University of New South Wales, Kensington, N.S.W., Australia*

Clive M. Harris (95), *School of Chemistry, The University of New South Wales, Kensington, N.S.W., Australia*

Stanley E. Livingstone (95), *School of Chemistry, The University of New South Wales, Kensington, N.S.W., Australia*

D. P. Mellor (1), *School of Chemistry, The University of New South Wales, Kensington, N.S.W., Australia*

R. S. Nyholm (51), *Department of Chemistry, University College, London, England*

J. N. Phillips (441), *Division of Plant Industry, C.S.I.R.O., Canberra, A.C.T., Australia*

A. M. Sargeson (183 and 237), *Biological Inorganic Chemistry Section, The John Curtin School of Medical Research, Australian National University, Canberra, A.C.T., Australia*

A. Shulman (383), *Department of Physiology, University of Melbourne, Melbourne, Victoria, Australia*

* Present Address: Department of Chemistry, Brown University, Providence, Rhode Island

† Present Address: Christian Brothers' Training College, Strathfield, N.S.W., Australia

‡ Deceased

Preface

The aim of this book is primarily to provide a reference work for senior students and research workers in the chemistry of metal chelates. It is, however, hoped that it will also be of interest to biologists and medical scientists whose growing preoccupation with metal-binding is clearly evident in the published proceedings of recent conferences.*

The emphasis in the book has been placed mainly on the structure and properties of metal chelates. An attempt has been made to cover in some detail aspects of the subject that have not been so treated previously. Some overlapping with earlier works is inevitable but it is hoped that there is no more than is essential for an understanding of the new material.

The chemistry of metal chelates is at present undergoing a period of rapid development and is engaging the attention of research workers in many diverse disciplines, both experimental and theoretical. Progress in this area of chemistry has received an added impetus because of its many applications to biology. It has been known for some time that metal chelates play essential roles in the chemistry of living matter: chlorophylls in photosynthesis, vitamin B_{12} in the formation of red cells and hemoproteins in respiration. While the structures and functions of these metal chelates are reasonably well known, there is a large number of metallo-proteins and other metal complexes of biological importance, whose structure and properties are much less completely understood. One of the most fruitful approaches to understanding the structures and properties of such complicated substances is almost certainly through a thorough knowledge of the structure and properties of simpler and possibly related substances.

The greater part of this book is devoted to relatively simple structures. It is only in the last few chapters that some of the biological aspects of metal chelation are considered.

February, 1964 D. P. Mellor

* "Biological Aspects of Metal Binding" Federation Proc. 20, Nr. 3 (1960); "Metal Binding in Medicine." Lippincott, Philadelphia.

Francis P. J. Dwyer (1910–1962)

With the passing of Francis Dwyer, Australia lost one of her outstanding chemists. He and his colleagues and students maintained for many years a steady stream of contributions to the chemistry of metal complexes.

Dwyer was one of a group of chemists that came into being at the University of Sydney in the early thirties under the guidance of G. J. Burrows, to whom this book is dedicated. Every contributor to this book is either a one-time colleague or student of Burrows or a student of a student. Since those early days, interest in metal complexes has spread to other Australian universities including the National University at Canberra. It was here that Dwyer occupied the Chair of Biological Inorganic Chemistry and developed a lively center of research. His interest in the biological importance of metal complexes developed rather late in his career and only after he had gained considerable insight into the structure and properties of the complexes themselves. One of his most notable experimental skills was the ability to resolve metal complexes into their optical isomers. He was, of course, not so much interested in the phenomenon of optical activity itself as in the light it could throw on the structure of metal complexes. His last published paper described a novel method for the resolution of amino acids. It will be evident in the following pages that many of his most important contributions to our knowledge of metal complexes centered around the phenomenon of optical isomerism.

Dwyer conceived the idea and plan of this book and up to the time of his death in June 1962, he was an active co-editor. The last chapter embodies results, unpublished at the time of his death, of work carried out by Dwyer and his colleagues in the departments of physiology, biochemistry, and biology in the Universities of Sydney, Melbourne, and Adelaide.

D.P.M.

Contents

5. OPTICAL PHENOMENA IN METAL CHELATES

A. M. SARGESON

6. OXIDATION-REDUCTION POTENTIALS AS FUNCTIONS OF DONOR ATOM AND LIGAND

D. A. BUCKINGHAM AND A. M. SARGESON

7. METAL CHELATES OF ETHYLENEDIAMINETETRA-ACETIC ACID AND RELATED SUBSTANCES

F. L. GARVAN

8. ENZYME–METAL ION ACTIVATION AND CATALYTIC PHENOMENA WITH METAL COMPLEXES

F. P. DWYER

9. METAL CHELATES IN BIOLOGICAL SYSTEMS

A. SHULMAN AND F. P. DWYER

10. PHYSICAL AND COORDINATION CHEMISTRY OF THE TETRAPYRROLE PIGMENTS

J. E. FALK AND J. N. PHILLIPS

CHAPTER 1

Historical Background and Fundamental Concepts

D. P. Mellor

School of Chemistry, University of New South Wales, Kensington, N.S.W., Australia

I. Introduction

A. Definitions

A complex has been defined as "a species formed by the association of two or more simpler species each capable of independent existence" (Rossotti and Rossotti, 1961). When one of the simpler species is a metal ion, the resulting entity is known as a metal complex. A characteristic feature of such a complex is that the metal atom occupies a central position in it, as exemplified by cobalt in hexamminecobalt(III) ion, platinum in tetrachloroplatinate(II) ion, and copper in bis(glycinato)copper(II). These are shown in (I), (II), and (III) respectively.

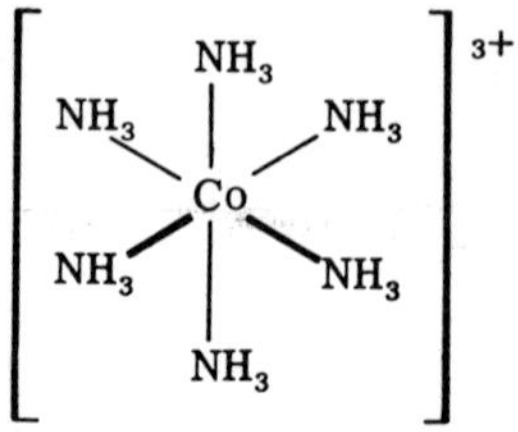

(I)

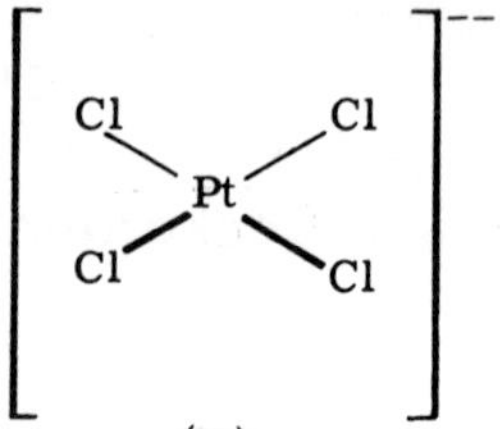

(II)

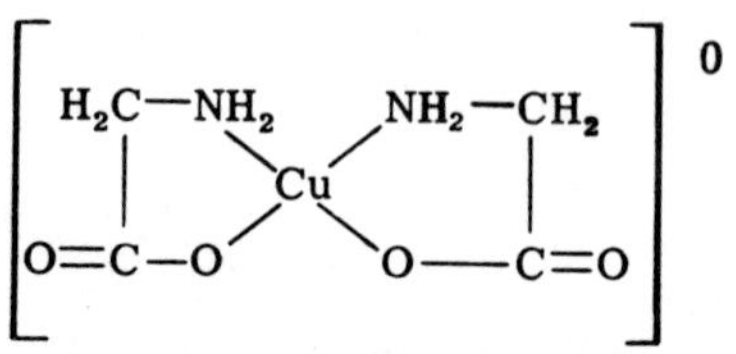

(III)

The metal-centered structure may carry a positive, negative, or zero charge. Complex ions are, of course, always associated with ions of opposite charge and the term metal complex is customarily applied to the compound itself.

Almost every kind of metal atom can serve as a central atom although some kinds do so more readily than others. Atoms of the transition series, for example, function in this way par excellence but atoms of the alkali metals are rarely found in this role.

When the central metal atom of a complex is bound to its immediate neighbors by covalent bonds formed as the result of the metal atom accepting an electron pair from each nonmetal atom, it is customary to call the latter the donor and the former the acceptor atom; alternatively, the nonmetal atom is called the coordinating atom and the bond between it and the metal atom a coordinate bond. However, metal atoms themselves sometimes contribute electrons to the bond as they do in π bonding. A more generally used convention which is preferable because it avoids any implication about the nature of the bond is to call any negative ion or polar molecule, bound to a metal atom, a ligand (L) and the bond between them a metal-ligand bond (M—L).

Since its introduction about 20 years ago, the term ligand has been used in two different senses. It is sometimes applied to the particular atom in a molecule by means of which the molecule is attached to a central metal atom, for example, the nitrogen atom in ammonia, or it may be applied to the molecule as a whole. This dual usage is well established and causes little or no confusion since the context generally indicates the sense in which it is being employed. Where there is any risk of ambiguity, it may be avoided by using the term ligand atom to denote the atom attached to a metal.

Some ligands are attached to the metal atom by more than one donor atom in such a manner as to form a heterocyclic ring. An example of such a ring is the one formed in (III) by the glycinate ion. This type of ring has been given a special name—chelate ring—and the molecule or ion from which it is formed is known as a chelating agent. The process of forming a chelate ring is known as chelation. The first chelating molecules discovered were those with two donor atoms and it was the caliperlike mode of attachment of the molecules to the metal atom that led Morgan and Drew (1920) to suggest the name *chelate*, which is derived from the Greek word χηλή meaning a lobster's claw.

Any metal complex in which one or more chelate rings are present is defined as a metal chelate[1] and it is with these and the molecules and ions that form such rings that this book is concerned.

It should be pointed out that not every author uses the term metal complex in the all-embracing sense indicated above. Some prefer to confine it to structures like (I) and (II), in which no chelate rings are present. On the other hand, use of the term metal complex as a synonym for metal chelate is fairly widespread. The essential point of the terminology used here is that a metal chelate is regarded as a special kind of metal complex. A minor disadvantage of this terminology is that there is no distinctive name for the class of complex exemplified by (I) and (II), though no confusion is likely to arise from calling them metal complexes.

In accordance with the definition given above, complexes like diaquo-8-hydroxyquinolinatocopper(II) ion (IV) and *trans*-dichlorobis(ethylenediamine)cobalt(III) ion (V) which contain ligands that are chelates as well as some that are not will be called metal chelates.

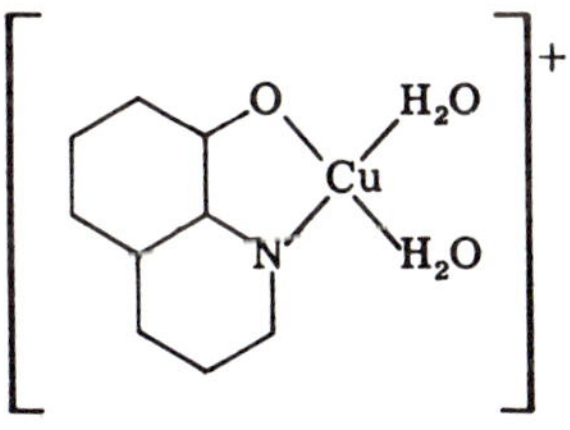

(IV)

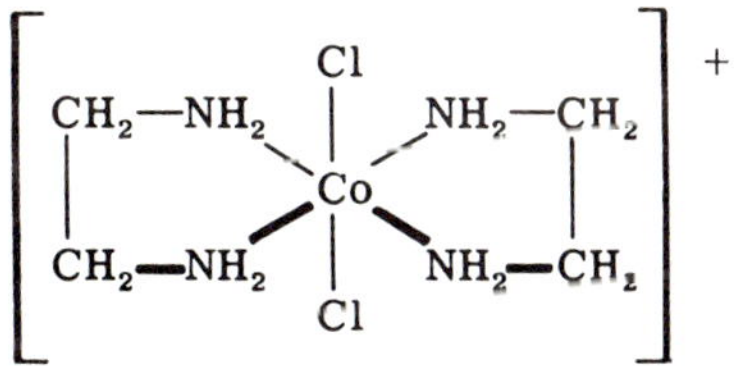

(V)

[1] When Morgan first introduced the word *chelate*, he used it as an adjective. Nowadays, it plays a more versatile role appearing as an adjective, verb, or noun. In this book, it will be used in all three ways, though the adjectival form will, as often as not, be the word *chelating* as used in the title of the book.

There is a good case for having a special name for metal chelates like (III) which are nonelectrolytes, since these compounds usually have characteristics that set them apart from metal chelate ions like (IV) and (V). Electrically neutral metal chelates were originally called "inner complex metallic salts," a name generally attributed to Ley (1904) who first used it in the title of a paper dealing with bis(glycinato)copper(II). Strangely enough, in this paper, he gave no explanation of the significance of the term "inner." It is presumed, however, that it originates from Werner's custom of referring to inner and outer spheres of coordination when describing a compound like $[Co(NH_3)_6]Cl_3$. Ammonia molecules belong to the inner sphere, ionized chlorine atoms to the outer sphere. An inner complex is therefore one that has no outer sphere; in other words, it is an electrically neutral metal complex or a nonelectrolyte. If this presumption is correct, it would be inconsistent to extend the name inner complex to include metal chelate ions like (IV) and (V) as advocated by some authors (Liebhafsky, 1946; West, 1960). On the grounds that there may be some confusion with the term inner orbital complex, some authors prefer to use the term neutral rather than inner complex. An objection may be raised to the use of the term neutral in this manner. Since neither term has any advantage over the other, they will be used interchangeably.

B. Nomenclature

The Commission on the Nomenclature of Inorganic Compounds set up by the International Union of Pure and Applied Chemistry (1957) has drawn up rules for the naming of metal complexes. Only a brief summary of the principal relevant rules will be given here.

When writing the formula of a metal complex the symbol for the central metal atom is placed first, and is followed by formulas of anionic and neutral ligands in that order. The formula of the whole complex entity (molecule or ion) is enclosed in square brackets.

However, in naming a metal complex, the opposite order is used. Ligands are cited first—anionic and neutral in that order—and then the metal, followed by a Roman numeral indicating its oxidation state. Thus (III) is written $[Cu(NH_2 \cdot CH_2 \cdot COO)_2]$ and named bis(glycinato)copper(II).

Some ligands are so constituted that there are alternative ways in which they may be attached to a metal atom. The dithiooxalato ion, for example, may be attached through either S and O or S or O. It is, in fact, attached through S and this is indicated by adding to the name the symbols *S,S′*.

When the central metal atom forms part of an anion, as in $K_2[Ni(C_2S_2O_2)_2]$, potassium bis(dithiooxalato-*S,S′*)nickelate(II), the ending *-ate* is used with the name of the metal. For a more detailed discussion of nomenclature, the reader should consult the Report of the Commission on the Nomenclature of Inorganic Compounds (1957).

II. Historical Background

A. Earliest Attempts to Discover the Structure of Metal Complexes

In order to understand how the phenomenon of chelation was discovered, it is necessary to glance briefly at the history of the early attempts to elucidate the structure of metal complexes. Since these attempts were made soon after the birth of structural organic chemistry, it is not surprising that the first formulations of the structure of metal complexes (metal ammines in particular) should have been in terms of chains and rings that had proved such common features in the structure of organic molecules. Thus Blomstrand, in 1869, proposed a structure for hexamminecobalt(II) chloride in which ammonia molecules were linked together in chains reminiscent of hydrocarbon chains:

$$Cl—NH_3—NH_3—NH_3—Co—NH_3—NH_3—NH_3—Cl$$

Copper(II) sulfate pentahydrate, on the other hand, was formulated as possessing a ring structure (VI).

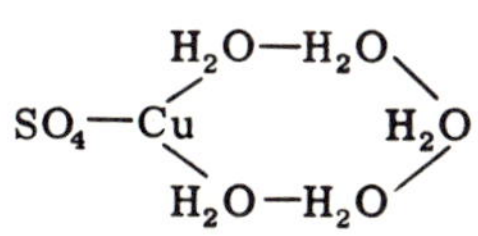

(VI)

One of the most notable exponents of the chain theory was Jorgensen (1887). Though he was able to modify and extend it to account for some of the chemical behavior and isomerism of metal ammines, the theory eventually broke under the weight of facts that he himself accumulated in the course of his extensive investigations. In his attempts to vindicate the chain theory Jorgensen provided most of the experimental evidence for the revolutionary structural theory introduced by Werner (1893).

About this time, many chemists regarded metal ammines as molecular compounds. For example, they wrote the formula for hexamminecobalt(III) chloride as: $CoCl_3 \cdot 6NH_3$. Their view was that in this compound cobalt has a valency of three as also has nitrogen. The question that greatly puzzled them was: by what mechanism did molecules in which the valencies of the atoms were, so it seemed, fully satisfied combine together to form such a highly stable compound as hexamminecobalt(III) chloride?

Werner was aware that certain structural entities—"complexes," he called them—persisted[2] through a series of chemical transformations; that the entity $[Co(NH_3)_6]^{3+}$, for example, remained intact throughout the following changes:

$$[Co(NH_3)_6]Cl_3 \xrightarrow{AgNO_3} [Co)NH_3)_6](NO_3)_3 \xrightarrow{H_2SO_4} [Co(NH_3)_6]_2(SO_4)_3$$

Having recognized the existence of this entity, his next step was to abandon the view that the valency of an atom and the number of bonds it can form are one and the same thing: that trivalent cobalt, for example, could only form three bonds. He introduced the idea of a metal-centered, octahedral structure. "If," he wrote, "we think of the metal atom as the center of the whole system, then we can most simply place the molecules bound to it at the corners of an octahedron" (Werner, 1893). This involved a complete break with contemporary ideas of valency and led him to the realization that it was necessary, as a modern chemist has put it, to "dissociate the concept of valency into several new concepts" (Pauling, 1948). This Werner did by distinguishing between what he called principal valency and auxiliary or "neben" valency. For cobalt in the hexamminecobalt(III) complex, these are three and six; for platinum in the tetrachloroplatinate(II) ion, they are two and four respectively. These concepts are, of course, still used though they go under other names—oxidation number and covalence—and are interpreted in terms of the electronic structure of atoms.

B. Substitution in Metal Complexes

Several years before Werner published his revolutionary ideas on the structure of metal complexes, Arrhenius (1887) had announced the theory of electrolytic dissociation. Werner was quick to see how measurements of electrolytic conductance could be used to support his views.

He was able to show, for example, that the result of the progressive substitution of ammonia molecules in the tetrammineplatinum(II) ion by chlorine ion is to reduce the charge on the complex ion by one for each substitution (Werner and Miolati, 1894).

By comparing the molecular conductivities of the above metal complexes with the molecular conductivities of simple salts like $BaCl_2$ and KCl, Werner reached the conclusion that the complexes yielded 3, 2, and 0 ions, respectively, when dissolved in water (Fig. 1). He introduced the convention of square brackets to indicate the entity which functioned as a complex. Fundamental to all Werner's structural thinking was the concept of coordination number which he defined as the number of atoms bound

[2] Persistence of structure was first recognized by von Liebig and Wohler (1832) in the course of their studies of the benzoyl radical. Werner recognized that the principle was also valid for inorganic compounds.

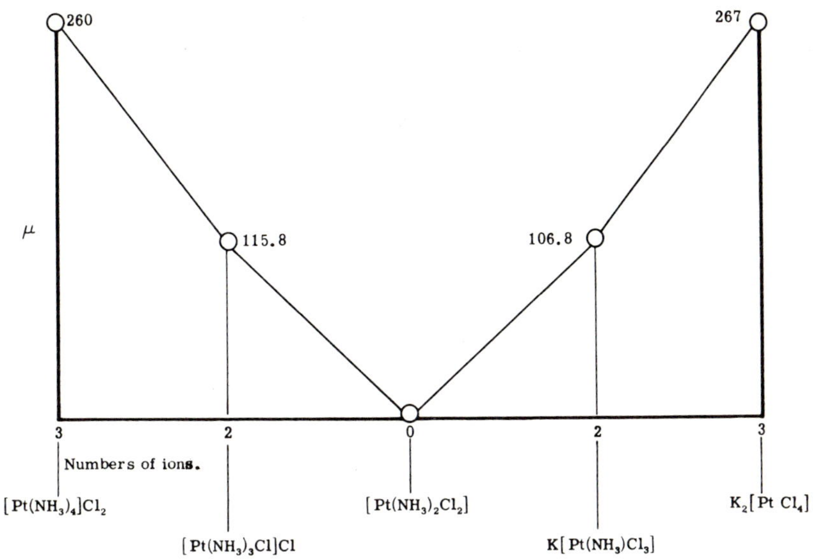

FIG. 1. Molecular conductivities (μ) of aqueous solutions (10^{-3} M) of platinum(II) ammines.

directly to the central metal atom. Throughout the transformations shown above, the number of atoms directly bound to platinum remains constant. It is, in fact, characteristic of platinum in the bivalent state to have a coordination number of four. A far more common coordination number is six which is often found to be independent of the principal valence of the metal. To illustrate this point, Werner cited as examples:

$$[Pt(NH_3)_6]Cl_4 \qquad [Co(NH_3)_6]Cl_3 \qquad [Ni(NH_3)_6]Cl_2$$

where the principal valences of the metal are 4, 3, and 2, respectively.

C. Discovery of Chelation

Two distinct steps were involved in this discovery. The first was the recognition of the fact that each metal has a characteristic coordination number which it exhibits in most, if not all, its compounds. For bivalent platinum, this number, as already stated, is four. Werner's insight into molecular structure led him unerringly to the concept of ring formation which he exemplified by discussing the structure of the compound $[Pt(C_2H_4(NH_2)_2)_2]Cl_2$. It was clear to him that in this compound two molecules of ethylenediamine had replaced the four molecules of ammonia in $[Pt(NH_3)_4]Cl_2$. He concluded that each molecule of ethylenediamine occupied two of the four coordination positions and that, in so doing, it formed a five-membered heterocyclic ring. It has been stated that "the idea of the ring structure in ethylenediamine complexes runs subconsciously through the early papers of Werner without being definitely expressed" (Diehl, 1937). In his classical paper of 1893, Werner was quite explicit about the structure of this ethylenediamine compound of platinum representing it as shown in (VII).

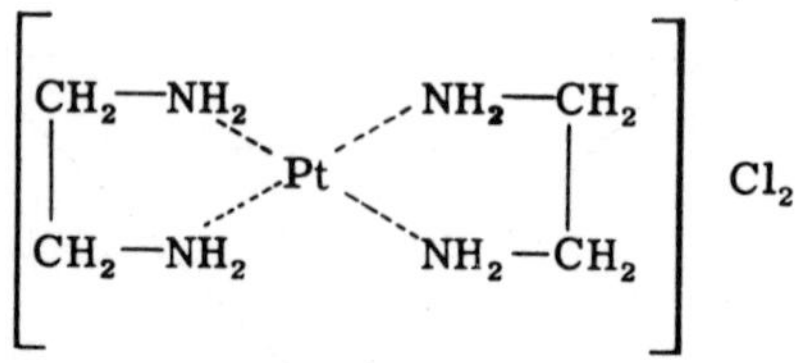

(VII)

At this point he made no mention of the stereochemistry of the platinum, but later in the same paper he put forward the view that the four bonds of platinum(II) are coplanar, that is, the metal is square-coordinated. He based this conclusion on the observation that dichlorodiammineplatinum(II) exists in two forms which he believed to be *cis* and *trans* geometrical isomers (VIIIa and b).

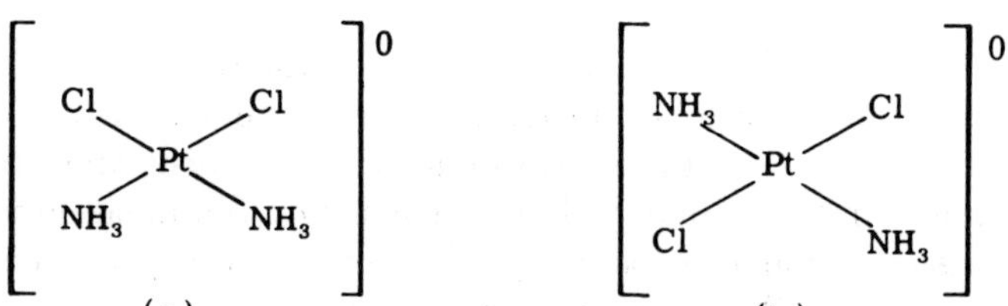

(a) (VIII) (b)

In the same paper, he also discussed the structure of the compound $[Co(en)_2X_2]X$ (en = ethylenediamine) which Jorgensen (1890) had prepared in two isomeric forms. Werner regarded the isomerism as evidence for the octahedral disposition of the six bonds about the cobalt atom and thus clearly indicated that he thought of ethylenediamine as a chelating agent.

D. Some of the First Chelating Molecules to Be Studied

The number and variety of molecules recognized as having the ability to chelate grew so rapidly after the discovery of the phenomenon that, in this brief introduction, it is possible to mention only a few of the most important ones.

Oxalic acid was probably the next molecule after ethylenediamine to be recognized as a chelating agent. Werner and Vilmos (1899) described the compound $[Co(en)_2C_2O_4]Cl$ which they prepared by treating $[Co(en)_2Cl_2]Cl$ with an alkali oxalate. They regarded the oxalate group as occupying the two (*cis*) coordination positions previously occupied by the chlorine atoms.

In their behavior as chelating molecules, there is an important difference between ethylenediamine and oxalic acid. The former is bound as such to a metal atom whereas oxalic acid loses two protons in the process. Many compounds function as chelating agents by the loss of one or more protons. Although it is the anion that combines with the metal atom, it is nevertheless customary to call the parent molecule a chelating agent.

Ever since their chelating properties were first discovered, ethylenediamine and oxalic acid have been widely used in studying the structure of metal complexes. They are among the simplest of all chelating molecules and the literature abounds with references to many hundreds of metal chelates derived from them.

The chelating ability of more complicated molecules was soon recognized. In 1901, Werner described a new type of compound which he prepared by treating potassium tetrachloroplatinate(II) with acetylacetone. The result of replacing the four chlorine ions of $[PtCl_4]^{--}$ with two acetylacetonate ions is a complex (IX) carrying no electrical charge—the first of its kind to be discovered. In the process of chelation each acetylacetone molecule loses a proton.

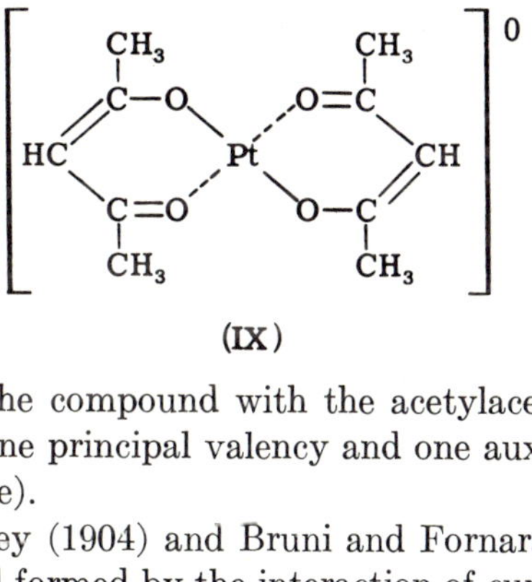

(IX)

Werner formulated the compound with the acetylacetonate ion attached to the platinum by one principal valency and one auxiliary valency (indicated by a dotted line).

Not long after, Ley (1904) and Bruni and Fornara (1904) studied the interesting compound formed by the interaction of cupric ion with glycine. That some interaction had taken place was inferred from the marked color change that followed the mixing of aqueous solutions of these substances. The electrical conductivity of the resulting solution was unusually low and the normal reactions of cupric ion were markedly affected. Analysis of the solid isolated from the solution showed a ratio of two molecules of glycine to one atom of copper. On the basis of these observations they attributed to the compound the structure shown in (III).

To this rather special and important type of complex, Ley, as already mentioned, gave the name "inner metallic complex salt." Oddly enough, this compound, which gave rise to the new name, is not typical of its class insofar as it is quite soluble in water, whereas the great majority of inner complexes are notable for their low solubility in that solvent. More nearly typical inner complexes are the metal glyoximes first prepared by Tschugaeff (1906).

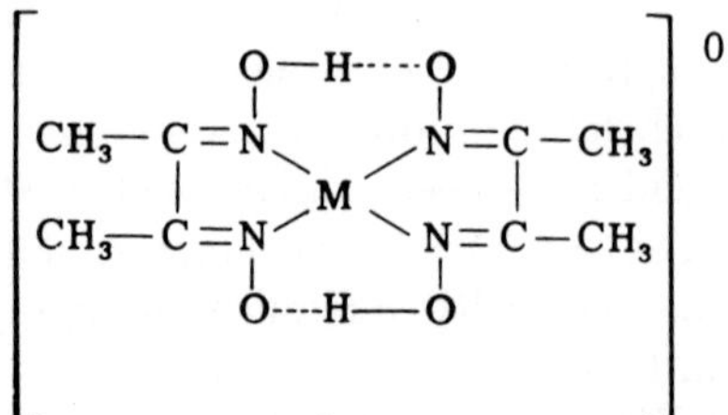

(X)

The best known metal chelate of this type (X) is, bis(dimethylglyoximato)nickel(II),[3] a compound frequently met in analytical work. Also important in analytical work and structural studies, but quite different in character, are the metal complexes of 2,2′-bipyridine and 1,10-phenanthroline. The iron complexes of these two chelating agents were first investigated by Blau in 1898; since then numerous metal chelates derived from these molecules have been described.

III. Fundamental Concepts

Metal chelates may be examined from at least three points of view. Attention may be focused on:

(1) the central metal atom,
(2) the chelating molecule,
(3) the nature of the bonds linking (1) and (2), and the influence of each and all of these on the behavior of the metal chelate as a whole.

A. The Central Metal Atom

1. *Some General Considerations*

As might be expected, the properties of a metal complex are influenced to a considerable extent by the nature and oxidation state of the central metal atom. One method of studying this influence is to compare the com-

[3] Tschugaeff formulated these compounds with oxygen as the donor atom and consequently with seven-membered chelate rings. All the evidence now available shows that chelation takes place through the nitrogen atoms and that nickel is square-coordinated.

pounds formed by a series of different metal atoms in a given oxidation state with a particular chelating agent. The central metal atom sometimes endows a metal chelate with an individuality which is quite striking.

A good illustration of this is to be found in the 2:1 chelates formed by copper and nickel with dimethylglyoxime. That there should be a marked difference between the absorption of light by these two compounds is not surprising. What is surprising is the fact that, despite the general similarity of the two molecules, the solubility in water of the two compounds is so different. The copper chelate is soluble in water; the nickel chelate is highly insoluble. The origin of the difference in solubility is to be found in subtle differences in the shapes of the two molecules and a consequent difference in the mode of packing of the two molecules in their respective crystals. X-ray analysis shows that in the nickel compound planar molecules are stacked together so that the nickel atoms are collinear and at 3.4 Å apart (Godycki and Rundle, 1953). The energy of the weak bond between the nickel atoms is about 12 kcal./mole^{-1}. In the molecule of the copper compound the chelate rings are not coplanar as they are in the nickel compound (Frasson *et al.*, 1959) but make a small angle (28°) with one another. Molecules of the copper compound are paired in the crystal in such a way that each copper atom has an oxygen atom from its associated molecule as one of its five immediate neighbors. The configuration of the copper atom is tetragonal pyramidal (Fig. 2). Ultimately the difference in the

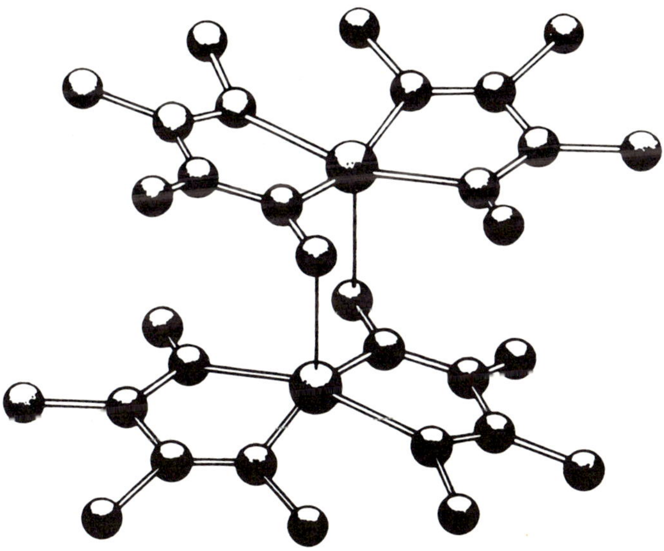

FIG. 2. The dimer of bis(dimethylglyoximato)copper(II).

behavior of the two compounds can be traced back to differences in the size and electronic structure of nickel and copper atoms.

Another example of the individuality of the chelate formed by a particular metal may be found among the chelates formed by bissalicylaldehydeethylenediimine with bivalent metals. Of these chelates (XI), one alone, namely, the cobalt compound, functions as an oxygen carrier.

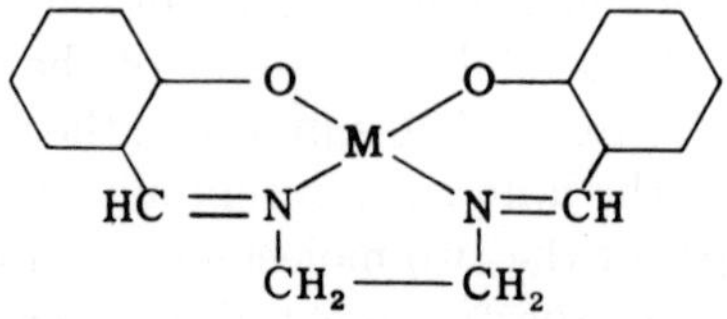

(XI)

Other examples of the highly individualistic nature of chelates are to be found in the living world. From all the wide variety of elements available to plants, magnesium is the metal selected by the processes of evolution as the most suitable for enabling chlorophyll to perform its function in photosynthesis; similarly iron in the hemoproteins and cobalt in vitamin B_{12} and the "cobamide coenzymes" endow their respective chelates with properties that enable them to perform their functions in living organisms. The origin of this high degree of specificity in the metal atoms is by no means fully understood.

2. *Oxidation Number*

The structure and properties of the complexes formed by a metal are influenced by its state of oxidation—in other words by its oxidation number. This number has been defined by Pauling (1948) as follows: "In a covalent compound of known structure the oxidation number of each atom is the charge remaining on the atom when each shared electron pair is assigned completely to the more electronegative of the two atoms sharing it. A pair shared by two atoms of the same element is split between them."

Transition elements often exhibit a wide range of oxidation numbers in their different compounds. Manganese can exist in no less than nine oxidation states, iron in four, and nickel in five. Table I shows some of the oxidation states that have been found for metals of the first transition series.

The oxidation number of an element is, of course, only another name for the classical or stoichiometric valence of an element—Werner's principal valence.

In some complexes, it is difficult to decide what the oxidation number of the metal atom is. This difficulty occasionally arises from the fact that a ligand appears to be capable of functioning in different ways. A good

TABLE I

OXIDATION STATES OF ELEMENTS IN THE FIRST TRANSITION SERIES

Sc	Ti	V	Cr	Mn	Fe	Co	Ni	Cu
—	—	−1	—	−1	—	−1	—	—
—	0	0	0	0	0	0	0	0
—	—	1	1	1	—	1	1	1
—	2	2	2	2	2	2	2	2
3	3	3	3	3	3	3	3	3
	4	4	5	4	4	4	4	
		5	5	5				
			6	6				
				7				

example of such a ligand is nitric oxide. By yielding an electron to a metal atom, it may enter the complex as the nitrosonium ion (NO^+); by removing an electron from a metal atom, it may enter as the nitrosyl ion (NO^-).

$$:N{\equiv}O:^{+} \xleftarrow{-e} :N\overset{\cdot}{=}O: \xrightarrow{+e} :N{=}\underset{\cdot\cdot}{O}:^{-}$$

This duality of function is the probable explanation of the existence of the two series of nitrosyl pentammine cobalt complexes $[Co(NH_3)_5NO]X_2$. One series is black, the other pink. The difficulty of deciding the oxidation number of cobalt may be illustrated by considering the black compound. Crystal structure analysis of the black chloride (Hodgkin and Dale, 1962) reveals that the complex ion has the form and dimensions shown below (XII).

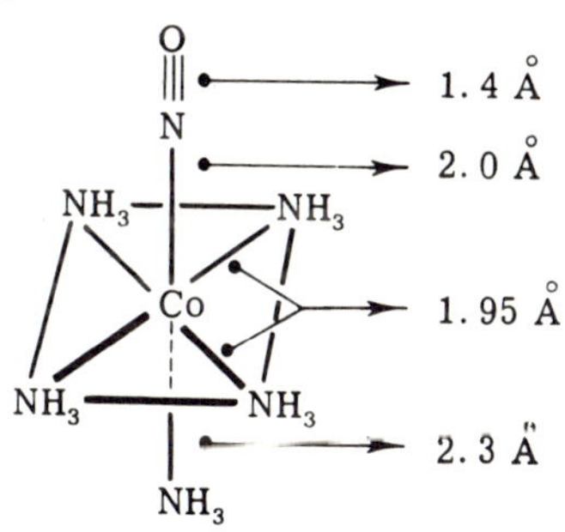

(XII)

A most significant feature of the structure is the linear group of atoms Co—N—O. The diamagnetism of the black series is consistent with an assignment of an oxidation number +1 or +3 to the cobalt atom but it does not enable a distinction to be drawn between the two possibilities. Its unusual color suggests that the black chloride is not a normal member of the pentammine series—$[Co(NH_3)X]^{++}$. That the three atoms Co, N, and O are collinear makes it reasonable to suppose that the structure of this part of the complex is Co—N≡O. In displacing an ammonia molecule from $[Co(NH_3)_6]^{++}$ nitric oxide reduces Co(II) to Co(I) and enters the complex as NO^+.

In their chemical and physical properties, the pink compounds closely resemble the normal members of the pentammine $[Co(NH_3)_5X]^{++}$. This suggests that in the formation of the pink compounds nitric oxide oxidizes Co(II) to Co(III) and in consequence enters the complex as NO^- to form a nonlinear structure:

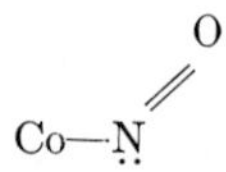

The crystal structure of a member of the pink series has yet to be determined.

The bent structure has been found in other metal complexes such as $[Ru(NH_3)_4(NO)(OH)]Cl_2$ (Parpiev and Porai-Koshits, 1959). It seems highly probable that the isomerism of the black and pink series of cobalt complexes arises from what might be termed "internal oxidation-reduction." In the black series, the oxidation states of cobalt and nitrogen are +1 and +3 respectively; in the pink series, they are +3 and +1.

That the isomerism arises from the different modes of attachment of nitric oxide to the metal—through nitrogen in one form and oxygen in the other—seems unlikely but this possibility has not yet been definitely excluded.

There has been some controversy over the assignment of the oxidation number of copper in the compound $K_2[Cu(Pc)]^{--}$ (Pc = phthalocyanine) which is formed by reducing Cu(II) Pc with potassium in liquid ammonia. Watt and Davies (1960) who first prepared the former compound suggested that the oxidation number of the copper is zero.

On the grounds that it is not possible to decide whether two electrons are added to the ligand or metal atom, Alexander and Pauson (1961) have concluded that the oxidation state of the central metal atom is "not only undetermined but indeterminate." However, electrons are delocalized to some extent in most complexes and, on the basis of the definition of oxidation number given at the beginning of this section, it is reasonable to assign an oxidation number zero to copper.

Despite the occasional difficulty in deciding the oxidation number of a metal atom in a complex, the concept is a useful one. The method of assigning the oxidation number is arbitrary and must not be interpreted as indicating the location of the charge on a complex ion. The charge on $[Co(NH_3)_6]^{3+}$ is not centered on the metal atom but distributed among the eighteen hydrogen atoms which form the periphery of the ion (Pauling,

1948). When $[Co(NH_3)_6]^{3+}$ is reduced to $[Co(NH_3)_6]^{++}$, there is a marked change in the electronic structure of the cobalt atom as evidenced by the profound change in its magnetic behavior. This justifies the selection of cobalt as the element that undergoes the change in valency (oxidation number) during the reduction.

Metal ions exist in aqueous solution as aquo complexes. It is not always certain how many water molecules are involved, though for many metals it is probably four or six. Replacement of these water molecules by other molecules (including chelating molecules) or ions may stabilize an otherwise unstable oxidation state. For example, $[Co(H_2O)_6]^{3+}$ is unstable in aqueous solution and is so powerful an oxidizing agent that it oxidizes water to oxygen; $[Co(NH_3)_6]^{3+}$ and the corresponding ethylenediamine complex are stable in aqueous solution. The hexaquo $[Co(H_2O)_6]^{++}$ is stable in aqueous solution but replacement of the water molecules by five cyanide ions results in a complex $[Co(CN)_5]^{3-}$ that is capable of reducing water to hydrogen. The ion $[Co(H_2O)_6]^{+}$ cannot survive in aqueous solution. However, cobalt can exist in aqueous solution in the +1 state if it is coordinated with three molecules of 2,2-bipyridine (Waind and Martin, 1958).

The quantitative assessment of the relative stabilizing or destabilizing effects of coordination and chelation is taken up in Chapter 6.

3. *Coordination Number*

This number which has already been defined as the number of donor or ligand atoms directly bound to a central metal atom is not necessarily fixed and invariable for each metal. Sometimes it depends on the nature of the donor atom; at other times it depends on the oxidation state of the metal. Atoms known to function as donors in metal complexes are shown in Table II. Those shown in boldface type have been observed as donor atoms in chelating molecules. The halogen atoms Cl, Br, and I participate in chelate ring formation in bridged, polynuclear complexes which are

TABLE II
DONOR ATOMS IN METAL COMPLEXES

			H
C	**N**	**O**	F
	P	**S**	**Cl**
	As	**Se**	**Br**
	Sb	Te	**I**

not usually thought of as metal chelates, though in fact they are according to the definition already given.

A number of general statements may be made about the coordination numbers of metals.

(*a*) The coordination number of some metals in a given oxidation state is, in their stable compounds, fixed and characteristic. With very rare exceptions, if any, the following metals have a coordination number of six: Co(III), Cr(III), Pt(IV), Pd(IV). This list is not intended to be a complete one.

(*b*) In contrast to the metals in (*a*) are those whose coordination number for a given oxidation state varies, usually with the nature of the ligand atoms. Examples include: Fe(II) (four with Cl^- and six with CN^-), Zn(II) (six with $NH_2{\cdot}CH_2{\cdot}CH_2{\cdot}NH_2$ and four with CN^-), and Cu(I) (two with Cl^- and four with $AsEt_3$).

(*c*) For some metals there is a correlation between oxidation state and coordination number, e.g., Mo(III) (six), Mo(IV) (eight), Pt(IV) and Pd(IV) (six), and Pd(II) (generally four). For other metals like osmium the coordination number is the same (six) for at least three different oxidation states: Os(II), Os(III), and Os(IV). Os(IV) can also be eight-covalent.

As a general rule, there is little difficulty in deciding, on the basis of the molecular formula, what the coordination number of a metal is in complexes where unidentate ligands only are present. This statement does not apply to compounds in which finite or infinite polynuclear structures are present. For example, the coordination of cadmium in $CsCdCl_3$ appears to be three. Crystal analysis shows that it is six (XIII). The cadmium atoms form part of an infinite complex.

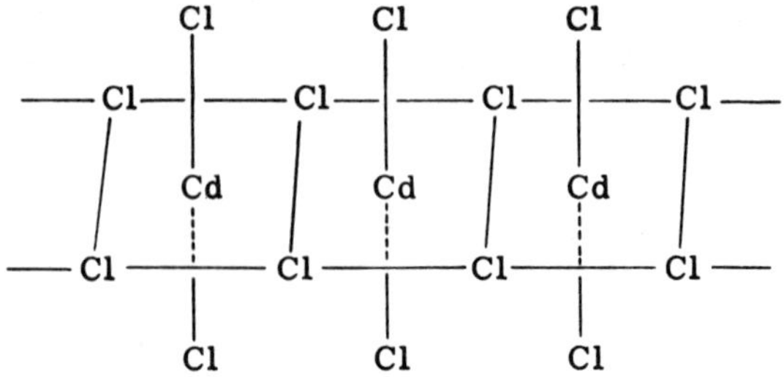

(XIII)

Occasionally a metal exists in two oxidation states in the one compound, each oxidation state having a different coordination number. In the compound $CsAuCl_3$ X-ray crystal analysis has revealed the presence of equal numbers of linear $[AuCl_2]^-$ and square $[AuCl_4]^-$ ions.

The coordination number of a metal in a metal chelate may generally be deduced from its stoichiometry. There are, however, instances where X-ray crystal analysis alone is able to provide the answer. This is especially the case where compounds of apparently square-coordinated metals are concerned [Ni(II), Cu(II), Pt(II), Pd(II)]. The square configuration for these metals is not nearly as universal as it was thought to be.

Copper(II), a metal once thought to have this configuration in most of its complexes, has a coordination number of four, five, or six depending on the nature of the ligands attached to it. There is probably more variety in the stereochemistry of copper(II) than in that of any other metal. The following configurations have been described for copper(II): square planar, distorted tetrahedral, distorted octahedral, tetragonal pyramidal, and trigonal bipyramidal.

Even after a detailed crystal analysis has been carried out on a compound, it is sometimes difficult to be certain of the coordination number of a metal. Thus, in the chelate bis(*o*-phenylenebisdimethylarsine)palladium(II) iodide there are two iodine atoms in the octahedral (1,6) positions at a distance of 3.5 Å from the palladium atoms (Stephenson, 1962). This distance is greater than the sum of the covalent radii of palladium and iodine but less than the nonbonded distance which is estimated to be approximately 4.2 Å. Assuming that there is a bond between palladium and iodine, then, of course, the coordination number of palladium is six.

B. Chelating Molecules

1. *Conditions Necessary for Chelation*

If a molecule is to function as a chelating agent, it must fulfill at least two conditions. First, it must possess at least two appropriate functional groups, the donor atoms of which are capable of combining with a metal atom by donating a pair of electrons. These electrons may be contributed by basic coordinating groups such as NH_2 or acidic groups that have lost a proton.

Some acidic groups that combine with metal atoms by the replacement of hydrogen are:

—COOH	—N(H)—H
$—SO_3H$	—N(R)—H
—OH (enolic and phenolic)	
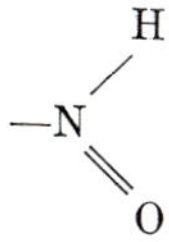	$—P(O)(OH)_2$
—SH	

Coordinating groups include:

=O	—O—R	—AsR_2
—NH_2	=NOH	
—NH	—OH (alcoholic)	—PR_2
—N=	—S— (thioether)	

Second, these functional groups must be so situated in the molecule that they permit the formation of a ring with a metal atom as the closing member. These two conditions are necessary, but they are not always sufficient for the formation of a chelate ring. Under some conditions, for example, in solutions of sufficiently low pH, a potential chelating molecule may attach itself to a metal atom through only one ligand atom. An example of this is compound (XIV) (Grinberg, 1962).

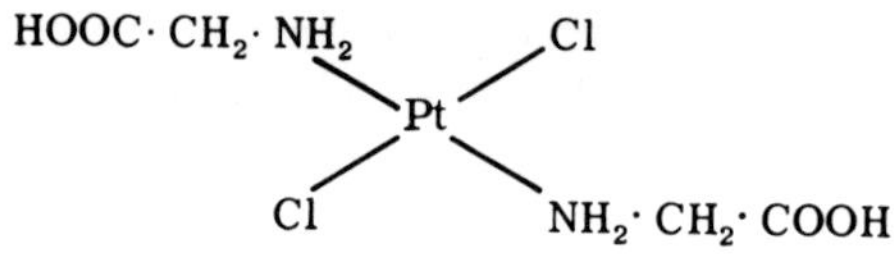

(XIV)

Steric factors occasionally influence chelation. One chelating molecule may attach itself to a metal atom readily enough, but the addition of a second and third is hindered or even prevented by the fact that there is a clash between the first and parts of the second and third when the latter move into the positions required for attachment. Whether such a clash will occur may sometimes depend on the size of the central metal atom; it may occur only with atoms below a certain radius. Irving and his co-workers (1949) have shown that substitution of a methyl group in the 2 position of 8-hydroxyquinoline prevents the formation of a tris complex with aluminum(III) but not with chromium(III) or iron(III). Of these three ions, aluminum is the smallest. However, the matter may not be as simple as it appears to be. Failure to form a precipitate with aluminum should not be taken to imply the absence of complex formation. As already pointed out, some stable inner complexes, for example bis(dimethylglyoximato)-copper(II) are soluble in water.

Steric factors are more clearly evident in the chelation of 2,9-dimethyl-1,10-phenanthroline with Cu(I) and Fe(II). The tetrahedral disposition of its valency bonds enables Cu(I) to form a stable 2,1 complex with this chelating agent. The octahedral disposition of its bonds allows Fe(II) to form a bis but not a tris complex (Fox *et al.*, 1962). Where the clash of chelating molecules is not so serious as to completely prevent chelation, it may reveal its influence by a lowering of the stability of the resulting complex. Considerable interest attaches to steric effects of this kind, because of their possible application to the design of selective chelating molecules.

In the process of chelation, there may be some modification of natural bond angles of the metal and the chelating agent to permit ring formation. The natural angle at octahedrally and square-coordinated metal atoms, 90°, is not the angle of regular or nearly regular five- and six-membered rings. Generally speaking, the deviation from 90° is not great though in extreme cases, in four-membered rings for example, it may be as large as 20°. Not only does chelation sometimes result in a modification of the bond angles of the chelating agent but it may also modify bond lengths and the chemical reactivity of the agent. Study of this last effect is a subject of growing importance and is taken up in Chapter 8.

2. *Classification of Chelating Molecules*

So far, only the simplest type of chelating molecule has been mentioned—the type that is attached to metal atoms by two donor or ligand atoms. To describe these, Morgan introduced the term bidentate chelating molecule. It will be evident in Chapter 3 that bidentate chelating agents are not only numerous but very varied in character. There are, however, many molecules with three or more donor atoms capable of combining with metal atoms and forming interlocking or fused chelate rings. According to the number of donor atoms capable of combining with a metal atom, these are called tridentate, quadridentate, quinquedentate, sexadentate, and octadentate chelating molecules, respectively. In forming a metal chelate, a multidentate chelating agent may not use all its donor atoms; for example, a potentially sexadentate chelating agent may function as a quinquedentate. In fact, most quinquedentates are of this character. Examples of tri-, quadri-, and sexadentate chelating agents are shown below: bis(3-dimethylarsinylpropyl)methylarsine (XV) (Barclay and Nyholm, 1953); porphin (XVI); and 1,8-bis(salicylidencamino)-3,6-dithiaoctane (XVII) (Dwyer and Lions, 1947).

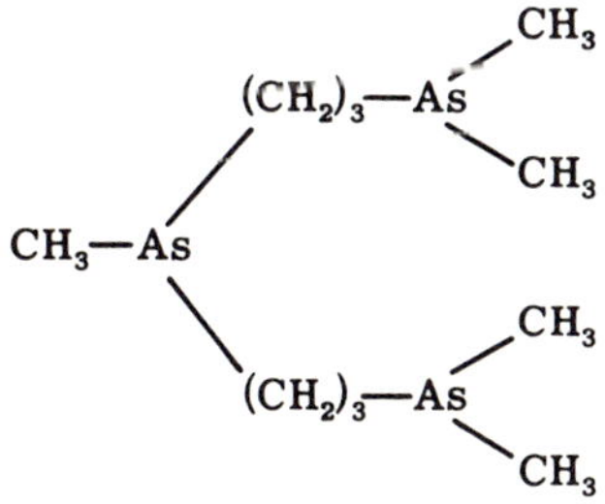

(XV)

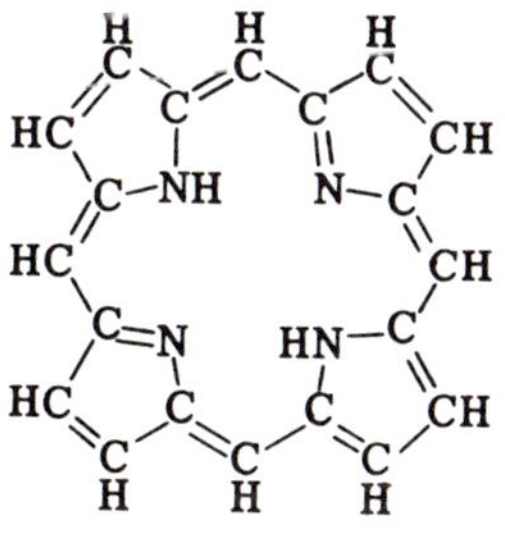

(XVI)

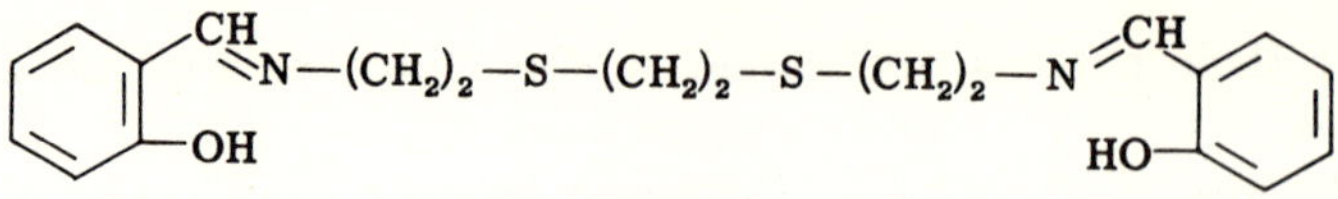

(XVII)

A more detailed classification of chelating agents (Diehl, 1937) is based on the nature of the functional groups which are described as: (1) basic or coordinating, e.g., NH_2; or (2) acidic, e.g., COOH. Bidentate chelates are thus divided into three groups, namely, those with:

(1) two acidic groups—oxalic acid;
(2) two basic groups—ethylenediamine;
(3) one acid and one basic group—glycine.

There are other ways of classifying chelating molecules. They may, for example, be classified as rigid or flexible. To the first group belong oxalic acid, 1,10-phenanthroline, porphin, and phthalocyanine; to the second belong ethylenediamine and ethylenediaminetetraacetic acid. Flexibility is an essential characteristic of all sexadentate chelating molecules.

Bidentate chelating molecules may be grouped as in Chapter 3—on the basis of their donor atoms: O,O; N,N; and N,O. This system is not generally used for higher multidentate molecules.

3. *Size of Chelate Rings*

Four-membered chelate rings are the smallest whose existence has been established with certainty. X-ray crystal analysis has disclosed the presence of such rings in carbonatotetramminecobalt(III) bromide and bis(*N*,*N*'-di-*n*-propyldithiocarbamate)nickel(II). The structure of these rings will be considered in more detail in the section on X-ray crystal analysis. Four-membered rings are comparatively rare mainly because their formation involves considerable strain.

Five- and six-membered rings are by far the most common among metal chelates. Unsaturated five- and six-membered rings in which resonance occurs are, as a rule, planar; saturated rings, on the other hand, are puckered.

Structures have been proposed for compounds in which the existence of larger chelate rings has been postulated, but very few have been studied by X-ray crystal analysis. Seven-, eight-, and higher membered rings are doubtless formed by the higher homologs of some sexadentate molecules, e.g., the higher homologs of ethylenediaminetetraacetic acid (XVIII).

Metal complexes of chelating agents in which n = 2, 3, 4, and 5 have been described.

X-ray crystal analysis has disclosed the presence of an unusually large chelate ring in the structure of vitamin B_{12} (Hodgkin *et al.*, 1957, 1959).

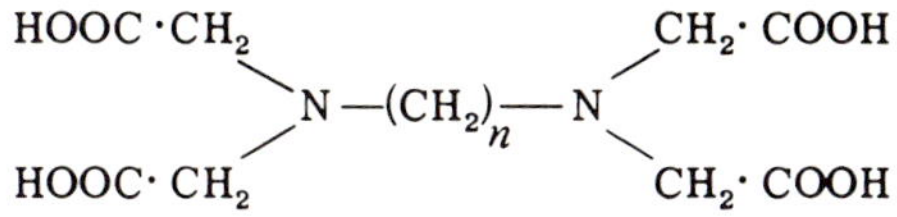

(XVIII)

On the basis of less direct evidence, Warner (1954) has suggested that a similarly large chelate ring is present in the structure of conalbumin.

4. *Symmetry of Chelate Rings*

Chelate rings are sometimes described as symmetrical or unsymmetrical depending on whether or not they are symmetrical about a plane perpendicular to the plane of the ring. Thus the ring formed by the oxalate ion is symmetrical while that formed by the *o*-aminophenolate ion is unsymmetrical.

Unsymmetrical rings are of special interest because they can give rise to *cis-trans* isomerism in square and octahedral complexes and optical isomerism in tetrahedral complexes.

The possible *cis* and *trans* isomers of bis(*o*-aminophenol)nickel(II) are shown in (XIX) and (XX):

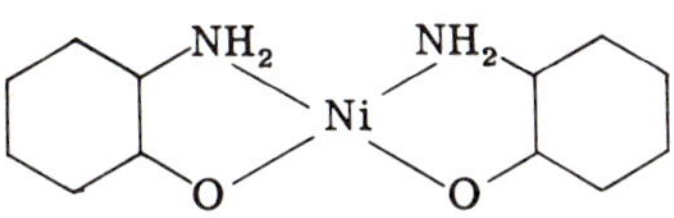

(XIX)

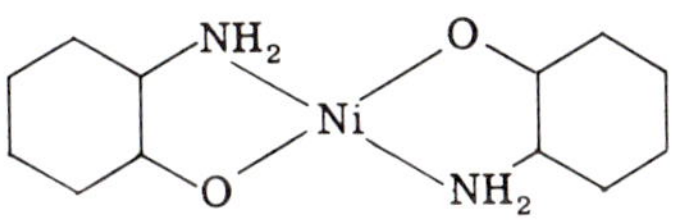

(XX)

It is important to note that the term unsymmetrical is here used in a restricted sense—to refer only to symmetry about a plane perpendicular to the plane of the ring. An unsymmetrical ring, so defined, is not necessarily asymmetric. Though no X-ray crystal analysis appears to have been done on the compound, it is almost certain that the chelate ring formed by *o*-aminophenol and nickel is planar. If this is so, the ring is symmetrical about a plane passing through the centers of the atoms of the ring. The stereochemistry of asymmetric chelate rings is taken up in Chapter 5.

C. The Metal-Ligand Bond

The strength and stability of this bond will, of course, depend on the nature of M and L—on their size and electronegativity; more specifically on their electronic structure. The ligand atom L may have other atoms attached to it, the number and nature of which will also influence the stability of the M—L bond.

Lewis (1916) and Langmuir (1919) were the first to identify a covalent bond with a pair of electrons—one from each atom, and both held in common. Some years later, Sidgwick (1927) developed an electronic interpretation of the structure of metal complexes in which one of the key ideas was the coordinate bond. The distinctive feature of this bond between M and L is that both electrons of the bond are contributed by one atom, L, known as the donor atom. For example, the nitrogen of the amino group in ethylenediamine contributes its lone pair to the metal atom, a process which Sidgwick symbolized by an arrow (XXI).

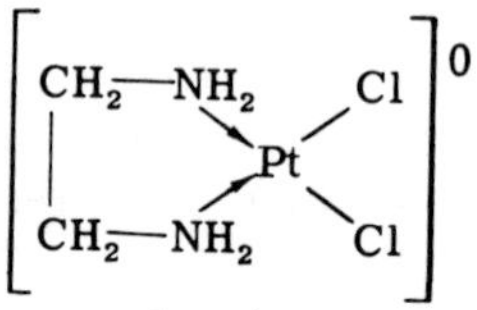

(XXI)

Though this symbol which replaced the dotted line introduced by Werner is still employed by some authors, its use is misleading if it suggests that the bond, once formed, is anything but a normal covalent bond. Those who use it claim as one of its advantages the fact that it enables a quick check of the oxidation state of the central atom.

The next important step in the development of the theory of the chemical bond was the application of quantum mechanical theory to the problem of molecular structure. During the 1930's three main lines of attack were developed: (1) the valence bond technique of Heitler, London, Slater, and Pauling; (2) the molecular orbital technique of Hund, Bloch, Mulliken, Lennard-Jones, and Huckel; and (3) the crystal field technique of Bethe, Kramers, and Van Vleck. The valence bond technique as expounded by Pauling (1939) in his book *The Nature of the Chemical Bond* exercised considerable influence on the development of structural chemistry and, although the emphasis in development has shifted to (2) and (3) during the last 10 years or so, (1) provides a good introduction to the theory of the metal-ligand bond. The theory of this bond is treated at length in Chapter 2.

Pauling introduced the concept of the hybridization of electronic orbitals, or wave functions. By this process, he was able to show that a set of orbitals, not all of which are geometrically equivalent, could be combined to produce a set of geometrically equivalent orbitals oriented in space differently from the original set. The one $2s$ and three $2p$ orbitals in the carbon atom, when hybridized, produce four equivalent sp^3 orbitals directed to the corner of a regular tetrahedron. Similarly one $4s$, three $4p$, and two $3d$ orbitals can be combined to give a set (d^2sp^3) directed towards the corners of a regular octahedron; one $4s$, two $4p$, and a $3d$ combined to give a set (dsp^2) directed towards the corners of a square. Thus he accounted for the observed stereochemistry of many different kinds of atoms.

In order to account for the structure of metal complexes like $Ni(CO_4)$ and $K_4[Fe(CN)_6]$, Pauling (1939) introduced the idea that "atoms of transition groups are not restricted to the formation of single bonds, but can form multiple bonds with electron-accepting groups by making use of the electrons and orbitals within the valence shell." This type of bonding later became known as π bonding. The nickel-carbon bond distances in nickel tetracarbonyl are consistent with a double-bonded structure (XXII) which resonates with the single-bonded structure (XXIII).

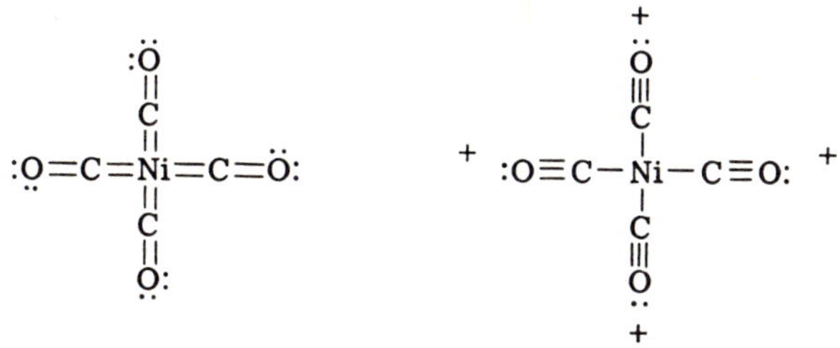

(XXII) (XXIII)

In (XXII) the nickel and other atoms are neutral whereas in (XXIII) a fourfold negative charge is placed on the nickel atom. The latter is contrary to what one would expect from an electropositive atom like nickel. Considerations of this kind later led Pauling (1948) to enunciate the principle of the essential electrical neutrality of atoms.

Pauling also applied the principles of magnetism to the investigation of the nature of the metal-ligand bond and, since there will be frequent mention of the magnetic properties of metal chelates in later chapters, a brief reference will be made to some aspects of magnetochemistry.

D. Magnetochemistry of Metal Complexes

Atoms, molecules, or ions may possess permanent magnetic dipole moments that arise in either or both of the following ways:

(1) from the orbital motion of electrons,

(2) from the spin of electrons.

Permanent magnetic dipoles tend to orient themselves in the direction of an applied magnetic field; the extent to which this is achieved depends upon the temperature since thermal motion tends to oppose orientation. The ratio of the number of magnetic dipoles in stable orientations with the field to the number in less stable orientations against the field decreases with increasing temperature. Paramagnetic polarization (of which magnetic susceptibility is a measure) is, therefore, temperature-dependent. In an ideal system where there are no dipole interactions, the molar susceptibility ψ_M can be represented by the equation

$$\psi_M = N\alpha + \frac{N\mu^2}{3kT}$$

where N is Avogadro's number; k, Boltzmann's constant; T, the absolute temperature; α, the diamagnetic susceptibility (a negative quantity); and μ, the permanent magnetic dipole moment.

The paramagnetic term is sometimes written C/T where C, the Curie constant of the substance, is $N\mu^2/3k$. The permanent dipole moment may be obtained from C by using the equation:

$$\mu = \sqrt{3kC/N} = 2.839 \sqrt{C} \text{ Bohr magnetons}$$

The resultant spin moment, μ, of n unpaired electrons is given by the expression:

$$\mu = \sqrt{n(n + 2)}$$

The magnetic dipoles of two electrons with opposed spins and occupying the same orbit cancel one another. The electron pair of a covalent bond therefore contributes nothing to the permanent magnetic dipole of a molecule.

Magnetic moments of ions of elements of the first transition series often approximate the value expected from the contribution of spin only. There are, however, instances where the orbital contribution to the moment is quite large. Except where the orbital contribution is very large, measurements of the molar susceptibility of metal complex enables one to deduce the number of unpaired electrons in a metal atom. From a knowledge of the number of unpaired electrons, conclusions of considerable interest may be drawn. They concern:

(1) the oxidation number of the metal atom;
(2) the stereochemistry of the metal atom;
(3) the nature of the metal ligand bond.

Pauling has developed a theory relating the number of unpaired electrons to these three factors. Predictions based on the theory are shown in Table III. The moments tabulated there are those arising from spin only.

Measurements of magnetic susceptibility often enable a check of the oxidation number of a metal in its complexes. Thus, Cu(I) can as a rule be distinguished from Cu(II) and Os(II) from Os(III). Caution must be used in applying this method. Occasionally, the orbital contribution to the moment of an ion may be so large that the observed moment is higher than the next spin-only moment. For example, the moment of a three unpaired

TABLE III
PREDICTED MAGNETIC MOMENTS OF COMPLEXES CONTAINING TRANSITION ELEMENTS

No.	The iron group	Number of electrons in 3*d* shell	For ionic or sp^3 (tetrahedral) bonds	For four dsp^2 (square) bonds	For six d^2sp^3 (octahedral) bonds
1	Ti(IV)	0	0.00	0.00	0.00
2	V(IV)	1	1.73	1.73	1.73
3	V(III), Cr(IV)	2	2.83	2.83	2.83
4	V(II), Cr(III), Mn(IV)	3	3.88	3.88	3.88
5	Cr(II), Mn(III), Fe(IV)	4	4.90	4.90	2.83
6	Mn(II), Fe(III), Co(IV)	5	5.91	3.88	1.73
7	Fe(II), Co(III)	6	4.90	2.83	0.00
8	Co(II), Ni(III)	7	3.88	1.73	—
9	Ni(II), Cu(III)	8	2.83	0.00	—
10	Cu(II)	9	1.73	—	—
11	Cu(I), Zn(II), Ge(IV)	10	0.00	—	—

spin cobalt(II) complex (theoretical spin-only value 3.88 Bohr magnetons—four unpaired spins 4.88) may be as high as 5.4 Bohr magnetons. Similarly, it is possible for an iron(II) complex to have a moment as high as 5.6 Bohr magnetons, e.g., bis(8-hydroxyquinolino)iron(II)monohydrate. This value (5.6) is so close to the value expected for ionic iron(III) (5.9) complexes as to make it unsatisfactory as a check on the oxidation state. A study of these large orbital contributions can be used to throw light on the stereochemistry of the metal atoms concerned but this is beyond the scope of this introduction.

Some years ago, Dwyer and Nyholm (1941) prepared a series of diphenylmethylarsine complexes of rhodium in which the stoichiometry seemed consistent with +2 oxidation state for the metal. However, all the complexes were diamagnetic, which indicated that the oxidation state could not be +2 (Mellor, 1943a). The anomaly was removed when Lewis *et al.* (1960) showed that the complexes were hydrides, for example: $HRhCl_2(AsMePh_2)_2$. In other words, the true oxidation state of the metal in these complexes is +3.

Departures of moments from the spin-only values predicted for particular oxidation states may occur for reasons other than the existence of an orbital contribution. The magnetic moments of a number of bridged copper(II) complexes are abnormally small because of electron-exchange demagnetization between the metal atoms (Harris *et al.*, 1962).

Pauling predicted that diamagnetic complexes of nickel(II) would be square-coordinated—a configuration that had not been observed for nickel up to that time. The correctness of the prediction was soon verified in a number of different ways. Sugden (1935) prepared *cis* and *trans* isomers of bis(methylbenzoylglyoxime)nickel(II); Cox and his associates (1935) demonstrated by X-ray crystal analysis the square bonding of nickel in potassium bis(dithiooxalato-*S,S′*)nickelate(II).

Other instances of square coordination predicted by Pauling—for Au(III), Ag(II), and Co(II)—were subsequently verified.

The complexes of iron(III) afford an illustration of how magnetic dipole moments are used to discriminate among different types of metal-ligand bonds. Electron distributions possible for iron(III) in its complexes are as shown in Scheme 1.

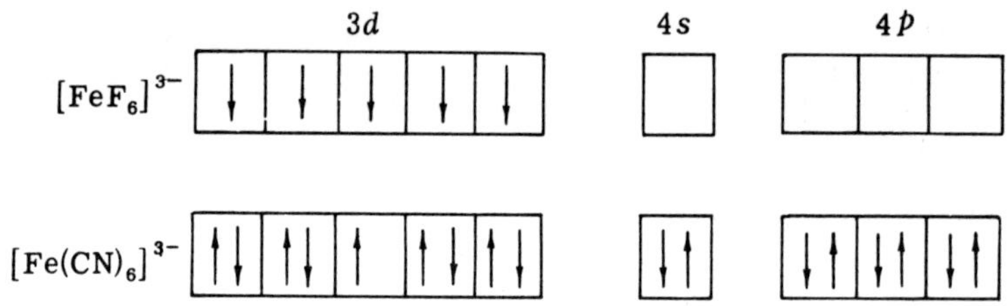

The magnetic moment of iron in K_3FeF_6 is 5.9 Bohr magnetons which is also the value expected for the free ferric ion. For $K_3[Fe(CN)_6]$, the moment is 1.9 Bohr magnetons. The numbers of unpaired electron spins in the two compounds are thus five and one respectively. In the former case, Pauling (1939) interpreted the result to mean that the bonds to iron are essentially ionic in character, in the latter, that they are covalent. The theory allowed no intermediate character for the bonds. Other interpretations have since been placed on the value of the higher moment. One is that outer *d* orbitals of iron are used when the ligand is one of high electronegativity like fluorine. The proposed electron distribution under these circumstances is shown in Scheme 2.

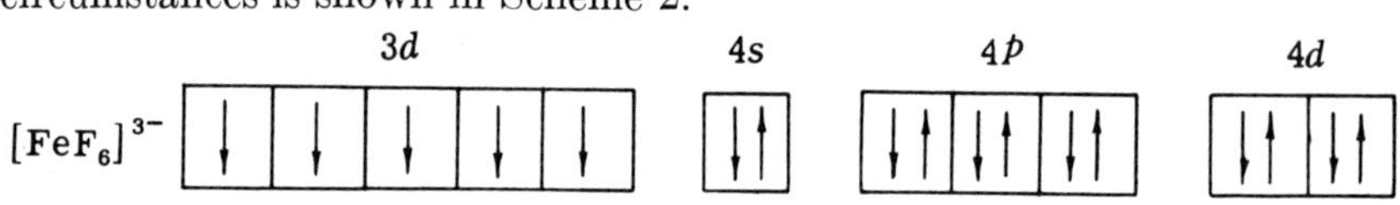

As a means of indicating this, the complex is described as an outer-orbital complex to distinguish it from $[Fe(CN)_6]^{3-}$ which, in contrast, is called an inner-orbital complex. There is, however, still some doubt whether outer *d* orbitals are used in the fluoride complex. For this reason, it is preferable to use the alternative terms "spin-free" and "spin-paired" to distinguish between these two types of complex. The present day trend is no longer to draw a sharp distinction between covalent and ionic bonding for the simple reason that it has not been possible to determine the degree of covalency by experimental techniques acceptable to the theoretical chemist (Orgel, 1960).

Despite its undoubted successes, Pauling's valence bond theory proved inadequate for a quantitative understanding of a number of aspects of the physical and chemical behavior of metal complexes. It failed to provide an adequate interpretation of their absorption spectra, free energies of formation, and oxidation-reduction potentials.

Over the last 10 years or so, the molecular orbital and crystal field techniques have been used to provide a more satisfactory explanation of these properties.

IV. Investigation of the Structure of Metal Chelates

A. Chemical Methods

1. *Octahedral Complexes*

The development of an understanding of the structure of metal chelates is so closely intertwined with that of the structure of metal complexes in general that a brief treatment of the latter is probably the best approach to the former. Chemical methods are based essentially on the study of two main types of isomerism—geometrical and optical.

a. Geometrical Isomerism. There are three possible types of regular arrangement of six atoms about a seventh:

(1) hexagonal (planar);
(2) trigonal prismatic;
(3) octahedral.

For structures of the type $[MA_2B_4]$, three isomers are possible for (1) and (2) and two for (3). Two only have been found. In this negative kind of way, isomer counts support the octahedral configuration.

b. Optical Isomerism. Evidence of a more positive character for the octahedral configuration has been obtained through a study of optical or mirror image isomerism.

Optical isomerism arises when a molecule or ion and its mirror image are not superposable.

The most general statement of the criterion for appearance of this type of isomerism is that the molecule or complex ion must not possess an axis of rotatory inversion. Of the axes of rotatory inversion, $\bar{1}$ is synonymous with a center of symmetry, $\bar{2}$ with a plane of symmetry, $\bar{3}$ implies the existence of a center of symmetry, and $\bar{6}$ the presence of a plane of symmetry. A fourfold axis of rotatory inversion is so rare among molecules that for all practical purposes it is generally safe to use the more familiar rule of the absence of a center or plane of symmetry.

There are two geometrical isomers of $[Co(en)_2Cl_2]Cl$—*trans* (XXIV) and *cis* (XXV).

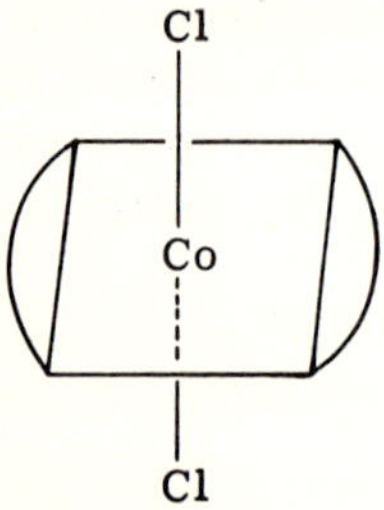

(XXIV)

(XXV)

The *cis* form has neither a plane nor a center of symmetry, and should therefore be capable of existing in *d* and *l* mirror image forms. Werner was the first to prove this point when he succeeded in resolving the complex $[Co(en)_2(NH_3)Cl]^{++}$ (1911) by means of the *d*-bromocamphor sulfonate. Within a short time, he also resolved $[Co(en)_2Cl_2]^+$, $[Co(en)_2(NO_2)_2]^+$, and $[Co(en)_3]^{3+}$ (1912).

At first, there was considerable reluctance to accept this evidence. Those who doubted the basis of Werner's structural theory believed that the optical activity in these compounds arose, in some way not specified, from the presence of carbon therein. In order to meet this criticism, Werner adopted the ingenious idea of using a non-carbon-containing metal complex as a chelate group. He chose *cis*-dihydroxytetramminecobalt(III) ion $[Co(NH_3)_4(OH)_2]^+$ for this purpose and succeeded in preparing compound (XXVI).

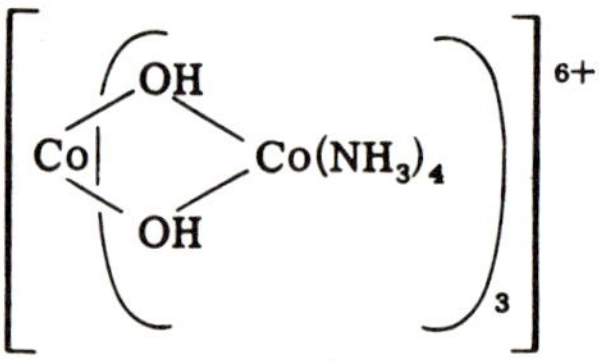

(XXVI)

Resolution of this polynuclear complex in 1914 was a remarkable vindication of Werner's structural theory. Many years later, the ethylenediamine analog was resolved by Goodwin and his co-workers (1958). The stereochemistry of the ethylenediamine analog is a good deal more complicated because the *cis*-dihydroxybis(ethylenediamine) chelate itself is capable of existing in *d* and *l* forms.

Demonstration of the existence of optically active isomers of complexes of cobalt(III) not only gave strong support to the theory of octahedral coordination, but also greatly strengthened belief in the existence of chelate rings. There is little doubt that progress in the study of optical isomerism among octahedral complexes would have been extremely slow had it not been for the use of chelating agents. Even today only one instance of a complex of the type [M(abcdef)] is known, and this has been separated into only a few of its thirty possible geometrical isomers (Gelman and Essen, 1950).

2. *Square Coordination*

The tetrahedral disposition of four valence bonds about the carbon atom was the first stereochemical configuration to be discovered, and so widespread did its occurrence eventually prove to be—not only among carbon compounds but also among the compounds of many other elements—that for a long time there was considerable reluctance among chemists to accept the view that any alternative arrangement—in particular, the direction of bonds towards the corners of a square—was possible.

It will be recalled that Werner suggested the square arrangement of bonds about the platinum atom to explain the existence of two isomeric forms of dichlorodiammineplatinum(II). For many years, the square configuration was the subject of controversy, and it was not until the late thirties that the question was finally settled. Considerable support was given to Werner's views by the resolution of a most ingeniously devised metal chelate, namely, bis(isobutylenediamine-*meso*-stilbenediamineplatinum(II) ion (XXVII).

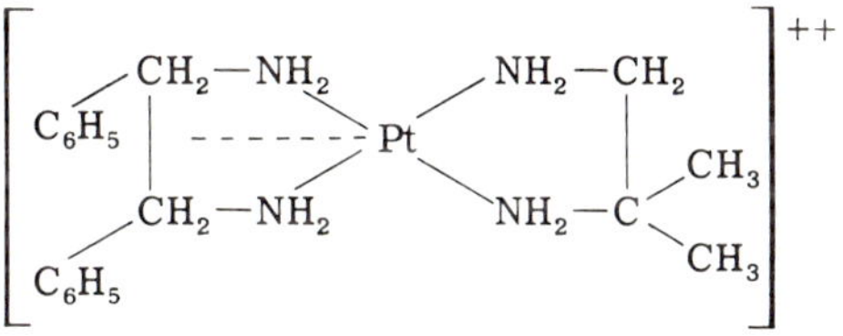

(XXVII)

The metal chelate was so designed by Mills and Quibell (1935) that it would be resolvable if platinum(II) were square-coordinated but not if it were tetrahedrally coordinated. The symmetry of the chelate rings is all important: the ring formed by isobutylenediamine is not symmetrical about any plane perpendicular to the ring, but is symmetrical about the plane of the ring itself.[4] The opposite is true of the ring formed by *meso*-

[4] The situation is more complicated than is indicated here. The chelate rings are not planar, as Mills and Quibell assumed, but puckered. However, the transition from one conformation of the ring to the other involves so little energy that the mean positions of the atoms produced a statistical plane of symmetry and the argument outlined above is not affected.

stilbenediamine. This ring is symmetrical about a plane perpendicular to the plane of the ring passing through the platinum atom and the middle of CH_2—CH_2 (XXVII). The ring is not symmetrical about the plane of the ring itself. If the planes of symmetry of the two chelate rings coincide, which will be so for tetrahedral coordination, the complex as a whole will have a plane of symmetry and will therefore not be resolvable. If, on the other hand, the planes of symmetry of the chelate rings are at right angles to one another, as they would be for square coordination, the complex as a whole will have no plane or center of symmetry and will therefore be resolvable. Experiment proved that the complex was resolvable. It should, however, be pointed out that the experimental result could also be explained if the bonds from the platinum atom were pyramidally arranged. The square configuration proved to be one that could not be established unequivocally by chemical means alone (Mellor, 1943b). Only by means of crystal structure analysis was it finally established beyond doubt.

3. *Tetrahedral Complexes*

Instances of optical isomerism in tetrahedrally coordinated metal complexes are uncommon. Of course, this does not mean that tetrahedral coordination does not occur among metals; far from it. It means that, although optical isomers may exist, it is for some reason—possibly because of their lability—difficult to separate them. A few resolutions have been effected with complexes of the type $M(a\text{–}b)_2$ where a–b is a ligand capable of forming an unsymmetrical chelate ring. When attached to a metal atom, the benzoylpyruvate ion forms a ring of this type, as also does the benzoylacetone ion. Successful resolutions have been reported for:

(1) dimethylammonium bis(benzoylpyruvato)beryllate(II);
(2) bis(benzoylacetonato)beryllium(II);
(3) potassium bis(8-hydroxyquinolinato-5-sulfonic acid)zincate(II).

B. Physical Methods

General Considerations

The physical methods used for studying molecular structure fall into two main classes:

(1) those based on X-ray, neutron, and electron diffraction techniques which yield detailed information about the whole structure of the molecule;
(2) those which yield fragmentary information—information about individual bonds or particular groups of atoms in a molecule.

The methods of class (2) include absorption spectroscopy of the region ranging from the microwave to the ultraviolet and the measurement of electric and magnetic dipole moments. The results of X-ray crystal analysis only will be briefly discussed here.

C. Crystal Structure Analysis

Although chemical methods of investigation left no doubt about the existence of chelate rings and about the general features of the structure of metal chelates, even of those formed with chelating agents as complicated as phthalocyanine, etioporphyrin, and ethylenediaminetetraacetic acid, confirmation of the correctness of these and other structures by X-ray crystal analysis greatly strengthened our knowledge of chelation. It introduced a metrical element into the understanding of chelate rings by revealing the length and angles of chemical bonds and other structural details that could not be gained by purely chemical methods.

The precise location of all the atoms within the unit cell of a crystal by X-ray analysis provides the strongest and most detailed evidence available for the existence of chelate rings. When X-ray methods were first applied to metal chelates, the analysis was often taken no further than determining the general shape and symmetry of the ion or molecule. Nowadays with the help of greatly improved techniques, including the use of electronic computers, structures as complex as that of vitamin B_{12} which involved the fixing of three parameters for each of one hundred and five atoms have been successfully solved.

In this section, a group of metal chelates chosen mainly because they were important in the development of the chemical theory of structure will be briefly examined with the object of showing the extent which the conclusions based on chemical theory have been confirmed. No attempt will be made to describe the way in which the molecules or ions are arranged in the unit cell of the crystal structure.

1. *Bidentate Ligands*

a. Four-Membered Planar Rings. One of the early crystal analyses of metal chelates was that of bis(*N*,*N*-di-*n*-propyldithiocarbamato)nickel(II) (Peryonel, 1941). Two different modes of chelation are possible with di-*n*-propyldithiocarbamate—either through two sulfur atoms, or nitrogen and sulfur. Crystal structure analysis shows that it is chelated through sulfur to produce a four-membered ring (XXVIII).

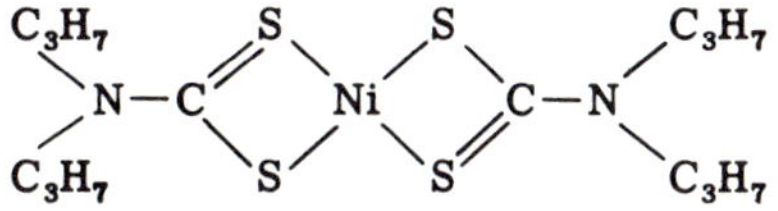

(XXVIII)

The dimensions of the ring were determined only approximately (Ni—S = 2.30 Å; C—S = 1.83 Å). The angle between the two Ni—S bonds was assumed to be 90°.

The ability of the carbonate ion to function as a chelating agent was recognized quite early by Werner and Vilmos (1899) and later used by them to prepare *cis* isomers which they did by replacing the chelated group with two unidentate ligands. Although few chemists doubted that the carbonate ion acts as a chelating agent in the ion $[Co(NH_3)_4CO_3]^+$, it was not until quite recently that this was verified by means of X-ray crystal analysis (Barclay and Hoskins, 1962). They showed that the four-membered chelate ring is, as would be expected, a highly strained one—the angle at the cobalt atom being 70.5° instead of the 90° expected for regular octahedral coordination (XXIX).

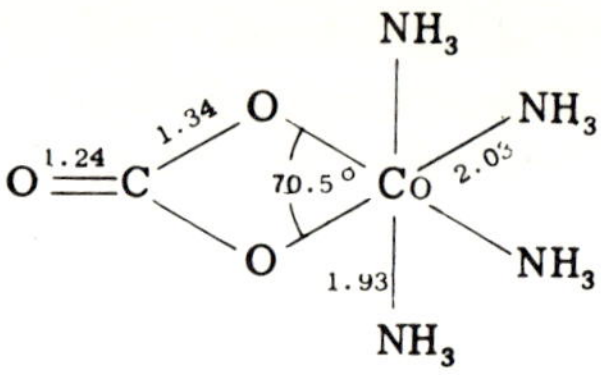

(XXIX)

The bond lengths of the chelated CO_3 group are interesting. The C—O bond not involved in chelation is significantly shorter than in the carbonate ion as it exists in calcite (1.24 as against 1.29). The C—O bonds in the chelate ring are significantly longer than in calcite. Also noteworthy is the fact that the length of the two Co—N bonds opposite the chelate ring is appreciably greater than that of the remaining Co—N bonds.

b. Five-Membered Planar Rings. X-ray structural analysis was first applied to metal chelates in which the central atom was square-coordinated. A notable example was the analysis of potassium bis(dithiooxalato-*S*,*S*′)-nickelate(II) (Cox *et al.*, 1935). The structural feature of principal interest is the large anion (XXX), in which nickel is square-coordinated and all the atoms are coplanar. It is also of interest to note that chelation takes place through the sulfur atoms.

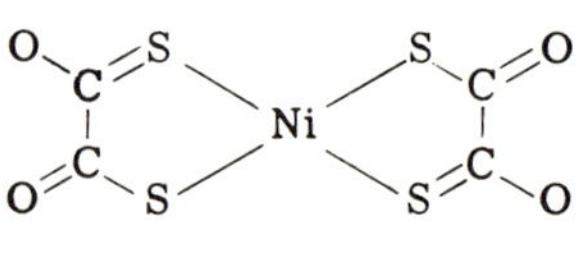

(XXX)

The only ring dimension measured was Ni—S (2.30 Å).

A crystal analysis of anhydrous oxalic acid carried out by Hendricks

(1935) showed that the molecule of this substance has a planar structure. One of the first detailed analyses of a metal chelate in which an oxalato group is present was made by Van Niekerk and Schoening (1951), who investigated *trans*-$K[Cr(C_2O_4)_2(H_2O)_2]3H_2O$. Their work left no doubt about the chelating function of the oxalate group and provided dimensions of the planar chelate ring, details of which are shown in (XXXI).

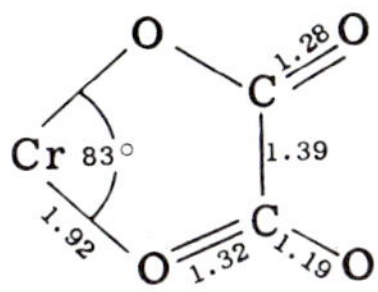

(XXXI)

Stosick's (1945) X-ray analysis of bis(glycinato)diaquonickel(II) revealed several structural features of interest. First, in confirmation of Ley and Bruni's original suggestion, he found that glycine functioned as a bidentate ligand; second, he found that the chelate ring formed by glycine was almost planar. This latter was not altogether surprising, since the earlier analysis of glycine by Albrecht and Corey (1939) had shown that the molecule of this substance is also approximately planar. In the nickel(II) complex, the metal atom forms coplanar and nearly square bonds to the nitrogen and two oxygen atoms of the glycine groups which are in *trans* positions. The two water molecules occupy 1,6 octahedral positions at a distance of 2.12 Å from the nickel atom. The dimensions of the chelate ring were not accurately determined.

An analysis of the structure of the historically interesting bis(glycinato)-aquocopper(II) molecule was not made until quite recently (Tomita and Nitta, 1961). A notable feature disclosed by their analysis is that two planar glycine rings are attached to copper with a *cis* configuration (III). This structure has since been confirmed by Freeman and Snow (1962) who began an analysis of the compound independently. The environment of the copper atom in the crystal consists of six atoms in a distorted octahedral arrangement shown in (XXXII). It is of interest to compare the bond lengths of the copper chelate determined by Freeman *et al.* (1962) with the corresponding bond lengths in glycine itself. The latter bond lengths, shown in parentheses, have been determined by Marsh (1958). It will be seen that the only significant differences are in the C—O bonds of the carboxyl group.[5]

[5] Strange as it may seem, the two chelate rings in bis(glycinato)copper(II) are not identical. X-ray analysis reveals small but definite differences in the dimensions of the two rings. The distances shown in (XXXII) refer to only one of the rings.

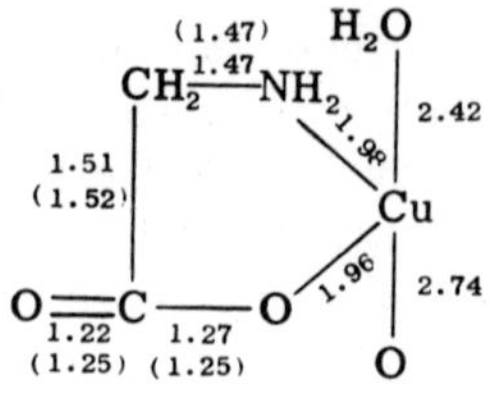

(XXXII)

c. Six-Membered Planar Rings. Nearly planar six-membered chelate rings have been found in diaquobis(acetylacetonato)cobalt(II), the crystal structure of which was recently analyzed by Bullen (1959). Since the rings are unsaturated, a planar ring structure would be expected. The dimensions of the ring are shown in (XXXIII).

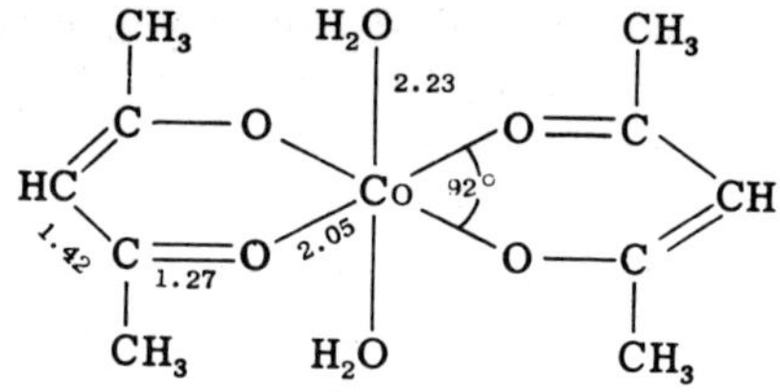

(XXXIII)

The atoms in the chelate ring are very nearly coplanar and so also are the two chelate rings. A curious feature of this structure is that the cobalt atom lies about 0.4 Å above the plane of the two rings. It will be seen that there is not much distortion of the O—Co—O angle.

d. Puckered Rings. Early investigations of metal chelates containing ethylenediamine were concerned mainly with the stereochemistry of the central metal atom. Little attention was paid to the structure of the five-membered ring formed by ethylenediamine which was generally assumed to be planar. All the five-membered rings in the structures so far considered are formed from chelating agents—thiooxalic and oxalic acids, glycine, and 8-hydroxyquinoline—which are themselves resonating planar molecules. However, some saturated five-membered rings are planar. There can, for example, be little strain in the planar, regular pentagonal ring of pentamethylene, since the ring angle of 108° closely approximates the tetrahedral angle of 109° 28′. The situation is different in the ring formed by ethylenediamine with a metal atom. The natural bond angle at the metal atom (N—M—N) for octahedral coordination is 90°, and the bond lengths C—C, C—N, and N—M are all different. For such a ring to assume planar configuration would involve strain.

One of the important results of the first complete X-ray crystal analyses of a trisethylenediamine chelate, namely $[Co(en)_3]Cl_3NaCl{\cdot}6H_2O$ (Nakatsu *et al.*, 1957), was that it focused attention on the conformation of the chelate ring. Nakatsu and his co-workers not only verified the octahedral configuration of cobalt(III), but they also determined the size and shape of the puckered chelate ring, the dimensions of which are summarized below:

C—C = 1.54 Å	<CoNC = 109.5°
C—N = 1.47 Å	<NCC = 109.6°
Co—N = 2.00 Å	<NCoN = 87.4°

The existence of puckering in chelate rings greatly complicates the stereochemistry of metal chelates. This subject is taken up in greater detail in Chapter 5.

2. *Bidentate Ligands That Exhibit Alternative Ways of Chelation*

Several examples have already been given of chelating molecules for which there are alternative ways of forming rings with metal atoms. In each case only one of the two possible ways was found to occur. The next chelating agent to be considered, namely, biuret, exhibits, in different compounds, different modes of chelation. Biuret may combine with a metal atom like copper by means of two nitrogen atoms, two oxygen atoms, or one nitrogen and one oxygen atom. This implies, of course, that the molecule is flexible—that rotation is possible about the C—N bonds. In the crystalline state, the molecule of biuret is approximately planar, and has the *trans* configuration shown in (XXXIV) (Hughes *et al.*, 1961). Freeman and his co-workers (1961), in the course of their studies of metal-peptide complexes, have carried out structural analyses of two biuret copper chelates:

(1) potassium bis(biureto)cuprate(II) tetrahydrate, $K_2[Cu(NH{\cdot}CO{\cdot}NH{\cdot}CO{\cdot}NH)_2]4H_2O$;

(2) bis(biuret)copper(II) chloride, $[Cu(NH_2{\cdot}CO{\cdot}NH{\cdot}CO{\cdot}NH_2)]Cl_2$.

The former compound is prepared in an alkaline solution. Structural analysis reveals that under these conditions each biureto ion is attached to the copper atom by means of two nitrogen atoms (XXXV). The four nitrogens are situated at the corners of an almost perfect square. That the biureto ion would chelate in this manner was suggested many years ago by Tschugaef (1907).

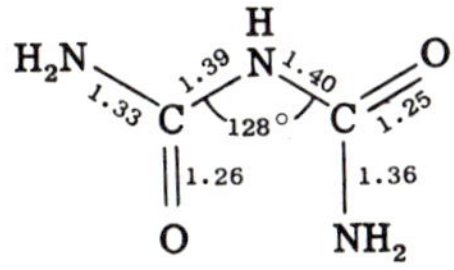

(XXXIV)

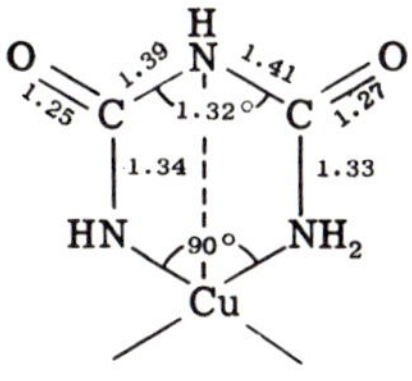

(XXXV)

The chelate ring is not quite flat, being slightly twisted about the axis—shown as a dotted line. An important point established by this analysis is that within the moderate limits of the accuracy of the biuret structure determination chelation produces no significant change in the linear and angular dimensions of the ligand which is taken to mean that chelation does not seriously disturb the bond orders and resonance of the ligand. In the crystal of the metal chelate, the molecules are packed in such a manner that each copper atom has an environment of six nitrogen atoms—four from its two chelating molecules, and one from each of two neighboring molecules. The former are at distances of 1.93 Å, the latter are at a distance of 3.33 Å and cannot therefore be regarded as bonded to the copper atom. The absorption spectrum of the crystalline chelate is almost identical with that of its aqueous solution which has an absorption maximum at 505 mμ. The chelate metal therefore preserves its structure in solution, since the maximum at 505 mμ is associated with copper coordinated with four peptide nitrogen atoms.

The second compound, bis(biuret)copper(II) chloride, is prepared from neutral or acid solution. Structural analysis reveals a most interesting difference in the manner of chelation (Freeman *et al.*, 1959). In this compound biuret is attached to a copper atom through four oxygen atoms.

The results of an analysis of bis(biuret)cadmium(II) chloride by Cavalca *et al.* (1959) illustrate the danger of drawing structural conclusions on the basis of analogous formulas. Surprisingly enough, in this cadmium compound biuret does not function as a chelating molecule, being attached to the metal atom through only one oxygen atom. The structure consists of infinite chains of the type shown in (XXXVI) (B = *trans* biuret, attached to cadmium through an oxygen atom).

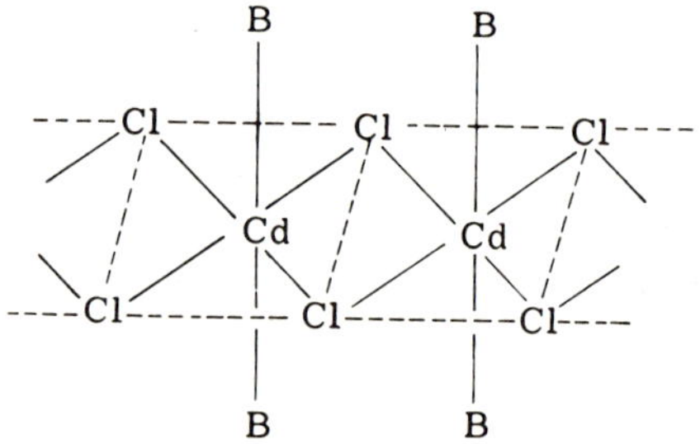

(XXXVI)

An even more extraordinary example of the failure of a molecule to chelate with a metal atom, despite its ability to do so, is to be found in the compound $K[Pt(acac)_2Cl]$ (acac = acetylacetonato group). X-ray analysis reveals that one of the acetylacetonato groups is attached to the platinum atom through carbon (Figgis *et al.*, 1962).

3. *Quadridentate Ligands*

One of the best known quadridentate chelating agents is phthalocyanine. All the chemical evidence (Linstead, 1934) is consistent with the view that the molecule in its combination with metal atoms functions as a quadridentate. Robertson's (1936) X-ray analysis of the chelating agent itself was an outstanding event in the history of structural chemistry. Writing of it, he said: "It is remarkable that the complex phthalocyanine molecule, governed by 60 independent parameters for the carbon and nitrogen atoms alone, should be the first organic structure to yield to an absolutely direct X-ray analysis which does not even involve any assumptions regarding the existence of discrete atoms in the molecule."

This notable achievement was made possible by the use of the heavy atom technique. For the application of this technique, the platinum derivative proved well suited. X-ray crystal analysis confirmed the structure put forward on the basis of chemical evidence. All the atoms in phthalocyanine are coplanar. The phthalocyanine molecule itself does not have a fourfold axis of symmetry; this is destroyed by the hydrogen bonding of two of the four nitrogen atoms. Between the four nitrogen atoms, there is sufficient space to accommodate a metal atom (XXXVII).

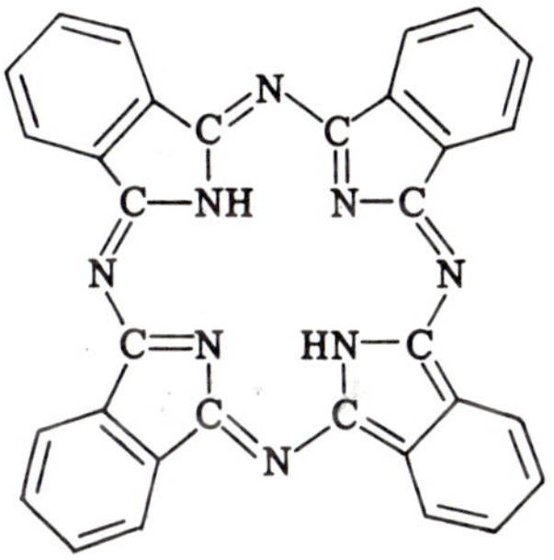

(XXXVII)

The distance from the center of the isoindole nitrogen atoms to the center of the molecule in free phthalocyanine is 1.90 ± 0.03 Å. If a covalent radius of 0.55 Å is assumed for nitrogen, there is a hole in the center of the molecule approximately 1.35 Å in radius into which metal atoms must fit. It might be expected that such a metal-binding site would be highly selective if not specific in its operation. In its ability to form metal chelates phthalocyanine can be ranked with acetylacetone and ethylenediaminetetraacetic acid. It is true that not all kinds of metal atoms are equally well accommodated in this site. Consequently, there is considerable variation in the stability of the metal chelates of phthalocyanine. Those of the alkali metals are among the least stable while those of nickel(II) and platinum(II) are among the most stable. The radii of the metal atoms of the phthalocyanine compounds whose structures have been examined by X-ray analysis range from 1.24 Å for nickel(II) to 1.38 Å for platinum(II).

Once a metal atom becomes attached to the four nitrogen atoms in phthalocyanine, the molecule assumes fourfold symmetry. No readjustment of bond angles is required for metal atoms that are normally square or octahedrally bonded, but for those that are tetrahedrally bonded large deformations of bond angles are inevitable. On an atom like beryllium(II) whose normal configuration is tetrahedral, phthalocyanine imposes its own steric requirements.

4. *Sexadentate Ligands*

That ethylenediaminetetraacetic acid (H_4EDTA) could function as a sexadentate group was first suggested by Brintzinger *et al.* (1943) who based their argument on the stoichiometry of Na[CoEDTA] which they prepared by heating the 4-hydrate at 150°C. A more substantial experimental basis for this view of the function of H_4EDTA was provided by Busch and Bailar (1953) and Dwyer *et al.* (1955) who resolved the compound Na[CoEDTA]-$4H_2O$.

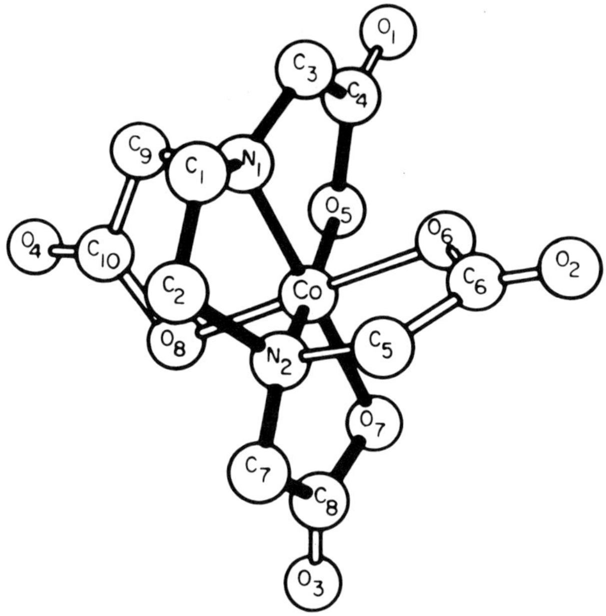

FIG. 3. The structure of $[CoEDTA]^-$ showing the mode of attachment of the chelating ion.

Weakliem and Hoard (1959) have confirmed the conclusions based on chemical evidence by means of a crystal analysis of this compound. The shapes and disposition of the five interlocking rings of the chelate ion are shown in Fig. 3.

The cobalt atom is bonded octahedrally to two nitrogen and four oxygen atoms. Three of the rings (including the ethylenediamine ring) form a girdle about the cobalt atom; the bonds in these three rings have been thickened. At right angles to the girdle and at right angles to one another are two almost planar rings.

Ethylenediaminetetraacetic acid does not always function as a sexadentate. In some metal chelates only four of the functional groups are attached to the central metal atoms; in others five of the groups are attached—the remaining coordination position being occupied by a water molecule or other unidentate ligand. In hydrogen ethylenediaminetetra-

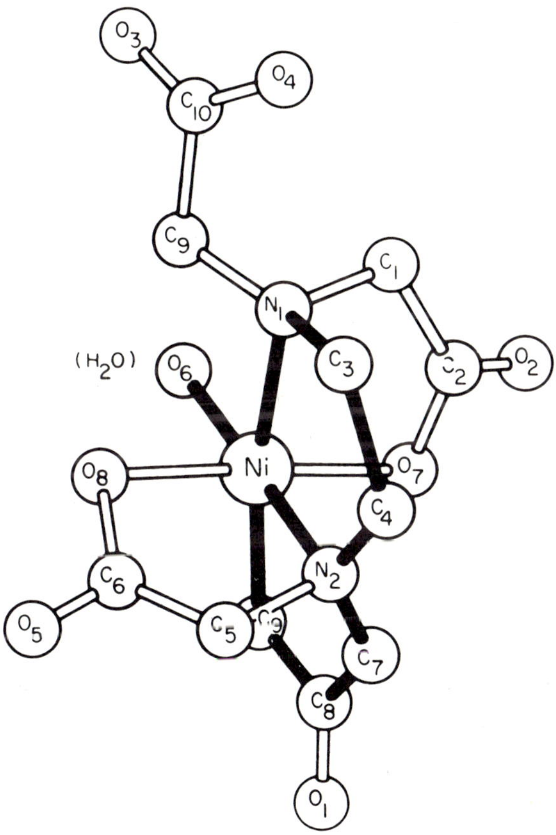

FIG. 4. The structure of $H[Ni(H_2O)HEDTA]$ in which $HEDTA^{3-}$ functions as a quinquedentate.

acetatoaquonickel(II), analyzed by Smith and Hoard (1959), EDTA functions as a quinquedentate group. From Fig. 4, which shows a model of $H[Ni(H_2O)HEDTA]$ in perspective, it will be seen that one of the "girdle" rings (shaded) has opened and the position previously occupied by a carboxyl oxygen is now filled by a water molecule.

V. The Stability of Metal Chelates

A. Definitions

The solution stability of a metal complex (as distinct from its thermal stability or its stability towards oxidation or reduction) is concerned with an equilibrium of the type:

$$M + nL \rightleftharpoons ML_n \tag{1}$$

By applying the law of mass action to Eq. (1), an equilibrium constant may be expressed as follows:

$$\beta_n{}^\circ = \frac{\{ML_n\}}{\{M\}\{L\}^n} \tag{2}$$

where $\{M\}$ is the activity of the species M and $\beta_n{}^\circ$ is the over-all thermodynamic stability constant for the reaction.

Reaction (1) almost always takes place stepwise:

$$M + L \rightleftharpoons ML \tag{3}$$

$$ML + L \rightleftharpoons ML_2 \tag{4}$$

$$ML_{n-1} + L \rightleftharpoons ML_n \tag{5}$$

The step stability constant $K_n{}^\circ$ for the last step is given by the expression

$$K_n{}^\circ = \frac{\{ML_n\}}{\{ML_{n-1}\}\{L\}} \tag{6}$$

The constant that is determined experimentally is usually the concentration or stoichiometric stability constant (β_n) in which concentrations of the species (indicated by square brackets) replace activities. Thus

$$\beta_n = \frac{[ML_n]}{[M][L]^n} \tag{7}$$

By applying the law of mass action to Eqs. (3), (4), and (5), we obtain the step constants

$$K_1 = \frac{[ML]}{[M][L]} \tag{8}$$

$$K_2 = \frac{[ML_2]}{[ML][L]} \tag{9}$$

$$K_n = \frac{[ML_n]}{[ML_{n-1}][L]} \tag{10}$$

$$\beta_n = K_1 \cdot K_2 \cdot K_3 \ . \ . \ . \ K_n \tag{11}$$

A comprehensive and thorough treatment of the theory and measurement of stability constants is to be found in the book by Rossotti and Rossotti (1961).

The reciprocal of β_n, known as the instability constant, which is sometimes used in the description of metal complexes (Yatsimirskii and Vasilev, 1960), is analogous to the dissociation constant of an acid; whereas in acid-base equilibria, a base combines with a hydrogen ion, in complex formation a ligand combines with a metal ion. It is therefore possible to devise metal-ion buffer solutions analogous to acid-base buffers.

Metal complexes may be roughly classified on the basis of the speed with which equilibrium between metal ion and ligands is attained. Those which attain it rapidly (practically instantaneously) are described as "labile," those which attain it slowly as "inert" or "robust." Both aspects of the equilibrium—position and speed of attainment—are important for a proper understanding of the behavior of metal complexes in solution. Some confusion has arisen at times from a failure to discriminate clearly between the two.

There is not necessarily any correlation between stability and lability or between stability and inertness. A complex may be very stable and, at the same time, quite labile. The stability constant involves the ratio of two velocity constants—one for the forward reaction and one for the backward reaction. Both of these may be large, yet their ratio may also be large.

Many of the most important discoveries of geometrical and optical isomerism that formed the experimental basis of the theory of the structure of metal complexes were made with inert complexes—complexes of Co(III), Cr(III), Pt(II), Pt(IV), and many others.

Metals that form labile complexes, e.g., Co(II), Cd(II), Mn(II), Zn(II), and Cu(II), are notable for the rarity with which they give rise to geometrical or optical isomers. Whether a metal will form labile or inert complexes depends upon its electronic configuration. Taube (1952) has examined inner and outer orbital octahedral complexes in the light of valence bond theory. Basing his classification on lability with respect to substitution, he has noted that among inner orbital complexes a striking discontinuity appears at the point where the inner d orbitals all become at least singly occupied. Such a transition occurs between vanadium(III) and chromium(III). Simple substitution reactions of the former are all rapid; those of the latter are all slow.

No such sharp distinction can be drawn between inert and labile outer orbital complexes on the basis of electronic structure. In general, however, outer orbital complexes are labile. For a detailed discussion of the inertness and lability of complexes, the reader is referred to the work of Basolo and Pearson (1958).

B. The Importance of Stability Constants

Just as a knowledge of the dissociation constants of acids and bases has done much to systematize our understanding of the behavior of these substances in all kinds of systems, so a knowledge of stability constants has done much to rationalize our understanding of the behavior of metal chelates in solution.

The conditions required for the maximal or complete formation of a complex may be predicted on the basis of its stability constant. Information about the concentration of different species of a metal complex in equilibrium mixtures is invaluable in the study of analytical separation procedures—solvent extraction, ion exchange, or chromatography. It is equally important in the study of the role of metal chelates in biological systems.

Since the stability constant refers to a system in equilibrium, it may be used to calculate the free energy change (ΔG) and other thermodynamic functions of the reaction involved by means of the equation:

$$-RT \ln K = \Delta G = \Delta H - T\Delta S$$

In theory, it is possible to determine the enthalpy change (ΔH) by measuring the stability constant at a series of temperatures; in practice, it is better to determine ΔH by direct calorimetric measurement.

A recent compilation (Bjerrum *et al.*, 1957) lists data for complexes formed by some sixty metals with fifty-six inorganic and 464 organic ligands.

C. The Chelate Effect

Metal chelates are inherently more stable than closely related non-chelate complexes. This statement which describes what is known as the chelate effect may be illustrated qualitatively by considering the two compounds $[Ni(en)_3]Cl_2$ and $[Ni(NH_3)_6]Cl_2$. The latter is rapidly hydrolyzed in water; the aqueous solution has a distinct odor of ammonia and at a concentration 0.01 M yields a precipitate of nickel hydroxide. Aqueous solutions of the ethylenediamine complex, on the other hand, are stable even at a concentration of 0.001 M. Numerous quantitative studies also support the above statement. Spike and Parry (1953) have made a detailed study of several complexes of cadmium. Table IV, which is taken from their paper, summarizes some of their results.

From Table IV, it will be seen that the stability of the metal chelate is

TABLE IV

CADMIUM COMPLEXES AND THE CHELATE EFFECT

Complex	Log stability constant[a]	ΔH (kcal./mole)	ΔF (kcal./mole)	$T\Delta S$ (kcal./mole)
$[Cd(NH_3)_2]^{++}$	4.950	7.12	6.75	0.37
$[Cd(NH_2CH_3)_2]^{++}$	4.808	7.02	6.56	0.46
$[Cd(en)]^{++}$	5.836	7.03	7.96	−0.93
$[Cd(NH_3)_4]^{++}$	7.44	12.7	10.16	2.53
$[Cd(NH_2CH_3)_4]^{++}$	6.55	13.7	8.94	4.77
$[Cd(en)_2]^{++}$	10.62	13.5	14.5	−0.98

[a] K_1 for 3, log β_2 for 1, 2, and 6, and log β_4 for 4 and 5.

in each case greater than that of the corresponding nonchelate complex. The strength of the metal-ligand bonds as indicated by ΔH is roughly the same in both types of complex.

There has been a good deal of argument about the origin of the chelate effect and there is no doubt that a number of factors are involved. A partial explanation may be given in terms of translational entropy (Orgel, 1960). Consider the case of two complexes ML_2 and M(LL) (e.g., $[Cd(NH_3)_2^{++}$ and $[Cd(en)]^{++}$).

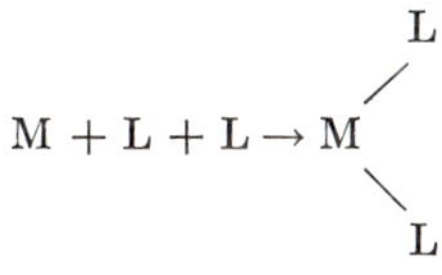

(12)

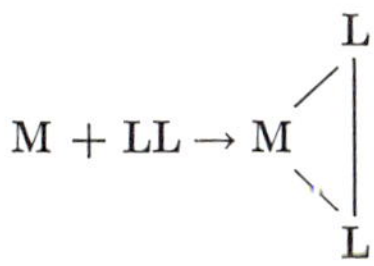

(13)

Before Reaction (12) occurs, there are three molecules moving independently but after the reaction is complete all three move together. The entropy change during this reaction includes a term which allows for the fact that the translational entropy of two molecules is lost. In Reaction (13), the translational entropy of only one molecule is lost. As Orgel points out this is a considerably oversimplified explanation because it ignores important degrees of freedom other than those of translation.

D. Factors Influencing the Stability of Metal Complexes

Two general approaches have been made to the problem of discovering the factors which influence the stability of a metal complex. The first involves the investigation of the stability of complexes formed by a particular metal with a series of different, but usually related, chelating molecules. The second, which is essentially complementary to the first, involves the investigation of the stability of complexes formed by a series of metal ions with a given chelating molecule.

In the first approach, the following factors are considered:

(1) The size of the chelate ring;
(2) The number of rings (fused rings in multidentate chelating molecules);
(3) The basic strength of the chelating molecule;
(4) Substitution in the chelating molecule which may modify basic strength or π bonding or introduce steric effects;
(5) The nature of the donor or ligand atoms.

1. *Ring Size*

Most of the observations on the effect of ring size on stability are qualitative, being based on preparative experience, but in the few instances that have been studied quantitatively, for example complexes of oxalate and malonate and of 8-hydroxyquinoline and 1-(*o*-hydroxyphenylisoquinoline), the five-membered rings formed by the first of each of these pairs of substances proved to be the more stable.

The comparison is only valid when other factors remain constant; in other words, it must be made between closely related chelate rings whose principal difference is their size.

2. *The Number of Rings*

Of two similar chelating agents, that which forms the greater number of chelate rings with a given metal forms the more stable complex. This is illustrated with the copper chelates whose stability constants are shown in Table V.

The effect of the number of rings in the chelating agent on stability may be seen by comparing β_2 (shown in parenthesis in the table) for the complex with ammonia with K_1 for the diamine, β_3 with K_1 for the triamine, and β_4 with K_1 for the tetraamine. These are further examples of the chelate effect.

TABLE V
STABILITY CONSTANTS OF COPPER CHELATES AS A FUNCTION OF THE NUMBER OF RINGS[a]

Chelating agent	Number of donor groups	Log stability constant K_1	K_2	K_3	K_4
NH_3	1	4.2	3.5 (7.7)	2.9 (10.6)	2.1 (12.7)
$NH_2 \cdot CH_2 \cdot CH_2 \cdot NH_2$	2	10.8	—	9.3	—
$NH(CH_2 \cdot CH_2 \cdot NH_2)_2$	3	16.0	—	—	5.4
$N(CH_2 \cdot CH_2 \cdot NH_2)_3$	4	18.8	—	—	—

[a] This table is based on one drawn up by Schwarzenbach (1961).

3. *The Basic Strength of the Chelating Molecule*

Both metal ions and hydrogen ions are Lewis acids, that is, electron acceptors. There is, therefore, a resemblance between the combination of a metal with a ligand and the neutralization of a base with hydrogen ion. Many ligands are the conjugate bases of acids. It would not be unreasonable then to expect some correlation between the basic strengths of a series of ligands and the stability of the complexes they form with a particular metal. Calvin and Wilson (1945) investigated this question by studying the copper chelates of a series of salicylaldehyde and β-diketone derivatives. On plotting log β_2 against pK_a, they obtained a number of roughly parallel straight lines—each line corresponding to a particular series of chelating agents. In other words, they showed that, for a series of closely related chelating agents, the greater the basic strength (pK_a), the greater the stability of the metal chelate.

4. *The Effect of Substitution in the Chelating Molecule*

Substitution in the chelating molecule may not only alter its basic strength and so alter the stability of any complexes it forms but it may also introduce steric effects which, as pointed out earlier, may in extreme cases prevent chelation altogether. Johnston and Freiser (1954) have shown that complexes of 8-hydroxyquinaldine are less stable than those of 8-hydroxyquinoline despite the fact that the former chelating molecule is the stronger base of the two. This they attribute to the effect of steric hindrance arising from the chelation of 8-hydroxyquinaldine.

5. *Nature of the Donor Atoms*

The question of the influence of the nature of the donor atom on stability is a complicated one since the donor atom must also be considered in relation to the acceptor atom. The relative affinities of ligand atoms for acceptor ions have been discussed in detail by Ahrland and his associates (1958). They found it convenient to divide acceptors into two main classes[6]:

(*a*) those which form their most stable complexes with atoms of the first element of each group of the periodic table, i.e., with N, O, and F; and

(*b*) those which form their most stable complexes with atoms of the second or subsequent elements (P, S, Cl . . .).

These classes are shown in Fig. 5.

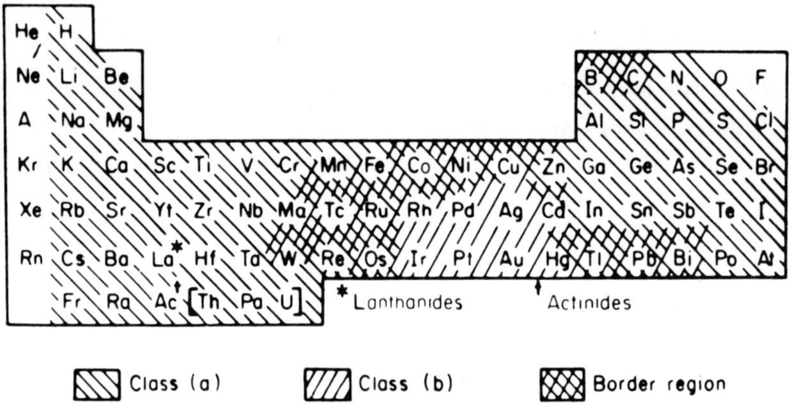

FIG. 5. The two main classes of acceptor atoms.

There is some overlapping of the two classes. The state of oxidation of the metal may well be a determining factor. For example, copper(I) belongs to class (*b*); copper(II) is on the borderline between the two classes. This classification is based for the greater part on unidentate ligands. The situation is more complicated in chelates where at least two donor atoms per ligand must be considered.

If the discussion is restricted to bidentate chelates, it is necessary to consider only two possibilities:

(1) chelating agents in which the donor atoms are identical as in oxalate and thiooxalate ion;

(2) chelating agents in which the donor atoms are different as in *o*-aminophenol and *o*-aminothiophenol.

In order to make a comparison under (2), it is essential that one donor atom, in this case nitrogen, be common to both chelating agents.

[6] This and an alternative classification based on the different types of nonbonding electronic "cores" present in acceptor metal atoms are discussed in detail in Chapter 2.

There are no data on the stability of the bisthiooxalatonickel(II) ion. However, the fact that the thiooxalato ion chelates with nickel through the sulfur rather than oxygen indicates that thiooxalato ion forms a more stable nickel chelate than the oxalato ion.

Measurements of the stability constants of the *o*-aminophenol and *o*-aminothiophenol complexes of zinc(II) show that the substitution of sulfur for oxygen does have a marked influence on stability, increasing it from $\log \beta_2 = 10.95$ for the former compound to $\log \beta_2 = 14.10$ for the latter.

6. *Influence of the Central Metal Atom*

This may be studied by comparing the stabilities of the complexes formed by a particular chelating molecule with a series of metal atoms in a given oxidation state. That the order of stability in such a series may be independent of the nature of the chelating agent was first discovered by Mellor and Maley (1948). For the bivalent metals, they found an order

$$\text{Pd} > \text{Cu} > \text{Ni} > \text{Co} > \text{Zn} > \text{Cd} > \text{Fe} > \text{Mn} > \text{Mg}$$

but were unable to decide what factor or factors determined this order. Very shortly afterwards, Irving and Williams (1948) noted the same order of stability in their studies on extraction. They showed, moreover, that if the stability constants of complexes of bivalent ions of the first transition series are plotted against atomic number, the stability increases to a maximum at copper. This restricted the series but put it on a more rational basis. The series then became

$$\text{Zn} < \text{Cu} > \text{Ni} > \text{Co} > \text{Fe} > \text{Mn}$$

It has since been tested and found to hold rigorously for upwards of eighty different ligands (Irving, 1959) in which the donor atoms are nitrogen and oxygen.

More recent investigations (Crabtree *et al.*, 1961) of complexes derived from metals in the first part of the first row of the transition series (Ti–Cr) have shown π bonding has a greater influence on their stability than it has on the stability of complexes of the metals from Mn–Cu. A consequence is that the order of stability of complexes of spin-free bivalent metals in the first part of the series is dependent on the nature of the ligand.

E. Specificity and Selectivity in the Formation of Metal Chelates

The results of stability constant measurements have made it abundantly clear that the ability of metal ions to form chelates is general, and with a given chelating agent only changes gradually from one metal to the next. On this evidence, it seems most unlikely that any chelating agent will be found that will combine with one kind of metal atom and no other. Absolu e specificity in this sense is unattainable. This does not mean that there is no such thing as a specific test where the specificity resides not in the reagent but the product of the test. Dimethylglyoxime, which combines readily with a number of metals, forms a highly insoluble red chelate with one metal only, namely, nickel. The specificity of this reaction lies in the color of the product; other metals like palladium and platinum form insoluble chelates with dimethylglyoxime which are yellow and blue, respectively, but nickel is the only metal to form a water-insoluble red compound. It may well be that the specificity of biologically important metals resides mainly in the unique physical and chemical properties of their chelate compounds.

While chelating agents do not exhibit absolute specificity in their reactions with metal ions, they may exhibit varying degrees of selectivity according to circumstances. A good example of this may be seen in the use of chelating agents as selective precipitants in analytical chemistry. 8-Hydroxyquinoline forms water-insoluble precipitates with upwards of thirty different metals. This number may be reduced by controlling the pH of the solution from which precipitation takes place. A still greater degree of selectivity may be brought about by the use of "masking" or sequestering reagents. Thus 8-hydroxyquinoline will precipitate very few metals from a buffer solution (ammonium acetate and acetic acid) containing ethylenediaminetetraacetic acid as a sequestering agent.

References

Ahrland, S., Chatt, J., and Davies, N. (1958). *Quart. Revs. (London)* **12,** 265.

Albrecht, G., and Corey, R. B. (1939). *J. Am. Chem. Soc.* **61,** 1087.

Alexander, W. A., and Pauson, P. L. (1961). *J. Inorg. & Nuclear Chem.* **17,** 186.

Arrhenus, S. (1887). *Z. physik. Chem. (Leipzig)* **1,** 631.

Barclay, G. A., and Hoskins, B. F. (1962). *J. Chem. Soc.* p. 586.

Barclayi, G. A., and Nyholm, R. S. (1953). *Chem. & Ind. (London)* p. 378.

Basolo, F., and Pearson, R. G. (1958). "Mechanisms of Inorganic Reactions." Wiley, New York.

Bjerrum, J., Schwarzenbach, G., and Sillen, L. G. (1957). "Stability Constants." *Chem. Soc. (London) Spec. Publ.* **No. 7.**

Blau, F. (1898). *Monatsh.* **19,** 647.

Blomstrand, C. W. (1869). "Chemie der Jetztzeit vom Standpunkte der electrochemischen Auffassung aus Berzelius Lehre entwickelt." Heidelberg.

Brintzinger, H., Thiele, H., and Müller, U. (1943). *Z. anorg. u. allgem. Chem.* **251,** 285.

Bruni, G., and Fornara (1904). *Rend. accad. nazl. Lincei* **13,** 26.

Bullen, G. J. (1959). *Acta Cryst.* **12,** 703.

Busch, D. H., and Bailar, J. C. (1953). *J. Am. Chem. Soc.* **75,** 4574.

Calvalca, L., Nardelli, M., and Fava, G. (1959). *Acta Cryst.* **13,** 594.

Calvin, M., and Wilson, K. W. (1945). *J. Am. Chem. Soc.* **67,** 2003.

Cox, E. G., Wardlaw, W., and Webster, K. C. (1935). *J. Chem. Soc.* p. 1475.

Crabtree, J. M., Marsh, D. W., Tomkinson, J. C., Williams, R. J. P., and Fernelius, W. C. (1961). *Proc. Chem. Soc.* p. 336.
Diehl, H. (1937). *Chem. Revs.* **21,** 39.
Dwyer, F. P., and Lions, F. (1947). *J. Am. Chem. Soc.* **69,** 2917.
Dwyer, F. P., and Lions, F. (1950). *J. Am. Chem. Soc.* **72,** 1545.
Dwyer, F. P., and Nyholm, R. S. (1941). *J. Proc. Roy. Soc. N.S.Wales* **75,** 127.
Dwyer, F. P., Gyarfas, E. C., and Mellor, D. P. (1955). *J. Phys. Chem.* **59,** 296.
Figgis, B. N., Lewis, J., Long, R. F., Mason, R., Nyholm, R. S., Pauling, J. P., and Robertson, G. B. (1962). *Nature* **195,** 1278.
Fox, D. B., Hall, J. R., and Plowman, R. A. (1962). *Australian J. Chem.* **15,** 235.
Frasson, E., Bardi, R., and Beggi, S. (1959). *Acta Cryst.* **12,** 201.
Freeman, H. C., and Snow, M. R. (1962). Private communication.
Freeman, H. C., Smith, J. E., and Taylor, J. C. (1959). *Nature* **184,** 707.
Freeman, H. C., Smith, J. E., and Taylor, J. C. (1961). *Acta Cryst.* **14,** 407.
Freeman, H. C., Nitta, I., Snow, M. R., and Tomita, K. (1962). Private communication.
Gelman, A. D., and Essen, L. N. (1950). *Doklady Akad. Nauk S.S.S.R.* **75,** 693.
Godycki, L. E., and Rundle, R. E. (1953). *Acta Cryst.* **6,** 487.
Goodwin, H., Gyarfas, E. C., and Mellor, D. P. (1958). *Australian J. Chem.* **11,** 426.
Grinberg, A. A. (1962). "The Chemistry of Complex Compounds." Pergamon Press, New York.
Harris, C. M., Kokot, E., and Lenzer, S. L. (1962). *Nature* **196,** 472.
Hendricks, S. B. (1935). *Z. Krist.* **91,** 48.
Hodgkin, D. C., and Dale, D. H. (1962). Private communication.
Hodgkin, D. C., Kamper, J., Lindsey, J., MacKay, M., Pickworth, J., Robertson, J. H., Shoemaker, C. B., White, J. G., Prosen, R. J., and Trueblood, K. N. (1957). *Proc. Roy. Soc.* **A242,** 228.
Hodgkin, D. C., Pickworth, J., Robertson, J. H., Prosen, R. J., Sparks, R. A., and Trueblood, K. N. (1959). *Proc. Roy. Soc.* **A251,** 306.
Hughes, E. W., Yakel, L., and Freeman, H. (1961). *Acta Cryst.* **14,** 345.
Irving, H., and Williams, R. J. P. (1948). *Nature* **162,** 746.
Irving, H., Butler, E. J., and Ring, M. F. (1949). *J. Chem. Soc.* p. 1489.
Irving, H. M. N. H. (1959). *Proc. Intern. Conf. on Coordination Chem., London* p. 13.
Johnston, W. D., and Freiser, H. (1954). *Anal. Chim. Acta* **11,** 201.
Jorgensen, S. M. (1887). *J. prakt. Chem.* **35,** 417.
Jorgensen, S. M. (1890). *J. prakt. Chem.* **41,** 440.
Langmuir, I. (1919). *J. Am. Chem. Soc.* **41,** 868, 1543.
Lewis, G. N. (1916). *J. Am. Chem. Soc.* **38,** 762.
Lewis, J., Nyholm, R. S., and Reddy, G. K. M. (1960). *Chem. & Ind. (London)* p. 1386.
Ley, H. (1904). *Z. Elektrochem.* **52,** 954.
Liebhafsky, H. A. (1946). *J. Chem. Ed.* **23,** 341.
Linstead, R. P. (1934). *J. Chem. Soc.* p. 1017.
Marsh, R. E. (1958). *Acta Cryst.* **11,** 654.
Mellor, D. P. (1943a). *J. Proc. Roy. Soc. N.S.Wales* **75,** 145.
Mellor, D. P. (1943b). *Chem. Revs.* **33,** 137.
Mellor, D. P., and Maley, L. (1948). *Nature* **159,** 370.
Mills, W. H., and Quibell, J. H. H. (1935). *J. Chem. Soc.* p. 839.
Morgan, G. T., and Drew, H. D. K. (1920). *J. Chem. Soc.* **117,** 1456.
Nakatsu, K., Shiro, M., Saito, Y., and Kuroya, H. (1957). *Bull. Chem. Soc. Japan* **30,** 158.
Nyholm, R. S. (1961). *Proc. Chem. Soc.* p. 273.

Orgel, L. E. (1960). "An Introduction to Transition-Metal Chemistry: Ligand-Field Theory." Wiley, New York.
Parpiev, N. A., and Porai-Koshits, M. A. (1959). *Soviet Phys. Cryst.* (*Engl. Transl.*) **4,** 26.
Pauling, L. (1939). "The Nature of the Chemical Bond." Cornell Univ. Press, Ithaca, New York.
Pauling, L. (1948). *J. Chem. Soc.* p. 1461.
Peyronel, G. (1941). *Z. Krist.* **103,** 157.
Robertson, J. M. (1936). *J. Chem. Soc.* p. 1195.
Rossotti, J. C., and Rossotti, H. (1961). "The Determination of Stability Constants." McGraw-Hill, New York.
Schwarzenbach, G. (1961). *Advances in Inorg. Chem. Radiochem.* **3,** 257.
Sidgwick, N. V. (1927). "The Electronic Theory of Valency." Oxford Univ. Press, London and New York.
Smith, G. S., and Hoard, J. L. (1959). *J. Am. Chem. Soc.* **81,** 556.
Spike, C. G., and Parry, R. W. (1953). *J. Am. Chem. Soc.* **75,** 2726, 3770.
Stephenson, N. C. (1962). *J. Inorg. & Nucl. Chem.* **24,** 797.
Stosick, A. J. (1945). *J. Am. Chem. Soc.* **67,** 365.
Sugden, S. (1935). *J. Chem. Soc.* p. 621.
Taube, H. (1952). *Chem. Revs.* **50,** 69.
Tomita, K., and Nitta, I. (1961). *Bull. Chem. Soc. Japan* **34,** 286.
Tschugaeff, L. A. (1906). *Ber. deut. chem. Ges.* **39,** 3197.
Tschugaeff, L. A. (1907). *Ber. deut. chem. Ges.* **40,** 1973.
Van Niekerk, J. N., and Schoening, F. R. L. (1951). *Acta Cryst.* **4,** 35; **5,** 196.
von Liebig, J., and Wohler, F. (1832). *Ann. Chem.* **3,** 249.
Waind, G. M., and Martin, B. (1958). "Chemistry of the Co-ordinate Compounds" (A Symposium sponsored by the Italian National Research Council and other bodies). Pergamon Press, New York.
Warner, R. C. (1954). *Trans. N. Y. Acad. Sci.* **16,** 182.
Watt, G. W., and Davies, J. W. (1960). *J. Inorg. & Nuclear Chem.* **14,** 32.
Weakliem, H. A., and Hoard, J. L. (1959). *J. Am. Chem. Soc.* **81,** 549.
Werner, A. (1893). *Z. anorg. u. allgem. Chem.* **3,** 267.
Werner, A. (1901). *Ber. deut. chem. Ges.* **34,** 2584.
Werner, A. (1911). *Ber. deut. chem. Ges.* **45,** 121.
Werner, A. (1912). *Ber. deut. chem. Ges.* **45,** 3065.
Werner, A. (1914). *Ber. deut. chem. Ges.* **47,** 3087.
Werner, A., and Miolati, A. (1894). *Z. physik. Chem.* (*Leipzig*) **14,** 506.
Werner, A., and Vilmos, Z. (1899). *Z. anorg. u. allgem. Chem.* **21,** 153.
West, B. O. (1960). *Revs. Pure and Appl. Chem.* (*Australia*) **10,** 207.
Yatsimirskii, K. B., and Vasilev, V. P. (1960). "Instability Constants of Complex Compounds" (translated from the Russian by D. A. Paterson). Pergamon Press, New York.

CHAPTER 2

The Nature of the Metal-Ligand Bond

D. P. Craig and R. S. Nyholm

Department of Chemistry, University College, London, England

I. Bond Types and Characteristic Properties

A. Introduction

The structure of a metal complex is well defined when the coordination number of the central metal, the stereochemistry, the conformation of the molecule or complex ion, and the nature of the bond between metal and ligand are all known. The nature of the bond itself is closely bound up with coordination number and stereochemistry, and an understanding of it and of its relationship with other properties is fundamental to the interpretation of the structure and reactivity of a metal complex.

We have based the discussion of metal-ligand bonds so far as possible on the fundamental atomic properties of electron configurations, the nature of the orbitals involved, their ionization potentials, and their overlapping powers, and have tried to show that many of the broad features of the bonds formed can be interpreted in those terms. Our discussion of the metal-ligand bond is not exhaustive. Some parts of the subject, such as the crystal and ligand field theories, have been treated fully in excellent and widely available works. We have chosen topics with a view to complementing existing sources as far as possible.

B. Bond Types

1. *General Classification*

Before analyzing bond properties that depend on the metal, it is helpful to set up categories of bond types in terms of which discussion may proceed. In the extreme ionic bond we suppose that a metal ion M^{n+} is attached to a certain number of ligands, either charged, like Cl^-, or uncharged, like NH_3. If the ligands are negatively charged the model is held together by electrostatic attraction between oppositely charged ions. If the ligands are neutral the attraction is between the M^{n+} ion and the dipole (static plus induced) on the NH_3 groups. In the extreme covalent bond, the binding is essentially of the type familiar in the covalent bonds between hydrogen atoms in H_2 or carbon atoms in hydrocarbons; however such a picture cannot apply at all exactly, because the bonded atoms in a complex are very different in size, and the better analogy would be the heteronuclear C—H bond. Also, the bonded atoms are of different electronegativities, so that in most cases the bond will have some polar character. With these points in mind we subdivide the covalent bond into three types: first, the σ bond in which the electron density is greatest along the internuclear axis, and formed in many cases, for example by the overlapping of a lone pair orbital of the ligand with a *dsp* hybrid orbital of the metal; second, the π bond analogous to the auxiliary bond in ethylene, formed by the overlapping of π-type orbitals on ligand and metal as in compounds of the type

$$(CH_3)_2\overset{-}{B} = \overset{+}{N}(CH_3)_2$$

and in the carbonyls of nickel and other transition metals; and, finally, other bonding situations that may broadly be described as multicenter bonds. The most familiar examples are in electron-deficient molecules such as B_2H_6 with a single pair of electrons concentrated in the region of overlap of an sp^3 hybrid orbital from each boron atom and the σ orbital of the hydrogen atom. Among metal complexes the metal cyclopentadienyls are examples of multicenter bonding, insofar as the bond forces cannot be treated as localized between particular atom pairs. The categories are, thus, as follows:

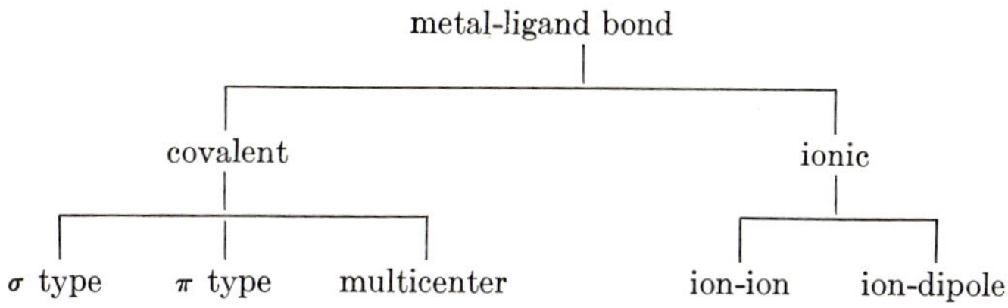

2. σ *Bonds*

The ionic bond can be viewed as an extreme case of the highly polar σ type and, since it is easier to separate the role of the metal and the ligand in this case, it forms a convenient starting point.

Given a metal ion M^{n+} and a ligand L (charged or uncharged) with one or more lone pairs available for bond formation, the attraction of the metal for lone pairs gives rise to several types of σ bond, as shown in the accompanying tabulation.

"SPECTRUM" OF σ BONDS

Most covalent ←				→ Most ionic
Equivalent covalent, e.g., SF_6. All bonds of same type and highly covalent.	*Mixed covalent*, e.g., PCl_5. Two types of covalent bond (axial and radial distinct in symmetry).	*Mixed ionic* and *covalent*, e.g., $HgCl_2 \cdot 2(py)$ strong covalent $6s6p$ linear Cl—Hg—Cl bonds with four weaker (largely ionic) bonds to Cl and Py groups.	*Partly covalent*, e.g., $[Fe(H_2O)_6]^{++}$. Essentially an Fe^{++} ion attached to six dipolar H_2O molecules but with a discrete complex recognizable.	*"Pure ionic"* as in crystalline $FeCl_3$, the Fe^{3+} ion being at center of six octahedrally disposed Cl^- ions. No discrete complex ions recognizable.

3. π *Bonds*

Two main types of π bonds in metal complexes are recognized. Consider the compound $(CH_3)_2B\overset{\leftarrow}{=}N(CH_3)_2$. This can be regarded as derived from the simple σ-bonded structure $(CH_3)_2B—N(CH_3)_2$ in which we have a boron atom with a sextet and a nitrogen atom with a lone pair occupying an orbital with the capability of overlap with the vacant $2p$ orbital of the boron atom. Studies by Becher and Goubeau (1952) of Raman and infrared spectra and electric dipole moments support the view that the structure is in fact $(CH_3)_2B\overset{\leftarrow}{=}N(CH_3)_2$. Ligands which have π lone pairs available including, for example, the O^{--} ion, should double-bond most readily with metals having vacant orbitals. The early transition metal ions, e.g.,

Ti^{4+}, V^{5+}, etc. readily form bonds with such ligands; infrared studies by Barraclough *et al.* (1959) and theoretical investigations by Gray and Ballhausen (1962) support the conclusion that double bonds between the metal and ligand are formed. The second type of π bond involves donation of a pair of electrons by the metal atom to the ligand. This requires the use of an empty p or d orbital on the ligand, or an anti-bonding π-type molecular orbital. No definite cases of the donation of a $p\pi$ pair of electrons by the metal atom have been discussed; but in principle they are feasible, and might be compared with the donation of $d\pi$ pairs of electrons to p orbitals which have been made available on the ligand (e.g., CO) or to empty d orbitals (e.g., in PPh_3). The possible types of π bonds are summarized below:

Types of π bonds in metal complexes

- (*i*) donation by metal
  - $p\pi$–$p\pi$ e.g., $[IPy_2]^+ClO_4^-$
  - $d\pi$–$p\pi$ e.g., $Ni(CO)_4$
  - $d\pi$–$d\pi$ e.g., $Ni(PF_3)_4$
- (*ii*) donation by ligand
  - $p\pi$–$p\pi$ as in $Me_2B \leftrightarrows NMe_2$

C. The Pauling Electroneutrality Principle

This principle is helpful in giving an indication of the number and nature of bonds between metal and ligands. As an example of its implications, it is on the face of it unreasonable to accept that a Fe^{3+} ion (requiring about 1200 kcal. for its formation by the removal of three electrons from an Fe atom) can be thermodynamically stable in the presence of ionizable electrons on neighboring ligands. The metal will tend either to withdraw entirely one or more electrons from the ligand, i.e., to effect oxidation (such as occurs with the I^- ion), or to polarize a larger number of ligands so as to acquire a small amount of negative charge from each. Thus, in the case of the Fe^{3+} ion this may be done by combining with four polarizable ligands to form the well known $[FeCl_4]^-$ ion or with six less easily polarized F^- ions giving the $[FeF_6]^{3-}$ ion. Whether Pauling's criterion of a charge limit of ± 0.5 e is correct or not, studies of the reduction of the spin-orbit coupling constant of metal ions in complex compounds can be plausibly interpreted to confirm that positive charges greater than $+1$ are unlikely to occur.[1]

[1] Thus spin-orbit coupling constants for the isoelectronic d^8 spin-free atoms or ions Fe^0, Co^{+1}, and Ni^{+2} in the gas phase are, respectively, -128, -228, and -315 cm.$^{-1}$.

When the electronic structure of both metal and ligand is allowed for, and the point-charge electrostatic model accordingly discarded, the electroneutrality principle can be given the following interpretation. We draw a sphere round the metal ion at a suitable radius and calculate the total charge within it—including both the electrons of the metal ion and those of the ligands. The principle is then stated in the form that the total charge must be close to electrical neutrality. The closer definition of the principle in this way also robs it of some precision: it is rather arbitrary to define the sphere by, for example, the ionic radius for the trivalent ion; one might have chosen the bipositive ionic radius or the covalent radius and so have included differing amounts of electronic charge, and thus have varied the total charge within the sphere. There is evidence, however, in the form of the actual electron distribution functions that any reasonable choice of radius still leaves the total charge within the range suggested by Pauling.

In utilizing the Electroneutrality Principle interest centers on the total electron population, and there is no conflict with the fact that in Ni(II) complexes spectra and magnetic behavior are conveniently interpreted in terms of a $d_\epsilon^6 d_\gamma^2$ subshell of electrons in a bipositive ion. At normal covalent bond radii the donor pairs of electrons on the ligands can be regarded as being shared so as to make the Ni(II) atom neutral *at that distance.* The fact that the spin-orbit coupling constant is reduced in the Ni(II) complexes below the value expected for a trivalent ion indicates that the bonding pairs penetrate the nonbonding $d_\epsilon^6 d_\gamma^2$ shell to some extent.

D. The Coordination Number

Although the variation in coordination number will be discussed again in connection with the electron configuration of certain metal atoms, it is helpful to point out here that the major determining factors are: (*a*) the oxidation number of the metal atom; (*b*) the effective nuclear charge on the metal atom (i.e., the electronegativity); (*c*) the number of available bonding orbitals. It is helpful for the purposes of classification to divide metal ion acceptors into classes as though they possessed initially (i.e., before complex formation) a positive charge equal to the oxidation state. This is then discharged to a greater or lesser degree by overlap of vacant orbitals with orbitals containing lone pairs of electrons on the ligands. The coordination number (C.N.) is seen mainly as a reflection of the electroneutrality principle. It is convenient to summarize certain generalizations concerning the preferred C.N. of a metal. (*i*) The higher

---

However, the spin-orbit coupling constant of Ni in $[Ni(NH_3)_6]^{++}$ is only -200 cm.$^{-1}$. The simple interpretation of this (Dunn, 1961) is that charge transfer from the H_2O molecules to the Ni^{++} atom has reduced its effective charge to about $+1$.

coordination numbers are usually associated with cationic complexes, the lower with anionic (c.f. $[Co(H_2O)_6]^{++}$ and $[CoCl_4]^{--}$). (*ii*) Higher coordination numbers arise with higher oxidation numbers, e.g., $[CuCl_2]^-$, $[CuCl_4]^{--}$, $[CuF_6]^{3-}$. (*iii*) Coordination number falls with increasing polarizing power of the metal atom, thus, Au(I) with I.P. (ionization potential) 9.22 e.v. forms $Et_3P \rightarrow Au—Cl$, involving two-coordinate Au(I), whereas Cu(I) with I.P. 7.72 e.v. forms the tetrameric $[Et_3P \rightarrow CuCl]_4$ in which the Cu(I) is four-coordinate. (*iv*) Coordination number falls as polarizability of the ligand increases; thus Cl^- tends to produce a lower C.N. with metal ions than F^-, e.g., FeF_6^{3-} and $FeCl_4^-$. A difficulty of interpretation arises because polarizability generally increases with increasing size of an anion and it has often been assumed that the difference between F^- and Cl^- could be simply due to size. We believe, however, that although steric effects can be important, especially with polydentate groups, polarizability is the major factor with monodentate ligands.

In the case of the light metals the maximum coordination is four, e.g., as in bisacetylacetonatoberyllium(II) or in the $[BF_4]^-$ ion. However it rises to six in the second row, e.g., as in the $[Mg(H_2O)_6]^{++}$ ion, and even to eight in later rows as in the complex $[Zr^{IV}(acetylacetone)_4]^0$. The limit of four for Li^+, Be^{++}, etc., in the first row can be explained in terms of the size of the metal ion or of the available orbitals. The four low-lying $2s2p^3$ orbitals form a tetrahedral set; alternatively one could say that only four bonding pairs of electrons can get close enough to the Li^+, Be^{++}, or B^{3+} ions without interelectronic repulsion between the σ-bonding pairs becoming prohibitively large.

Until recently it was generally assumed that the coordination number six was not exceeded in the first long period but now several cases of seven and eight coordination in the first transition series have been reported. It is presumably significant that they all arise either with chelates and/or weak field ligands; thus the ligand ethylenediaminetetraacetic acid forms seven-coordinate complexes with Fe(III) (Hoard *et al.*, 1961a) and Mn(II) (Hoard *et al.*, 1961b), whereas eight coordination has been observed in the complex $TiCl_4 \cdot 2$diarsine and in the corresponding quadrivalent vanadium complex (Clark and co-workers, 1962).

E. Stereochemical Types

The stereochemistry expected for metal atoms with spherically symmetrical nonbonding shells is well understood. The arrangements (see Gillespie and Nyholm, 1957; Gillespie, 1961) anticipated for the various coordination numbers in terms of maximum symmetry arising from repulsion of bonding pairs and lone pairs of electrons are shown in Table I.

Only in the case of eight coordination is there any real uncertainty as

TABLE I
STEREOCHEMISTRY AND COORDINATION NUMBER
(SPHERICALLY SYMMETRICAL CORE)

Coordination number	Number of nonbonding pairs	Hybridization involved (including lone pairs)	Example	Shape
2	0	sp	$HgCl_2$	Linear
	1	sp^2	$SnCl_2$	V-shaped
	2	sp^3	H_2O	V-shaped
	3	$sp^3d_{z^2}$	$[ICl_2]^-$	Linear
	4	$sp^3d_{z^2}d_{x^2-y^2}$	?	Linear
3	0	sp^2	BCl_3	Trigonal
	1	sp^3	NH_3	Pyramidal
	2	$sp^3d_{z^2}$	ClF_3	T-shaped
	3	$sp^3d_{z^2}d_{x^2-y^2}$	?	T-shaped or pyramidal
4	0	sp^3	CH_4	Tetrahedral
	1	$sp^3d_{z^2}$	SF_4	Irregular tetrahedral
	2	$sp^3d_{z^2}d_{x^2-y^2}$	$[ICl_4]^-$ [a]	Square
5	0	$sp^3d_{z^2}$	PF_5	Trigonal bipyramidal
	1	$sp^3d_{z^2}d_{x^2-y^2}$	IF_5	Square pyramidal
6	0	$sp^3d_{z^2}d_{x^2-y^2}$	SF_6	Octahedral
	1	sp^3d^2	$[SbF_6]^-$	Distorted octahedral
7	0	sp^3d^3	IF_7	Pentagonal bipyramidal
8	0	sp^3d^4	$[TaF_8]^{3-}$	Square antiprism
			$TiCl_4{\cdot}2$ diarsine	Dodecahedral

[a] The recently reported compound XeF_4 is also presumably square planar (Claasen *et al.*, 1962).

to the expected stereochemistry provided that we are dealing with an isolated (ideally gaseous) MX_8 complex. In solid compounds lattice forces are frequently comparable with those arising from bond pair–lone pair repulsion and some deviations from regularity occur, e.g., in ions like ZrF_7^{3-} and TaF_8^{3-}. Further, there may be changes from the gas or solution phase on passing over to a solid, as occurs with $PCl_5 \rightarrow [PCl_4]^+[PCl_6]^-$, with $PBr_5 \rightarrow [PBr_4^+]Br^-$, and with $NbCl_5$ which becomes an octahedral dimer with halogen bridges in the solid state (I) (Zalkin and Sands, 1958).

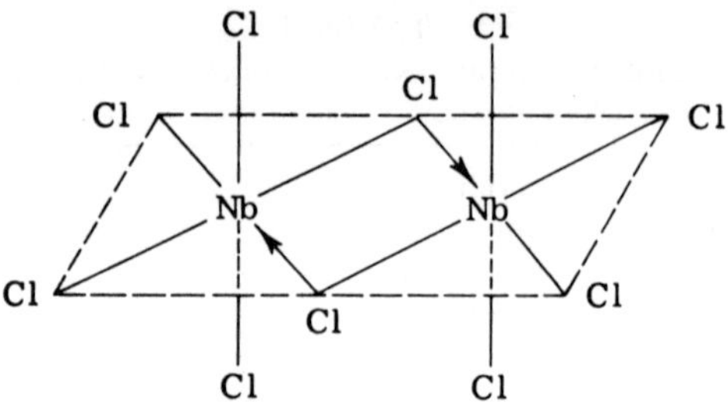

(I)

II. The Classes of Acceptor Metal Atoms

A. Classification

1. *Classification in Terms of the Nonbonding Core*

We distinguish different types of nonbonding electronic cores in acceptor metal atoms. Several of these have been recognized implicitly in other terms, e.g., the "light" metals, and where possible these names are retained. The various types are given in Table II.

TABLE II
Types of Nonbonding Core

Nonbonding core	Name
$1s^2$	Light metals, Li^+, Be^{++}, B^{3+}
ns^2np^6	Pretransition metals, e.g., Na^+, Ca^{++}, Sc^{3+}
$(n-1)d^{10}$	Posttransition metals, e.g., Cu^+, Zn^{++}, Ga^{3+}
$(n-1)d^{10}ns^2$ [a]	Inert pair metals, e.g., Tl^+, Pb^{++}
$(n-1)d^1 \rightarrow (n-1)d^9$	Transition metals
$(\mathrm{n}-1)(f^1\text{–}f^{13})ns^2np^6$	Lanthanide metals ($n = 4$) and actinide metals ($n = 5$)

[a] The configuration $ns^2np^6s^2$ (e.g., Al^+) is so rare that it can be discussed with $(n-1)d^{10}ns^2$.

2. *Class A and Class B Metals*

a. Basis of the Classes. An alternative division of metals into two classes according to their affinity for ligands in aqueous solution has been recognized for a long time (see Ahrland and Chatt, 1958). The class A metals are distinguished by an order of affinity in water towards the halogens $F^- > Cl^- > Br^- > I^-$, and by the fact that they generally coordinate better with the more electronegative ligands. The Be^{++} and Mg^{++} ions show this type of behavior. In sharp distinction one might take the Hg^{++} ion which coordinates much more readily with I^- than F^- in aqueous solution.

These comparisons of the affinity of a metal ion for different anions in aqueous solution refer to free energy changes. The statement that $F^- > I^-$ towards Al^{3+} means that $-\Delta G$ for the reaction

$$-\mathrm{Al}^{3+} \leftarrow \mathrm{OH_2} + \mathrm{F}^- \rightleftharpoons -\mathrm{Al}^{3+}\mathrm{F}^-$$

is greater than for the reaction with I^-. The relation between the ΔH values of association is the same insofar as entropy changes are constant. Broadly, it is found that the light, the pretransition, the lanthanide, and the actinide metals fall into class A.

The posttransition and usually the inert pair elements generally fall into class B. The transition elements show a gradual change from class A to class B behavior passing along the series $d^1 \rightarrow d^9$: thus Ti, Zr, and Hf usually show class A behavior, whereas palladium and platinum belong to class B; nickel can show class A behavior. The change along the series is nicely illustrated by the behavior of bivalent transition metal ions towards chloride ions. With elements on the right-hand side of the series complex formation occurs readily even in aqueous solution as in $[CuCl_4]^{--}$ and $[CoCl_4]^{--}$.[2] For the ions in the center, namely Cr^{++}, Mn^{++}, Fe^{++}, complex formation takes place in alcohol as solvent, but in the elements at the left-

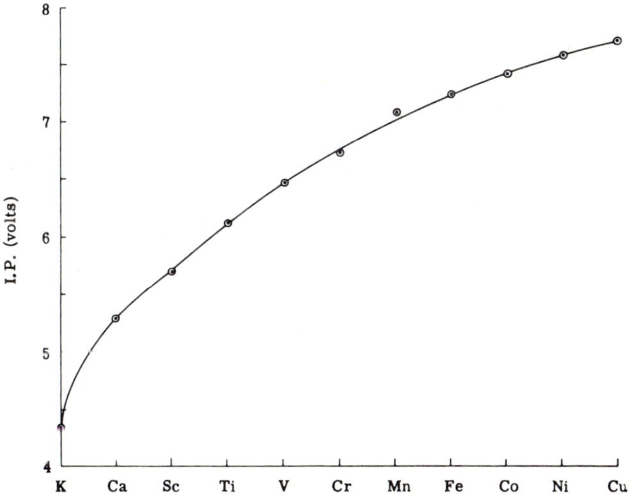

FIG. 1. Ionization potentials for the process $3d^n4s^1 \rightarrow 3d^n$ in the first long period.

[2] The apparently anomalous behavior of Ni^{++} in not forming $[NiCl_4]^{--}$ readily in water can be explained in terms of the large hydration energy arising partly from the crystal field stabilization energy (Gill and Nyholm, 1959).

hand side, Ti^{++} and V^{++}, complex formation can be induced only in the presence of other less readily coordinating solvents or melts. Class A metals have nonbonding cores with high shielding power, i.e., they have s^2p^6 or s^2 outermost shells or only a few nonbonding d electrons. The class B elements, however, have either filled or nearly filled d^{10} cores. As an illustration of the shielding power of these cores it is useful to show the first ionization potential (I.P.) value from potassium to zinc in the first long period. These values have been assembled from published data (Moore, 1952) to apply to the removal of a 4s electron (Fig. 1), e.g., the, I.P. value for Ni^0 (ground state $3d^84s^2$) is the energy required for the process $3d^94s^1 \rightarrow 3d^9$. This gives a measure of the relative shielding powers of the cores $3s^23p^6$, $3s^23p^63d^n$, and $3d^{10}$. The steadily increasing I.P. from K (d^0s^1) to Ni (d^9s^1) is noteworthy as is the high value for the d^{10} core shown by Cu ($d^{10}s^1$).

The class A metals are those with lower values of external field, on account of the core shielding. Their polarizing power at short interionic distances is less than that of metals with, for example, d^{10} cores. Their ability to combine with the less polarizable F^- more strongly than with OH^-, and with the heavier halogens, suggests that the bonds formed are electrostatic ion-ion in type. Class B metals, with stronger short range electrostatic fields polarize ligands more, and should form bonds in which electrostatic ion-induced-dipole and to some extent covalent forces are important. Their greater affinity for the more polarizable halide ions is thus understandable. It is compatible with this emphasis on the importance of the polarizing power that class B metals should form complexes with neutral ligands such as NH_3 more readily than do class A.

b. A Simple Index of Class A and Class B Character. By basing the discussion on the simplest possible models of covalent and ionic binding we may define a quantity, which will be called R, to measure the characteristics of the bonds likely to be formed by a given metal.

Let us consider a hydrated metal ion M^{n+} in aqueous solution reacting with hydrated halide ions X^-. If we concern ourselves with the formation of only one $M^{n+} \cdot\cdot\cdot X^-$ ionic bond in the reaction

$$[M(H_2O)_y]^n + X^-\cdot(H_2O)_z \rightarrow [M(H_2O)_{y-1}X]^{n-1} + (z+1)H_2O$$

the energy quantities involved are:

(1) ΔH_1, the energy required to remove part of the solvation shell of the $[M(H_2O)_x]^{n+}$ ion;

(2) ΔH_2, the energy required to remove the hydration shell[3] of the X^- ion.

[3] Strictly speaking this should read "part of the hydration shell" because the X^- ion in an M^+X^- ion pair will still be partly hydrated.

(3) ΔH_3, the (electrostatic) energy involved when the two point ions M^+ and X^- are brought together from infinity.

This equals, in absolute units $ne^2/(r_{M^+} + r_{X^-})$ where n is the number of unit positive charges (e) on the metal ion, e is the charge on the anion and r_{M^+} and r_{X^-} are the ionic radii of the metal ion and the anion given in angstroms, respectively. This expression may be simplified to

$$\frac{331.3}{r_{M^+} + r_{X^-}} \text{ kcal./mole}$$

Given the following ionic radii: F^-, 1.36 Å; Cl^-, 1.81 Å; Br^-, 1.95 Å; I^-, 2.16 Å, and Li^+, 0.60 Å, we find that the energies (ΔH_3) evolved in forming the ion pairs Li^+F^-, Li^+Cl^-, Li^+Br^-, and Li^+I^- are 169, 138, 130, and 120 kcal./mole, respectively. Thus, if we were to ignore hydration energies entirely, the order $F^- > Cl^- > Br^- > I^-$ towards class A metals in terms of simple ionic bonding could be understood. If we take into account the heat of hydration of the anion $F^- = 117.4$, $Cl^- = 85.4$, $Br^- = 73.8$, $I^- = 72.4$ kcal.-gm.-ion, it is obvious that, in certain instances (e.g., $Cl^- \rightarrow Br^-$), the difference in hydration energy is greater than the difference in electrostatic energy evolved in ion pair formation. Since towards class A metals, Cl^- is in fact in water a better coordinating ligand than Br^-, we must conclude that only part of the solvation energy of the halide ion is lost during "ionic" complex formation.

Now for the class B metal ions such as Cu^+, Hg^{++}, etc., which prefer the order $I^- > Br^- > Cl^- > F^-$ let us postulate, as a first approximation, that a covalent bond is formed. For this type of bond the energy evolved during the reaction

$$[M(H_2O)_x]^n + X^-(H_2O)_y \rightarrow [M(H_2O)_{x-1}X]^{n-1}$$

can be regarded very approximately as the sum of four terms: (*i*) the ionization potential (I.P.) of the metal ion; (*ii*) the electron affinity (E.A.) of the anion; (*iii*) the two terms involving the loss of hydration energy of the cation and anion respectively (ΔH_1 and ΔH_2 discussed above). As a rule, differences in the E.A. (for the halogens) are small and hence we find that a rough measure of the tendency of the metal ion to form covalent bonds is given by the total ionization potential.

The main terms of interest in comparing the tendency of a metal ion to form a covalent or an electrostatic bond are, respectively, I.P.[4] and $ne^2[/(r_{M^+} + r_{X^-})$. For convenience we shall henceforth refer to the ratio R, given by the expression

$$R = \frac{\text{I.P.} \times (r_{M^+} + r_{X^-})}{ne^2}$$

[4] We neglect the E.A. as being small (compared with the I.P.) and because it is nearly the same for all four halogens.

This is further simplified to

$$R = \frac{\text{I.P.} \times (r_{M^+} + 1.82)}{n}$$

where r_{M^+} is given in angstroms and the I.P. in volts. The constant value 1.82 Å for r_{M^+} has been chosen for convenience simply because it is the average of the radii of the F^-, Cl^-, Br^-, and I^- ions. It is emphasized that R is purely empirical; no significance should be attached to the absolute value.

Calculated values of R are given for common uni-, bi-, ter-, and quadrivalent ions in Table III.

Taking each of the different charge types of the ions separately we see that a low value of R is generally associated with class A behavior whereas a high value correlates with class B behavior. The main factor responsible for a high value of R is the ionization potential; this reflects a strong external field and polarizing power.

TABLE III

VALUES OF $R = \frac{\text{I.P.} \times (r_{M^{n+}} + 1.82)^a}{n}$

Metal class[b]	Univalent cations Ion	R	R′	Bivalent cations Ion	R	R′	Tervalent cations Ion	R	R′	Quadrivalent cations Ion	R	R′
A												
	Li^+	13.0	13.0	Be^{2+}	29.3	14.7	B^{3+}	48.0	16.0	C^{4+}	72.9	18.2
	Na^+	14.2	14.2	Mg^{2+}	28.0	14.0	Al^{3+}	41.2	13.7	Si^{4+}	57.6	14.4
	K^+	13.6	13.6	Ca^{2+}	25.3	12.3	Sc^{3+}	38.5	12.8	Ge^{4+}	61.3	15.3
	Rb^+	13.8	13.8	Sr^{2+}	24.5	12.3	Cr^{3+}	45.0	15.0	Sn^{4+}	58.1	14.5
	Cs^+	13.6	16.6	Ba^{2+}	24.0	12.0	Fe^{3+}	44.9	15.0	Pb^{4+}	64.2	16.1
							La^{3+}	35.8	11.9			
B												
							Co^{3+}	45.8	15.3			
	Cu^+	21.5	21.5	Zn^{2+}	35.2	17.6	Ga^{3+}	46.4	15.5	Ti^{4+}	57.3	14.3
	Ag^+	23.3	23.3	Cd^{2+}	36.0	18.0	In^{3+}	46.3	15.4	Zr^{4+}	52.1	13.0
	Au^+	29.5	29.5	Hg^{2+}	42.5	21.8	Tl^{3+}	52.0	17.3	Hf^{4+}		
	Tl^+	20.0	20.0									

[a] The figures in parentheses give the values of $R' = R/n$ discussed below.
[b] Division approximate only and depends upon ligand.

In order to compare metal ions having different charges we list also $R' = R/n$. Theoretical justification for this is not easy to adduce. One finds that R' correlates quite readily with class A and class B behavior.

In general an R' value greater than 19 indicates a class B metal; similarly, a value less than 15 indicates class A. Values in the range 15–19 are found with metals whose properties indicate that they may behave as type A or B depending on the ligands used for complex formation.

3. *Classification by Behavior toward π Acceptor Ligands*

The division of the metals into class A and class B is not a useful one in discussing behavior towards ligands such as CO and NO and others capable of forming π bonds. There is evidence, to be discussed below, that the ability to form π donor bonds is confined to the transition metals, and that it is best developed in the center of the series, i.e., in the elements Cr, Mn, and Fe which show behavior intermediate between that of class A and class B.

B. Complexes of the Acceptor Metals

1. *The Light Elements*

In going from the light elements through the pretransition metals to the lanthanides and actinides we pass from the use of only $s\sigma$ and $p\sigma$ bond orbitals to complexes involving outer d orbitals as well. In the transition metals a gradual change in behavior from an essentially ionic bond to a covalent or ion-induced dipole bond then becomes apparent. The situation where π bond formation is likely will be emphasized.

For complex formation the light elements are Li^+, Be^{++}, B^{3+}, and, formally, C^{4+}. The coordination number does not exceed four and in general these typical class A ions display a strong affinity for highly electronegative ligands of low polarizability. The tendency to utilize all four $2s2p^3$ orbitals to complete the octet is so strong that it is commonly found that either π bond formation occurs using an extra lone pair of the ligand (as in BF_3) or there are multicenter bonds as in BeH_2 and B_2H_6. The contrast between the light and heavy elements is illustrated by Hg^{II}. This very often attains a coordination of two only as compared with Be^{II} which tends to utilize all four $2s2p^3$ orbitals and frequently polymerizes in order to do so.

2. *Pretransition Elements*

These are in many ways similar to the light elements. Examples of four coordination include sodium acetylacetone·H_2O (II) and the $[AlCl_4]^-$ ion. The behavior of the pretransition elements is distinguished from that of the light elements by a much greater reluctance to form double bonds and multicenter bonds, and by the ability to increase the coordination number from four to five or six by using outer d orbitals.

The observation that double bonding is of much wider occurrence in

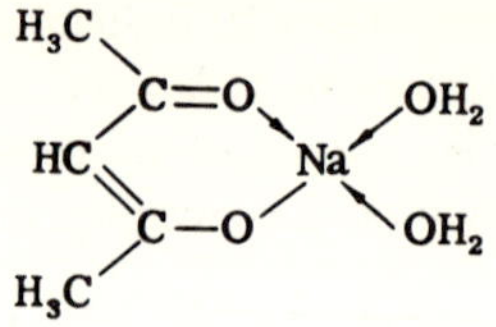

(II)

the first row than in later rows of the Periodic Table is not well understood. It is probably connected with the fact that with the larger and more polarizable atoms there is an energetic advantage in having a larger rather than a smaller number of surrounding ligands, especially if they are highly electronegative and therefore highly polarizing. In the later elements of the second row the number of ligands rises in many cases to five and six, the extra bonds perhaps being formed with the help of *d* orbitals; but in the earlier elements the higher coordination numbers are less common (e.g., as in AlF_6^{3-}) and the *d* orbitals, even when significantly populated, still make a contribution to the polarizability, and thus favor attachment of four ligands as in Al_2Cl_6 instead of three ligands as in BF_3. The importance of *d* orbitals in second row elements is discussed more fully below.

As usual, the tendency towards six coordination increases as the charge on the atom rises, i.e., $Na^+ < Mg^{++} < Al^{3+}$, and as the polarizability of the ligand decreases, thus F^- gives rise to AlF_6^{3-} whereas Cl gives $[AlCl_4]^-$.

3. *Lanthanides and Actinides*

These elements are logically considered next because the unfilled *f* orbits are so well shielded by the outer s^2p^6 shell that they have very little effect upon the coordination of ligands. Accordingly they are intermediate in properties between the pretransition elements and the early transition metals and indeed behave more like the former. Successive additions of *f* electrons affect the shielding and lead to a steady decrease in size of the rare earth ions (lanthanide contraction) but the effects on complex formation are much less than is observed in the stepwise addition of *d* electrons in the transition metal series. It was, of course, the close similarity in properties of the rare earth metals in the trivalent state that made separation so difficult before ion exchange techniques were developed. As seen in Table III the R value for La^{3+} is the lowest of all the trivalent metals quoted and the R' value is less than that of the alkali or alkaline earth metals. Thus, the ligands with greatest affinity for these elements are the anions of low polarizability. Thus we find that oxyligands, the fluoride ion, and polydentates with oxygen and nitrogen as donors form the most stable complexes. The spectral and magnetic properties of certain of the ions are influenced by the crystal field to some extent, but in many cases the magnetism is largely independent of the anion or ligands and the spectra are ion-like. This is in marked contrast to examples next to be discussed, in which the unpaired electrons are subject to strong influences from the ligands.

4. *The Transition Metals*

We assign "transition metal behavior" to those metals which have a d^1–d^9 electron configuration. The tendency to covalent character increases as the oxidation state increases; and, for the same oxidation state, as the number of d electrons increases. Metal atoms with a d^0 configuration can be regarded as of the pretransition type, except that as the oxidation state of the atom rises the polarization of the ligand increases and the degree of covalent or ion-induced dipole bonding becomes greater.

In order to discuss transition metal ions, it is convenient to start with the crystal field model. In ligand field theory the effect of covalent bonding is also included. These theories are by now well known, and available in standard works (Griffith, 1961; Orgel, 1960; Jorgensen, 1962). We shall do no more than outline the terms to be used.

In the case of a transition metal ion with n nonbonding d electrons in a perfect octahedron of negatively charged ligands (or ion dipoles), the d orbitals are split into the (upper) d_γ doublet and the (lower) d_ϵ triplet. The separation between these levels is referred to as 10 Dq or Δ; in hexahydrated bivalent metal complexes Δ is about 10,000 cm.$^{-1}$ and in tervalent complexes about 20,000 cm.$^{-1}$. The interaction between the ligand field and the d electrons depends upon:

(*i*) the strength of electrical field due to the ligands, measured by Δ,
(*ii*) the magnitude of the interelectronic repulsions.

Depending upon the relative magnitude of Δ and the interelectronic repulsion spin-pairing may occur, as shown in Scheme 1.

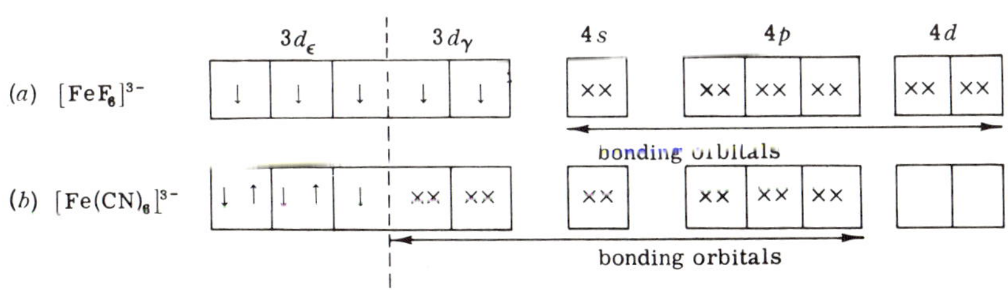

As an illustration of this Fig. 2 shows how the first ionization potential of the neutral first transition series metal atoms, corrected to apply to the transition $d^n \rightarrow d^{n-1}$, increases from $d^1 \rightarrow d^9$ except for the break at d^5 and d^6 arising from the intervention of the half-filled d^5 configuration.

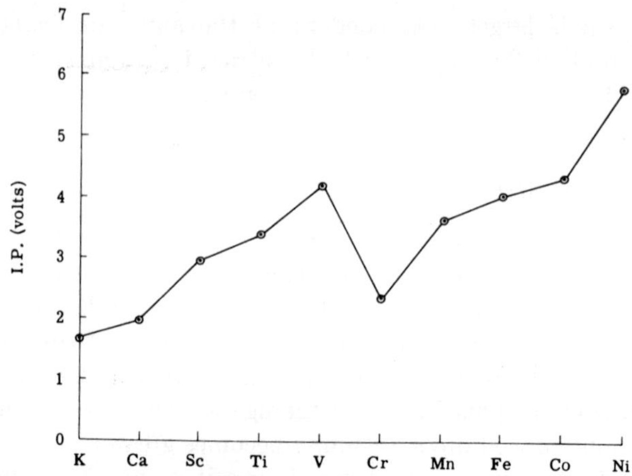

FIG. 2. Ionization potentials for the process $3d^n \rightarrow 3d^{n-1}$ in the first long period.

This increase in I.P. is a reflection of the decrease in shielding power of each successive d electron added, hence of the greater polarizing power of the metal ion as we pass from $d^1 \rightarrow d^9$. These values should be compared with those for the $3d^n4s^1 \rightarrow 3d^n$ I.P. values shown in Fig. 1 where there is a steady increase from d^0s^1 to $d^{10}s^1$. This indicates the field at a ligand outside the d^n configuration. As a result the behavior of a transition metal ion is expected to vary between two extremes:

(*a*) The situation arising in complexes when the number of nonbonding d electrons is small. In this instance the R value is relatively small (for Ti^{++}, 27.5) and essentially ionic, or type A complexes are expected;

(*b*) The situation which arises when the number of d electrons approaches ten. Here the R value is much larger (for Cu^{++}, 35.5). The higher effective nuclear charge leads not only to a greater polarization of the lone pairs of the ligands (and hence more covalency in the metal-ligand bond) but also interactions between nonbonding electrons increase as the number of d electrons increases.

The purely electrostatic model being considered first, the stereochemistry expected for each of the spin-free and spin-paired configurations for d^0 to d^{10} is given in Table IV.

Table V shows the shapes expected for the corresponding spin-paired configurations.

To appreciate the importance of an increase in polarizing power of the metal atom in causing departures from the behavior predicted in Tables

TABLE IV
STEREOCHEMISTRY OF SPIN-FREE COMPLEXES[a]

Number of nonbonding *d* electrons	Unpaired electrons	Four coordination	Six coordination
0	0	Regular tetrahedral	Regular octahedral
1	1	Almost regular tetrahedral	Almost regular octahedral
2	2	Regular tetrahedral	Almost regular octahedral
3	3	Distorted tetrahedral	Regular octahedral
4	4	Square planar	Tetragonal
5	5	Regular tetrahedral	Regular octahedral
6	4	Almost regular tetrahedral	Almost regular octahedral
7	3	Regular tetrahedral	Almost regular octahedral
8	2	Distorted tetrahedral	Regular octahedral
9	1	Square planar	Tetragonal
10	0	Regular tetrahedral	Regular octahedral

[a] It must be emphasized that for this simplified table the effects of lattice forces, covalent bonding, spin-orbit coupling, etc., are ignored.

IV and V, it is convenient to consider an example. Let us take the Ni^{++} (d^8) ion. Starting with a six-coordinate, hexaaquo ion, crystal field theory

TABLE V
STEREOCHEMISTRY OF SPIN-PAIRED COMPLEXES

Number of nonbonding *d* electrons	Unpaired electrons	Four coordination	Six coordination
3	1	Almost regular tetrahedral	—
4	0	Regular tetrahedral	—
5	1	Distorted tetrahedral	Almost regular octahedral
6	0	Distorted tetrahedral	Regular octahedral
7	1	Square planar	Tetragonal
8	0	Square planar	Tetragonal

suggests maximum stability for a perfect octahedron in which the configuration is $d_\epsilon^6 d_\gamma^2$. That this is not correct is shown first by magnetic data, the μ_{eff} of the $[Ni(H_2O)_6]^{++}$ ion being about 3.2 B.M. instead of the expected spin-only value of 2.83 B.M.; X-ray studies support the conclusion that some distortion from regular octahedral symmetry occurs. This can be described by allowing for mixing in of some of the $d_\epsilon^5 d_\gamma^3$ configuration via spin-orbit (s.o.) coupling. The extent to which this mixing in occurs depends on the relative magnitudes of Δ, the separation between d_ϵ and d_γ orbital sets in an octahedral field of negative charges, and the energy separation between the lowest and next J values.

Passing next to tetrahedral Ni(II) complexes, e.g., the $[NiCl_4]^{--}$ ion, the value of Δ is only about 3500 cm^{-1}. and in this instance λ operates in the opposite sense, i.e., it tends to diminish the distortion from a regular tetrahedral arrangement (as shown in Table IV, the spin-free d^8 $(d_\gamma{}^4 d_\epsilon{}^4)$ configuration is expected to lead to a considerably distorted tetrahedral arrangement). In addition to the effects of λ and Δ, covalent binding will tend to favor a regular tetrahedral arrangement, because it will reduce mutual repulsions between the four σ-bonding pairs of electrons.

An illustration of the way in which the opposed forces determine the stereochemistry is given by the tetrachloro- and tetrabromocopper(II) anions. Simple crystal field theory shows that a square planar arrangement is expected for a four-coordinated d^9 ion. This expectation is indeed borne out with the more electronegative ligands like NH_3 and H_2O (strictly speaking one should recognize that there are usually two more distant ligands on either side of the square which give rise effectively to a tetragonal arrangement). The effect of the highly polarizing cupric ion on the more or less easily polarized chloride and bromide ions acts to enhance the repulsion between the σ lone pair on the halogen atoms. As a compromise distortion occurs, the $[CuCl_4]^{--}$ and $[CuBr_4]^{--}$ ions assuming a stereochemistry intermediate between a square and a tetrahedral arrangement (Helmholtz and Kruh, 1952; Morosin and Lingafelter, 1960).

The large values for the electron affinities of the ions of the posttransition elements coupled with considerations of orbital overlap (see Section III,D,1) lead us to conclude that $d\pi$ bond formation by these metals is less likely than has often been assumed. The fact that they have a notable affinity for ligands such as tertiary arsines, the S^{--} ion, and other ligands of this type seems to arise from their polarizing power rather than their tendency to form double bonds.

5. *The Posttransition Elements*

Since each of these ions has a d^{10} nonbonding core, a characteristic feature is their relatively high polarizing power. This is reflected in the high electron affinities of the ions, a fact conveniently illustrated by comparing electron affinities of isovalent ions occurring in the Periodic Table before and after the transition series.

Thus the electron affinities of K^+ and Cu^+ are 4.34 and 7.72 e.v., respectively; those of Ca^{++} and Zn^{++} are 11.87 and 18.0 e.v., respectively. The higher values are associated with greater polarizing power and larger s–p separations leading to a decrease in preferred coordination numbers. This is particularly evident in comparisons of elements before and after the lanthanide series and the third transition series. Thus the ionization potentials of Cs and Au are 31,407 and 74,410 $cm.^{-1}$ and the s–p separations 11,178 $cm.^{-1}$ and 37,359 $cm.^{-1}$. Although it has been generally accepted that one can explain the preferred coordination number of two for univalent gold in terms of this large s–p promotion energy, an explanation based on the mixing of the $6s$ orbital with the $5d_{z^2}$ leading to asymmetry in the nonbonding core has been proposed by Orgel (1958). It seems likely that the two explanations are essentially complementary.

6. *Inert Pair Elements*

Finally, we refer briefly to a small group of ions showing a degree of adaptability to chemical circumstance that suggests the term chemical chameleon. They have one pair of s electrons outside the d^{10} shell, as for example in univalent Ga, In, and Tl and in bivalent Sn and Pb. The energy required to remove the one or two electrons to leave a $d^{10}s^2$ core is considerably less than that required to remove the electrons from $d^{10}s^1$ or $d^{10}s^2$ atoms to leave d^{10} cores. As the figures in the accompanying tabulation show, the behavior to be expected is intermediate between that of a pre- and a posttransition element. In fact these elements behave in two

	Element		
First I.P. (e.v.)	Rb 4.18	Ag 7.27	In 5.79
Sum of first and second I.P. (e.v.)	Sr 17.01	Cd 24.90	Sn 22.05

ways: in essentially ionic compounds, e.g., thallous salts, such as TlCl, and also $PbSO_4$, they tend to resemble pretransition metals. The s^2 pair in such cases seems simply to contribute a spherical component to the ionic core. However with a polarizable ligand the lone pair acquires some p character and in Sn(II) and Pb(II) forms strong bonds towards S^{--} and other sulfur ligands.

III. Consideration of Orbitals and Orbital Properties

A. Effective Nuclear Charge

1. *Effective Nuclear Charge and Screening*

As has been discussed, there are many complexes in which the attraction between metal and ligand can be interpreted in terms of an essentially electrostatic interaction, perhaps modified by polarization of the ligand. Thus, for the complexion $[FeF_6]^{3-}$, a model based upon an Fe^{3+} ion and six F^- ions is convenient; one assumes that the ferric ion acts as a source of potential corresponding to the full charge of three units. In complexes with more covalent character other procedures are needed to represent the external field of the ions. The point is conveniently illustrated by comparing univalent copper and sodium. The univalent ionic radii are almost identical (~0.96 Å), but the difference in I.P. values (Cu = 7.72 e.v., Na = 5.14 e.v.) shows that the attraction by the nucleus for the outermost s electron is greater in copper than in sodium. The reason for this is that the penultimate shell in the case of sodium ($2s^22p^6$) shields the nucleus more effectively than the more diffuse $3d^{10}$ subshell in copper. The potential of Na^+ thus drops off more quickly with distance than in Cu^+, and the ionization potentials of the atoms reflect this. In a loose way it is said that the effective charge is greater in Cu^+ than in Na^+, but there is some confusion in the literature over the significance of the effective nuclear charge (Z^* or Z_{eff}) in heavy atoms, as well as over the best values to use. We first evaluate the situation and indicate the procedure to be followed.

The essential distinction concerned in the concept of effective nuclear charge is between the effective charge as a parameter in a Slater-type atomic orbital, and the effective charge measured at a *point* describing the electric field of an atom. In a complex atom the latter electric potential is the difference between the potential of the nucleus, which varies according to Coulomb's law $V_{nucl} = -Ze^2/r$, Z being the (true) nuclear charge, and r the radial distance, and the opposing potential of the electron cloud surrounding the nucleus. Because the amount of electron charge lying within the sphere of radius r from the nucleus increases quickly as r increases, the net potential drops far more rapidly than $1/r$. Thus it is impossible to describe the potential of a complex atom by one single charge parameter. Let us instead define $\xi(r)$ to be the effective charge at a distance r from the nucleus, according to $\xi(r) = -V(r)r/e^2$, where $V(r)$ is the net potential. If the potential obeyed Coulomb's law ξ would be a constant, independent of distance. In practice it varies rapidly from a value for small r equal to the true nuclear charge Z, to zero at large r for an atom, or to $+n$ for a positive ion of charge n. Thus the value of ξ at a given radial distance is the actual charge which, if placed at the nucleus, would give the potential at that radius in a Coulomb field. A second effective charge $\xi'(r)$ applicable to atomic fields is sometimes used which similarly refers to the *force* acting on a charged particle instead of to the potential. The definition $\xi'(r) = r^2e^{-2}dV/dr$ shows that ξ' is the actual charge which, if placed at the nucleus, would give the force on a negative charge at that radius in a Coulomb field.[5] ξ and ξ' are different from one another, especially where the field is varying rapidly.

[5] Values of ξ and ξ' are often available in calculations of self-consistent field wave functions for atoms (see Hartree, 1957).

2. *Slater's Rules*

While ξ and ξ' apply to a chosen distance from the nucleus, and vary from point to point, the Slater effective nuclear charge Z^* is a constant, and is a special type of average over the whole nuclear field. In a hydrogenic (one electron) system Z^* equals the actual nuclear charge, but in complex atoms it has meaning only in relation to the Slater radial orbital (Slater, 1930), which has the functional form shown in Eq. (1)

$$R(n,Z^*) = Nr^{n-1}e^{-Z^*r/n} \tag{1}$$

where N is the normalizing coefficient and n the principal quantum number· Values of Z^* to be used to get approximate atomic wave functions are given by Slater's rules with surprising accuracy for small atoms ($n < 3$), and for the inner electrons of larger ones.

In atoms with one electron outside a closed shell core, such as the alkali metals, and the $d^{10}s^1$ configurations of Cu, Ag, and Au, another empirical estimate of the effective charge can be made by employing a relation between ionization potential and nuclear charge strictly applicable to hydrogenic atoms, namely, I.P. = 13.595 Z^{*2}/n^2 e.v. Yet another estimate can be obtained by equating the I.P. to one-half the potential energy at the atomic radius ρ: I.P. = 7.20 $\xi^{\dagger}/\rho$ e.v., in which case the calculated nuclear charge is, strictly, applicable to a single radial distance and is not an average in the sense of Z^* and Z. In Table VI these three quantities are compared in a series of representative examples.

TABLE VI

s^1 atom	Z^* calculated from Slater's rules	$Z\dagger$ calculated assuming I.P. = 13.595 $Z\dagger^2/n^2$	ξ^a calculated using I.P. = 7.20 ξ^*/r	I.P. (e.v.)
H	1.00	1.00	1.00	13.595
Li	1.30	1.26	1.16	5.39
Na	2.20	1.84	1.35	5.14
K	2.20	2.22	1.41	4.34
Rb	2.20	2.77	1.44	4.18
Cs	2.20	3.21	1.44	3.90
Cu	2.50	3.02	1.37	7.72
Ag	2.50	3.73	1.51	7.57
Au	2.50	4.94	1.84	9.22

[a] The radii of these atoms are estimated from the metallic radii (Pauling, 1961).

3. *Screening in the Class A and Class B Metals*

It was noted earlier, in Section II,A,2,*b*, that the distinction between class A and class B metal ions is correlated with a difference of the ionization potential, in configurations with one electron outside a core, and therefore with a difference of shielding power of this core. In a case of an ion with a perfectly shielding core, the potential field at all distances up to the radius of the core would continue to be that of a unit charge, and two such ideally screened ions, with opposite unit charges, would attract mutually according to Coulomb's law and approach to a touching distance determined by the nature of the short range repulsion operating. Real ions do not behave like this, although various forms of hard sphere approximation based on ion-ion attraction are quite successful with ions (such as the alkali halide ions) having good core screening. In an ion with imperfect shielding, the potential field changes gradually from point charge character, at long range, to the field of several charge units, as the ill-defined edge of the diffuse core electrons begins to be penetrated. A neutral or anionic ligand may approach so that a part of its electronic charge lies within the strong zone of the potential field of the metal, and the induced dipole moment can play an important part in the binding. In general, therefore, poorly shielded metal ions have higher polarizing power at short range than well shielded ions, even though at long range their fields are identical.

B. *d*-Orbitals

1. *Atomic Orbitals in Acceptor Atoms*

In class A acceptors the detailed character of the metal orbitals has not received much attention; it is presumed, especially for the lighter elements, that Slater orbitals are adequate for approximate calculations and that self-consistent field wave functions (Hartree, 1957) can be used otherwise. In class B metals, and transition metals generally, it is quite clear that the properties in bond formation depend rather intimately on the nature of the *d* orbitals available in each. This is true both in the phenomena broadly described as crystal field effects, and in those in essentially covalently bound complexes with π bonding. In cases of this kind, the *d* orbitals are "inner" orbitals, 3*d* in the first transition series.

Thus in the present state of knowledge theoretical understanding of even the gross features of many classes of complex calls for discussion of *d* orbitals in three connections:

(1) The use of outer *nd* orbitals hybridized with *ns* and *np* orbitals in σ bond formation.
(2) The use of inner *d* orbitals $(n - 1)d$ for σ bond formation with *ns* and *np* orbitals;
(3) The use of filled $(n - 1)d$ orbitals for π bond formation with vacant ligand orbitals.

The preoccupation of following sections with d orbitals rather than with atomic orbitals generally thus reflects their importance in current theory.

2. *The Sizes and Forms of 3d Orbitals in the Transition Metals*

a. Slater Orbitals. The size of an orbital is a useful but somewhat elusive concept, in that it can be defined only in relation to a given orbital form, i.e., in relation to a given dependence of the electron density function ψ^2 on the radial distance r. Even then there are several possibilities, conveniently introduced by referring to the Slater radial wave function (Eq. 1), which is the simplest in common use. In examples with azimuthal quantum number l equal to its maximum value of $(n - 1)$ the Slater function is identical with the hydrogen-like wave function of the same nuclear charge. This is true of the $3d$ orbitals in the first transition series.

The most commonly used definition of orbital size is the value of the *radial maximum* r_M. This is the distance at which the radial distribution function $\psi^2 r^2$ takes its maximum value. It is the radius of maximum probability for an electron in the orbital. For a Slater orbital (Eq. 1) $r_M = 9/Z$, measured in atomic units of length, 1. a.u. = 0.529 Å. Alternatively, the size may be specified by the *mean radius* $\bar{r}$, defined by Eq. (2):

$$\bar{r} = \int r\psi^2 d\tau \tag{2}$$

In the familiar $2p$ orbitals of carbon ($Z = 3.25$) $r_M = 0.65$ Å and $\bar{r} = 0.81$ Å, compared with the covalent radius in sp^3 hybridization of about 0.77 Å. Calculated for a Slater $3d$ orbital $\bar{r} = 10.5/Z$. The ratio $\bar{r}/r_M$ is 7/6 in this example, but differs in orbitals of other kinds. In self-consistent field orbitals, which usually make the radial maximum fall relatively much closer to the nucleus, the ratio is nearer to $\bar{r}/r_M = 1.6$ (Hartree, 1957).

In Slater's method the effective nuclear charge Z in the wave function (Eq. 1) is related to the atomic number N, according to $Z = N - \sigma$, where σ is a number characteristic of the electron configuration of the atom. σ, the screening number, is found according to a simple formula given in standard textbooks (e.g., Eyring *et al.*, 1944). The values for the d electrons of the transition metal bivalent ions found in this way are given in Table VII. The configuration is $(3d)^{n-2}$. The increase of effective charge

TABLE VII

SLATER EFFECTIVE NUCLEAR CHARGES FOR $3d$ ELECTRONS

Ti^{++}	V^{++}	Cr^{++}	Mn^{++}	Fe^{++}	Co^{++}	Ni^{++}	Cu^{++}	Zn^{++}
3.65	4.30	4.95	5.60	6.25	6.90	7.55	8.20	8.85

along the series is a constant 0.65 for each unit increment in atomic number, corresponding to a steady reduction in size from Ti to Zn. Moreover, exactly the same values apply to configurations of the neutral atom $(3d)^{n-2}(4s)^2$, because Slater's rules allow no screening by electrons outside the group concerned.

In free atom configurations $(3d)^n$, where two electrons have been added to the $3d$ orbitals, the effective charges are less in each case by 0.70 charge units, up to and including Ni, in which the d^{10} configuration is attained. According to the formulas for r_M and $\bar{r}$, this corresponds to an expansion of the d orbitals as electrons are added. This feature, the variations of orbital size across the series, and other general behavior trends in free atoms and ions are correctly given by Slater's rules. However in individual cases Slater orbitals, with exponents from Slater's rules, do not approximate closely to the correct forms, so far as these are known from self-consistent field studies. This is illustrated in a typical example in Fig. 3, which shows the $3d$ radial

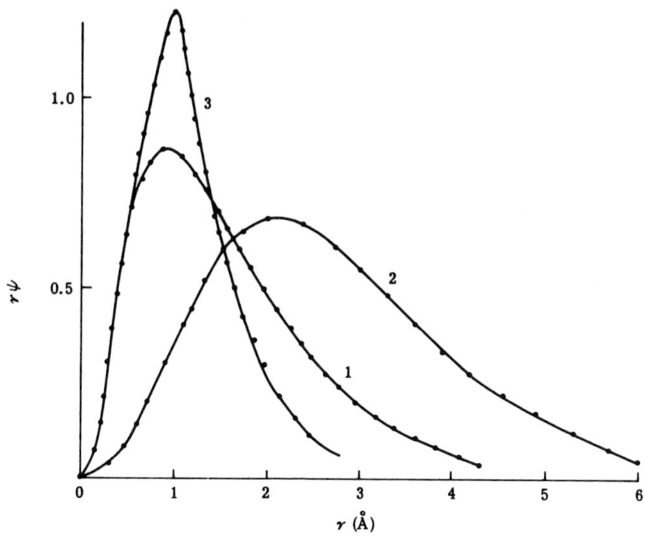

FIG. 3. Comparison of $3d$ radial wave functions for V^{++}: (1) Self-consistent field wave function according to Hartree; (2) Slater wave function with exponent found from Slater's rules; (3) Slater wave function with exponent adjusted to give the same radial maximum as (1).

wave function in V^{++}. The self-consistent field wave function (Hartree, 1956) has its radial maximum at 0.5 Å, the Slater orbital with exponent 4.3 taken from Table VII has maximum r_M = 2.1 a. u. = 1.1 Å. It is evident that properties calculated for such an orbital would bear little relation to the real system. It is not only the orbital exponent that is wrong, for if we retain the Slater form of orbital, but choose a new exponent to make the radial maximum fall at 0.5 Å (the required increase is from 4.3 to 10), we get the radial wave function (3) of Fig. 3. This wave function is seriously incorrect in form. It rises to too sharp a maximum, and falls off too quickly outside it.

The detailed features of the wave functions of individual atoms can best be had from self-consistent field (S.C.F.) calculations. These also show clear trends along the series of transition metals. As in the Slater scheme there is a steady size reduction across the series from Ti to Zn, but the radial maxima fall much closer to the nucleus at 0.5–0.6 Å, and the electron distribution is more diffuse outside the radial maximum. The range of variation along the series of transition metals is less than in the Slater scheme, and the effect of this is especially marked among the lighter atoms. The effect of adding electrons to the *d* shell is greater in the S.C.F. functions, as to both the greater diffusion of the electron density over a wider range of *r* and the greater change in the position of the radial maximum. One can say that the Slater scheme places the 3*d* radial maxima unrealistically far from the nucleus and does so more seriously in configurations of a smaller number of *d* electrons.

b. Analytical Self-Consistent Field Wave Functions. Self-consistent field wave functions calculated by the Hartree method lead to results for the radial dependence expressed in numerical form, and thus not so easily visualized as Slater radial functions or other simple analytical expressions. In the transition metal atoms a superior procedure has been developed by Watson (1960) who has calculated self-consistent wave functions directly in analytical form, each one being expressed as a linear combination of four (in the case of 3*d*) Slater functions. These are probably the best transition metal wave functions so far available, but their use for even simple molecular calculations leads to very complex expressions. Recently a significant advance has been made (Richardson *et al.*, 1962) by the proposal that good fits to S.C.F. 3*d* wave functions can be obtained by linear combination of two Slater functions with exponents chosen to optimize the overlap with the more exact functions. The 3*d* radial functions are then in the form

$$R(3d) = c_1R(3d,\alpha_1) + c_2R(3d,\alpha_2) \tag{3}$$

α_1 and α_2 are the exponents (Z/n) in Slater functions (Eq. 1) and the coefficients c_1 and c_2 determine the relative importance of the two terms and take care of the normalization. The values of the α's and c's for the first transition series metals are given in Table VIII, abstracted from the more extensive tabulation of Richardson *et al.* (1962). The precise exponent

TABLE VIII
EXPONENTS AND COEFFICIENTS IN TWO PARAMETER 3*d* FUNCTIONS[a]

		Ti	V	Cr	Mn	Fe	Co	Ni	Cu
α_1		4.55	4.75	4.95	5.15	5.35	5.55	5.75	5.95
α_2	I[b]	1.40	1.50	1.60	1.70	1.80	1.90	2.00	—
	II	1.60	1.70	1.80	1.90	2.00	2.10	2.20	2.30
	III	1.80	1.90	2.00	2.10	2.20	2.30	2.40	2.50
c_1	I	0.4206	0.456	0.4876	0.514	0.5366	0.555	0.5683	—
	II	0.4391	0.476	0.5060	0.532	0.5505	0.568	0.5817	0.5933
	III	0.4623	0.498	0.5247	0.547	0.5661	0.582	0.5959	0.6062
c_2	I	0.7839	0.752	0.7205	0.693	0.6678	0.646	0.6292	—
	II	0.7397	0.706	0.6750	0.649	0.6260	0.606	0.5890	0.5744
	III	0.6910	0.655	0.6289	0.605	0.5860	0.565	0.5497	0.5371

[a] From Richardson *et al.* (1962).
[b] I, II, and III refer to the neutral atom, and the uni- and bivalent ions, respectively.

values given in Table VIII are not critical, insofar as the criterion of maximum overlap with the more exact functions is somewhat insensitive. However we may discern some significant trends. The behavior of the wave function near the nucleus is largely determined by the first term, with the larger value of the exponent α_1. This also determines the position of the radial maximum, and one sees as before that this maximum falls closer to the nucleus in the elements to the right of the transition series. In general the inner region of the 3*d* electron distribution is not of much importance in overlap. As far as the chemical bonds to ligands are concerned, the second term belonging to the lower value of α is much the more important because it determines the character of the wave function at greater radial distances.

Two distinct trends appear in the exponents for the second term. First, the 3*d* electron occupies an orbital which becomes more strongly bound in moving across the series from Ti to Cu. The second feature is concerned with the dependence of the wave functions on the state of ionization. If we compare exponents in the series M(I), M(II), and M(III), referring in order to the neutral atom in the configuration $3d^n$, the univalent ion $3d^{n-1}$, and the bivalent ion $3d^{n-2}$, we see that there is rather strong contraction of the orbital as 3d electrons are removed. This is a consequence of reduction in the "inner screening" of one 3*d* electron by another. As to outer screening, we recall the fact, originally noted by Hartree, and confirmed by Watson, that the 3*d* wave functions in the configuration $3d^{n-2}4s^2$ of the atom and in $3d^{n-2}$ of the bipositive ion are practically the same; a small allowance for outer screening by the $4s^2$ electron pair ought in principle to be made, especially in the earlier transition elements, but it is unimportant compared with inner screening, and with the changes that take place as a result of the change in nuclear charge in going from one transition metal to its neighbors in the series.

The qualitative features of transition metal $3d$ orbitals which are made apparent by such detailed studies of individual atoms may be summarized under three headings:

(a) *The orbital size*, whether judged by the position of the radial maximum or by mean radius, diminishes steadily from left to right across the transition series, i.e., with increasing atomic number.
(b) *Inner screening.* $3d$ orbitals are expanded by increased $3d$ population in a given element. The expansion affects the outer part of the charge most, the position of the maximum being practically unaltered.
(c) *Outer screening.* $3d$ orbitals are almost unaffected[6] by the presence of $4s$ or $4p$ electrons in the same element.

In the two-parameter orbitals of Table VIII, the radial maximum r_M is the value for which the product rR attains its maximum, and this depends strongly on the contribution by the term with the larger α. Thus r_M is only a little larger than the value calculated for the first term alone. For example, in the d^6 configuration of Fe^{++} the two term $3d$ wave function has a radial maximum about 0.1 a.u. greater than the radial maximum for the first term alone (3/5.35 = 0.56 a.u.). For bond formation this strong dependence on the inner regions of the orbital makes r_M an undesirable measure, because the outer parts are more important in their contribution to the overlap with bonded atoms. Thus r_M values are of little interest, and we shall consider the mean radii $\bar{r}$ [see Eq. (2)]. It may readily be shown that the mean radius of the two parameter wave function is given by Eq. (4):

$$\bar{r} = 3.5c_1^2/\alpha_1 + 3.5c_2^2/\alpha_2 + 7c_1c_2(\gamma^7/\beta^8) \qquad (4)$$

where γ is the geometrical mean and β the arithmetical mean of the two exponents α_1 and α_2, i.e., $\gamma = (\alpha_1\alpha_2)^{1/2}$ and $\beta = (\alpha_1 + \alpha_2)/2$. The values of r found in this way from the values in Table VIII are shown in Fig. 4.

C. Influence of Ligand Atoms

1. *Influence of the Molecular Environment*

Two factors that change the effective nuclear charge in a systematic way have already been noted. The effective charge is smaller (giving orbitals capable of better overlap with acceptors) in the lighter atoms, and in electronic structures of the same atomic number it drops rapidly as the number

[6] There is a small expansion, affecting mainly the outermost region of the charge distribution.

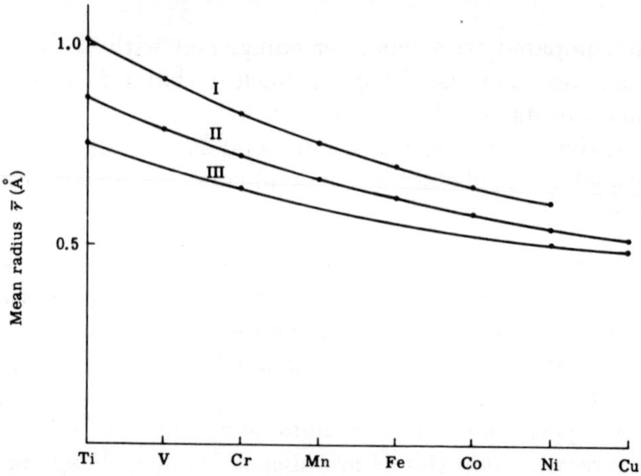

FIG. 4. Mean radii of 3d orbitals of transition metal atoms (I), unipositive ions (II), and bipositive ions (III) according to the wave functions of Richardson *et al.* (1962).

of d electrons goes up. These both apply to free atoms and ions. In molecules still other factors enter. In a crude sense, it is obvious that electron transfer into the d orbitals of the donor as in σ bond formation will reduce the effective charge for the 3d electrons. The meaning of this is that, for example, the $3d\pi$ electrons are screened from their own nucleus by electrons belonging to the ligand atom, usually its lone pair electrons. Electron transfer from the ligand is not required in this process, though it may of course occur. All that is necessary is that ligand electrons penetrate the region occupied by the metal d electrons and so partially screen them. If the ionization potential of the ligand lone pair electrons is high, so that the electron distribution is not too seriously perturbed by the field of the metal, it should be possible to estimate the extent of the penetration of lone pair electrons by calculating overlap integrals of the lone pair orbital with a $3d_{z^2}$ orbital of the metal. Typical values of overlap integral are given in Table IX. They are based upon Slater orbitals both for the metal and the

TABLE IX
LONE PAIR–$3d_{z^2}$ OVERLAPS

$Z(3d)$	$Z(2p,2s)$	r (Å)	S	Example
7.5	3.25	1.84	0.201	M—C≡O
7.5	3.9	1.95	0.174	M—NH_3
7.5	3.9	1.77	0.218	M—N≡O
7.5	4.55	2.06	0.120	M—OH_2
6.0	3.25	1.84	0.30	M—C≡O
6.0	4.55	2.06	0.20	M—OH_2

ligand; because Slater orbitals, as discussed above, are inaccurate in form, the values have no absolute significance but should be relatively in the correct order. The metal $3d$ exponent is 2.5, the Slater value for Ni^{++}, and the ligand exponents are also Slater values.

The variation in the overlaps shows a strong dependence on the bond length, which emerges as the most important variable, being much more important than the ligand exponent in determining the overlap. The last two values in the table are those for a lower $3d$ exponent; they confirm that the relationship between the overlaps in the several cases do depend only a very little on the $3d$ exponent.

The degree of penetration into the metal's d orbitals, and thus the inner screening, appears to depend mainly on the bond distance, and is much greater for carbonyls and nitrosyls than for ammines and hydrates at the observed bond distances. We thus can expect that the expansion of $d\pi$ orbitals in a given metal will be far greater for CO as ligand than for H_2O. The expansion will also increase the more ligands there are attached to the central atom.

There is a further mechanism that is also expected to cause an increase in d orbital size. It is recognized that in problems such as that of the d electrons in SF_6, very diffuse d orbitals are contracted (Craig and Magnusson, 1956) by ligands, particularly electronegative ones, and enabled to take part in binding. Contrariwise, very tightly bound orbitals may be expanded by ligands, by a mechanism essentially different from inner screening. In physical terms this is best explained by considering first a single, tightly bound d electron surrounded by negative ions. Analysis of the electrostatic energy terms shows that the d electron energy is reduced when the orbital is expanded, so that a larger proportion of the electronic charge lies outside the shell of negative charge. The degree of expansion is greater for loosely rather than tightly bound electrons, and is greater for orbitals directed away from the charges than for those directed towards them; i.e., in the example of octahedrally arranged charges, the d_ϵ orbitals are more expanded than the d_γ orbitals.

The effect is far more general than this description implies, because the expansion can be produced by ligands with dipole moments, even when they carry no net charge. Since this must be true of virtually all ligands, the expansion should be a rather general phenomenon. Much interest attaches to the magnitude of the effect, and of its variation from one ligand to another, because this will be one of the factors affecting the strength of the bonds formed. A number of trial calculations of dipole-induced d orbital changes have been made and briefly reported (Craig and Magnusson, 1958). They refer to a d electron in a Slater orbital (Eq. 1) with typical exponents given by Slater's rules for a transition metal. The perturbing field is that of six dipoles, with their negative ends inward, of 2.5 debyes. Results in a few cases are given in Table X. The changes are very sensitive

TABLE X
EXPONENT CHANGES IN 3*d* ORBITALS

Bond distance (Å)	Free atom effective nuclear charge	$Z(3d_\epsilon)$	$Z(3d_\gamma)$
1.85	5.55	5.0	5.65
1.85	5.7	5.15	5.7
2.4	5.55	5.4	5.55
2.4	5.7	5.6	5.7

to the bond distance: the over-all effect, averaged over d_ϵ and d_γ orbitals, is an expansion; the contraction of the d_γ orbitals is slight, and swamped by the larger expansion of the d_ϵ orbitals, which are alone of interest in the formation of π bonds in an octahedral field. The change in effective charge in the d_ϵ orbitals of 0.55 units corresponds roughly to moving one place to the left in the transition series; i.e., one could say that a *d* electron in Fe^{++} in an octahedral field of this kind behaves like a *d* electron in a free Mn^{++} ion. Physically, orbital expansions of this kind are analogous to, though generally larger than, outer screening expansions in free atoms, i.e., the small expansion that distinguishes *d* orbitals in the d^{n-2} and $d^{n-2}s^2$ configurations.

A third influence is that of *d* electron loss from the metal to the ligand in donor-acceptor π bonding. This results in a reduction of inner screening of the remaining electrons, and so acts in opposition to the increased inner screening due to penetration by lone pair ligand electrons. There are, thus, three ways in which free atom *d* orbitals may be modified by ligand atoms; (*a*) by inner screening due to ligand lone pairs penetrating the 3*d* shell leading to expansion; (*b*) by outer screening due to the electrostatic influence of ligands considered as point charges or dipoles also leading to expansion; and (*c*) by π electron loss to ligands in π bonding, giving orbital contraction, by reduction of *atomic* inner screening.

2. *Nonempirical Treatments*

Before going on to consider changes in orbital size in specific cases, we mention briefly recent attempts to deal in a more exact way with the characteristics of *d* orbitals in molecular fields. Perhaps the most significant are those which use a crystalline field model, and give S.C.F. wave functions for a metal surrounded by point charges (Freeman and Watson, 1960). Such calculations show the characteristic crystal field splitting, and also indicate a net orbital expansion (for negative charges). The detailed effects agree with those based on Slater functions (Craig and Magnusson, 1958) in giving an expansion of *d* orbitals with density maxima falling in regions free of negative ionic charges, and contraction otherwise. It proves difficult to go one stage further in elaboration by including a realistic potential for the ligand ions, because of the problem of including exchange terms in the potential. If the exchange terms are not included, the repulsions between *d* electrons and inner shell ligand electrons are seriously underestimated, and overlarge orbital expansions are found (Phillips, 1959). No entirely successful way round this difficulty has yet been found (Marshall and Stuart, 1961).

3. *Outer d Orbitals and Isovalent Hybridization*

Complexes of metals in high oxidation states, such as the hexaaquoferric ion $[Fe(H_2O)_6]^{3+}$, may be described as ionic, or, alternatively, if their properties in some respects suggest covalent character, the bonds can be said to be covalent, formed using hybrids of 4*s*, 4*p*, and 4*d* orbitals. Such a hybridization scheme implies that the outer (4*d*) orbitals, which in the free element are too diffuse and weakly bound to contribute to binding, are polarized sufficiently in the molecular environment to give them useful binding ability. Since in any realistic assessment of the polarization of these orbitals they will still be large compared with the 4*s* and 4*p* orbitals, electrons in them will have a high probability of being located in the outer regions of the electron distribution of the metal, and therefore close to the ligands. Thus the electron density will be high near the ligands whether the ionic bond picture is used, in which the electrons are located wholly on the ligand, or the covalent picture, in which the electrons formally belong to the central metal but occupy orbitals extending out to the vicinity of the ligands. There is, thus, no hard and fast distinction between the ionic and covalent viewpoints as to electron density.

D. Double Bonding in Carbonyls and Related Molecules

1. *Overlapping Power of Metal 3d Orbitals*

If the forms of the 3*d* orbitals in metals are more or less accurately known, overlap values have some quantitative usefulness, and may give a guide to the feasibility of $d\pi$—$p\pi$ metal-ligand bonding in particular cases, as well as giving relative values from one metal to another.

An example of great practical interest is the overlap concerned in double bonding between metal and first row ligand atoms, as in metal carbonyls or nitrosyls, in which one or more $2p\pi$ ligand orbitals form bonds with a 3*d* metal orbital. This is measured by the overlap integral between two orbitals φ_A and φ_B:

$$S_{AB} = \int \varphi_A \varphi_B d\tau$$

In actual bonding orbitals the value of S usually lies in the range 0.1–0.3; it depends on the distance between the centers A and B and on the characteristics of the orbitals. By using the two parameter $3d$ wave functions for the metal atoms, with values taken from Table VIII and suitable ligand orbitals, a set of overlap integrals may be calculated. The values given in Fig. 5 refer to the metal-carbon π bond in carbonyls, and are obtained for a metal carbon distance of 1.83 Å, and an exponent 1.625 for the carbon $2p\pi$ orbital.

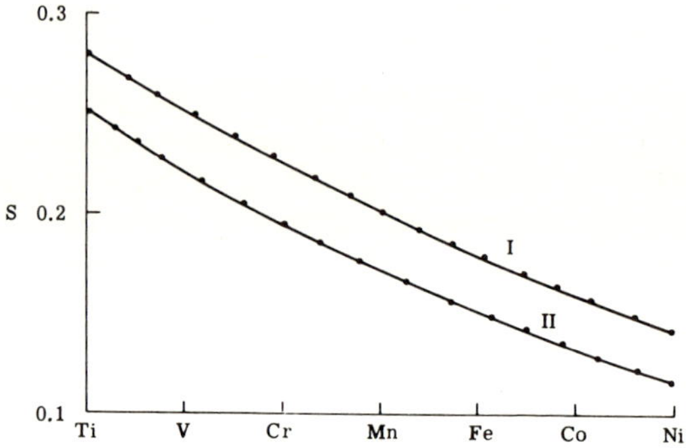

FIG. 5. Overlap integrals for $3d\pi$—$2p\pi$ bonding. The $2p\pi$ orbital is a Slater type ($Z = 3.25$); the $3d\pi$ orbitals are those of Richardson *et al.* (1962) for neutral atoms (I) and unipositive ions (II). The internuclear distance is 1.83 Å.

For the elements in a given ionization state the orbital sizes diminish from left to right of the series, from Ti to Cu; Fig. 5 shows that the overlap integrals with a common $2p\pi$ orbital (appropriate to carbon) also decrease. In the univalent and bivalent ions the orbital sizes, and the overlaps, are successively less, in agreement with the physical picture that, as d electrons are removed, the inner screening is reduced, and the remaining d electrons are more tightly held.

The overlaps throughout are large enough to support the idea that significant double bonding may occur; the overlap in the π bond of ethylene is in the range 0.25–0.30, and in general overlaps of 0.1 and above are compatible with π bonds. The sensitivity of overlap to changes in d orbital population indicates that external influences on the population due to ligands will affect the overlapping power of the metal orbitals; even small structural changes capable of changing the effective nuclear charge of the metal $d\pi$ orbitals might lead to an important increase or decrease in π bonding.

2. *Double Bonding in Carbonyls and Nitrosyls*

a. Theoretical Preliminaries. It is clear from the overlap integrals in Fig. 5 that the $p\pi$–$d\pi$ overlap falls off to small values at the end of the first transition series, so that conditions become steadily less favorable to π bonding. In the earlier transition elements, where the overlap is large, there are insufficient d electrons in the metal orbitals; there are, thus, quite specific limitations to the degree of metal-ligand π bonding possible at the two extremes of the transition series. This leads to a prima facie expectation that conditions are most favorable near the middle of the series, in the elements Cr, Mn, and Fe. There is evidence referred to below to support this conclusion. Also, as long as the size of the $d\pi$ orbital is the controlling factor, conditions are best in the neutral elements, and successively worse in the uni- and bivalent ions. Under conditions of molecule formation there will often be some degree of inner screening of the metal $d\pi$ electrons by ligand lone pairs; the effect of inner screening is favorable to π bonding, in that it increases the size of the $d\pi$ orbitals, and increases the $p\pi$–$d\pi$ overlap. The degree of inner screening itself has some dependence on overlap, namely that of the lone pair orbital of the ligand, with a metal d orbital; and the effect of this is that the variation of $d\pi$–$p\pi$ double bonding upon the size of the d orbitals is sharper than Fig. 5 shows. The double bonding to nickel is thus much less than with cobalt or iron and negligible in zinc.

b. Experimental Evidence for Metal-Ligand π Bonding. Although in the preceding discussion we have made considerable use of the hypothesis of double bonding between transition metals and certain ligands, it must be emphasized that this is still very much presumptive. The usual kind of bond length evidence adduced for double bonding in the first row of the Periodic Table (e.g., for C=C, N=O, or C=N bonds) is not applicable for carbonyls, for example, partly because of the paucity of experimental data, partly because of uncertainty as to just what radius is "normal" for zerovalent metal, and partly because the metal-carbon distance in metal carbonyls appears to be relatively insensitive to different transition metal atoms.

The arguments in favor of the hypothesis are summarized below. Taken individually they cannot be regarded as providing strong support but *in toto* they do enable us to accept the proposition as a good working hypothesis. Much of the argument relates to metal carbonyls.

(1) Metal carbonyls so far as is known at present are formed only by transition metals which have d electrons available for π bond formation.

(2) The ligands which most easily displace some or all of the CO groups

from metal carbonyls are those which are at least capable of accepting $d\pi$ electrons from a transition metal atom. These ligands are of two kinds: (*a*) those having as a donor atom a first row element which can make a *p* orbital available by rearrangements of the electrons in the ligand molecule (e.g., N in pyridine); (*b*) those in which the donor has vacant *d* orbitals available (e.g., P or As in tertiary phosphines or arsines).

It seems significant that ligands like NH_3 which cannot form π bonds by either of the above methods will displace CO from carbonyls of the early transition metals, e.g., from $Cr(CO)_6$; it seems that the remaining CO groups readily accept all of the π bonding available if they have the chance to do so.

(3) It is to be expected that π bonding will be enhanced as the energy required to remove a *d* electron from the nonbonding *d* shell diminishes. Three cases need to be distinguished.

(*a*) The effect of formal charge. In the sequence $[Fe(CO)_4]^{2-}$, $[Co(CO)_4]^-$, and $[Ni(CO)_4]$ we have d^{10} metal atoms with steadily increasing (positive) nuclear charges; this should operate in the sense of tending to diminish the availability of $d\pi$ electrons. Infrared stretching frequencies of the C—O bonds support this view; the average C—O stretching frequencies being $[Fe(CO)_4]^{--}$ (1788 cm.$^{-1}$); $[Co(CO)_4]^-$ (1901 cm.$^{-1}$); $Ni(CO)_4$ (2087 cm.$^{-1}$). If we accept the view that the two extreme structures are

$$M \leftarrow C \equiv O \quad \text{and} \quad M{=}C{=}O$$

then a *decrease* in C—O stretching frequency indicates a decrease in triple bond character of the C—O bond and hence by inference an increase in *double bond* character of the M—C bond.

(*b*) Similar conclusions follow from a study of the effect of groups of different electronegativity on the C—O stretching frequency in substituted metal carbonyls. Thus, along the sequence I → Br → Cl in compounds such as $[Fe(CO)_4(hal)_2]^0$ and $[PtCO(hal)_2]_2$ the mean C—O stretching frequency gradually increases as the electronegativity of the halogen increases, as shown in Table XI.

This is simply interpreted to mean that in the sequence Cl → Br → I the positive charge on the metal atom decreases enabling readier donation

TABLE XI

INFRARED DATA FOR SUBSTITUTED METAL CARBONYLS (CM.$^{-1}$)

Compound	$[PtCOCl_2]_2$	$[PtCOBr_2]_2$	$[PtCOI_2]_2$	$[Fe(CO)_4Br_2]^0$	$[Fe(CO)_4I_2]^0$
Mean C—O stretching frequency	2152	2130	2112	2109	2091

to the metal atom. The triple bond character in the C—O bond is decreased. [For detailed references see Nyholm (1961).]

We also observe that with the same metal atom carbon monoxide and π-bonded complexes generally occur most readily when the metal atom has a low oxidation state. Thus PF_3, PCl_3, and similar poor donors will stabilize zerovalent nickel but not higher oxidation states. Similarly, simple carbonyl halides are restricted to oxidation states of two or less.

Finally, it is of interest to compare the relative stabilities of zerovalent metal carbonyls having different ionization potentials for the nonbonding core. For the very stable $Cr(CO)_6$ the I.P. of the d^6 spin-paired nonbonding core is estimated at 4.0 e.v.—similar to that of the d^8 core in $Fe(CO)_5$ (3.8 e.v.). However, the much less stable $Ni(CO)_4$ with a d^{10} nonbonding core has an I.P. of 5.87 whereas the (unknown) compounds $Pd(CO)_4$ and $Pt(CO)_4$ have a d^{10} nonbonding core for which the I.P. values are, respectively, 8.3 and 8.2 e.v. It is tempting to attribute the fact that these two compounds have not been prepared to their poor π-bonding capacity in the zerovalent state.

(4) Bond length data have been used to justify the existence of double bonding in metal carbonyls, but many more accurate bond lengths are needed before these can be interpreted with any confidence.

3. *The Rare Gas (Eighteen Electron) Rule*

a. General. The composition of certain classes of transition metal complexes, especially those of the metals in their zerovalent states, appears to be determined by the condition that the number of electrons in the valence shell is eighteen. For example, in $Ni(CO)_4$ the number is made of ten electrons in the 3*d* and 4*s* orbitals of nickel plus two lone pair electrons from each carbonyl group. In such molecules the attainment of the inert gas electron configuration of krypton seems to lead to stability, and is therefore intimately concerned in the problem of the metal-ligand bond.

Experimentally, it is found that the composition of a large number of transition metal carbonyls, nitrosyls, aryls, isocyanides, and other complexes can be correctly predicted by using the rare gas rule. There are also many exceptions, like the seventeen electron $V(CO)_6$, showing that obedience to the rule is only one of the factors leading to high stability in a complex.

The analogous regularities in electron configurations leading to stable molecules are the octet rules for first and second row elements. In the first row especially the octet rule applies very widely, the main systematic exceptions being confined to Be and B, in which the neon configuration requires the addition of six and five electrons respectively. Even in B, as noted above, the tendency to complete the octet by double bonding as in BF_3 or by coordination with a donor is strong.

In the second row an octet rule again applies to some extent in the elements silicon to chlorine, with reference to the attainment of the electron configuration of argon, but exceptions occur much more widely.

The analogous rule for the first long period, including the transition series of elements, applies to the completion of the electron configuration of krypton, which requires a further eighteen electrons. As might be anticipated, this rule has an even narrower range of application.

To the right-hand side of the period, in the elements from germanium to bromine, the krypton configuration is attained in compounds with elements of low electronegativity, but is easily exceeded otherwise, for example, in pentacovalent arsenic. To the left of the period, in the transition elements, the failure of the rule is in the opposite sense, namely that the number of electrons is less than required for the krypton configuration, leading to complexes with incompletely filled $3d$ shells, often with paramagnetism. The krypton configuration appears to be energetically significant only in rather special circumstances, when the attached elements are weakly electronegative, and when double bonding is at least a formal possibility.

b. The Modified Potential Field in Metal Complexes. In the potential field of a point positive charge, as in the hydrogen atom, all the orbitals of a given principal quantum number, for example $3s$, $3p$, and $3d$, have the same energy. If the field falls off more slowly than coulombic, we should find that the more projecting orbitals are relatively more tightly bound than the less projecting, leading to an order of stability $3d > 3p > 3s$. However, in practice, all complex atoms have fields which fall off faster than coulombic, because the field of the nucleus is more and more screened by inner electrons at larger distances. We then encounter the characteristic result that the more projecting orbitals are the less strongly bound, giving the order $3d < 3p < 3s$. The physical reason for this is that the electrons in the outer regions are in a very weak field, and make a smaller contribution to binding than electrons nearer to the nucleus: the orbital with the more projecting character is the less strongly held. In the transition series this results in the $4s$ orbitals being occupied before the $3d$ are completely filled. Now if the ligands can change the potential field sufficiently, in the sense of making it decrease less rapidly in the region important for $3d$ electron binding, conditions may be established leading to stability of the krypton configuration. This possibility has been recently examined (Craig and Doggett, 1963).

The electrostatic potential may be calculated from the S.C.F. orbitals given by Watson (1960) and compared with the potential for krypton calculated from results of Worsley (1958). Because the nuclear charge of Cr is less by 10 units than that of Kr, its potential close to the nucleus is much less negative, and it is found that the difference persists at all distances that are significant for 3*d* orbitals with radial maxima at between 0.25 and 0.5 Å.

The possibility to be investigated is whether the potential of Cr(d^5s) and of other transition metals can be so modified by ligands that the rare gas configuration is stabilized. In practice the problem is whether the potential is brought closer to that of krypton when the effects of ligand atoms are allowed to modify the metal potential. With first row elements as ligand atoms, as in carbonyls and nitrosyls, the ligand fields may readily be investigated, both as to the extent to which the accurately spherical character of the atomic field is destroyed, and as to how far the radial dependence is altered. It is found by Doggett (1961) that an octahedral array of carbon atoms at the observed metal-carbon distance gives less than 1% deviation from spherical symmetry up to a distance of 0.6 Å. The region within a sphere of this radius contains most of the *d* electron density.

The more difficult question is that of the change in radial dependence of the potential caused by adding the ligand potential to that of the metal. Neither in CO nor in other ligands with inward-facing lone pairs is there a close resemblance to the krypton potential. However, by supposing that $d\pi$ electrons are to some extent removed by double bonding with the ligands, so reducing inner screening, the potential may be made to fall off more quickly near the nucleus, and to approach significantly closer to the krypton potential in the critical region. There is no good way of estimating the degree to which the central metal loses electrons by π bonding; it may only be said that electron withdrawal by ligand groups is an efficient mechanism for reducing inner screening of the *d* electrons, decreasing the rate of fall-off of the potential, and so stabilizing the rare gas electron configuration. It gives definite support for the view that, where the σ bond component is formed by lone pairs from the ligands, the eighteen electron rule will hold only if electrons are removed from the *d* shell by double bond formation.

Where the bonds are not formed by lone pairs, but as in aryl and cyclopentadienyl compounds by atoms with singly occupied valence orbitals, the modified potential shows an approach to kryptonlike character, without electron donation from the *d* shell.

We conclude that in carbonyls and nitrosyls, and with other lone pair–bonded ligands, the conditions for stable rare gas configurations are essentially the conditions for strong double bond formation, discussed in Section III,D,2. Briefly, breakdown of the eighteen electron rule is most likely: (*i*) when the ligand is an insufficiently strong π acceptor, as in NH_3;

(*ii*) when the π overlap between metal and ligand is small, as in the late transition elements; or (*iii*) when the metal nuclear charge is so low that electron withdrawal has an insufficiently large effect, as in the early transition metals.

c. *Departures from the Rare Gas Rule.* The rare gas rule can only be expected to apply, as we have seen above, under rather narrow conditions, namely in metals in low or zero oxidation states and with ligands that form double bonds or have other special characteristics such as olefins, aryls, and cyclopentadienyls. Thus, the question of likely deviations from the rare gas rule is an interesting one only if we restrict it to deviations within this class of complex compounds.

The empirical observations of the breakdown of the rule may be arranged in a rough sequence. In the elements Cr, Mn, and Fe, in the center of the first transition series, the rule applies rather widely, with ligands CO, NO, PX_3, CN^-, etc. In neutral Mn, with an odd number of electrons, at least one of the ligands must be different from the rest, and must supply an odd number of electrons. We find NO fulfilling this function in $Mn(CO)_4NO$; another possibility is a ligand supplying one electron, such as hydrogen, to form a hydride, or a second Mn atom as in $(CO)_5Mn$—$Mn(CO)_5$, or as another possibility the complex can ionize to $[Mn(CO)_5]^-$. In Mn complexes of the type $Mn(CO)_4L$, small structural changes in L may cause a shift in the monomer-dimer equilibrium, so that in one case (e.g., L = PPh_3) the seventeen electron monomer and in another, in L = $P(OPh)_3$, the eighteen electron dimer is the more stable. In Cr and Fe departures from the rare gas rule are less common.

In the early elements of the transition series attainment of a potential of the kind to stabilize the rare gas configuration requires more drastic modification to the free atom potential, and departures are likely always to be in the sense of fewer electrons than the stable complement of eighteen. Vanadium hexacarbonyl, with seventeen electrons, is in this category. In the later elements of the transition series, the situation is quite different, in that the π donor capacity is reduced, and the nature of the metal-donor bonds changed progressively toward pure σ bonding, depending largely on the polarization of ligand lone pairs by the metal. Since such bonds are largely electrostatic in character there is no limitation to the total number of electrons imposed by the electronic structure of the metal and the number will rise above eighteen. Thus, the deviation at the end of the transition series will be in the sense of electron excess.

In practice the deviations fall into four main classes. Let us consider a d^n metal ion M^{t+}. As the value of n decreases from 10 to 0, so the number of σ bonding pairs[7] of electrons needed to reach the rare gas number increases

[7] The rare gas configuration is sometimes achieved using both σ and π pairs of elec-

from four (e.g., in $Ni(CO)_4$) to nine; no definite examples of the latter, with the possible exception of $[Nd(H_2O)_9]$ $(BrO_3)_3$ (Helmholtz, 1939), are known.

The first kind of deviation arises in complexes in which all of the five $(n - 1)d$, ns, and three np orbitals are used but some may contain unpaired electrons. In the d^4 configuration three possible situations can arise in all of which the nine [nd, $(n + 1)s$, and np] orbitals are used. First, we have diamagnetism with seven coordination as is observed in $[Mo(diarsine)(CO)_3I_2]^0$ (Nigam, Nyholm, and Stiddard, 1960) or $[Re(diarsine)_2(CO)I_2]I$ (Kirkham *et al.*, 1962) which obey the 18 electron rule; second, we have paramagnetism with two unpaired electrons, leaving six orbitals for σ bond formation. Complexes of quadrivalent Fe, Ru, and Os, e.g., $[Fe(diarsine)_2Cl_2][ClO_4]_2$ (Nyholm and Parish, 1956), tervalent Mn, Tc, and Re, e.g., $[Re(diarsine)_2Cl_2]Cl$ (Curtiss, Fergusson, and Nyholm, 1958), and bivalent Cr, Mo, and W, e.g., $Mo(diarsine)_2Cl_2$ (Lewis, Nyholm, and Smith, 1962) are examples. Finally, there are a few complexes known with four unpaired electrons and a coordination number of five. The compound $[Et_4N]_2MnCl_5$ appears to be an example of this, but it is not yet certain that the anion is not polymeric (Gill, 1961). Examples of paramagnetic complexes with coordination numbers 8–2 are given in the accompanying tabulation.

Coordination number	Number of non-bonding electrons: Total	Number of non-bonding electrons: Unpaired	Example
8	1	1	$[Mo^{V}(CN)_8]^{3-}$
7	3	1	$[Mo^{III}(diarsine)(CO)_2Br_3]^0$
6	5	1	$[Fe^{III}(CN)_6]^{3-}$
5	7	1	$[Fe^{I}(diarsine)(CO)_2I]^0$
4	9	1	$[CuCl_4]^{2-}$
3	$d^{10} + s'$	1	$C^{III}Ph_3$
2	$d^{10} + s^2p'$	1	$Cl^{IV}O_2$

The second type of deviation involves the use of fewer than the number of available orbitals for bonding. Thus the $[Ni(CN)_4]^{--}$ ion has eight non-bonding $3d$ electrons, using four d orbitals; another four are used for bonding to the CN groups. This leaves one vacant, presumably p orbital (Scheme 2). An examination of data for d^8 complexes reveals that the tendency for

trons from the ligand: the coordination of NO can be regarded at least formally as involving one σ bond and one π bond, three electrons coming from the ligand.

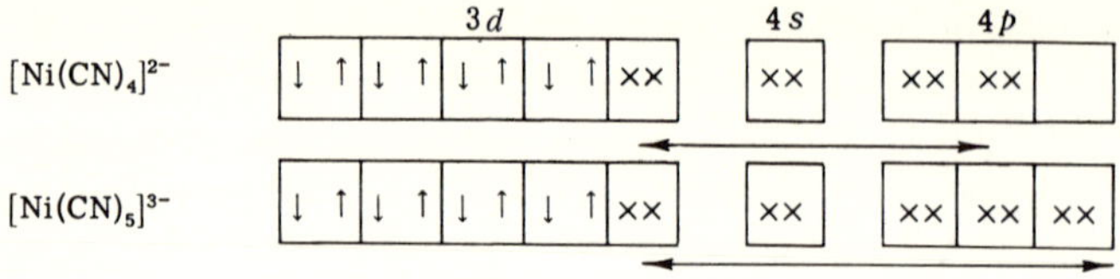

four coordination to arise increases as (*i*) the formal (positive) charge on the d^8 metal increases and (*ii*) the effective nuclear charge in a vertical sequence increases. As an example of (*i*) we note the sequence $[Cr(CO)_5]^{--}$, $[Mn(CO)_5]^-$, $[Fe(CO)_5]^0$, $[Co(PhNC)_5]^+$ but with Ni(II) and Cu(III) four coordination is common. This has been discussed elsewhere (Nyholm, 1961) and can be correlated with the increasing $3d$–$4p$ separation as the positive charge increases. Similarly we note in regard to (*ii*) that in their tendency to increase the coordination number from four to five Ni(II) > Pd(II) > Pt(II) which is the order of increasing $(n-1)d$–np separation for bipositive ions.

The third variation involves coordination numbers exceeding the number of vacant orbitals available. Several of the complexes of d^8 elements, e.g., Ni^{II}, form complexes in which the coordination number is six. An example is the compound $[Ni(II)(diarsine)_2I_2]^0$ shown by Stephenson (1962) to involve square coordination to the four arsenic atoms but with two unusually long bonds normal to the square, as shown in (III). It is con-

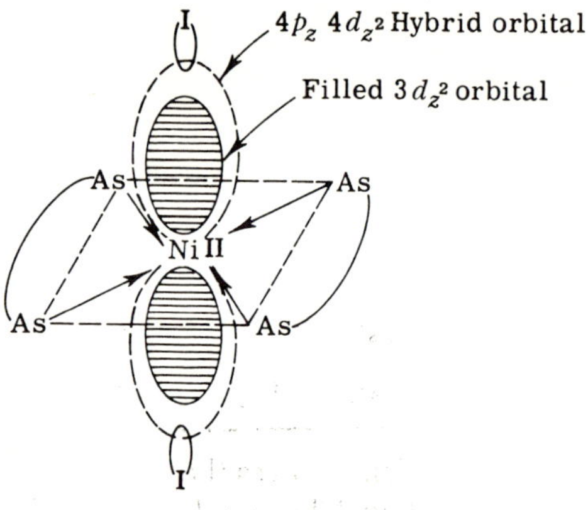

(III)

venient to assign to this complex four square $3d_{x^2-y^2}4s4p_x4p_y$ bonds and two longer more ionic bonds. It is notable that the absorption spectrum of the $[Ni(diarsine)_2]^{++}$ ion changes markedly when the two iodine atoms are attached. Some degree of covalent binding is implicit; the use of $4p_z4d_{z^2}$ hybrid orbitals seems feasible. This deviation is only a variation on the second type discussed above, insofar as the important feature is a structure with four $3d4s4p^2$ bonds.

A fourth type of variation needs to be recognized with largely ionic binding: in compounds of the type $[TiF_6]^{--}$ and $[FeF_6]^{3-}$, one must regard the structure as based essentially on an ionic model involving Ti^{4+} or Fe^{3+} ions and F^- ions. The latter will arrange themselves at the corners of an octahedron. Some covalent contribution to the binding undoubtedly occurs but marked deviations from the nine orbital rule take place. In TiF_6^{--} only six of nine possible orbitals are used; in $[FeF_6]^{3-}$ five $3d$ orbitals are occupied by the five unpaired electrons and six more orbitals would be needed for covalent binding, two in excess of those expected from the rare gas configuration.

IV. Metal-Metal Bonds in Complex Compounds

Metal-to-metal bonding in complex compounds has been recognized during recent years in an increasing number of examples of widely varying type. The occurrence of stable metal-metal bonds depends on the nature of the attached ligands, on the electron configuration of the metal atom, and on its oxidation state.

In Table XIII we provide a summary of the main types of metal-metal bonds which occur in elements and chemical compounds. We should really add a fifth type, "Ionic," since in certain alloys the loss of an electron by one metal and its gain by another is so nearly complete that a salt-like compound is formed. The 1:1 "alloy" between caesium and gold (Cs^+Au^-) is an example of this type of binding.

Some apparently simple compounds, e.g., $MoCl_2$, are complex and involve a group of metal atoms bound together by a multicenter bond. The compound $MoCl_2$, for example, has as its structural basis a $[Mo_6Cl_8]^{4+}$ unit, in this there is an octahedron of molybdenum atoms which are conveniently placed at the centers of the six faces of a regular cube. The Mo metal atoms are also bound together by eight chlorine atoms located at the corners of the tube.

The covalent type of metal-metal bond is more important in complex compounds. It occurs most frequently with metal atoms in low oxidation states; the effect of this low formal charge is presumably to increase the size of the orbital containing the unpaired electron, thereby facilitating overlap and metal-metal bond formation. Thus for d^9 metal atoms the tendency towards dimer formation increases as one passes along the sequence $Cu^{2+} \rightarrow Ni^+ \rightarrow Co^0 \rightarrow Fe^{-I}$. Except in its carboxylic acid derivatives, Cu^{2+} shows little tendency towards M—M bond formation; however, Ni^+ in $K_4Ni_2(CN)_6$ gives rise to a diamagnetic compound, presumably owing to the formation of a Ni—Ni bond. The only known examples of

Co^0 (e.g., $Co_2(Co)_8$) and Fe^{-I} (e.g., $[Fe_2(CO)_8]_2^-$) are dimeric and diamagnetic.

Examples are also known in which metal-metal bond formation occurs between metal atoms with different initial electron configurations. Thus, complexes of a $d^{10}s^1$ atom (Au^0) with metal atoms having d^9 (e.g., Co^0) d^8 (e.g., Fe^0) d^7 (e.g., Mn^0) and d^5 (W^I) have recently been prepared by Coffey, Lewis, and Nyholm. By reacting the sodium salt of the carbonyl anion with $Ph_3P \rightarrow AuCl$, the following have been isolated: $Ph_3P \rightarrow Au—Co(CO)_4$; $(Ph_3P \rightarrow Au)_2$-$Fe(CO)_4$; $Ph_3P \rightarrow Au—Mn(CO)_5$, and $Ph_3P \rightarrow Au$-$W(CO)_3CP$. Many other combinations such as d^7–d^5 and d^7–d^9 are possible. There is little doubt that by attaching appropriate groups to the metal atoms to confer upon it an appropriate effective electronegativity a wide variety of such compounds can be prepared.

TABLE XII
METAL-METAL BONDS

Type I	Type II	Type III	Type IV
Solid metals (multicenter bonds)	Concentrated metal compounds, the metal having a small number of *d* electrons, e.g., $d^1(NbI_4)_n$ $d^2(MoO_2)_n$-$(Nb_6Cl_{12})_{unit}^{6+}$ $d^3(MoI_3)_n$ $d^4Mo_6Cl_8{}_{unit}^{4+}$ Bonding usually multicenter but not necessarily so.	Compounds with one or more *covalent* bonds: (*a*) Diatomic metals in gas phase, e.g., Li_2,Au_2. (*b*) Metal atoms with same electron configuration, e.g., $Mn_2(CO)_{10}$. (*c*) Metal atoms with different electron configuration, e.g., $HgFe(CO)_4$	Metal donor type, e.g., the weak Ni—Ni bond in bisdimethylglyoximatonickel(II)

References

Ahrland, S., and Chatt, J. (1958). *Quart. Revs.* (*London*) **12,** 265.

Barraclough, C., Lewis, J., and Nyholm, R. S. (1959). *J. Chem. Soc.* p. 3552.

Becher, H. J. (1952). *Z. anorg. Chem.* **270,** 273.

Becher, H. J., and Goubeau, J. (1952). *Z. anorg. Chem.* **268,** 131.

Clark, R. J. H., Lewis, J., Nyholm, R. S., Pauling, P., and Robertson, G. B. (1961). *Nature* **192,** 222.

Clark, R. J. H., Lewis, J., and Nyholm, R. S. (1962). *J. Chem. Soc.* p. 2460.

Classen, H. H., Selig, H., and Malm, J. G. (1962). *J. Am. Chem. Soc.* **84,** 3593.

Coffey, C. E., Lewis, J., and Nyholm, R. S. (1963). *J. Chem. Soc.* (in press).

Craig, D. P., and Doggett, G. (1963). *J. Chem. Soc.* (in press).

Craig, D. P., and Magnusson, E. A. (1956). *J. Chem. Soc.* p. 4895.

Craig, D. P., and Magnusson, E. A. (1958). *Discussions Faraday Soc.* **26,** 116.
Craig, D. P., and Zauli, C. (1962). *J. Chem. Phys.* **37,** 601.
Curtis, N. F., Fergusson, J., and Nyholm, R. S. (1958). *Chem. & Ind.* (*London*) p. 625.
Doggett, G. (1961). Ph.D. Thesis, University of London, England.
Dunn, T. (1961). *Trans. Faraday Soc.* **57,** 1441.
Eyring, H., Walter, J., and Kimball, G. E. (1944). "Quantum Chemistry," p. 89. Wiley, New York.
Freeman, A. J., and Watson, R. E. (1960). *Phys. Rev.* **118,** 1168.
Gill, N. S. (1961). *Chem. & Ind.* (*London*) p. 989.
Gill, N. S., and Nyholm, R. S. (1959). *J. Chem. Soc.* p. 3997.
Gillespie, R. J. (1961). *Can. J. Chem.* **39,** 2336.
Gillespie, R. J., and Nyholm, R. S. (1957). *Quart. Revs.* (*London*) **11,** 339.
Gillespie, R. J., and Nyholm, R. S. (1958). *Progr. in Stereochem.* **2,** 261.
Gray, H. B., and Ballhausen, C. J. (1962). *Inorganic Chemistry* **1,** p. 111.
Griffith, J. S. (1961). "The Theory of Transition Metals Ions." Cambridge Univ. Press, London and New York.
Hartree, D. R. (1956). *J. Opt. Soc. Am.* **46,** 350.
Hartree, D. R. (1957). "The Calculation of Atomic Structures." Wiley, New York.
Helmholtz, L. (1939). *J. Am. Chem. Soc.* **61,** 1544.
Helmholtz, L., and Kruh, R. F. (1952). *J. Am. Chem. Soc.* **74,** 1176.
Hoard, J. L., Lind, M., and Silverton, J. V. (1961a). *J. Am. Chem. Soc.* **83,** 2770.
Hoard, J. L., Pedersen, B., Richards, S., and Silverton, J. V. (1961b). *J. Am. Chem. Soc.* **83,** 3533.
Irvine, J., and Wilkinson, G. (1951). *Nature* **168,** 514; *Science* **113,** 742; *J. Am. Chem. Soc.* **73,** 5501.
Jorgensen, C. K. (1962). "Absorption Spectra and Chemical Bonding in Complexes." Pergamon, New York.
Kirkham, W. J., Nyholm, R. S., and Sandhu, S. (1962). Unpublished observations.
Lewis, J., Nyholm, R. S., and Smith, P. W. (1962). *J. Chem. Soc.* p. 2592.
Marshall, W., and Stuart, R. (1961). *Phys. Revs.* **123,** 2048.
Moore, C. E. (1952). "Atomic Energy Levels." *Natl. Bur. Standards* (*U.S.*) *Circ.* **No. 467,** Vols. I, II, and III.
Morosin, B., and Lingafelter, E. C. (1960). *Acta Cryst.* **12,** 744.
Nigam, H., Nyholm, R. S., and Stiddard, M. H. B. (1960). *J. Chem. Soc.* p. 1806.
Nyholm, R. S. (1961). *Proc. Chem. Soc.* (Tilden Lecture), p. 273.
Nyholm, R. S. and Parish, R. V. (1956). *Chem. & Ind.* (*London*) p. 470.
Orgel, L. E. (1958). *J. Chem. Soc.* p. 4186.
Orgel, L. E. (1960). "An Introduction to Transition Metal Chemistry." Methuen, London.
Pauling, L. (1961). "Nature of the Chemical Bond." Cornell Univ. Press, Ithaca, New York.
Phillips, J. C. (1959). *J. Phys. Chem. Solids* **11,** 226.
Richardson, J. W., Nieupoort, W. C., Powell, R. R., and Edgell, W. F. (1962). *J. Chem. Phys.* **36,** 1057.
Slater, J. C. (1930). *Phys. Rev.* **36,** 57.
Stephenson, N. (1963). In preparation.
Watson, R. E. (1960). *Phys. Rev.* **118,** 1036; **119,** 1934.
Worsley, M. (1958). *Proc. Roy. Soc.* (*London*) **A247,** 390.
Zalkin, A., and Sands, D. E. (1958). *Acta Cryst.* **11,** 615.

CHAPTER 3

Bidentate Chelates

CLIVE M. HARRIS AND STANLEY E. LIVINGSTONE

School of Chemistry, The University of New South Wales, Kensington, N.S.W., Australia

I. Introduction

Because of the vast number of metal complexes of bidentate chelating agents it is impossible to review this field exhaustively in a single chapter. This discussion is mainly concerned with certain aspects which have not been reviewed and with the more recent evidence for the structures of the metal chelates. Olefine complexes are not discussed.

The bidentate chelating agents and their metal complexes are classified in this chapter according to the donor atoms of the ligand.

II. Oxygen Chelates

A. Oxyanions Which Form Four-Membered Rings

Although the occurrence of four-membered rings is comparatively rare, Werner (1920) listed a number of cationic complexes, e.g., $[Co(NH_3)_4CO_3]^+$

and $[Co(en)_2CO_3]^+$, in which the CO_3^{--} ion was presumed to be bidentate as in (I). Anionic complexes $[M(CO_3)_2]^{--}$ (M = Co, Ni, Cu) and $[Co(CO_3)_3]^{3-}$

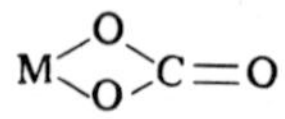

(I)

are also known (Applebey and Lane, 1918; Mori *et al.*, 1956). The carbonato group can be unidentate as in the $[Co(NH_3)_5CO_3]^+$ ion. Coordination of the CO_3^{--} ion lowers its symmetry from D_{3h} to C_{2v}, irrespective of whether it is unidentate or bidentate. This change is accompanied by the appearance in the infrared spectra at ~1050 cm.$^{-1}$ of a band, which is inactive for the free CO_3^{--} ion, and by the splitting of the degenerate C—O asymmetric stretching mode at ~1450 cm.$^{-1}$ (Gatehouse *et al.*, 1958a); a similar effect has been observed in nitrato complexes (Gatehouse *et al.*, 1957, 1958b). The bidentate nature of the carbonato group in the compound $[Co(NH_3)_4CO_3]Br$ has been confirmed by Barclay and Hoskins (1962) by means of an X-ray structure determination (see Chapter 1).

Other oxyanions such as SO_4^{--}, SO_3^{--}, SeO_4^{--}, SeO_3^{--}, CrO_4^{--}, MoO_4^{--}, and PO_4^{3-} have been reported as functioning as chelating agents (Diehl, 1937; Daniel and Salmon, 1957). The SO_4^{--} ion has T_d symmetry which is changed by coordination to C_{3v} if the sulfato group is unidentate and to C_{2v} if it is bidentate. Thus, infrared measurements can distinguish between a free SO_4^{--} ion and a unidentate and a bidentate sulfato group, since the band at ~1100 cm.$^{-1}$, due to the S—O stretching vibration, splits into two bands in unidentate and into three bands in bidentate complexes. However, it is probably impossible to distinguish between a bridging and a chelate sulfato group from infrared spectral data alone (Nakamoto *et al.*, 1957; Baraclough and Tobe, 1961). It has been inferred from infrared measurements that the nitrato group is bidentate in certain uranyl complexes (Gatehouse and Comyns, 1958).

B. Alkanoate Ions

Although Lowry and French (1924) proposed that the acetato groups in Cu(II) acetate form four-membered rings, the hydrate exists as the binuclear molecule $Cu_2(OAc)_4{\cdot}2H_2O$, in which the two Cu atoms are bridged by four acetato groups as in (II) (Van Niekirk and Schoening, 1953). The structures of $Cr_2(OAc)_4{\cdot}2H_2O$ (Van Niekirk *et al.*, 1953a) and the pyridine adduct of copper acetate, $Cu_2(OAc)_4{\cdot}2(py)$ (Barclay and Kennard, 1961b), are similar to (II).

Measurements of the visible and near ultraviolet spectra of anhydrous and hydrated Cu(II) alkanoates have led to the suggestion that these com-

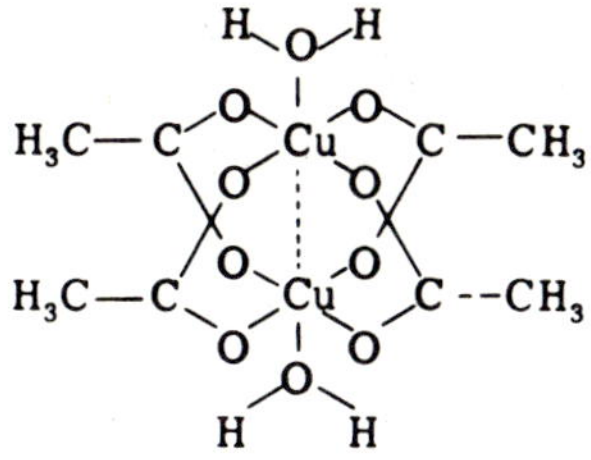

(II)

pounds, with the exception of Cu(II) formate, possess structures similar to Cu(II) acetate, involving Cu-Cu interaction (Tsuchida and Yamada, 1955, 1958; Tsuchida *et al.*, 1956, 1958; Yamada *et al.*, 1957, 1958). These compounds exhibit anomalous paramagnetic behavior and their magnetic moments vary with temperature, due to exchange interactions between the pairs of adjacent Cu atoms. This leads to the formation of a lower diamagnetic singlet level and an excited paramagnetic triplet level for each pair of Cu atoms (Guha, 1951; Bleaney and Bowers, 1952; Martin and Waterman, 1957, 1959a,b; Martin and Whitley, 1958). The exchange demagnetization results from a δ-bond between adjacent Cu atoms due to lateral overlap of their $3d_{x^2-y^2}$ orbitals (Figgis and Martin, 1956; Ross, 1959; Ross and Yates, 1959; Figgis and Lewis, 1960).

X-ray structural investigations have shown that Cu(II) formate and Cu(II) formate tetrahydrate are polymeric with formate bridges linking the Cu atoms (Kariyama *et al.*, 1954; Barclay and Kennard, 1961a). Three different types of bridging arrangements of the carboxy group have been observed: these have been classified by Martin and Waterman (1959a,b) as *syn-syn* (III), *anti-anti* (IV), and *anti-syn* (V) (R = H or Me). The

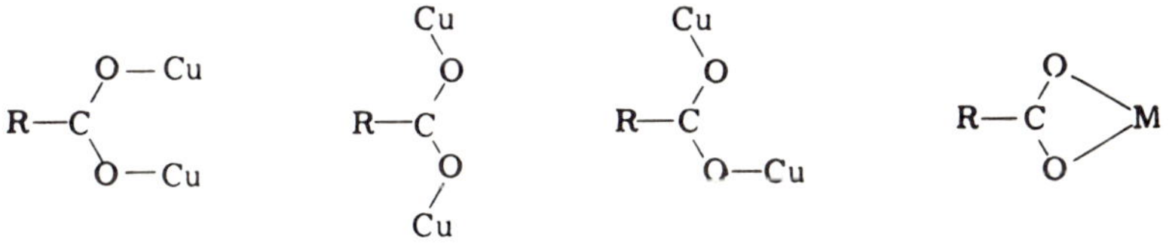

(III) (IV) (V) (VI)

acetates, $M_2(OAc)_4{\cdot}2H_2O$ (M = Cu, Cr), have the *syn-syn* bridging arrangement, Cu(II) formate tetrahydrate the *anti-anti* bridging arrangement, and anhydrous Cu(II) formate the *anti-syn* bridging arrangement. Zinc acetate dihydrate has an octahedral structure containing two four-membered rings as in (VI) (Van Niekirk *et al.*, 1953b). Silver perfluorobutyrate possesses the dimeric structure (VII) in which the carboxy

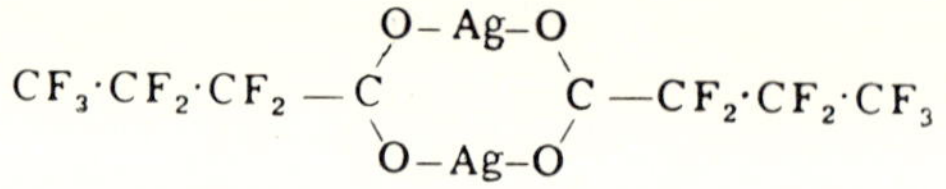

(VII)

groups bridge two Ag atoms. The Ag · · · Ag distance (2.90 Å) is nearly identical with the interatomic distance in metallic silver (Blakeslee and Hoard, 1956).

C. Dicarboxylate Ions

1. *The Oxalate Ion*

The oxalate ion functions as a bidentate with the formation of the five-membered chelate ring (VIII). Complex oxalates are known with all

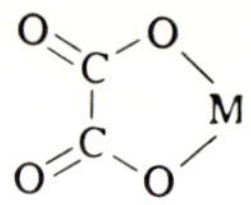

(VIII)

the elements of Groups IIA (except Ra) to VIIA (except Tc), VIII, IB to IVB (except C and Si), and VB (except N and P), and with all the lanthanides and with the actinides from Th to Am. Mono-, bis-, tris-, and tetrakisoxalato complexes are known and have been reviewed by Krishnamurty and Harris (1961). *Cis-trans* isomerism has been established in some bisoxalato complexes of certain six-coordinated metals. The trisoxalato complexes $[M^{III}(C_2O_4)_3]^{3-}$ have played an important role in coordination theory because of their dissymmetry, which has permitted resolution into optically active isomers in many instances (Diehl, 1937).

X-ray investigations of *trans*-$K[Cr(C_2O_4)_2(H_2O)_2]{\cdot}3H_2O$ and $K_3[Cr(C_2O_4)_3]$ have shown that the oxalato group is bidentate and approximately planar (Van Niekirk and Schoening, 1951, 1952). The oxalato group functions as a bidentate to two metal atoms simultaneously in the Au complex (IX) (Gibson and Weller, 1941) and in the Pd complex (X) (Chatt *et al.*, 1938). A different bonding arrangement occurs in silver oxalate, which has the chain structure (XI) (Griffith, 1943).

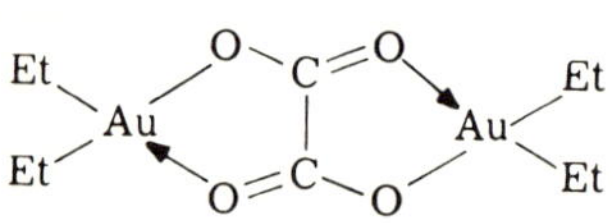

(IX)

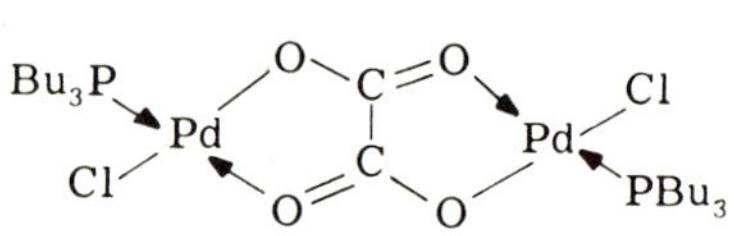

(X)

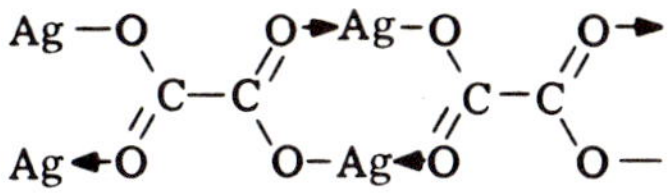

(XI)

The infrared spectrum of the oxalate ion displays three strong bands at 1627, 1338, and 768 cm.$^{-1}$ (Murata and Kawai, 1956). The symmetry changes from D_{2h} in the oxalate ion to C_{2v} in the chelated oxalato group (Fujita *et al.*, 1957). The oxalato complexes display strong doublet bands at ~1700 cm.$^{-1}$, assigned to the stretching frequency of the modified C=O groups, and a strong band at ~1400 cm.$^{-1}$, attributed to the C—O stretching vibration and corresponding to the symmetric OCO stretch at 1338 cm.$^{-1}$ in the oxalate ion (Schmeltz *et al.*, 1957). The shift in the frequency of the band at ~1400 cm.$^{-1}$ in the complexes of Cu(II), Co(II), and Ni(II) has been correlated with the stability order—Cu > Co > Ni. It is assumed that, as the M—O bond becomes stronger, the C—O bond becomes longer and exhibits a lower stretching frequency (Fujita *et al.*). The frequency shifts observed for the trisoxalato complexes $[M^{III}(C_2O_4)_3]^{3-}$ (M = Al, Fe, Cr, Co), compared with the stretching vibrations found in the spectra of sodium oxalate (regarded as fully ionic) and dimethyl oxalate (regarded as fully covalent), have been used to calculate the degree of covalency in the M—O bond. The degree of covalency was found to be 50% (Schmeltz *et al.*).

The visible and ultraviolet spectra of many oxalato complexes have been examined and exchange reactions involving oxalato complexes have been extensively studied (Krishnamurty and Harris, 1961).

2. *Other Dicarboxylate Ions*

The malonate ion forms well-defined complexes which have six-membered chelate rings (XII), but otherwise resemble the oxalato complexes. The higher homologs, if chelate, would form rings with more than six atoms. Few well-defined metal complexes are known with the higher homologous dicarboxylic acids, although homophthalic acid has been reported to form a complex (XIII) with one eight-membered ring (Duff, 1921).

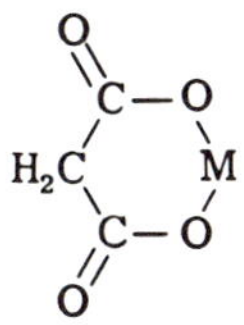

(XII)

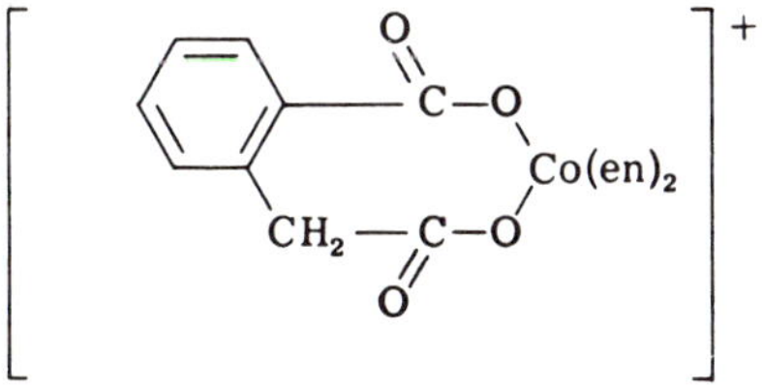

(XIII)

With the exception of Cu(II) oxalate and malonate, the anhydrous and hydrated Cu(II) salts of α,ω-dicarboxylic acids and their pyridine adducts display similar magnetic behavior to Cu(II) acetate (Kokot, 1961).

D. β-Diketones

Acetylacetone undergoes keto-enol tautomerism:

$$CH_3\underset{\underset{O}{\|}}{C}\,CH_2\underset{\underset{O}{\|}}{C}\,CH_3 \rightleftarrows CH_3\underset{\underset{OH}{|}}{C}{=}CH\underset{\underset{O}{\|}}{C}\,CH_3$$

By loss of a proton it can form six-membered chelate rings (XIV) with a

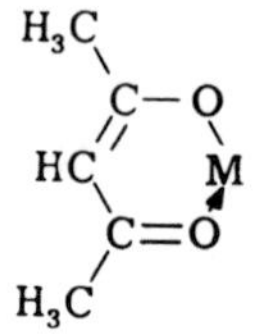

(XIV)

large number of metals. Most of the metal chelates are neutral complexes of the types $[M^{II}(acac)_2]$ (M^{II} = Be, Cu, Zn, Cd, etc.; acac = acetylacetonate), $[M^{III}(acac)_3]$ (M^{III} = Al, V, Cr, Mn, Fe, Co, etc.), and $[M^{IV}(acac)_4]$ (M^{IV} = Zr, Hf, Ce, Th, U, Pu). Many of these acetylacetonates are readily soluble in organic solvents, some, e.g., those of V(III), Cr(III), Mn(III), Fe(III), and Co(III), can be vaporized, whereas those of Al, Sc, and In can be distilled. Acetylacetonates of certain bivalent metals readily take up two extra ligands such as water, alcohol, ammonia, or organic amines to form stable octahedral complexes. Unstable adducts, $2M(acac)_3{\cdot}NH_3$, $3M(acac)_3{\cdot}2NH_3$, occur with the rare earths. Much of the extensive literature on metallic acetylacetonates, published prior to 1914, was reviewed by Morgan and Moss (1914).

The chelates of Si(IV), Ti(IV), and B(III) are of interest, since they are rare examples of these elements in cationic complexes. The first two elements have a maximum coordination number of six and cannot form neutral chelates $[M(acac)_4]$ as can the heavier quadrivalent elements, Zr, Ce, and Th, which can display a coordination number of eight. Consequently, Si(IV) and Ti(IV) form the complexes $[M(acac)_3]X$ (X = a complex ion such as $FeCl_4^-$) (Dilthey, 1906). Similarly, B(III), which has a maximum covalency of four, cannot form a complex $[M(acac)_3]$, but forms the neutral complex $[B(acac)F_2]$ (Morgan and Tunstall, 1924) and the cationic complex $[B(acac)_2]X$ (X = $FeCl_4$, $AuCl_4$, etc.) (Dilthey and Schumacher, 1906). The complex $[Si(acac)_3]Cl{\cdot}HCl$, by its resolution into optical enantiomers, has been proved to possess an octahedral structure with three chelate rings (Dhar *et al.*, 1959). Partial resolution of the neutral complexes [M(acac)$_3$] (M = Cr, Co) has been effected (Dwyer and Gyarfas, 1951; Moeller and Gulyas, 1958).

The anionic complex Na[Co(acac)$_3$] is known (Rosenheim and Garfunkel, 1911). Werner (1901) isolated the neutral complex [Pt(acac)$_2$] and two anionic complexes, K[PtCl$_2$(acac)] and K[PtCl(acac)$_2$]. Grinberg and Chapurskii (1959), from a reinvestigation of these compounds, concluded that one of the acetylacetonato groups in K[PtCl(acac)$_2$] is unidentate.

Bivalent metal acetylacetonates follow the normal stability sequence: Mn < Fe < Co < Ni < Cu > Zn (Izatt *et al.*, 1954).

Other β-diketones of general formula R·CO·CH$_2$·COR′, e.g., benzoylacetone and dibenzoylmethane, form metallic complexes similar to those formed by acetylacetone. It has been found that when one of the H atoms of the central methylene group of acetylacetone is substituted by an isopropyl or *sec*-butyl group, the β-diketone does not give the usual colored metal chelates with Cu(II) and Fe(III). However, α-*n*-alkyl– and α-isobutyl–substituted acetylacetones give normal complexes (Morgan and Thomasson, 1924). It would seem that steric hindrance between the isopropyl or *sec*-butyl groups and the H atoms of the terminal methyl groups of the β-diketone interferes with the planar configuration of the chelate ring and thereby lowers the stability of the complex (Martell and Calvin, 1952b). The stability constants of the complexes of several methylene-substituted β-diketones with a number of bivalent metals appear to depend upon the type of substitution, and, in the case of cyclic compounds, upon the type of ring system. In general, the stability of the metal derivatives is related to the acid dissociation constants of the β-diketones (Martin and Fernelius, 1959). Fluorine substitution on the terminal C atoms of acetylacetone increases the amount of the enol form relative to the keto form in aqueous solution (Reid and Calvin, 1950; Belford *et al.*, 1956).

Uranium complexes of β-diketones of the types [U(R·CO·CH:CO·R′)$_4$] and [UO$_2$(R·CO·CH:CO·R′)$_2$] have been described (Gilman *et al.*, 1956; Jones *et al.*, 1956a,b,c) and the infrared spectra of uranyl β-diketonates have been examined (Comyns *et al.*, 1958; Sacconi *et al.*, 1958a,b).

Replacement of the acetylacetonato groups in the complexes [M(acac)$_3$] (M = Fe, Ru, Rh) by other β-diketones has been achieved by the heating of the complexes with the β-diketone in a high boiling solvent (Wolf *et al.*, 1960).

The literature on the infrared spectrum of acetylacetone has been summarized critically by Cotton (1960). A band at 1720 cm.$^{-1}$ indicates the presence of a small amount of the diketo form (XV) or possibly of the keto-enol tautomer (XVI). However, the absence of a band at 1675 cm.$^{-1}$, characteristic of conjugated aliphatic ketones, suggests that (XVI) is an unlikely structure. Resonance between the forms (XVII) and (XVIII) has been suggested by Rasmussen *et al.* (1949) to account for: (*a*) the absence of absorption at ~7000 cm.$^{-1}$, where overtones of normal hydrogen-bonded OH groups usually occur; (*b*) the OH absorption at ~2700 cm.$^{-1}$; (*c*) a very strong broad band in the range 1640–1530 cm.$^{-1}$. The proton resonance spectrum is consistent with this postulate (Duval *et al.*, 1952).

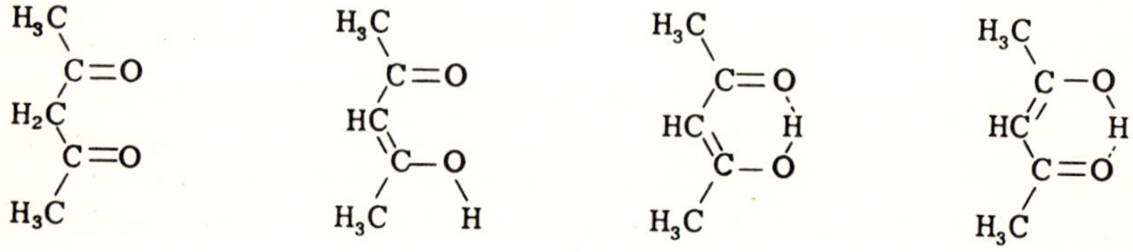

(XV) (XVI) (XVII) (XVIII)

The infrared spectral investigations of metal acetylacetonates and related complexes, published prior to 1959, have been reviewed by Cotton (1960), who remarked on the uncertainty in the assignments of the intense bands at ~1580, 1520, and 1380 cm.$^{-1}$. Since the appearance of Cotton's review some important work has been published. Nakamoto *et al.* (1959) examined a number of β-diketone complexes and observed that the bands ranging from 480–420 cm.$^{-1}$ shift to higher frequency if the metal is changed in the order of complex stability: Co(II) < Ni(II) < Cu(II) < Pd(II). They suggested that this band might well be the M—O stretching mode. A normal coordinate treatment on the chelate ring of [Cu(acac)$_2$] was used by Nakamoto and Martell (1960) to calculate the frequencies in the range 1700–350 cm.$^{-1}$ and good agreement was obtained with the observed bands. The results indicate: (1) the band at 1580 cm.$^{-1}$, previously assigned to the asymmetric C═O stretch, is the C═C stretch; (2) the bands at 684 and 654 cm.$^{-1}$, previously assigned to the Cu—O stretch, are Cu—O stretching vibrations coupled with other modes; (3) the band at 455 cm.$^{-1}$ is the *sym* Cu—O stretching mode. Nakamoto *et al.* (1961a) applied perturbation theory to the normal coordinate treatment of the Cu(II) complex in order to calculate the frequencies for [M(acac)$_3$] (M = Co, Cr, Al, Fe). They concluded that it is almost impossible to assign the bands between 1600 and 1400 cm.$^{-1}$ by means of the empirical approach used by earlier workers since: (*a*) the bond orders of C═C and C═O are similar; (*b*) the relative positions are sensitive to a change in M; (*c*) coupling in such a chelate ring system is serious and the concept of group frequency is not generally acceptable. The bands in the spectra of the M(III) chelates have been assigned and those of greatest interest are:

1590–1572 cm.$^{-1}$	C=C stretching mode (ν_8)
1554–1526 cm.$^{-1}$	C=O stretching mode (ν_1)
490– 434 cm.$^{-1}$	M—O stretching mode (ν_5)

The force constant for the M—O stretching mode is much less for Fe(III) than for Al(III), Co(III), and Cu(II), indicating that the Fe—O bond is relatively weak.

A study of the infrared spectra of the complexes $M(acac)_2$ (M = Be, Co, Ni, Cu, Zn, Pd) reveals that the M—O stretching force constants increase in the order Zn ≈ Co < Ni < Cu < Pd and the C=O force constants decrease in the same order. This is what would be expected from the stability order. However, assignments of bands between 1600 and 1400 cm.$^{-1}$ are still obscure. It is suggested that strong metal chelates have the structure (XIX) and weak chelates the structure (XX). In (XIX)

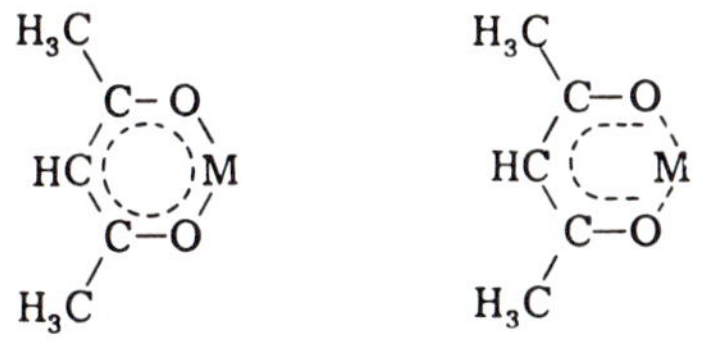

(XIX) (XX)

π-electrons as well as d-electrons of the metal tend to be more or less localized in the chelate ring, whereas in (XX) the π-electrons are localized more in the ligand (Nakamoto and co-workers, 1961b).

Double bonding between metal and oxygen in some acetylacetonates has been postulated from a study of their visible ultraviolet absorption spectra (Basu and Chatterji, 1958).

An X-ray structural analysis of $[Fe(acac)_3]$ has shown that there is an octahedral arrangement of Fe—O bonds and that the chelate rings are planar (Roof, 1956). The individual chelate rings have C_{2v} symmetry and the two M—O bonds in the ring are of equal length. The ring C—C distances (1.39 Å) are equal to the C—C distance in benzene and both C—O distances (1.28 Å) are intermediate between single and double bond lengths. This is consistent with the resonance structures (XIX) or (XX), although the benzenoid structure (XIX) has been questioned by Holm and Cotton (1958), who found no evidence for it from the nuclear magnetic resonance spectra of diamagnetic acetylacetonates. From a comparison of the M—O distances (1.90 Å) in the Ni(II), Cu(II), and Co(III) complexes, Shkolnikova and Shugam (1960) concluded that strong π-bonding between the metal and the ligand does not occur.

Cu(II) acetylacetonate has been found from X-ray studies to be square-planar (Cox and Webster, 1935; Shugam, 1951; Kogama *et al.*, 1953) and electron diffraction measurements on the vapor confirm the square-planar structure (Shibata and Sone, 1956). Cu(II) acetylacetonate and ethylacetoacetate react with heterocyclic bases, such as pyridine, in nondonor solvents to form adducts which, from the form of the equilibrium constants, are assumed to contain five-coordinate Cu(II) (Graddon, 1959). This assumption is supported by evidence from infrared measurements; the stability constants of adducts of this type have been determined by Graddon and Watton (1960, 1961a).

The complex $Co(acac)_2 \cdot 2H_2O$ has a tetragonally distorted octahedral configuration with the water molecules in the apical sites. The two chelate rings are almost planar but the Co atom is out of the plane (Bullen, 1959). The analogous Ni(II) complex is isomorphous and isostructural with the Co(II) complex (Holm and Cotton, 1961). However, the crystalline anhydrous Ni(II) complex is trimeric with the Ni atoms bound together in linear chains of three by the acetylacetonato O atoms in such a way that each Ni atom is surrounded by a distorted octahedron of O atoms as in (XXI) (Bullen, 1956; Bullen *et al.*, 1961). The compound is also trimeric in benzene (Graddon and Watton, 1961b) but an electron diffraction study on the monomer in the vapor state indicates an approximately square-planar configuration (Shibata *et al.*, 1957).

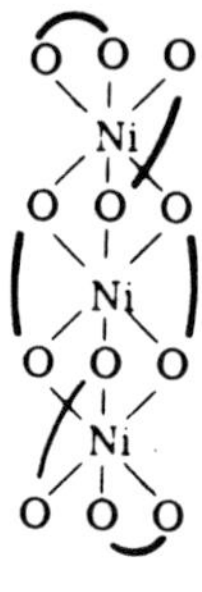

(XXI)

The visible absorption spectra of some anhydrous Ni(II) β-diketonates were studied in nonaqueous solution over a range of temperature by Fackler and Cotton (1961) who interpreted the results in terms of monomer-polymer equilibria. Both the anhydrous compounds bis(3-phenyl-2,4-pentanediono)nickel(II) and bis(dibenzoylmethano)nickel(II) have been obtained in a green paramagnetic and a red diamagnetic form.

The acetylacetonato group can act as a unidentate ligand by coordination to the metal atom through the active methylene carbon atom. This type of coordination occurs in the dimeric complex acetylacetonyltrimethylplatinum(IV) (XXII) (Hazell *et al.*,1959) and 2,2′-bipyridylacetylacetonyltrimethylplatinum(IV) (XXIII) (Swallow and Truter, 1961). Harris and Livingstone (1962) suggested that one of the organic ligands in the complex K[PtCl(acac)$_2$] may be bound in this way.

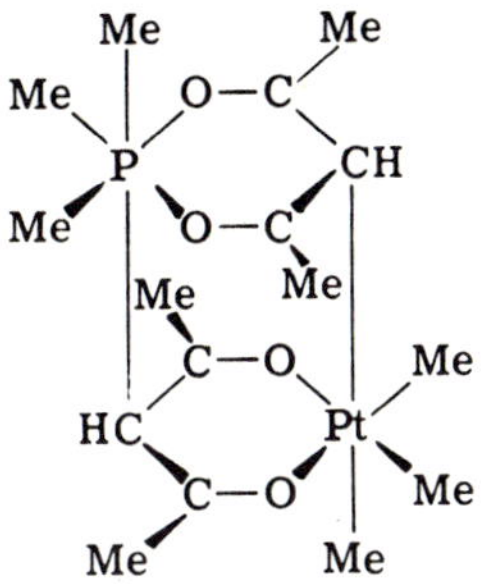

(XXII)

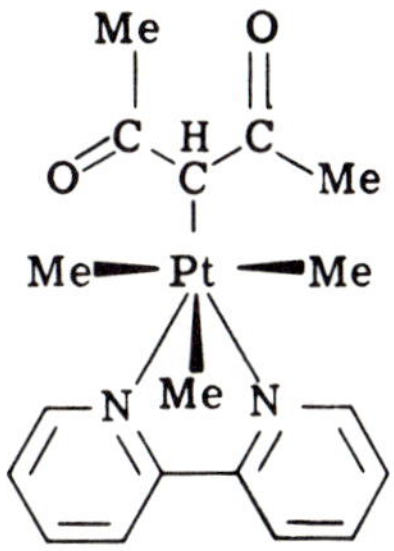

(XXIII)

Direct substitution of the H atom on the central C atom of the chelate ring in metal chelates produces complexes of the general type (XXIV)

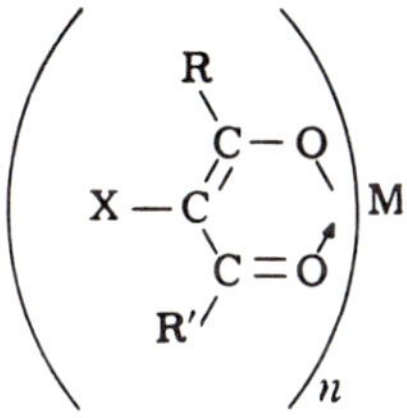

(XXIV)

(n = 2 or 3; X = Cl, Br, I, SCl, SCN) (Kluiber, 1961). The action of nitrous acid on Ni(acac)$_2$ gives a red diamagnetic complex, which is thought to have the structure (XXV) and the action of N_2O_4 on Cu(acac)$_2$ in

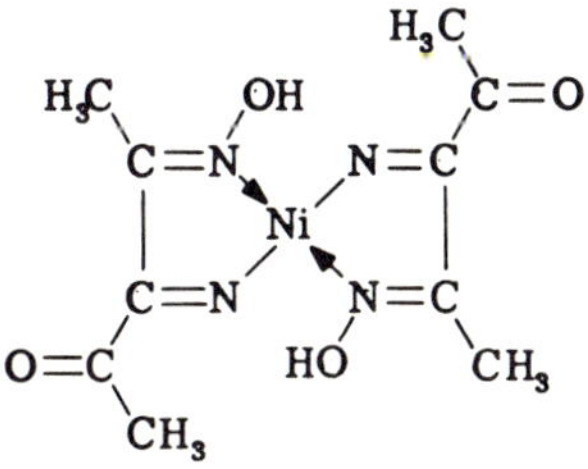

(XXV)

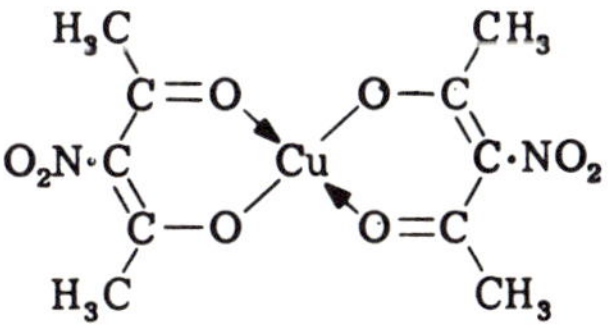

(XXVI)

benzene solution gives a complex to which is assigned the structure (XXVI) (Djordjevic *et al.*, 1959).

E. α-Hydroxycarboxylic Acids

The carboxyl and hydroxyl groups of α-hydroxycarboxylic acids can each lose a proton and coordinate to a metal ion to form a metal chelate. Salicylic acid forms anionic complexes such as $[M(O \cdot C_6H_4 \cdot CO_2)_2]^{--}$ (M = Be, Pd) and $[VO(O \cdot C_6H_4 \cdot CO_2)_2]^{--}$. The Cu(II) complexes of salicylic acid (log β_2, 16.9) are more stable than those of the aliphatic carboxylic acids, glycolic (log β_2, 3.7) and lactic (log β_2, 2.7) (Bjerrum *et al.*, 1957).

Boron forms stable complexes with a number of α-hydroxy-acids (Rosenheim and Vermehren, 1924). The bis(salicylato)borate and the bis(α-hydroxybutyro)borate ions have been resolved, indicating a tetrahedral configuration for boron in these complexes (Boeseken and Meulenhoff, 1924; Boeseken *et al.*, 1926).

F. o-Diphenols

Both phenolic groups in catechol (*o*-dihydroxybenzene) can lose a proton to give a doubly negatively charged chelate group which will form complexes with elements having a high affinity for oxygen (e.g., Ti, Zr, Hf, V, Nb, Cr, Mo, As, Sb). The following complexes are known: $M_2^I[M^{IV}(cat)_3] \cdot nH_2O$ (M^I = K, NH_4; M^{IV} = Ti, Zr, Th; $catH_2$ = catechol) (Rosenheim and Sorge, 1920; Rosenheim *et al.*, 1931); $(NH_4)_3[V(cat)_3] \cdot 2H_2O$; $(NH_4)_3[NbO(cat)_3] \cdot 9H_2O$ (Sidgwick, 1950b); and $M_3^I[MoO(cat)_3] \cdot nH_2O$ (Rosenheim and Nernst, 1933). The green hydrated complexes $M_3^I[Cr(cat)_3] \cdot nH_2O$ turn red on dilution of their solutions, presumably due to the formation of $M^I[Cr(cat)_2(H_2O)_2]$ (Weinland and Walter, 1923).

Arsenic forms the complexes $H[As(cat)_2] \cdot 4H_2O$ and $H[As(cat)_3] \cdot 5H_2O$; the latter was resolved through its alkaloid salts. The NH_4, K, and Ba salts and the free acid were each obtained in optically active forms, thus showing that the As atom was octahedrally coordinated (Rosenheim and Plato, 1925). Arsenic was the first element shown to be capable of acting as a center of both tetrahedral (Mills and Raper, 1925) and octahedral symmetry. Antimony is reported as forming the complexes $M^I[Sb(cat)_3]$, $M^I[Sb(cat)_2]$, and $M_3^I[SbO(cat)_3]$ [Weinland and Scholder, 1923).

The chelate-forming reaction between some *o*-diphenols and anions of Mo(VI), W(VI), V(V), Sn(IV), and Bi(III) has been studied (Halmekoski, 1959).

G. Salicylaldehyde

Salicylaldehyde (*o*-hydroxybenzaldehyde) is a weak acid and its phenoxide ion acts as a negatively charged chelate group with many metal ions. Mellor and Maley (1947) observed that the sequence for bivalent metals was the same for salicylaldehyde, glycine, and 8-quinolinol, viz., Mg < Mn < Fe < Cd < Zn < Co < Pb < Ni < Cu < Pd. Irving and Williams (1948) further correlated stability data and found that for a number of ligands the stability sequence for the first row bivalent transition elements is Mn < Fe < Co < Ni < Cu > Zn. A correlation has been observed between the values of log β_2 and the carbonyl stretching frequency for a number of salicylaldehyde complexes of bivalent metals (Bellamy and Branch, 1954).

Anhydrous bis(salicylaldehyde)nickel(II) is paramagnetic and it has been suggested that the compound is a trimer similar to bis(acetylacetonato)nickel(II) (Lyle *et al.*, 1959; Miller and Sharpe, 1961). The dihydrate has been shown to be octahedral (Stewart *et al.*, 1961).

Addition compounds of Cu(II) complexes of salicylaldehyde and its derivatives with pyridine and 2,2′-bipyridyl have been prepared (Muto, 1958). They are of the types of Cu $py_2(chel)_2$ and Cu bipy$(chel)_2$ (py = pyridine; bipy = 2,2′-bipyridyl; chelH = salicylaldehyde or a derivative such as 5-bromo-3-nitrosalicylaldehyde) and apparently contain six-covalent Cu(II) but nothing is known of their structure.

H. Diols and Polyols

Glycol and glycerol can act as bidentates, coordinating through the two *vicinal* hydroxyl groups (Bailar and Busch, 1956). Polyhydric aclohols such as glycerol, mannitol, and sorbitol can act as dibasic acids and form chelate complexes, notably with boron and iron (Martell and Calvin, 1952a).

The stereochemistry of boron complexes of 1,2- and 1,3-diols has been recently discussed (Dale, 1961). The latter are assumed to have six-membered nonplanar chelate rings. The essentially planar structure of the five-membered rings in the 1,2-diol complexes is indicated by the fact that 1,2-cyclopentanediol forms a much stronger complex than 1,2-cyclohexanediol and that *racemic* butane-2,3-diol, with *trans* disposition of Me groups, forms a stronger complex than the *meso* isomer which has two Me groups on the same side.

III. Nitrogen Chelates

A. Alkyl and Aryl Diamines

The best known member of this class of compound is ethylenediamine, $H_2NCH_2CH_2NH_2$ (en), which in general forms "high spin" complexes with the metals of the first transition series having the configurations d^4, d^5, d^6, d^7, and d^8, where "low spin" complexes are possible. However, the complex $[Co(en)_3]^{3+}$ is diamagnetic like all Co(III) complexes except K_3CoF_6, and anhydrous $[Ni(en)_2](ClO_4)_2$ is diamagnetic also, but the complexes $[Ni(en)_2X_2]$ (X = Cl, Br, I, SCN) are paramagnetic (Glaser and Pfeiffer, 1939; Asmussen, 1944).

The occurrence of *cis-trans* isomerism in octahedral bis complexes of ethylenediamine with Co(III) and Cr(III) has been reviewed by Basolo (1953), and substitution reactions, involving complexes of this ligand, have been extensively studied (Basolo and Pearson, 1958). The tris complexes $[M(en)_3]^{n+}$ [M = Zn(II), Cd(II), Cr(III), Co(III), Rh(III), Pt(IV)] and some bis complexes, e.g., *cis*-$[Co(en)_2NH_3Cl]^{++}$, have been resolved into their optical antipodes (Basolo, 1953).

Investigations on the complexes $Pt(en)X_3$ ($X_3 = Cl_3, Br_3, I_3, BrI_2$) have shown that these compounds do not contain Pt(III) but exist in the solid state as alternate $Pt(en)X_4$ and $Pt(en)X_2$ groups bridged through a halogen atom (Watt and McCarley, 1957). These complexes are similar to the compound $Pt^{IV}(NH_3)_2Br_4 \cdot Pt^{II}(NH_3)_2Br_2$ which has a structure consisting of chains of alternate Pt(IV) and Pt(II) atoms bridged by Br atoms (Brosset, 1948). Basolo *et al.* (1958) suggested that the catalysis of the slow $[Pt(en)_2Cl_2]^{++}$—Cl^- exchange by the $[Pt(en)_2]^{++}$ ion is due to the formation of the chloro-bridged intermediate $[ClPt(en)_2ClPt(en)_2Cl]^{3+}$.

Although ethylenediamine usually functions as a chelating agent, there are complexes in which it bridges two metal atoms: (XXVII) (Helman, 1943; Powell and Sheppard, 1959b); (XXVIII) (Schwarzenbach *et al.*,

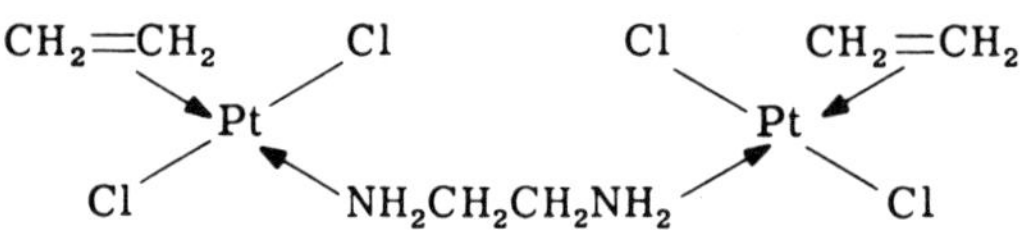

(XXVII)

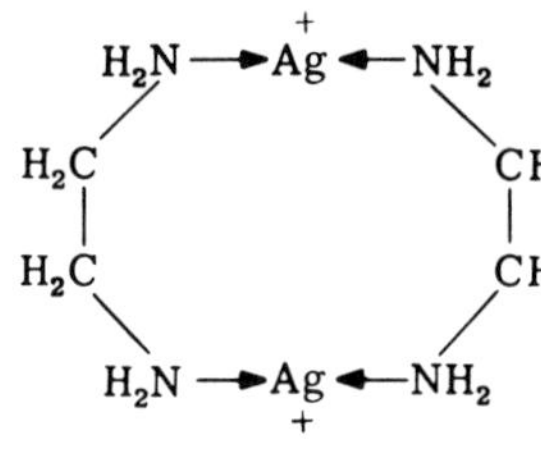

(XXVIII)

1952); and (XXIX) (Truter and Cox, 1956). In the complex $Sb_2Cl_6 \cdot (en)$ the ligand probably forms a bridge between two Sb atoms, since *p*-phenylenediamine, which cannot act as a chelate group, forms a similar compound (Prasad *et al.*, 1959). Ethylenediamine is unidentate in the Ag(I) complex (XXX) (Schwarzenbach *et al.*, 1952). Pfeiffer *et al.* (1952) postu-

$$(CH_3)_3(en)Pt^{IV} \leftarrow NH_2CH_2CH_2NH_2 \rightarrow Pt^{IV}(en)(CH_3)_3$$

(XXIX)

$$NH_2CH_2CH_2NH_2 \rightarrow \overset{+}{Ag} \leftarrow NH_2CH_2CH_2NH_2$$

(XXX)

lated that the complexes $AgClO_4 \cdot L$ and $M(ClO_4)_2 \cdot 2L$ (M = Cu, Zn, Hg; L = $H_2N(CH_2)_nNH_2$) contain 7-, 8-, and 9-membered rings when n = 4, 5, and 6, severally, but dimeric structures similar to (XXVIII) are possible.

Ethylenediamine, in common with aliphatic compounds of general formula YCH_2CH_2Y, has the more stable *trans* configuration (XXXI) rather than the *cis* arrangement (XXXII). However, in the complexes which have been examined by X-ray diffraction, the ligands have the *gauche* arrangement (XXXIII) (Wilkins and Williams, 1960). The configurations shown represent a view down the line of the C—C bond.

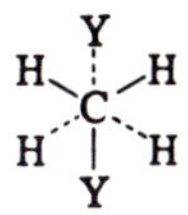

(XXXI)

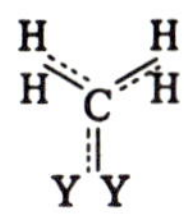

(XXXII)

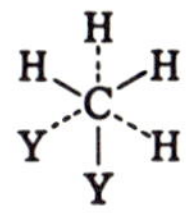

(XXXIII)

X-ray investigations have shown that the *gauche* configuration (XXXIII) occurs in $[Co(en)_3]Cl_3 \cdot 3H_2O$ (Nakatsu *et al.*, 1956, 1957), *trans*-$[Co(en)_2Cl_2]Cl \cdot HCl \cdot 2H_2O$ (Nakahara *et al.*, 1952), and *trans*-$[Co(en)_2Br_2]Br \cdot HBr \cdot 2H_2O$ (Ooi *et al.*, 1959) and have established the existence of the cation $[H_2O \cdot H \cdot OH_2]^+$ in the latter two complexes. Infrared measurements support the assignment of a *gauche* configuration for the ligand in $[Co(en)_3]Cl_3$ (Mizushima and Quagliano, 1953) and in $[M(en)_3]^{3+}$ (M = Co, Cr, Rh), $[M(en)_2]^{++}$ (M = Pd, Pt), and *trans*-$[Co(en)_2Cl_2]^+$ (Powell and Sheppard, 1959a). Although they suggest a *cis* arrangement in $[Ni(en)_3]^{++}$ and $[M(en)_2]^{++}$ (M = Cu, Zn), a structure determination on $[Ni(en)_3](NO_3)_2$ by Watanabe and Atoji (1951) is at variance with this hypothesis.

The stabilities of the complexes of ethylenediamine with the bivalent elements of the first transition series follow the Irving–Williams sequence: Mn < Fe < Co < Ni < Cu > Zn. However, the values of K_1 and K_2 obtained for Cr(II) place Cr between Fe and Co in the stability sequence. The displacement of the position of maximum absorption of $[Cr(en)_2]^{++}$ relative to the aquo species $[Cr(H_2O_6]^{++}$ is 4300 cm.$^{-1}$, which is almost as great as the corresponding displacement for the Cu(II) complex (5600 cm.$^{-1}$). Hence the degree of tetragonal distortion is of the same order in these two $[M(en)_2]^{++}$ complexes and this is the reason for the very small values of K_3 (Pecsok and Bjerrum, 1957). Salts of the ion $[Cu(en)_3]^{++}$ have been isolated (Sidgwick, 1950a) and, despite their low stability, the sulfate

$[Cu(en)_3]SO_4 \cdot H_2O$ has been separated into optical antimers by hand-picking of the crystals; transient rotations were observed for the two isomers in concentrated ethylenediamine solution (Gordon and Birdwhistell, 1959).

1,3-Diaminopropane forms complexes similar to those of ethylenediamine. The lower stability associated with the six-membered ring is reflected in the values of log β_2 for the Cu(II) and Ni(II) complexes. These values are approximately three log units less than those of the corresponding ethylenediamine complexes (Poulsen and Bjerrum, 1955).

The stability constants of the Ni(II) and Cu(II) complexes of some C-substituted ethylenediamines have been determined (Basolo and Murmann, 1952; Basolo *et al.*, 1953, 1954). The ligands have approximately the same base strength and the values of log β_2 for the Ni(II) and Cu(II) complexes show no pronounced trend. However, the values of log K_3 for the Ni(II) complexes fall with increase in C-substitution, which favors the low spin arrangement in the Ni(II) complexes since the ions $[Ni(chel)_2(H_2O)_2]^{++}$ (chel = en or pn) are paramagnetic but the bis complexes of *racemic*, and *meso*-2,3-diaminobutane, $H_2N \cdot CH(Me) \cdot CH(Me) \cdot NH_2$, 1,2-diamino-2-methylpropane, $H_2N \cdot C(Me)_2 \cdot CH_2 \cdot NH_2$, and 2,3-diamino-2,3-dimethylbutane, $H_2N \cdot C(Me)_2 \cdot C(Me)_2 \cdot NH_2$, are diamagnetic. These four complexes are square-planar in the solid state and in solution this diamagnetic yellow square-planar form is in equilibrium with a paramagnetic violet octahedral form (Basolo *et al.*, 1954). Wilkins (1957) has shown by exchange studies that the rates of dissociation in aqueous acid of complexes of Cu(II) and Ni(II) with C-alkyl–substituted ethylenediamines decrease as alkyl substitution increases.

N-alkyl- and *N,N'*-dialkyl–substituted ethylenediamines have basic strength comparable to, but coordinating ability poorer than, ethylenediamine. From measurements of the stability constants of the Ni(II) and Cu(II) complexes of a number of these ligands it has been concluded that steric hindrance by the alkyl groups lowers the stability of the complexes (Basolo and Murmann, 1952; Basolo *et al.*, 1953, 1954).

Complexes of C-aryl–substituted ethylenediamines have been investigated by Lifschitz *et al.* (1939). Of particular interest are the Ni(II) complexes of stilbenediamine, $H_2N \cdot CH(Ph) \cdot CH(Ph) \cdot NH_2$ (stien), and phenylethylenediamine, $H_2N \cdot CH(Ph) \cdot CH_2 \cdot NH_2$ (phenen). These complexes are of two types: (*a*) yellow and diamagnetic—$Ni(stien)_2X_2$ (X = Cl, NO_3, ClO_4, Cl_2CHCO_2, Cl_3CCO_2), $Ni(phenen)_2X_2$ (X = NO_3, ClO_4); (*b*) blue and paramagnetic—$Ni(stien)_2X_2$ (X = $PhCO_2$, CH_3CO_2, $ClCH_2CO_2$, Cl_2CHCO_2, Cl_3CCO_2), $Ni(phenen)_2(NO_3)_2$. Some of the complexes are known in both blue and yellow forms but, with the exception of $Ni(phenen)_2(NO_3)_2$, those which are salts of strong mineral acids form a blue modification only in the hydrated state, e.g., $Ni(phenen)_2(ClO_4)_2 \cdot 2H_2O$, if at all. All the stien complexes dissolve in water to give yellow solutions and in chloroform or benzene to give blue or green solutions. Furlani and Sartori (1958) have investigated the spectra of these complexes. The yellow forms are either square-planar or tetragonal, while the blue forms are probably octahedral with two univalent anions coordinated in the *trans*-octahedral positions. Dissolution in water solvates the anions and converts the blue complexes to the yellow form. *Racemic*- and *active*-stien tend to form blue complexes while *meso*-stien tends to form mainly yellow complexes (Lifschitz and Boz, 1940). The different behavior is attributed to the steric repulsion by *meso*-stien towards ligands in the octahedral sites. Nyburg *et al.* (1961) have shown that the blue complex $[Ni(meso\text{-stien})_2(H_2O)_2](Cl_2CHCO_2)_2$ has a *trans*-octahedral structure. It should be emphasized that this diaquo ion is not stable in aqueous solution and changes to the yellow form. A discussion of the yellow and blue complexes in terms of weak and strong tetragonal distortion, depending on the ligand field, has been given by Ballhausen and Liehr (1959).

N,N,N'-Triethylethylenediamine and *N,N,N',N'*-tetramethylethylenediamine do not form bis complexes due to steric obstruction by the *N*-alkyl groups but the bridged diol complex (XXXIV) and the mono complexes $[M(Me_2N{\cdot}CH_2{\cdot}CH_2{\cdot}NMe_2)X_2]$ (M = Pd, Pt, Ir; X = Cl, Br) are known (Pfeiffer and Glaser, 1938; Mann and Watson, 1958). The hydrolysis constants of the aquated ions $[Cu(chel)_2]^{++}$ (chel = *N,N*-dimethylethylenediamine or *N,N,N',N'*-tetramethylethylenediamine) have been interpreted by Gustafson and Martell (1959) in terms of the formation of the aquated hydroxo species, $Cu(chel)OH^+$ and $Cu(chel)(OH)_2$, and the binuclear $\mu\mu'$-diol species similar to (XXXIV). X-ray analysis by Hassel and Pedersen

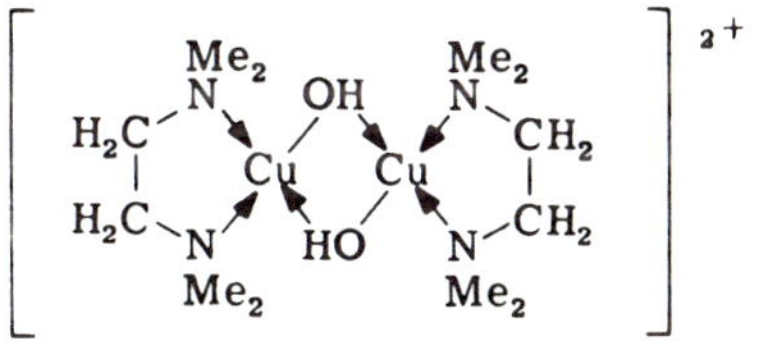

(XXXIV)

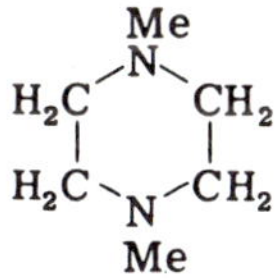

(XXXV)

(1959) has confirmed the boat configuration of 1,4-dimethylpiperazine (XXXV) in the Pd(II) complex of the series $[M(C_6H_{14}N_2)Cl_2]$ (M = Pd, Pt, Ir), described by Mann and Watson (1958).

o-Phenylenediamine, $o\text{-}C_6H_4(NH_2)_2$, is a poor coordinating agent; unstable complexes are known with bivalent Co, Ni, Cu, Zn, and Cd (Hieber *et al.*, 1929). The coordinating ability of the tetramethyl derivative $o\text{-}C_6H_4{\cdot}(NMe_2)_2$ is also poor and only the mono complexes $[M(chel)X_2]$ (M = Pd,

Cu; X = Cl, Br) are known (Stewart, 1958; Lockyer, 1959) because steric hindrance by the Me groups prevents the formation of bis complexes.

Complexes are known in which ethylenediamine has lost a proton and acts as a negatively charged ligand. Block and Bailar (1951) prepared the Au(III) complex [Au(en)(en–H)]Br_2 and Dwyer and Hogarth (1953, 1955) have isolated the following osmium complexes: [Os(en–H)$_2$(en)]X_2, [Os(en–H)(en)$_2$]X_3 (X = Br, I), [Os(en–H)$_2$(en)$_2$]I_2. The last compound apparently contains eight-covalent Os(IV) and can be oxidized to compounds of osmium in higher oxidation states. The acid dissociation constants of some metal chelates of ethylenediamine and propylenediamine have been determined by Grinberg *et al.* (1959).

B. Aromatic Heterocyclic Bases

1. *1,10-Phenanthroline and 2,2′-Bipyridyl*

a. General. 1,10-Phenanthroline (phen) (XXXVI) and 2,2′-bipyridyl (bipy) (XXXVII) were first used as coordinating agents by Blau (1888,

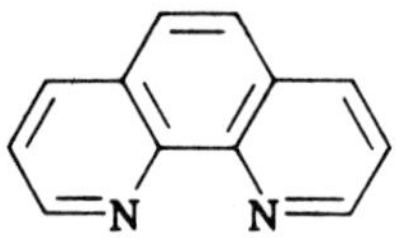

(XXXVI)

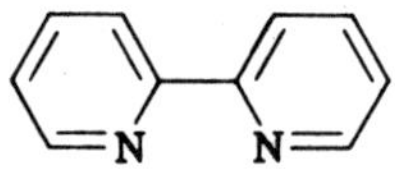

(XXXVII)

1898) and the metal chelates of these and related ligands were reviewed by Brandt *et al.* (1954). As well as being good σ-donors because of the presence of the sp^2 donor orbital on the N atoms, these ligands also function as acceptors due to the presence of delocalized π-orbitals associated with their aromatic ring systems. They stabilize both low and high valence states, e.g., V(−I) in $[V(bipy)_3]^-$ and Ag(II) in $[Ag(bipy)_2]^{++}$ (Chatt, 1958; Orgel, 1960).

2,2′-Bipyridyl forms "high spin" tris complexes with V(II), Mn(II), Co(II), Ni(II), and Cu(II) but "low spin" tris complexes with Cr(II) and Fe(II) (Burstall and Nyholm, 1952; Perthel, 1959; Harris *et al.*, 1961b). The magnetic moments of the $[M(phen)_3]^{++}$ (M = Mn, Fe, Co, Ni, Cu) are similar to those of their 2,2′-bipyridyl analogs. From a reinvestigation of the stability constants of the complexes formed by these two ligands with the bivalent metals of the first transition series, Banks and Bystroff (1959) found that, although the 1,1 complexes follow the Irving–Williams stability sequence, the 1,2 and 1,3 complexes depart radically from it.

b. Manganese Complexes. Bis and tris complexes of Mn(II) have been isolated although the $[Mn(chel)_3]^{++}$ (chel = phen or bipy) species predominates in solution (Miller and Brandt, 1955). The mono complexes Mn(chel)Cl_2 can be prepared from dimethylformamide solution (Broomhead and Dwyer, 1961). From infrared measurements Schilt and Taylor (1959) inferred that [Mn(phen)$_4$](ClO_4)$_2$ contains octocovalent Mn(II) but this requires confirmation. The reluctance of Mn(II) to form "low spin" complexes with these ligands is attributed to its low electronegativity and the stability of the d^5 configuration. A binuclear oxygen-bridged structure involving Mn(III) and Mn(IV) has been proposed for the cation in [Mn-(bipy)$_2$O](ClO_4)$_3$·2H_2O and its low magnetic susceptibility is attributed to metal-metal interaction (Nyholm and Turco, 1960).

c. Iron Complexes. The complexes [Fe(chel)$_3$]$^{++}$ are anomalously high in the stability sequence and $K_3 > K_1$ and K_2 for the 2,2′-bipyridyl system. Irving and Williams (1953) suggested that this arises from orbital stabilization on the formation of the diamagnetic tris species. The paramagnetism of the mono complexes Fe(chel)Cl_2 (Broomhead and Dwyer, 1961) and of the bis complexes Fe(chel)$_2Cl_2$ (Basolo and Dwyer, 1954) confirms the presence of ligand-field stabilization in the diamagnetic tris complexes.

The dibasic character exhibited by the complexes [Fe(chel)$_2$(CN)$_2$] (Schilt, 1960a,b) is attributed to the formation of a protonated species which Hamer and Orgel (1961) have shown to possess an isocyanide structure [(chel)$_2$Fe(C≡NH)$_2$]$^{++}$, analogous to the methyl derivative [(phen)$_2$-Fe(C≡N·CH_3)$_2$]$^{++}$. Although spectral and magnetic measurements have indicated protonation of the Fe(chel)$_3$]$^{++}$ ions in perchloric acid solution, no compounds could be isolated (Healy and Murmann, 1957).

The Fe(III) complexes [Fe(chel)$_3$]$^{3+}$, [Fe(chel)$_2$(CN)$_2$]$^+$, and [Fe(chel)-(CN)$_4$]$^-$ are of the "low spin" type (Schilt, 1960a) whereas the chloro complexes [Fe(phen)$_2Cl_2$]$^+$ and [Fe(phen)Cl_4]$^-$ are "high spin" (Harris and Lockyer, 1958). The derivatives of $Fe_2(CO)_9$ were shown by Hieber and Floss (1957) to be the salts [Fe(phen)$_3$][$Fe_2(CO)_8$] and [Fe(phen)$_3$][$Fe_4(CO)_{13}$].

d. Cobalt Complexes. Although the [Co(chel)$_3$]$^{++}$ ions are of the "high spin" type the complexes [Co(chel)$_2$(CN)$_2$] exhibit spin-pairing (Cambi and Paglia, 1956). The cyanide complexes of Co(II) and Co(III) have been investigated by Cambi and Paglia (1958) and measurements of the oxidation potentials of the [Co(chel)$_3$]$^{++}$/[Co(chel)$_3$]$^{3+}$ systems confirm the greater stability of the Co(II) complexes (Paglia and Sironi, 1957). Exchange studies have been made on the Co(II)–(phen) system and the rate constants for the dissociation of the [Co(phen)$_3$]$^{++}$ and [Co(phen)$_2$]$^{++}$ ions have been evaluated (Ellis and Wilkins, 1959). The exchange of ligand with the [Co(chel)$_3$]$^{3+}$ ions is catalyzed by Co(II) impurities and attempts to resolve the tris complexes of Co(III) were unsuccessful (Ellis *et al.*, 1957). Cabani (1960) investigated polarographically the complexes formed by the action of oxygen on solutions of Co(II) and 2,2′-bipyridyl.

e. Nickel Complexes. Among the bivalent metals of the first transition series Ni(II) forms the most stable bis and tris complexes with 1,10-phenanthroline and the $[Ni(chel)_3]^{++}$ ions have been optically resolved. Harris and McKenzie (1961) have investigated the following types of complex: $[Ni(chel)_2X_2]$, $[Ni(chel)_2(H_2O)X]X$, $[Ni(chel)_2(H_2O)X]ClO_4$, $[Ni(chel)_2(H_2O)_2]X_2$, the isomeric $[Ni(bipy)_2I_2]$ and $[Ni(bipy)_3][Ni(bipy)I_4]$, and $[Ni(chel)_3][NiX_4]$ (X = Cl, Br, I, SCN). The lilac isothiocyanato complex $[Ni(bipy)_2(NCS)_2]$ reverts to a yellowish brown isomer on standing and the lilac diaquo complex $[Ni(bipy)_2(H_2O)_2](ClO_4)_2 \cdot H_2O$ dehydrates without change in color or magnetic moment to the perchlorato complex $Ni(bipy)_2(ClO_4)_2$ which is a uni-univalent electrolyte in nitrobenzene (Barker *et al.*, 1961). The cyanide complexes of Ni(II) with 1,10-phenanthroline have been studied by Schilt (1957). Rate constants for the dissociation of $Ni(phen)^{++}$ and $Ni(phen)_2^{++}$ species have been determined (Wilkins and Williams, 1957; Ellis *et al.*, 1959). The mono complexes $[Ni(chel)X_2]$ (X = Cl, Br) have magnetic moments in the range 3.3–3.5 B.M. (Broomhead and Dwyer, 1961).

f. Copper Complexes. Mono, bis, and tris complexes are known in solution and in the solid state. The bis complexes $Cu(chel)_2X_2$ (X = Cl, Br, I) function as uni-univalent electrolytes in nonaqueous solvents and the five-coordinated Cu(II) complexes $[Cu(chel)_2X]ClO_4$ (X = Cl, Br, I, NCS), $[Cu(chel)_2NH_3](ClO_4)_2$, $[Cu(phen)_2(py)](ClO_4)_2$, and $[Cu(chel)_2ClO_4][PF_6]$ have been isolated (Harris *et al.*, 1961b; Barker *et al.*, 1961). Barclay and Kennard (1961c) have shown by an X-ray structural analysis that the Cu atom in $[Cu(bipy)_2I]I$ has an approximately trigonal bipyramidal configuration (XXXVIII) and that both optical enantiomers are present in the unit cell.

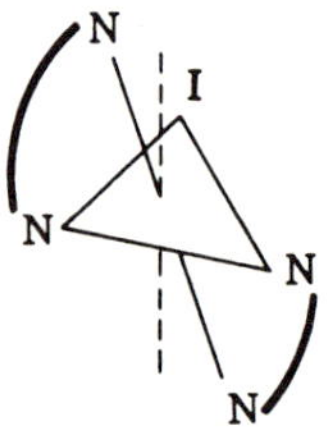

(XXXVIII)

Banks and Bystroff (1959) suggested that the $Cu(phen)_2^{++}$ complex exists in solution as the *cis*-diaquo species $[Cu(phen)_2(H_2O)_2]^{++}$ but the complex may well be the mono-aquo species since $[Cu(phen)_2H_2O](ClO_4)_2$ is obtained from the recrystallization of $[Cu(phen)_2ClO_4]ClO_4$. From solution spectra Jørgensen (1955) has concluded that the $[Cu(chel)_3]^{++}$ ions possess cubic symmetry. This conclusion is supported by the isomorphism of $[Cu(bipy)_3]SO_4$ and $[Ni(bipy)_3]SO_4$ (Jacobs and Speeke, 1955). James and Williams (1961) believe that the greater stability of the $Ni(phen)_2^{++}$ and $[Ni(phen)_3]^{++}$ ions compared with the stability of the analogous Cu(II) complex ions is due to the absence of the Jahn–Teller stabilization which is normally associated with Cu(II) complexes. In the mono complexes the greater polarizing power of the Cu^{++} ion produces the normal stability order Cu > Ni.

g. The Stabilization of Low Oxidation States. The complex $[Cr(bipy)_3](ClO_4)_2$ can be oxidized to $[Cr(bipy)_3](ClO_4)_3$ and reduced to $[Cr(bipy)_3]ClO_4$, which is a "low spin" complex (Hein and Herzog, 1952). The Cr(0) complex $[Cr(bipy)_3]^0$ has been obtained by Na reduction of $[Cr(bipy)_3]Br_2$ in tetrahydrofuran (Herzog *et al.*, 1957; Herzog and Schön, 1958) and by the action of Cr(II) acetate on 2,2′-bipyridyl (Herzog and Renner, 1959). It is diamagnetic and isoelectronic with the $[Fe(bipy)_3]^{++}$ ion. The stepwise oxidations

$$[Cr(bipy)_3]^0 \xrightarrow{-e} [Cr(bipy)_3]^+ \xrightarrow{-e} [Cr(bipy)_3]^{++} \xrightarrow{-e} [Cr(bipy)_3]^{3+}$$

can be effected by careful iodine-oxidation in dimethylformamide and the three-step reduction of

$$[Cr(bipy)_3]^{3+} \longrightarrow [Cr(bipy)_3]^0$$

has been confirmed polarographically (Vlček, 1961). The vanadium complexes $[V(bipy)_3]I_2$, $[V(bipy)_3]I$, $[V(bipy)_3]^0$, and $Li[V(bipy)_3]$ have also been isolated (Herzog, 1956, 1958; Herzog and Taube, 1956).

The magnetic moments of the Ti, Cr, and V complexes of 2,2′-bipyridyl have been measured by Perthel (1959) (see Table I). Orgel (1961) has

TABLE I
MAGNETIC MOMENTS (μ IN B.M.) OF TRIS-2,2′-BIPYRIDYL COMPLEXES OF TITANIUM, VANADIUM, AND CHROMIUM

Titanium complexes	Vanadium complexes	Chromium complexes	Electronic configuration
—	$[V(bipy)_3]^{++}$ green; μ, 3.7 3.8	$[Cr(bipy)_3]^{3+}$ yellow; μ, 3.8	$3d_\epsilon^3$
$[Ti(bipy)_3]^0$ blue; μ, 0	$[V(bipy)_3]^+$ red; μ, 2.8	$[Cr(bipy)_3]^{++}$ red; μ, 2.9	$3d_\epsilon^4$
$[Ti(bipy)_3]^-$ blue; μ, 1.7	$[V(bipy)_3]^0$ blue; μ, 1.7–2.0	$[Cr(bipy)_3]^+$ blue; μ, 2.0–2.1	$3d_\epsilon^5$
	$[V(bipy)_3]^-$ red; μ, 0	$[Cr(bipy)_3]^0$ red; μ, 0	$3d_\epsilon^6$

discussed the diamagnetism of the Ti(0) complex $[Ti(bipy)_3]^0$ in terms of "high spin" and "low spin" arrangements of $(d_\epsilon)^4$ systems in trigonal fields and he has suggested that the *d*-orbital splitting favors the "low spin" configuration because of strong π-bonding effects.

The existence in solution of $[Co(bipy)_2]^+$ and $[Co(bipy)_3]^+$ ions has been demonstrated polarographically (Vlček, 1957a; Martin and Waind, 1958a) and both complexes have been isolated as their deep blue diamagnetic perchlorates; Rh(I) and Rh(II) complexes were also reported (Martin and Waind, 1958b). Polarographic evidence for the formation of unstable complexes of nickel in low oxidation states has been given by Vlček (1957b).

2. *Substituted Derivatives of 1,10-Phenanthroline and 2,2′-Bipyridyl*

Many substituted derivatives have been synthesized and their analytical utility has been discussed (Brandt *et al.*, 1954; Schilt and Smith, 1957, and references therein). Substitution in these ligands may lead to enhanced stability of the metal complexes owing to increased basic strength or π-acceptor properties of the ligand. On the other hand, substitution can cause a lowering of stability because of the ligand's lower basic strength or by virtue of steric hindrance caused by the substituent groups (Brandt *et al.*, 1954; Williams, 1955; Tomkinson and Williams, 1958; Banks and Bystroff, 1959; James and Williams, 1961; James *et al.*, 1961).

3. *Other Ligands Containing the —N=C—C=N— Grouping*

Ligands containing this grouping have been reviewed by Brandt *et al.* (1954). Since then metal chelates of pyridine derivatives containing this grouping (XXXIX) have been investigated (Busch and Bailar, 1956;

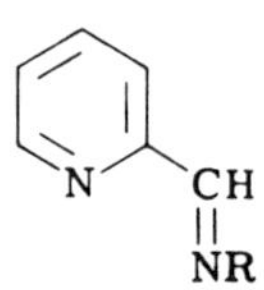

(XXXIX)

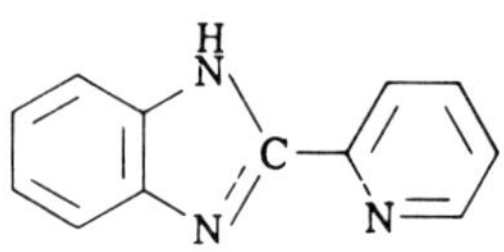

(XL)

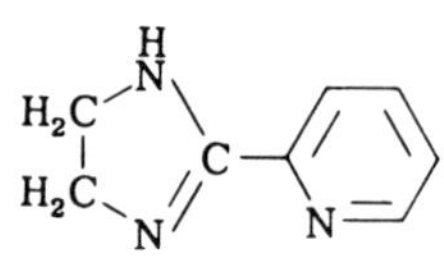

(XLI)

Stouffer and Busch, 1956; Bähr and Döge, 1957; Krause and Busch, 1958). Complexes of similar bidentates containing quinoline, isoquinoline, or thiazole have been studied (Breckenridge, 1954). The preparation, analytical aspects, and infrared spectra of the metal chelates of imidazole derivatives, such as 2-(2-pyridyl)-benzimidazole (XL) and 2-(2-pyridyl)-imidazoline (XLI), have been reported (Walter and Freiser, 1954; Harkins and Freiser, 1956; Harkins *et al.*, 1956). In addition to forming metal chelates similar to those of 2,2′-bipyridyl, 4,2′-pyridylimidazole and 2,2′-biimidazole, by loss of their imidazole-imino hydrogen, form inner complexes (Holmes *et al.*, 1961).

C. Other Nitrogen-Nitrogen Chelates

Dimethylglyoxime behaves as a weak monobasic acid and forms metal chelates (M = Ni, Pd, Pt) in which the metal is square-coordinated to four nitrogen atoms (XLII); however, the Cu(II) complex is dimeric and

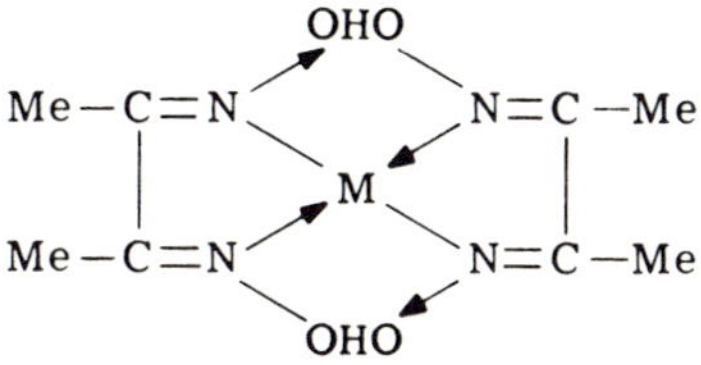

(XLII)

five-coordinated. The structure and solid state spectra of the insoluble Ni(II) and Pd(II) complexes indicate metal-metal interaction in the octahedral sites. The extensive published work on these complexes and those of other *vic*-dioximes has been recently discussed (Miller, 1961; Harris and Livingstone, 1962). The Pd(II) complex of benzoylmethylglyoxime has been isolated in two forms and this is one of the few instances where *cis-trans* isomerism has been claimed for Pd(II) (Dwyer and Mellor, 1935).

Biguanide (XLIII), guanylurea (XLIV), and their derivatives form

(XLIII)

(XLIV)

stable complexes with many metals (Rây, 1961). Because of the several possible ways in which these ligands can act as bidentate chelating agents, the structures of the complexes have not been established with certainty and it is likely that a particular ligand may coordinate in more than one way, depending on the metal, its valency, and the type of complex formed. However, it has been shown by Freeman *et al.* (1961) that in the biuret (XLV) complex, $K_2[Cu(NH{\cdot}CO{\cdot}NH{\cdot}CO{\cdot}NH)_2]{\cdot}4H_2O$, the Cu atom is

(XLV)

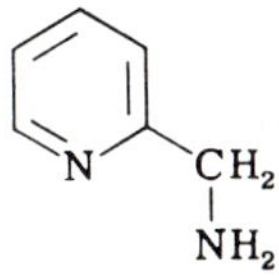

(XLVI)

square-coordinated by four terminal N atoms (at 1.93 Å), belonging to two (biuret)$^{--}$ residues.

Complexes of 2-picolylamine (XLVI) with Mn(II), Fe(II), Co(II), Co(III), Ni(II), Cu(II), Cd(II), and In(III) have been investigated by Sutton (1960a,b,c; 1961a,c).

D. Nitrogen-Oxygen Chelates

1. *Aminocarboxylic Acids*

Glycinate, anthranilate, and other α- and β-aminocarboxylate ions act as chelating agents to form complexes with many metal ions. The structures of several complexes of aminocarboxylic acids have been established by X-ray diffraction studies. Investigations of α- and β-aminobutyrates of Cu(II) show that the ligands are bidentate (Stosick, 1945; Bryan *et al.*, 1961). Bis(*dl*-proline)copper(II)dihydrate (proline = 2-pyrrolidinecarboxylic acid) has an octahedral arrangement about the Cu atom with the two water molecules in the apical positions (Welsh and Mathieson, 1953); however, anhydrous bis(1-aminocyclopentanecarboxylato)copper(II) has a square-planar arrangement with the carbonyl oxygens located 3.15 Å from the Cu atom in the octahedral sites (Barclay and Stephens, 1963). The infrared spectra of amino acid complexes have been reviewed by Cotton (1960).

2. *o-Aminophenols*

It has been suggested that certain *o*-aminophenols are carcinogenic because of their ability to coordinate with metals in the body (Boyland and Watson, 1956). The metal chelates of *o*-aminophenol are more stable than those of β-diketones and substituted salicylaldehydes (Charles and Freiser, 1952). The stability constants of some metal chelates of several *o*-aminophenols have been reported (Sims, 1959).

3. *8-Quinolinol*

The use of 8-quinolinol (XLVII), also known as 8-hydroxyquinoline

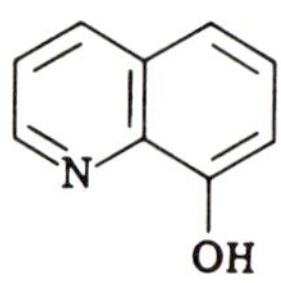

(XLVII)

and oxine, for the determination of metals has been the subject of more than five hundred papers since 1927 and its reactions have been reviewed by Phillips (1956). Oxine and its derivatives act as bactericides and fungicides and it has been suggested that the fungicidal action of oxine is due, at least in part, to its ability to render inactive trace metal ions by chelation (Zentmyer, 1944; Albert *et al.*, 1945, 1947). Normal oxinates (HQ = oxine) are known of the following types: $M^{II}Q_2$ (M^{II} = Be, Mg, Ca, Sr, Ba, Zn, Cd, V, Mn, Fe, Co, Ni, Pd, Cu); $M^{III}Q_3$ (M^{III} = Al, La, Ce, Pr, Ga, In, Tl, V, Cr, Fe, Sb, Bi); $M^{IV}Q_4$ (M^{IV} = Th, Pu). Other oxinates of more complicated formulae are known, e.g., $ScQ_3{\cdot}QH$, $ThQ_4{\cdot}QH$, $UO_2Q_2{\cdot}QH$, $PuO_2Q_2{\cdot}QH$.

By loss of a proton, 8-quinolinol forms inner complexes with metal ions, but it can also function as a neutral bidentate ligand as in the ion $[Ag(QH)_2]^+$, which has been resolved into optical isomers via its α-bromo-*d*-camphorsulfonate (Hein and Regler, 1936, 1941). The addition of hydroxyl ions to the $[Ag(QH)_2]^+$ ion results in the formation of the complex [AgQ(QH)] (Block *et al.*, 1951; Wendlandt and Haschke, 1962). A compound thought to be the Ag(II) complex AgQ_2 has been reported, but its existence has not been confirmed (Block *et al.*, 1951).

The stability constants of bivalent metal oxinates follow the Irving–Williams series (Johnston and Freiser, 1952; Irving and Rossotti, 1954). The effect of steric hindrance on stability has been investigated for chelates of substituted oxines (Irving *et al.*, 1949). The absorption spectra of the oxinates of bivalent metals display a bathochromic shift of the principal band of the ligand and this shift is greatest for the most stable complex (Sone, 1953; Moeller and Cohen, 1950; Moeller and Pundsack, 1954).

X-ray diffraction studies of $ZnQ_2{\cdot}2H_2O$ (Merritt, 1953; Merritt *et al.*, 1954) and $CuQ_2{\cdot}2H_2O$ (Kruh and Diggins, 1955) show that these complexes possess a *trans*-planar arrangement of the two chelate groups about the metal atom with two water molecules completing a six-covalent structure. The compounds $MQ_2{\cdot}2H_2O$ (M = Zn, Cd, Co, Ni, Pb) are isomorphous (Merritt *et al.*, 1954). Basolo and Matoush (1953) suggested that in the paramagnetic complex NiQ_2 the Ni atom is tetrahedral but it is probably octahedral as in the paramagnetic isomer of bis(*N*-methylsalicylaldimine)-nickel(II) (Harris *et al.*, 1958).

4. *Salicylaldimine Derivatives*

Salicylaldimine derivatives, *o*-$HO{\cdot}C_6H_4{\cdot}CH{:}NR$ (R = H, OH, and alkyl), form Ni(II) chelates $[Ni(O{\cdot}C_6H_4{\cdot}CH{:}NR)_2]$. The Ni(II) complexes (R = H, OH, and Me) and the analogous Cu(II) complex (R = H) have been shown to be *trans*-planar (Merritt *et al.*, 1956; Frasson and Sacconi, 1958; Stewart and Lingafelter, 1959).

Willis and Mellor (1947) observed that some of these diamagnetic Ni(II) chelates develop paramagnetism in organic solvents and the complex

(R = Me) has been obtained in diamagnetic and paramagnetic forms (Harris *et al.*, 1958; Sacconi *et al.*, 1958c). The phenomenon of solution paramagnetism has been investigated by a number of researchers and their work has been summarized by Harris *et al.* (1961a). One explanation assumes a planar structure in solution and a temperature-dependent population of a singlet ground state and a thermally accessible upper triplet state. However, recent work indicates that the paramagnetism may be dependent on association in solution (Holm, 1961).

IV. Sulfur Chelates

A. Chelating Agents with Two Sulfur Atoms as Donors

Sulfur when coordinated can be classified as follows: (*a*) covalently bound sulfur derived from an ion such as mercaptide or xanthate; (*b*) coordinately bound sulfur from a thioether. In the former the sulfur atom is bicovalent and has a V-shaped distribution of bonds, whereas in the latter it is tricovalent and pyramidal. Thioethers do not coordinate very strongly to metals apart from Pd, Pt, and Hg, but their coordinating ability is enhanced by chelate ring formation. Mercaptans (thiols) by losing a proton coordinate more strongly, especially to Pd(II), Pt(II), Cd(II), and Hg(II). Williams (1959) has suggested that the principal difference between thiols and thioethers as ligands is that the former are more highly polarizable but not as effective d_π electron acceptors as the latter.

The thioether 1,2-dimethylthioethane, $MeS{\cdot}CH_2CH_2{\cdot}SMe$, forms chelate complexes with Pt(II) (Tschugaeff and Kobljanski, 1913), Pd(II), Cu(I), Au(I), Au(III), Zn(II), Cd(II), Hg(II), Sn(IV), and Bi(III) (Morgan and Ledbury, 1922), but those with Pd(II), Pt(II), and Hg(II) are apparently the most stable. Complexes of Cu(II) with the diethyl and diisobutyl analogs, of Pt(II) with the diethyl analog (Tschugaeff and Subbotin, 1910; Tchougaeff and Fraenkel, 1912), and of Pd(II) and Hg(II) with the diphenyl and di-*p*-tolyl analogs have been reported (Tschugaeff, 1908; Bennett *et al.*, 1930). Although the aliphatic dithioethers $RS{\cdot}CH_2CH_2{\cdot}SR$ (R = Me, Et) form six-covalent paramagnetic Ni(II) chelates, 4-methyldithioveratrole (XLVIII) shows little tendency to complex with Ni(II) (Tschugaeff, 1908; Backhouse *et al.*, 1957).

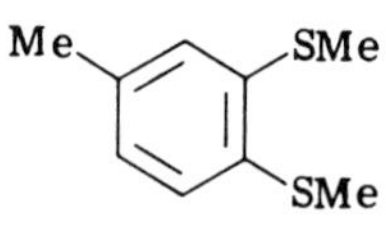

(XLVIII)

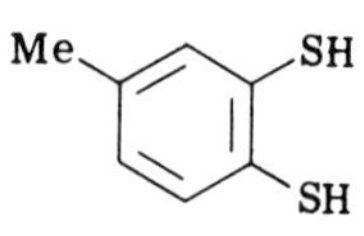

(XLIX)

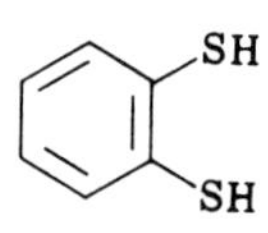

(L)

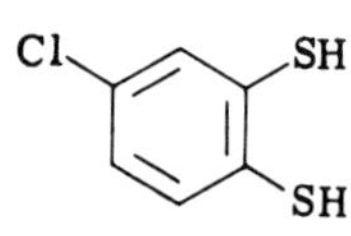

(LI)

Vasil'eva and Nedopekin (1959) have shown that a number of α-dithiols of general formula $HS{\cdot}CH_2(SH)CH_2{\cdot}CH_2X$ form strong chelate complexes with Cu(II), Ag(I), Hg(II), Sn(II), Pb(II), As(III), Bi(III), Fe(III), Co(II), Ni(II), and Pd(II), but they do not form stable complexes with Mg, Ca, Sr, Ba, Al, Zr, Ce, Th, V, Cr, Mo, W, and Mn. Ethanedithiol, $HSCH_2CH_2SH$ (esH_2), gives very stable nickel complexes $[Ni(es)_2]^{--}$ and $[Ni_2(es)_3]^{--}$ (Leussing and Alberts, 1960). Toluene-3,4-dithiol (XLIX) has been used for the colorimetric determination of Sn and Mo. Molybdenum complexes have been recently studied by Gilbert and Sandell (1960) and the Zn complex [Zn(phen)(tdth)] ($tdthH_2$ = toluene-3,4-dithiol) has been isolated (Wallenfels and Sund, 1957). Chelate complexes of antimony with *o*-benzenedithiol (L) have been described by Brown and Austin (1941). 4-Chlorobenzene-1,2-dithiol (LI) forms a very stable bridged Pd(II) complex (LII) (Chatt and Mann, 1938).

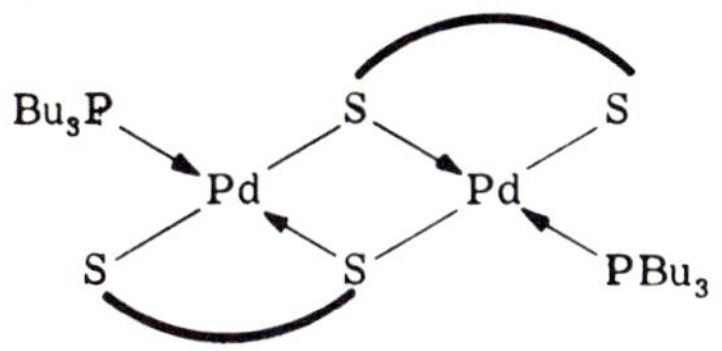

(LII)

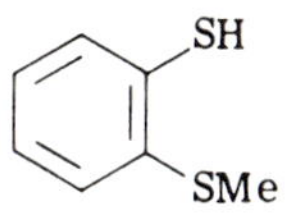

(LIII)

$EtSCH_2CH_2CH_2SH$

(LIV)

Complexes are known with chelating ligands which contain a thiol and a thioether group: *o*-methylthiobenzenethiol (LIII) forms square-planar inner complexes with Ni(II), Pd(II), and Pt(II), and 3-ethylthiopropane-1-thiol (LIV) forms a "low-spin" complex with Co(II) (Livingstone, 1956a,b). With Pd(II) the ligands (LIII) and (LIV) form thiolo-bridged complexes [(LV): M = Pd; X = Cl, Br, NO_2, SCN] and (LIV) is known to give similar Pt(II) complexes [(LV): M = Pt, X = Cl, Br, I] (Livingstone, 1956c,d). With this ligand several binuclear complexes containing another metal atom in addition to Pt have been isolated [(LVI): when M = Hg, X = Br; when M = Pd, X = Cl, Br, I]. If ammonia is passed into a solution of dichlorobis(3-ethylthiopropane-1-μ-thiolo)diplatinium(II) [(LV): M = Pt; X = Cl] in benzene the solution becomes decolorized and deposits the colorless chloride (LVII) which can be reconverted to the original neutral-bridged complex by heating (Livingstone, 1956d).

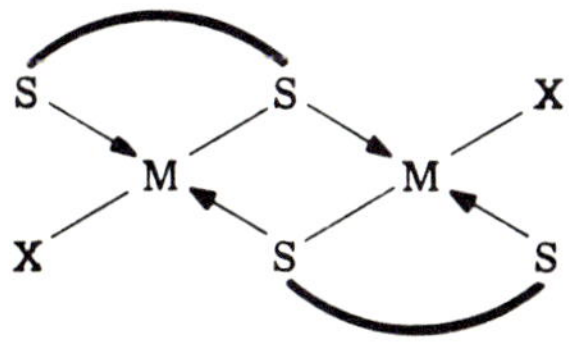

(LV)

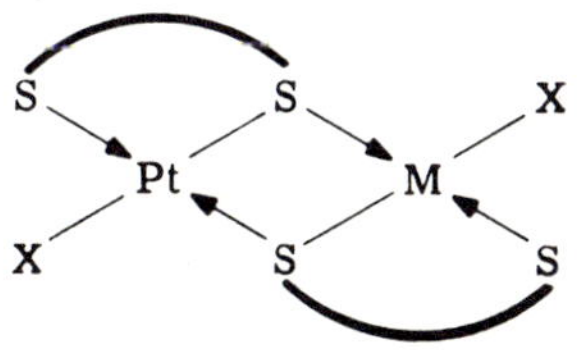

(LVI)

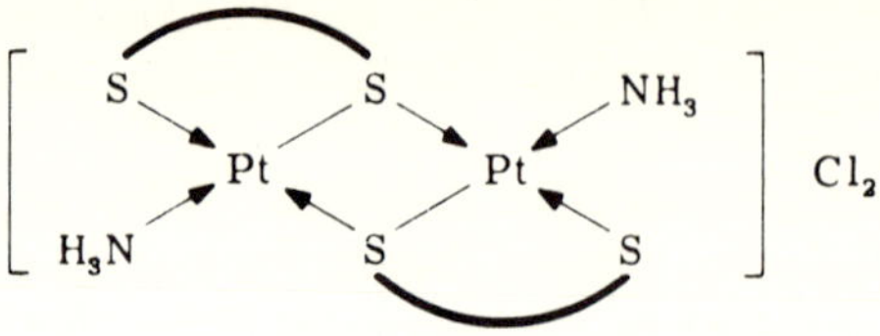

(LVII)

The dithio-oxalate ion forms anionic metal chelates $[M(C_2O_2S_2)_2]^{2-}$ (M = Ni, Pd) and $[M(C_2O_2S_2)_3]^{3-}$ (M = Co, Rh), which are intensely colored and very stable (Jones and Tasker, 1909; Robinson and Jones, 1912). The bis complexes $K_2[M(C_2O_2S_2)_2]$ (M = Ni, Pd, Pt) have been shown to contain square-planar anions with the structure (LVIII) (Cox

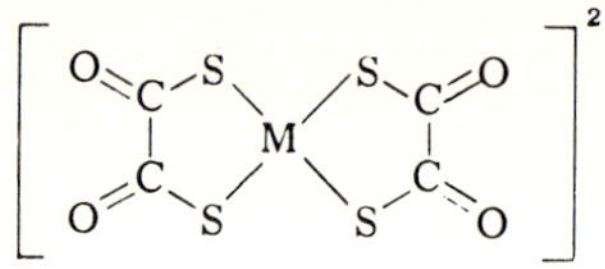

(LVIII)

et al., 1935) and the tris complex ions $[M(C_2O_2S_2)_3]^{3-}$ (M = Cr, Co, Rh) have been resolved (Dwyer and Sargeson, 1959).

Four-membered chelate rings are formed by xanthates (LIX), dithio-

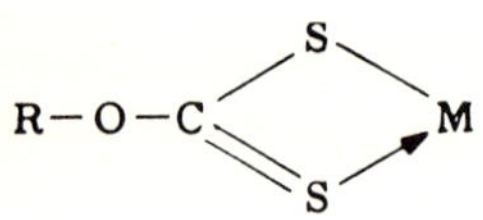

(LIX)

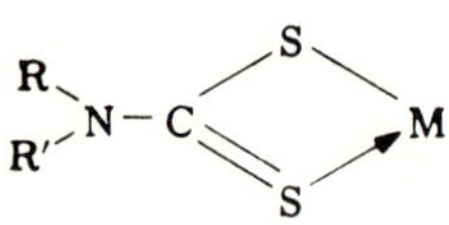

(LX)

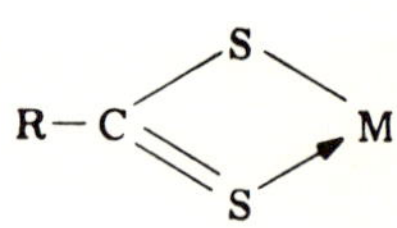

(LXI)

carbamates (LX), and dithiocarboxylates (LXI). The ethyl xanthate ion forms the neutral diamagnetic complexes $[Ni(EtOCS_2)_2]$ and $[Co(EtOCS_2)_3]$ (Cambi and Szego, 1931). Alkyl xanthates and dithiocarbamates form intensely colored binuclear complexes $Mo_2O_3(ROCS_2)_4$ which are diamagnetic and probably contain Mo(IV) and Mo(VI) but their structures are not known with certainty (Malatesta, 1939a,c). Alkyl xanthate and dithiocarbamate complexes of As(III) are known; evidence from dipole moment measurements suggests that the ligands are unidentate (Malatesta, 1939b). The *N*-alkyl dithiocarbamates of Cu(II) are unstable and are easily transformed into Cu(I) complexes whereas only Cu(I) complexes can be obtained with dithiocarboxylic acids. The *N*,*N*-dialkyl dithiocarbamates of Cu(II) are not isomorphous with the corresponding Ni(II) complexes (Cambi and Coriselli, 1936).

N-Substituted and N,N-disubstituted dithiocarbamate complexes of Ni(II) and Co(III) are diamagnetic while Mn(III) forms only "high-spin" complexes (Cambi and Szego, 1933) and the Cr(III) complexes with these ligands have a magnetic susceptibility approximately the same as that of the Cr^{3+} ion (Malatesta, 1939c). *N,N*-Dialkyl dithiocarbamates of Ru(III), Rh(III), and Pd(II) and two diamagnetic nitroso Ru(IV) dialkyl dithiocarbamates are known (Malatesta, 1938; Cambi and Malatesta, 1938). The diamagnetic nitroso complex $[Co(Me_2NCS_2)_2NO]$, first reported by Malatesta and Cambi (1940), has been shown by Alderman *et al.* (1962) to possess a five-coordinate rectangular-pyramidal configuration; the Co atom lies at 0.54 Å above the plane containing the four S atoms while the NO group is at the apex and inclined at 139° to the vertical axis of the pyramid.

The magnetic properties of four types of Fe(III) complexes have been investigated: (*a*) $Fe(ROCS_2)_3$; (*b*) $Fe(RCS_2)_3$; (*c*) $Fe(RNH{\cdot}CS_2)_3$; (*d*) $Fe(RR'N{\cdot}CS_2)_3$. The complexes of types (*a*) and (*b*) are spin-paired, those of type (*c*) spin-free, while those of type (*d*) have magnetic susceptibilities which lie between the values expected for one and five unpaired electrons and vary with temperature in a complex manner (Cambi and Szego, 1931, 1933). In a more detailed study of the *N,N*-dimethyl and *N,N*-di-*n*-butyl dithiocarbamates of Fe(III), Kokot (1961) has shown that the reciprocal of the magnetic susceptibility rises steeply with temperature then falls to a minimum and rises again. For the former compound the $\chi_M^{-1} - T$ curve below 120°K. approaches the theoretical curve for one unpaired electron while for the latter compound the curve above 300°K. runs almost parallel to the theoretical curve for five unpaired electrons. The ligand field in these compounds must be close to the critical value which determines the transition $3d_\epsilon^5 \rightarrow 3d_\epsilon^3 3d_\gamma^2$ of the Fe^{3+} ion (spin-paired → spin-free transition). The nature of the substituent on the N atom is apparently able to alter the ligand field sufficiently to change the magnetic behavior of the compound.

The infrared spectra of some *N*-alkyl and *N,N*-dialkyl dithiocarbamates display a strong band at 1512 1480 $cm.^{-1}$, indicating partial double bond character in the C—N bond and that the contribution of form (LXII) is of considerable importance. The complex xanthates do not absorb in the "double bond" region and it has been inferred that the canonical form (LXIII) contributes little to the structure (Chatt *et al.*, 1956). A structure

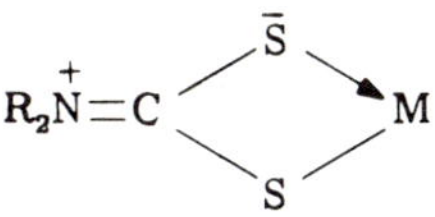

(LXII)

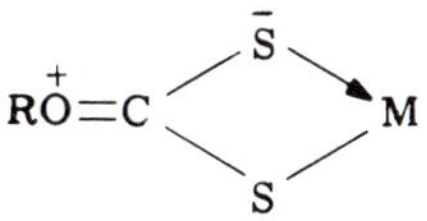

(LXIII)

analysis of Ni(II) *N,N*-diethyl dithiocarbamate by Shugam and Levina (1960) indicates: (*a*) square coordination of the four S atoms about the Ni atom; (*b*) the C═S double bond is delocalized in the ring; (*c*) the C—N and C—S distances are shorter than expected for single bonds. Approximately square-planar coordination of the four S atoms about the Cu atom occurs in Cu(II) *N,N*-di-*n*-propyl dithiocarbamate (Peyronel and Pignedoli, 1959).

B. Sulfur-Oxygen Chelates

Ligands in this class include the following ions: thioglycolate, $^-SCH_2CO_2^-$; *S*-alkylthioglycolate, $RS{\cdot}CH_2CO_2^-$; *o*-mercaptobenzoate (thiosalicylate), $^-S{\cdot}C_6H_4{\cdot}CO_2^-$; and *o*-alkylmercaptobenzoate, $RS{\cdot}C_6H_4{\cdot}$-CO_2^-. The bivalent ions M^{++} (M = Mn, Co, Ni, Zn) react with thioglycolate ion to form mono and bis complexes in solution. The changes in the order of stability are similar to those observed for oxygen and nitrogen ligands (Leussing, 1958). With thioglycolate ion Ni(II) forms polynuclear complexes, in which the Ni atoms are apparently bridged by two tricovalent S atoms (Leussing *et al.*, 1960). Thioglycolate and ethyl thioglycolate complexes of Fe(II) have been prepared in the absence of oxygen and the role of these complexes in the catalytic oxidation of thiol acids has been investigated (Schubert, 1932). The metal chelates of β-mercaptopropionic acid, having six-membered rings, are much less stable than those of mercaptoacetic (thioglycolic) acid which forms five-membered chelate rings (Fernando and Freiser, 1958). Esters of thioglycolic acid RO_2CCH_2SH (R = Et, *n*-hexadecyl) form inner complexes with Ni(II). The diamagnetism of these complexes and their dipole moments in benzene solution support the postulate of a square-planar configuration (Draney and Cefola, 1954).

o-Methylmercaptobenzoic acid forms inner complexes with Pd(II), Cu(II), Cd(II), and Hg(II) but not with Pt(II) (Livingstone *et al.*, 1950; Livingstone and Plowman, 1950, 1951). Treatment of the Pd complex (LXIV) with hydrochloric acid causes the chelate rings to open with the consequent formation of the dichloro compound (LXV) which can be con-

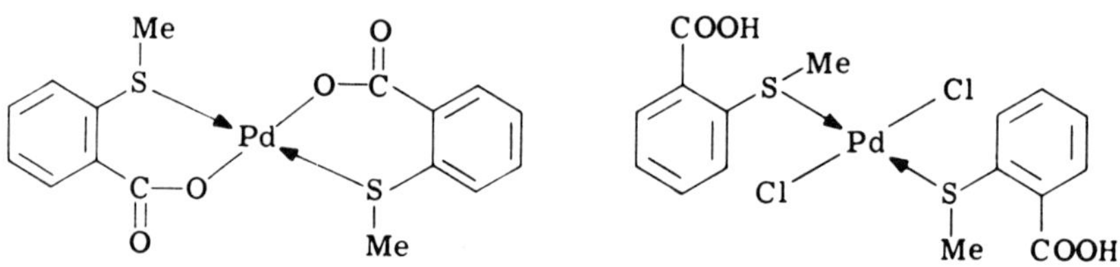

(LXIV) (LXV)

verted to the original complex (LXIV) by treatment with alkali (Livingstone *et al.*, 1950).

The complex (LXIV) reacts with K_2PdX_4 ($X = Cl, Br$) to give halogen-bridged complexes (LXVI) (Livingstone and Plowman, 1950); similar bridged complexes are known with other *o*-alkylmercaptobenzoic acids (Livingstone, 1956c). Binuclear complexes (LXVII) containing two different metal atoms have also been reported (Livingstone and Plowman, 1951).

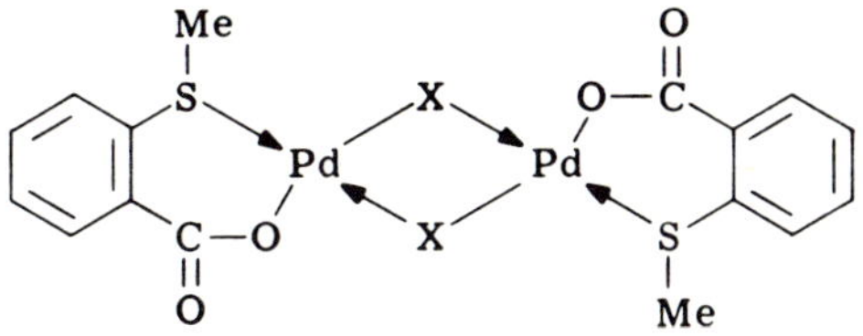

(LXVI; X = Cl, Br)

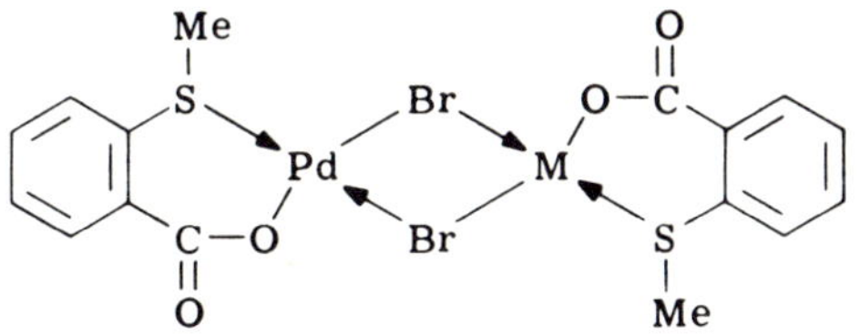

(LXVII; M = Hg, Cu)

The determination of the stability constants of a number of metal chelates of *S*-ethylthioglycolic, 3-(*S*-ethyl)mercaptopropionic, and *o*-ethylmercaptobenzoic acids showed that for each of these ligands the "natural order" for $\log K_1$, $Co < Ni < Cu > Zn$, is maintained but Cd and Pb fall out of line in the Mellor–Maley series, which is usually valid for oxygen and nitrogen ligands. It was also found that for some metals $\log K_2 > \log K_1$, in contrast to that found for O—O, N—N, N—O, and N—S chelates (Irving and Fernelius, 1956).

C. Sulfur-Nitrogen Chelates

The sulfur-nitrogen chelating agents can also be divided into two classes: the thiols and the thioethers. 2-Mercaptoethylamine is the simplest representative of the first class; the stability constants for its chelates with Co(II), Ni(II), Zn(II), Cd(II), and Pb(II) and those of methyl-α-amino-β-mercaptopropionate, $HS \cdot CH_2 \cdot CH(NH_2) \cdot CO_2 \cdot CH_3$, with Zn(II), Cd(II), and Pb(II) have been determined (Bjerrum *et al.*, 1957). The Au(III) complex (LXVIII) of 2-mercaptoethylamine can be alkylated to yield the complex (LXIX) (X = picrate or Br) but the S atom in dithio-oxamido-

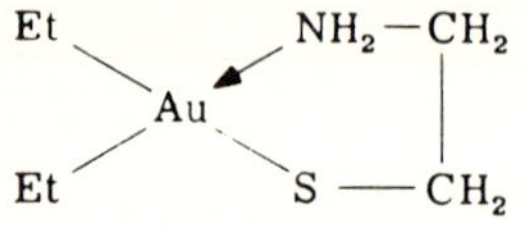

(LXVIII)

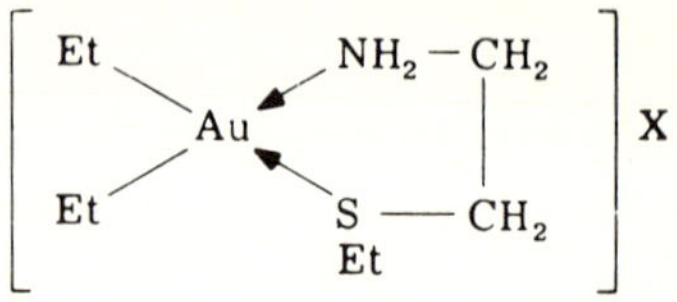

(LXIX)

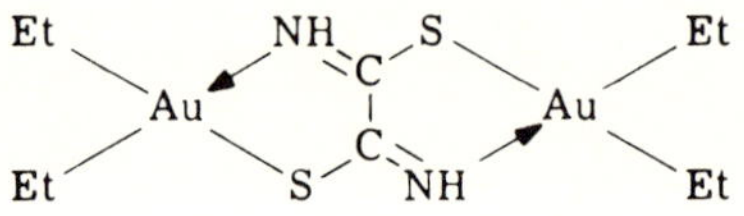

(LXX)

tetraethyldigold(III) (LXX) has none of the reactivity associated with (LXVIII) and it has been suggested that the chelate ring system in (LXX) is stabilized by single-double bond resonance (Ewens and Gibson, 1949).

Dithio-oxamide (rubeanic acid) forms with Ni(II) and Cu(II) insoluble chelates which are doubtless polymeric and probably possess the structure (LXXI) (Jensen, 1944; Ewens and Gibson, 1949). The structures of these metal chelates and those of *N*,*N*′-disubstituted thio-oxamides have been discussed by Hurd *et al.* (1961).

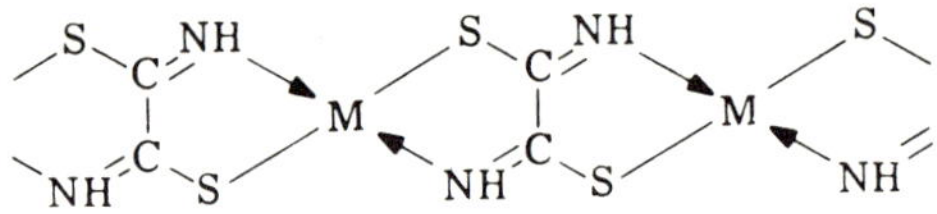

(LXXI)

The chelates of Zn(II) and Pb(II) with *o*-aminobenzenethiol and 6-mercaptopurine are more stable than the corresponding chelates with *o*-aminophenol and 6-hydroxypurine (Charles and Freiser, 1952; Cheney *et al.*, 1959). *o*-Aminobenzenethiol forms "low-spin" inner complexes with Co(II), Co(III), Ni(II), and Pd(II) (Hieber and Brück, 1949, 1952; Livingstone, 1956a,b). The yellow Ni(II) complex can be oxidized by air in strongly alkaline medium to the deep blue μ-dioxo-Ni(IV) complex (LXXII). A similar μ-dithio-Ni(IV) complex (LXXIII) of dithiobenzoic acid is known (Hieber and Brück, 1949, 1952).

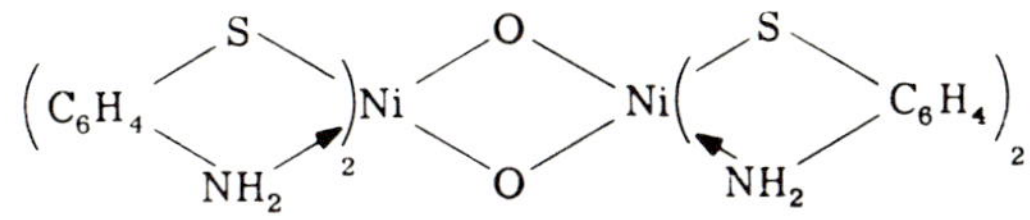

(LXXII)

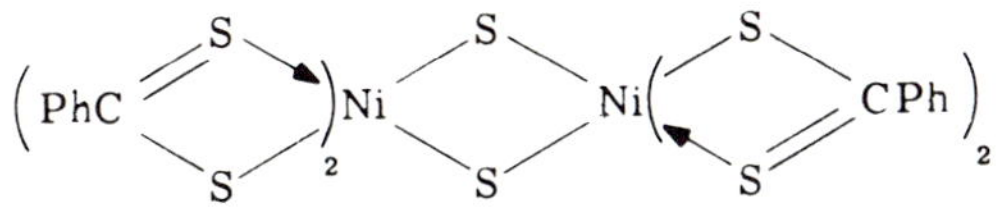

(LXXIII)

The relative stabilities of chelates of a large number of metals with 8-mercaptoquinoline have been reported by Bankovskis *et al.* (1960). Thiosemicarbazide behaves both as a neutral ligand (LXXIV) and, by loss of a proton, as a negatively charged ligand (LXXV), forming chelate

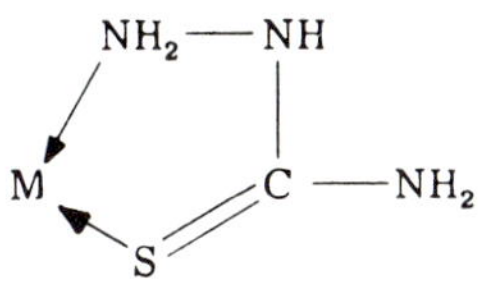

(LXXIV)

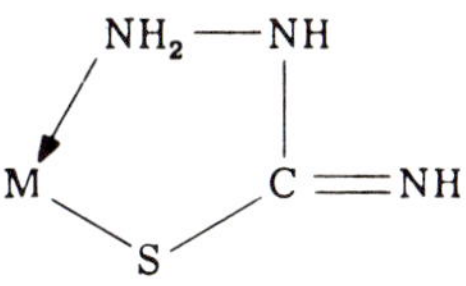

(LXXV)

complexes with Ni(II), Pd(II), and Pt(II) (Jensen, 1934).

3-Mercapto-1,5-diphenylformazan (dithizone = HDz) has found use in analytical chemistry since it forms highly colored inner complexes with those elements which tend to form M—S bonds in preference to M—O bonds. The structures of these compounds are still in dispute and the interpretation of their spectra is far from complete. An X-ray examination (Harding, 1958) of $HgDz_2 \cdot 2py$ shows that the Hg atom is bonded to one S atom of each dithizone group at the normal covalent distance but the N atom of the azo group is weakly coordinated at a distance 0.4–0.5 Å greater than expected for a covalent Hg—N bond giving the molecule a distorted tetrahedral arrangement (LXXVI). From infrared and continuous variation studies Duncan and Thomas (1960) suggested that in the complex $CuDz_2$ the ligand is bonded through the S atom only but a recent X-ray determination shows that the Cu atom is square-planar and the ligand is coordinated through N and S (Bryan and Knopf, 1961).

The metal chelates of guanylthiourea (LXXVII) have been studied only

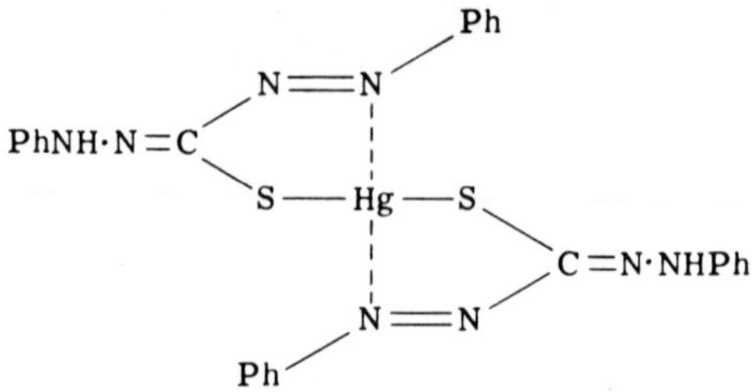

(LXXVI)

(LXXVII)

to a limited extent but complexes of Co(III), Ni(II), Cu(II), and Pd(II) have been reported. Complexes are known containing neutral and deprotonated guanylthiourea and it is highly probable that coordination takes place through the S and the imino N atoms (Rây, 1961). Bandyopadhayay (1955) reported that dithiobiuret (LXXVIII) fails to form stable complexes.

(LXXVIII)

However, inner complexes of Ni(II), Pd(II), Pt(II), Cu(II), and Cd(II) have been prepared; the nickel complex is diamagnetic (Livingstone, 1962).

The chelating ability of some β-aminothioethers such as methyl-2-aminoethylsulfide, $MeS{\cdot}CH_2CH_2{\cdot}NH_2$, has been studied (Gonick *et al.*, 1954). The free $—NH_2$ group of β,β'-diaminodiethylsulfideplatinum (IV) (LXXIX) is capable of adding acids and Mann (1930) resolved the com-

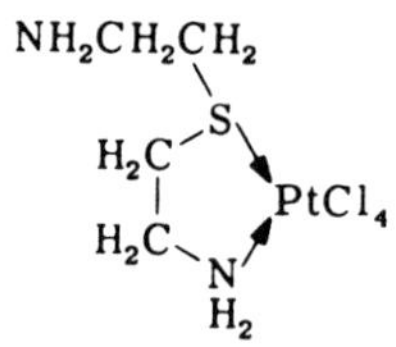

(LXXIX)

pound by isolating its *d*-camphorsulfonate and showed that the asymmetry is due to the tricovalent S atom. Similar optical activity had previously been shown to occur in the sulfoxides.

V. Phosphorus and Arsenic Chelates

The diarsines (LXXX) and (LXXXI) were shown by Chatt and Mann

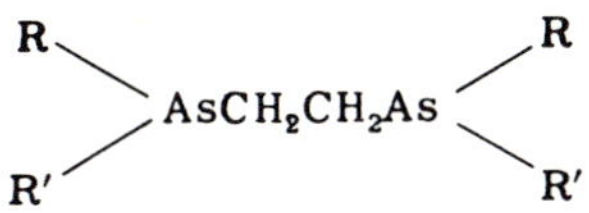

(LXXX); R = *n*-Bu, Ph; R′ = *n*-Bu, Ph

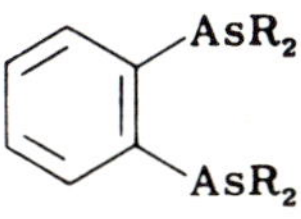

(LXXXI); R = Me, *n*-Bu

(1939) to form Pd(II) chelates. Complexes of *o*-phenylenebisdimethylarsine (As–As) [(LXXXI): R = Me] are known with most of the transition metals and with Sn, P, As, Sb, and Bi and these are listed in Table II.

In all cases where either a "low-spin" or a "high-spin" arrangement is possible magnetic moment determinations have shown that the ligand (As–As) favors the "low-spin" arrangement, except with Mn(II) and Mn(III) (Nyholm and Sutton, 1958b). In many instances this diarsine stabilizes an uncommon oxidation state—e.g., Tc(V), Re(V), Fe(IV), Ni(0), Ni(III), and Ni(IV)—or a high coordination number—e.g., six for Pd(II), Pt(II), and Au(III), seven for Mo(II), and eight for Ti(IV), V(IV), and Tc(V).

The diarsine can also replace some of the CO groups in metallic carbonyls and indeed all the CO groups in $[Ni(CO)_4]$ have been replaced with the formation of the Ni(0) complex $[Ni(As\text{–}As)_2]$ (Nyholm and Ramana Rao, 1960).

It is evident that *o*-phenylenebisdimethylarsine is effective in stabilizing both high and low oxidation states. Chatt (1958) has observed that low valency stabilization is a π-bond effect associated with ligands having suitable π-type orbitals (i.e., atomic *p*, *d*, *dp* hybrid or aromatic π-orbitals) which accept electrons from the filled *d*-orbitals in the penultimate shell of the metal atom. The function of the ligands in the stabilization of low valency states is to facilitate the entry of electrons to the complex, thereby permitting reduction of the metal. On the other hand, high valency stabilization is a σ-bond effect associated with the electronegativity of the donor atom. Carbon monoxide and isonitriles are good π-electron acceptors, whereas F^- and O^{--} ions stabilize high valency because of their lone pairs' repulsion towards the *d*-electrons of the metal atom, causing these electrons to be more easily removed by oxidation. Ligands such as the

TABLE II

METAL COMPLEXES OF *o*-PHENYLENEBISDIMETHYLARSINE

Complex	Metal	Complex	Metal
$[M^0(As\text{–}As)_2]$	Ni[a]	$[M^{II}(As\text{–}As)(CO)_3X_2]$	Mo[x]
$[M^0(As\text{–}As)(CO)_2]$	Ni[b,c]	$[M^{II}(As\text{–}As)_2(CO)_2X]^+$	Mo[x]
$[M^0(As\text{–}As)(CO)_3]$	Fe[d]	$[M^{III}(As\text{–}As)_2]^{3+}$	Au[g]
$[M^0(As\text{–}As)_2CO]$	Fe[d]	$[M^{III}(As\text{–}As)X_3]$	Ni[b]P[y]As[y]Sb[y]Bi[z]
$[M^0(As\text{–}As)(CO)_4]$	Cr[e]Mo[e]W[e]	$[M^{III}(As\text{–}As)H_2OCl_2]^+$	Mn[o]
$[M^0(As\text{–}As)_2(CO)_2]$	Cr[e]Mo[e]W[e]	$[M^{III}(As\text{–}As)_2I]^{++}$	Au[g]
$[M^I(As\text{–}As)_2]^+$	Cu[f]Ag[g]	$[M^{III}(As\text{–}As)X_4]^-$	Cr[aa]Ru[t]
$[M^I(As\text{–}As)(CO)_2I]$	Fe[d]	$[M^{III}(As\text{–}As)H_2OX_3]$	Ti[bb]Cr[aa]
$[M^{II}(As\text{–}As)X_2]$	Ni[b]Pd[h]Zn[i] Cd[i]Hg[i]	$[M^{III}(As\text{–}As)_2X_2]^+$	Cr[aa]Tc[q]Re[r]Fe[s,cc,dd] Ru[t]Os[u]Co[v]Rh[ee] Ni[j]Au[g]As[y]
$[M^{II}(As\text{–}As)_2]^{++}$	Ni[j,k]Pd[l]Pt[k] Zn[i]Cd[i]Hg[i]	$[M^{III}(As\text{–}As)_3]^{3+}$	Bi[z]
$[M^{II}(As\text{–}As)_2X]^+$	Ni[k,m]Pd[l,m]Pt[k]	$[M^{IV}(As\text{–}As)X_4]$	Ti[bb]Mo[x]Sn[ff]
$[M^{II}(As\text{–}As)_2ClO_4]^+$	Ni[n]	$[M^{IV}(As\text{–}As)_2X_2]^{++}$	Fe[dd]Os[u]Ni[gg]Pd[k]Pt[k]
$[M^{II}(As\text{–}As)(CO)_2X_2]$	Fe[d]	$[M^{IV}(As\text{–}As)_2Cl_4]$	Ti[hh],Zr[hh],Hf[hh],V[hh]
$[M^{II}(As\text{–}As)_2X_2]$	Mn[o]Tc[p,q]Re[r] Fe[s]Ru[t]Os[u] Co[v]Ni[j,k,m] Pd[k,l,w]Pt[k]	$[M^V(As\text{–}As)_2Cl_4]^+$	Tc[q]Re[ii]

X = univalent anion such as Cl, Br, I, SCN, NO_2. [a] Nyholm and Ramana Rao (1960); [b] Nyholm (1952); [c] Nyholm and Short (1953); [d] Nigam *et al.* (1959); [e] Nigam *et al.* (1960a); [f] Kabesh and Nyholm (1951); [g] Harris and Nyholm (1957); [h] Chatt and Mann (1939); [i] Sutton (1961b); [j] Nyholm (1950c); [k] Harris *et al.* (1960); [l] Harris and Nyholm (1956); [m] Harris *et al.* (1956b); [n] Barker *et al.* (1961); [o] Nyholm and Sutton (1958b); [p] Fergusson and Nyholm (1959); [q] Fergusson and Nyholm (1960); [r] Curtis *et al.* (1958); [s] Nyholm (1950a); [t] Nyholm and Sutton (1958c); [u] Nyholm and Sutton (1958d); [v] Nyholm (1950d); [w] Harris *et al.* (1956a); [x] Nigam *et al.* (1960b); [y] Sutton (1958b); [z] Sutton (1958a); [aa] Nyholm and Sutton (1958a); [bb] Sutton (1959); [cc] Lockyer (1959); [dd] Nyholm and Parish (1956); [ee] Nyholm (1950b); [ff] Allison and Mann (1949); [gg] Nyholm (1951); [hh] Clark *et al.* (1961); [ii] Fergusson and Nyholm (1958).

diarsine (As–As) can stabilize low oxidation states in the manner described above, and also high valency by forming "low-spin" complexes of the metal in a normal valency state with an effective atomic number greater than that of the next inert gas; electrons are forced into a higher energy level, whence they can be removed by oxidation. The stabilization of low valency states has been discussed further by Orgel (1959).

Metal chelates of other *o*-phenylene compounds (LXXXII)–(LXXXVI) have been investigated. Mono and bis complexes of Pd(II) with the amine-arsine (LXXXII) and the phosphine-arsine (LXXXIII) have been reported (Mann and Stewart, 1955; Jones and Mann, 1955). The four-coordinate complexes $[M(chel)_2]^+$ (M = Cu, Ag, Au) of the phosphine-arsine

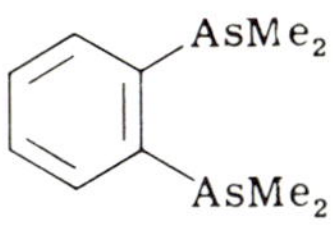

(LXXXII)

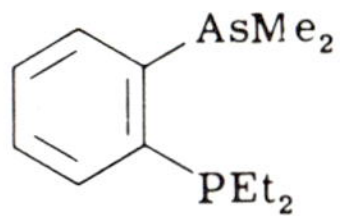

(LXXXIII)

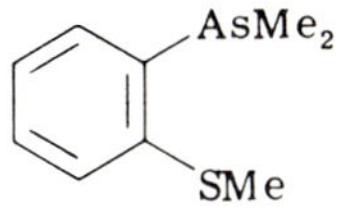

(LXXXIV)

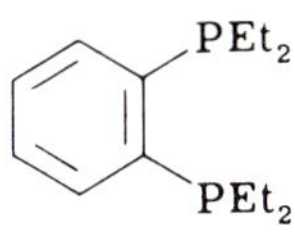

(LXXXV)

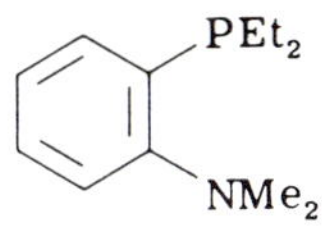

(LXXXVI)

(LXXXIII) were prepared by Cochran *et al.* (1957), who showed that the Au(I) complex has a distorted tetrahedral arrangement. Metal chelates of the arsine-thiether (LXXXIV) with Co(II), Rh(III), Ir(III), Ni(II), Pd(II), Pt(II), Cu(I), Ag(I), and Au(I) have been studied (Livingstone, 1957, 1958; Chiswell and Livingstone, 1959, 1960a,b,c). Diamagnetic Ni(II) complexes of the diphosphine (LXXXV) and Pd(II) complexes of the amine-phosphine (LXXXVI) are known (Chatt and Hart, 1960; Mann and Watson, 1957).

Harris and Livingstone (1962) have discussed the bisdiarsine complexes of Ni(II), Pd(II), Pt(II), and Au(III) which exhibit coordination numbers greater than four. The complexes $M(As\text{–}As)_2X_2$ (M = Ni, Pd, Pt; X = Cl, Br, I, SCN) function as uni-univalent electrolytes in nitrobenzene and a structure determination of the isomorphous iodides $M(As\text{–}As)_2I_2$ has shown a square-planar distribution of the four As atoms about the metal atom with the two I atoms completing a distorted octahedral configuration. The elongation of the M—I bonds in these spin-paired d^8 complexes is due to repulsion of the I^- ions by the filled $(n - 1)d_{z^2}$-orbital of the metal atom. A similar arrangement probably obtains in the six-coordinate $[Au(As\text{–}As)_2X_2]^+$ ion and in the Pd(II) complexes $Pd(chel)_2X_2$ of all the above ligands (LXXXII)–(LXXXVI). X-ray structural analysis has indicated that the complex $[Ti(As\text{–}As)_2Cl_4]$ possesses a dodecahedral structure (Clark *et al.*, 1961).

The aliphatic ligands (LXXXVII) and (LXXXVIII) show little tendency to stabilize high oxidation states or high coordination numbers

$Me_2AsCH_2CH_2CH_2SMe$

(LXXXVII)

$Et_2PCH_2CH_2PEt_2$

(LXXXVIII)

(Chiswell and Livingstone, 1961; Wymore and Bailar, 1960). However, the diphosphines (LXXXVIII) and (LXXXIX) are effective in stabilizing

$$Ph_2PCH_2CH_2PPh_2$$

(LXXXIX)

low valence states; e.g., Mo(0) complexes, [Mo(chel)$_3$] (Chatt and Watson, 1960), and Ni(0) complexes, [Ni(chel)$_2$] and [Ni(chel)(CO)$_2$] (Chatt and Hart, 1960), are known with both diphosphines.

Bis complexes of α-picolyldimethylarsine (XC) with Cu(I), Ag(I), Pd(II), Pt(II), Ru(II), and Ru(III) have been described by Goodwin and Lions (1959).

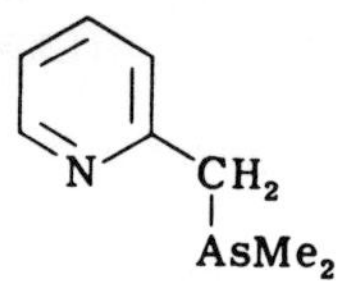

(XC)

References

Albert, A. A., Goldacre, R. J., Davey, M. E., and Stone, J. D. (1945). *Brit. J. Exptl. Pathol.* **26,** 160.

Albert, A. A., Rubbo, S. D., Goldacre, R. J., and Balfour, B. G. (1947). *Brit. J. Exptl. Pathol.* **28,** 69.

Alderman, P. R. H., Owston, P. G., and Row, J. M. (1962). *J. Chem. Soc.* p. 668.

Allison, J. A. C., and Mann, F. G. (1949). *J. Chem. Soc.* p. 2915.

Applebey, M. P., and Lane, K. W. (1918). *J. Chem. Soc.* **113,** 610.

Asmussen, R. W. (1944). "Magnetokemiske Undersögelser over Uoraniske Komplesforbinelser." Gillerups Forlag, Copenhagen.

Backhouse, R., Foss, M. E., and Nyholm, R. S. (1957). *J. Chem. Soc.* p. 1714.

Bähr, G., and Döge, H. G. (1957). *Z. anorg. u. allgem. Chem.* **292,** 119.

Bailar, J. C., and Busch, D. (1956). *In* "Chemistry of the Coordination Compounds" (J. C. Bailar, ed.), p. 24. Reinhold, New York.

Ballhausen, C. J., and Liehr, A. D. (1959). *J. Am. Chem. Soc.* **81,** 538.

Bankovskis, J., Ievinš, A., and Liepina, Z. (1960). *Zhur. Anal. Khim.* **15,** 4.

Bandyopadhayay, D. (1955). *J. Indian Chem. Soc.* **32,** 651.

Banks, C. V., and Bystroff, R. I. (1959). *J. Am. Chem. Soc.* **81,** 6153.

Barclay, G. A., and Hoskins, B. F. (1962). *J. Chem. Soc.* p. 586.

Barclay, G. A., and Kennard, C. H. L. (1961a). *J. Chem. Soc.* p. 3289.

Barclay, G. A., and Kennard, C. H. L. (1961b). *J. Chem. Soc.* p. 5244.

Barclay, G. A., and Kennard, C. H. L. (1961c). *Nature* **192,** 425.

Barclay, G. A., and Stephens, F. S. (1963). *J. Chem. Soc.* p. 2027.

Barker, N. T., Harris, C. M., and McKenzie, E. D. (1961). *Proc. Chem. Soc.* p. 335.

Barraclough, C. G., and Tobe, M. L. (1961). *J. Chem. Soc.* p. 1993.

Basolo, F. (1953). *Chem. Revs.* **52,** 459.
Basolo, F., and Dwyer, F. P. (1954). *J. Am. Chem. Soc.* **76,** 1454.
Basolo, F., and Matoush, W. R. (1953). *J. Am. Chem. Soc.* **75,** 5663.
Basolo, F., and Murmann, R. K. (1952). *J. Am. Chem. Soc.* **74,** 5243.
Basolo, F., and Pearson, R. G. (1958). "Mechanisms of Inorganic Reactions." Wiley, New York.
Basolo, F., Chen, Y. T., and Murmann, R. K. (1953). *J. Am. Chem. Soc.* **75,** 1478.
Basolo, F., Chen, Y. T., and Murmann, R. K. (1954). *J. Am. Chem. Soc.* **76,** 956.
Basolo, F., Wilks, P. H., Pearson, R. G., and Wilkins, R. G. (1958). *J. Inorg. & Nuclear Chem.* **6,** 161.
Basu, S., and Chatterji, K. K. (1958). *Z. physik. Chem. (Leipzig)* **209,** 360.
Belford, R. L., Martell, A. E., and Calvin, M. (1956). *J. Inorg. & Nuclear Chem.* **2,** 11.
Bellamy, L. J., and Branch, R. F. (1954). *J. Chem. Soc.* p. 4491.
Bennett, G. M., Mosses, A. N., and Statham, F. S. (1930). *J. Chem. Soc.* p. 1668.
Bjerrum, J., Schwarzenbach, G., and Sillén, L. G. (1957). "Stability Constants of Metal-ion Complexes," Part I. Special Publication No. 6, Chemical Society, London.
Blakeslee, A. E., and Hoard, J. L. (1956). *J. Am. Chem. Soc.* **78,** 3029.
Blau, F. (1888). *Ber.* **21,** 1077.
Blau, F. (1898). *Monatsch. Chem.* **19,** 647.
Bleaney, B., and Bowers, K. D. (1952). *Proc. Roy. Soc.* **A214,** 451.
Block, B. P., and Bailar, J. C. (1951). *J. Am. Chem. Soc.* **73,** 4722.
Block, B. P., Bailar, J. C., and Pearce, D. W. (1951). *J. Am. Chem. Soc.* **73,** 4971.
Boeseken, J., and Meulenhoff, J. (1924). *Koninkl. Ned. Akad. Wetenschap., Proc.* **27,** 174.
Boeseken, J., Muller, H. D., and Japhonjouw, R. T. (1926). *Rec. trav. chim.* **45,** 919.
Boyland, E., and Watson, G. (1956). *Nature* **177,** 837.
Brandt, W., Dwyer, F. P., and Gyarfas, E. C. (1954). *Chem. Revs.* **54,** 959.
Breckenridge, J. G. (1954). *Can. J. Chem.* **32,** 512, 641.
Broomhead, J. A., and Dwyer, F. P. (1961). *Australian J. Chem.* **14,** 250.
Brosset, C. (1948). *Arkiv. Kemi. Mineral. Geol.* **25,** No. 19.
Brown, H. P., and Austin, J. A. (1941). *J. Am. Chem. Soc.* **63,** 2054.
Bryan, R. F., and Knopf, P. M. (1961). *Proc. Chem. Soc.* p. 203.
Bryan, R. F., Poljak, R. J., and Tomita, K. (1961). *Acta Cryst.* **14,** 1125.
Bullen, G. J. (1956). *Nature* **177,** 537.
Bullen, G. J. (1959). *Acta Cryst.* **12,** 703.
Bullen, G. J., Mason, R., and Pauling, P. (1961). *Nature* **189,** 291.
Burstall, F. H., and Nyholm, R. S. (1952). *J. Chem. Soc.* p. 3570.
Busch, D. H., and Bailar, J. C. (1956). *J. Am. Chem. Soc.* **78,** 1137.
Cabani, S. (1960). *Gazz. chim. ital.* **90,** 1410.
Cambi, L., and Coriselli, C. (1936). *Gazz. chim. ital.* **66,** 779.
Cambi, L., and Malatesta, L. (1938). *Rend. ist. lombardo sci.* **71,** 118.
Cambi, L., and Paglia, E. (1956). *Atti accad. nazl. Lincei Rend., Classe sci. fis., mat. e. nat.* **21,** 372.
Cambi, L., and Paglia, E. (1958). *J. Inorg. & Nuclear Chem.* **8,** 249.
Cambi, L., and Szego, L. (1931). *Ber.* **64,** 2591.
Cambi, L., and Szego, L. (1933). *Ber.* **66,** 656.
Charles, R. G., and Freiser, H. (1952). *J. Am. Chem. Soc.* **74,** 1385.
Chatt, J. (1958). *J. Inorg. & Nuclear Chem.* **8,** 515.
Chatt, J., and Hart, F. A. (1960). *J. Chem. Soc.* p. 1378.
Chatt, J., and Mann, F. G. (1938). *J. Chem. Soc.* p. 1949.
Chatt, J., and Mann, F. G. (1939). *J. Chem. Soc.* p. 1622.

Chatt, J., and Watson, H. R. (1960). *Proc. Chem. Soc.* p. 243.
Chatt, J., Mann, F. G., and Wells, A. F. (1938). *J. Chem. Soc.* p. 2086.
Chatt, J., Duncanson, L. A., and Venanzi, L. M. (1956). *Suomen Kemistilehti* **29,** 75.
Cheney, G. E., Freiser, H., and Fernando, Q. (1959). *J. Am. Chem. Soc.* **81,** 2611.
Chiswell, B., and Livingstone, S. E. (1959). *J. Chem. Soc.* p. 2931.
Chiswell, B., and Livingstone, S. E. (1960a). *J. Chem. Soc.* p. 97.
Chiswell, B., and Livingstone, S. E. (1960b). *J. Chem. Soc.* p. 1071.
Chiswell, B., and Livingstone, S. E. (1960c). *J. Chem. Soc.* p. 3181.
Chiswell, B., and Livingstone, S. E. (1961). *J. Inorg. & Nuclear Chem.* **23,** 37.
Clark, R. J. H., Lewis, J., Nyholm, R. S., Pauling, P., and Robertson, G. B. (1961). *Nature* **192,** 222.
Cochran, W., Hart, F. A., and Mann, F. G. (1957). *J. Chem. Soc.* p. 2816.
Comyns, A. E., Gatehouse, B. M., and Wait, E. (1958). *J. Chem. Soc.* p. 4655.
Cotton, F. A. (1960). *In* "Modern Coordination Chemistry" (J. Lewis and R. G. Wilkins, eds.), p. 301. Wiley (Interscience), New York.
Cox, E. G., and Webster, K. C. (1935). *J. Chem. Soc.* p. 731.
Cox, E. G., Wardlaw, W., and Webster, K. C. (1935). *J. Chem. Soc.* p. 1475.
Curtis, N. F., Fergusson, J. E., and Nyholm, R. S. (1958). *Chem. & Ind. (London)* p. 625.
Dale, J. (1961). *J. Chem. Soc.* p. 922.
Daniel, S. S., and Salmon, J. E. (1957). *J. Chem. Soc.* p. 4207.
Dhar, S. K., Doron, V., and Kirschner, S. (1959). *J. Am. Chem. Soc.* **81,** 6372.
Diehl, H. (1937). *Chem. Revs.* **21,** 39.
Dilthey, W. (1906). *Ann.* **344,** 304.
Dilthey, W., and Schumacher, F. J. (1906). *Ann.* **344,** 300, 326.
Djordjevic, C., Lewis, J., and Nyholm, R. S. (1959). *Chem. & Ind. (London)* p. 122.
Draney, J. J., and Cefola, M. (1954). *J. Am. Chem. Soc.* **76,** 1975.
Duff, J. C. (1921). *J. Chem. Soc.* **119,** 1982.
Duncan, J. F., and Thomas, F. G. (1960). *J. Chem. Soc.* p. 2814.
Duval, C., Freymann, R., and Lecompte, J. (1952). *Bull. soc. chim. France* **19,** 106.
Dwyer, F. P., and Gyarfas, E. C. (1951). *Nature* **168,** 29.
Dwyer, F. P., and Hogarth, J. W. (1953). *J. Am. Chem. Soc.* **75,** 1008.
Dwyer, F. P., and Hogarth, J. W. (1955). *J. Am. Chem. Soc.* **77,** 6152.
Dwyer, F. P., and Mellor, D. P. (1935). *J. Am. Chem. Soc.* **57,** 605.
Dwyer, F. P., and Sargeson, A. M. (1959). *J. Am. Chem. Soc.* **81,** 2335.
Ellis, P., and Wilkins, R. G. (1959). *J. Chem. Soc.* p. 299.
Ellis, P., Wilkins, R. G., and Williams, M. J. G. (1957). *J. Chem. Soc.* p. 4456.
Ellis, P., Hogg, R., and Wilkins, R. G. (1959). *J. Chem. Soc.* p. 3308.
Ewens, R. V. G., and Gibson, C. S. (1949). *J. Chem. Soc.* p. 431.
Fackler, J. P., and Cotton, F. A. (1961). *J. Am. Chem. Soc.* **83,** 2818, 3775.
Fergusson, J. E., and Nyholm, R. S. (1958). *Chem. & Ind. (London)* p. 1555.
Fergusson, J. E., and Nyholm, R. S. (1959). *Nature* **183,**, 1039.
Fergusson, J. E., and Nyholm, R. S. (1960). *Chem. & Ind. (London)* p. 347.
Fernando, Q., and Freiser, H. (1958). *J. Am. Chem. Soc.* **80,** 4928.
Figgis, B. N., and Lewis, J. (1960). *In* "Modern Coordination Chemistry" (J. Lewis and R. G. Wilkins, eds.), p. 436. Wiley (Interscience), New York.
Figgis, B. N., and Martin, R. L. (1956). *J. Chem. Soc.* p. 3837.
Frasson, E., and Sacconi, L. (1958). *J. Inorg. & Nuclear Chem.* **8,** 443.
Freeman, H. C., Smith, J. E. W. L., and Taylor, J. C. (1961). *Acta Cryst.* **14,** 407.
Fujita, J., Nakamoto, K., and Kobayashi, M. (1957). *J. Phys. Chem.* **61,** 1014.
Furlani, C., and Sartori, G. (1958). *J. Inorg. & Nuclear Chem.* **8,** 126.

Gatehouse, B. M., and Comyns, A. E. (1958). *J. Chem. Soc.* p. 3965.
Gatehouse, B. M., Livingstone, S. E., and Nyholm, R. S. (1957). *J. Chem. Soc.* p. 4222.
Gatehouse, B. M., Livingstone, S. E., and Nyholm, R. S. (1958a). *J. Chem. Soc.* p. 3138.
Gatehouse, B. M., Livingstone, S. E., and Nyholm, R. S. (1958b). *J. Inorg. & Nuclear Chem.* **8,** 75.
Gibson, C. S., and Weller, W. T. (1941). *J. Chem. Soc.* p. 102.
Gilbert, T. W., and Sandell, E. B. (1960). *J. Am. Chem. Soc.* **82,** 1087.
Gilman, H., Jones, R. G., Bindschadler, E., Blume, D., Karmas, G., Martin, G. A., Nobis, J. F., Thirtle, J. R., Yale, H. L., and Yoeman, F. A. (1956). *J. Am. Chem. Soc.* **78,** 2790.
Glaser, H., and Pfeiffer, P. (1939). *J. prakt. Chem.* **153,** 300.
Gonick, E., Fernelius, W. C., and Douglas, B. E. (1954). *J. Am. Chem. Soc.* **76,** 4671.
Goodwin, H. A., and Lions, F. (1959). *J. Am. Chem. Soc.* **81,** 311.
Gordon, G., and Birdwhistell, R. K. (1959). *J. Am. Chem. Soc.* **81,** 3567.
Graddon, D. P. (1959). *Nature* **183,** 1610.
Graddon, D. P., and Watton, E. C. (1960). *Nature* **187,** 1021.
Graddon, D. P., and Watton, E. C. (1961a). *J. Inorg. & Nuclear Chem.* **21,** 49.
Graddon, D. P., and Watton, E. C. (1961b). *Nature* **190,** 907.
Griffith, R. L. (1943). *J. Chem. Phys.* **11,** 499.
Grinberg, A. A., and Chapurskii, I. N. (1959). *Zhur. Neorg. Khim.* **4,** 314.
Grinberg, A. A., Vrubleskaja, L. V., Gil'dengershel, Kh. I., and Stetsenko, A. T. (1959). *Zhur. Neorg. Khim.* **4,** 1018.
Guha, J. (1951). *Proc. Roy. Soc.* **A206,** 353.
Gustafson, R. L., and Martell, A. E. (1959). *J. Am. Chem. Soc.* **81,** 525.
Halmekoski, J. (1959). *Ann. Acad. Sci. Fennicae, Ser A II* **96.**
Hamer, N. K., and Orgel, L. E. (1961). *Nature* **190,** 439.
Harding, M. M. (1958). *J. Chem. Soc.* p. 4136.
Harkins, T. R., and Freiser, H. (1956). *J. Am. Chem. Soc.* **78,** 1143.
Harkins, T. R., Walter, J. L., Harris, O. E., and Freiser, H. (1956). *J. Am. Chem. Soc.* **78,** 260.
Harris, C. M., and Livingstone, S. E. (1962). *Revs. Pure and Appl. Chem.* (*Australia*) **12,** 16.
Harris, C. M., and Lockyer, T. N. (1958). *Chem. & Ind.* (*London*) p. 1231.
Harris, C. M., and McKenzie, E. D. (1961). *J. Inorg. & Nuclear Chem.* **19,** 372.
Harris, C. M., and Nyholm, R. S. (1956). *J. Chem. Soc.* p. 4375.
Harris, C. M., and Nyholm, R. S. (1957). *J. Chem. Soc.* p. 63.
Harris, C. M., Nyholm, R. S., and Stephenson, N. C. (1956a). *Rec. trav. chim.* **75,** 687.
Harris, C. M., Nyholm, R. S., and Stephenson, N. C. (1956b). *Nature* **177,** 1127.
Harris, C. M., Lenzer, S. L., and Martin, R. L. (1958). *Australian J. Chem.* **11,** 331.
Harris, C. M., Nyholm, R. S., and Phillips, D. J. (1960). *J. Chem. Soc.* p. 4379.
Harris, C. M., Lenzer, S. L., and Martin, R. L. (1961a). *Australian J. Chem.* **14,** 420.
Harris, C. M., Lockyer, T. N., and Waterman, H. (1961b). *Nature* **192,** 424.
Hassel, O., and Pedersen, B. F. (1959). *Proc. Chem. Soc.* p. 394.
Hazell, A. C., Swallow, A. G., and Truter, M. R. (1959). *Chem. & Ind.* (*London*) p. 564.
Healy, E. A., and Murmann, R. K. (1957). *J. Am. Chem. Soc.* **79,** 5827.
Hein, F., and Herzog, S. (1952). *Z. anorg. u. allgem. Chem.* **267,** 337.
Hein, F., and Regler, H. (1936). *Ber.* **69B,** 1962.
Hein, F., and Regler, H. (1941). *Ber.* **74B,** 1926.
Helman, A. D. (1943). *Compt. rend. acad. sci. U.R.S.S.* **38,** 243.
Herzog, S. (1956). *Naturwissenschaften* **43,** 35.

Herzog, S. (1958). *Z. anorg. u. allgem. Chem.* **294,** 155.
Herzog, S., and Renner, K. (1959). *Chem. Ber.* **92,** 872.
Herzog, S., and Schön, W. (1958). *Z. anorg. u. allgem. Chem.* **297,** 323.
Herzog, S., and Taube, R. (1956). *Naturwissenschaften* **43,** 349.
Herzog, S., Renner, K., and Schön, W. (1957). *Z. Naturforsch.* **12b,** 809.
Hieber, W., and Brück, R. (1949). *Naturwissenschaften* **36,** 312.
Hieber, W., and Brück, R. (1952). *Z. anorg. u. allgem. Chem.* **269,** 13.
Hieber, W., and Floss, J. G. (1957). *Chem. Ber.* **90,** 1617.
Hieber, W., Schleiszmann, C., and Ries, K. (1929). *Z. anorg. u. allgem. Chem.* **180,** 89.
Holm, R. H. (1961). *In* "Advances in the Chemistry of the Coordination Compounds" (S. Kirschner, ed.), p. 341. Macmillan, New York.
Holm, R. H., and Cotton, F. A. (1958). *J. Am. Chem. Soc.* **80,** 5658.
Holm, R. H., and Cotton, F. A. (1961). *J. Phys. Chem.* **65,** 321.
Holmes, F., Jones, K. M., and Torrible, E. G. (1961). *J. Chem. Soc.* p. 4790.
Hurd, R. N., De La Mater, G., McElheny, G. C., and McDermott, J. P. (1961). *In* "Advances in the Chemistry of the Coordination Compounds" (S. Kirschner, ed.), p. 350. Macmillan, New York.
Irving, H., and Rossotti, H. S. (1954). *J. Chem. Soc.* p. 2910.
Irving, H., and Williams, R. J. P. (1948). *Nature* **162,** 746.
Irving, H., and Williams, R. J. P. (1953). *J. Chem. Soc.* p. 3192.
Irving, H., Butler, E. J., and Ring, M. F. (1949). *J. Chem. Soc.* p. 1489.
Irving, R. J., and Fernelius, W. C. (1956). *J. Phys. Chem.* **60,** 1427.
Izatt, R. M., Haas, C. G., Block, B. P., and Fernelius, W. C. (1954). *J. Phys. Chem.* **58,** 1133.
Jacobs, G., and Speeke, F. (1955). *Acta Cryst.* **8,** 67.
James, B. R., and Williams, R. J. P. (1961). *J. Chem. Soc.* p. 2007.
James, B. R., Parris, M., and Williams, R. J. P. (1961). *J. Chem. Soc.* p. 4630.
Jensen, K. A. (1934). *Z. anorg. u. allgem. Chem.* **221,** 6, 11.
Jensen, K. A. (1944). *Z. anorg. u. allgem. Chem.* **252,** 227.
Johnston, W. D., and Freiser, H. (1952). *J. Am. Chem. Soc.* **74,** 5239.
Jones, E. R. H., and Mann, F. G. (1955). *J. Chem. Soc.* p. 4472.
Jones, H. O., and Tasker, H. S. (1909). *J. Chem. Soc.* **95,** 1904.
Jones, R. G., Karmas, G., Martin, G. A., and Gilman, H. (1956a). *J. Am. Chem. Soc.* **78,** 4285.
Jones, R. G., Bindschadler, E., Karmas, G., Yeoman, F. A., and Gilman, H. (1956b). *J. Am. Chem. Soc.* **78,** 4287.
Jones, R. G., Bindschadler, E., Karmas, G., Martin, G. A., Thirtle, J. R., Yeoman, F. A., and Gilman, H. (1956c). *J. Am. Chem. Soc.* **78,** 4289.
Jørgensen, C. K. (1955'. *Acta Chem. Scand.* **9,** 1362.
Kabesh, A., and Nyholm, R. S. (1951). *J. Chem. Soc.* p. 38.
Kariyama, R., Ibamato, H., and Matsuo, K. (1954). *Acta Cryst.* **7,** 482.
Kluiber, R. W. (1961). *J. Am. Chem. Soc.* **83,** 3030, and references therein.
Kogama, H., Saito, Y., and Kuroya, H. (1953). *J. Inst. Polytech. Osaka City Univ. Ser.* **C4,** 43; *Chem. Abstr.* **48,** 3097 (1954).
Kokot, E. (1961). Ph.D. Thesis, University of New South Wales, Australia.
Krause, R. A., and Busch, D. H. (1958). *Nature* **181,** 1529.
Krishnamurty, K. V., and Harris, G. M. (1961). *Chem. Revs.* **61,** 213.
Kruh, R., and Diggins, C. W. (1955). *J. Am. Chem. Soc.* **77,** 806.
Leussing, D. L. (1958). *J. Am. Chem. Soc.* **80,** 4180.
Leussing, D. L., and Alberts, G. S. (1960). *J. Am. Chem. Soc.* **82,** 4458.

Leussing, D. L., Laramy, R. E., and Alberts, G. S. (1960). *J. Am. Chem. Soc.* **82,** 4826.
Lifschitz, I., and Bos, J. G. (1940). *Rec. trav. chim.* **59,** 407.
Lifschitz, I., Bos, J. G., and Dijkema, K. M. (1939). *Z. anorg. u. allgem. Chem.* **242,** 97.
Livingstone, S. E. (1956a). *J. Chem. Soc.* p. 437.
Livingstone, S. E. (1956b). *J. Chem. Soc.* p. 1042.
Livingstone, S. E. (1956c). *J. Chem. Soc.* p. 1989.
Livingstone, S. E. (1956d). *J. Chem. Soc.* p. 1994.
Livingstone, S. E. (1957). *Chem. & Ind. (London)* p. 143.
Livingstone, S. E. (1958). *J. Chem. Soc.* p. 4222.
Livingstone, S. E. (1962). Unpublished work.
Livingstone, S. E., and Plowman, R. A. (1950). *J. Proc. Roy. Soc. N.S.Wales* **84,** 188.
Livingstone, S. E., and Plowman, R. A. (1951). *J. Proc. Roy. Soc. N.S.Wales* **85,** 116.
Livingstone, S. E., Plowman, R. A., and Sorenson, J. (1950). *J. Proc. Roy. Soc. N.S.Wales* **84,** 28.
Lockyer, T. N. (1959). M.Sc. Thesis, University of New South Wales, Australia.
Lowry, T. M., and French, H. S. (1924). *Proc. Roy. Soc.* **A106,** 489.
Lyle, F. K., Morosin, B., and Lingafelter, E. C. (1959). *Acta Cryst.* **12,** 938.
Malatesta, L. (1938). *Gazz. chim. ital.* **68,** 195.
Malatesta, L. (1939a). *Gazz. chim. ital.* **69,** 408.
Malatesta, L. (1939b). *Gazz. chim. ital.* **69,** 629.
Malatesta, L. (1939c). *Gazz. chim. ital.* **69,** 752.
Malatesta, L., and Cambi, L. (1940). *Gazz. chim. ital.* **70,** 734.
Mann, F. G. (1930). *J. Chem. Soc.* p. 1745.
Mann, F. G., and Stewart, F. H. C. (1955). *J. Chem. Soc.* p. 1269.
Mann, F. G., and Watson, H. R. (1957). *J. Chem. Soc.* p. 3950.
Mann, F. G., and Watson, H. R. (1958). *J. Chem. Soc.* p. 2772.
Martell, A. E., and Calvin, M. (1952a). "Chemistry of the Metal Chelate Compounds," pp. 35, 135. Prentice-Hall, Englewood Cliffs, New Jersey.
Martell, A. E., and Calvin, M. (1952b). "Chemistry of the Metal Chelate Compounds," p. 175. Prentice-Hall, Englewood Cliffs, New Jersey.
Martin, B., and Waind, G. M. (1958a). *J. Inorg. & Nuclear Chem.* **8,** 551.
Martin, B., and Waind, G. M. (1958b). *Proc. Chem. Soc.* p. 169.
Martin, B. B., and Fernelius, W. C. (1959). *J. Am. Chem. Soc.* **81,** 2342.
Martin, R. L., and Waterman, H. (1957). *J. Chem. Soc.* p. 2545.
Martin, R. L., and Waterman, H. (1959a). *J. Chem. Soc.* p. 1359.
Martin, R. L., and Waterman, H. (1959b). *J. Chem. Soc.* p. 2960.
Martin, R. L., and Whitley, A. (1958). *J. Chem. Soc.* p. 1394.
Mellor, D. P., and Maley, L. (1947). *Nature* **159,** 370.
Merritt, L. L. (1953). *Anal. Chem.* **25,** 718.
Merritt, L. L., Cady, R. T., and Mundy, B. W. (1954). *Acta Cryst.* **7,** 473.
Merritt, L. L., Guare, C., and Lessor, A. E. (1956). *Acta Cryst.* **9,** 253.
Miller, J. R. (1961). *J. Chem. Soc.* p. 4452.
Miller, J. R., and Sharpe, A. G. (1961). *J. Chem. Soc.* p. 2594.
Miller, R. R., and Brandt, W. W. (1955). *J. Am. Chem. Soc.* **77,** 1384.
Mills, W. H., and Raper, R. (1925). *J. Chem. Soc.* **127,** 2479.
Mizushima, S., and Quagliano, J. V. (1953). *J. Am. Chem. Soc.* **75,** 6084.
Moeller, T., and Cohen, A. J. (1950). *J. Am. Chem. Soc.* **72,** 3546.
Moeller, T., and Gulyas, E. (1958). *J. Inorg. & Nuclear Chem.* **5,** 245.
Moeller, T., and Pundsack, F. L. (1954). *J. Am. Chem. Soc.* **76,** 617.
Morgan, G. T., and Ledbury, W. (1922). *J. Chem. Soc.* **121,** 2882.

Morgan, G. T., and Moss, H. W. (1914). *J. Chem. Soc.* **105,** 189.
Morgan, G. T., and Thomasson, R. W. (1924). *J. Chem. Soc.* **125,** 754.
Morgan, G. T., and Tunstall, R. B. (1924). *J. Chem. Soc.* **125,** 1963.
Mori, M., Shibata, M., Kyuno, E., and Adachi, T. (1956). *Bull. Chem. Soc. Japan* **29,** 883.
Murata, H., and Kawai, K. (1956). *J. Chem. Phys.* **25,** 589.
Muto, Y. (1958). *Bull. Chem. Soc. Japan* **31,** 56.
Nakahara, A., Saito, Y., and Kuroya, H. (1952). *Bull. Chem. Soc. Japan* **25,** 331.
Nakamoto, K., and Martell, A. E. (1960). *J. Chem. Phys.* **32,** 588.
Nakamoto, K., Fujita, J., Tanaka, S., and Kobayashi, M. (1957). *J. Am. Chem. Soc.* **79,** 4904.
Nakamoto, K., McCarthy, P. J., and Martell, A. E. (1959). *Nature* **183,** 459.
Nakamoto, K., McCarthy, P. J., Ruby, A., and Martell, A. E. (1961a). *J. Am. Chem. Soc.* **83,** 1066.
Nakamoto, K., McCarthy, P. J., and Martell, A. E. (1961b). *J. Am. Chem. Soc.* **83,** 1272.
Nakatsu, K., Saito, Y., and Kuroya, H. (1956). *Bull. Chem. Soc. Japan* **29,** 428.
Nakatsu, K., Shiro, M., Saito, Y., and Kuroya, H. (1957). *Bull. Chem. Soc. Japan* **30,** 158.
Nigam, H. L., Nyholm, R. S., and Ramana Rao, D. V. (1959). *J. Chem. Soc.* p. 1397.
Nigam, H. L., Nyholm, R. S., and Stiddard, M. H. B. (1960a). *J. Chem. Soc.* p. 1803.
Nigam, H. L., Nyholm, R. S., and Stiddard, M. H. B. (1960b). *J. Chem. Soc.* p. 1806.
Nyburg, S. C., Wood, J. S., and Higginson, W. C. E. (1961). *Proc. Chem. Soc.* p. 297.
Nyholm, R. S. (1950a). *J. Chem. Soc.* p. 851.
Nyholm, R. S. (1950b). *J. Chem. Soc.* p. 857.
Nyholm, R. S. (1950c). *J. Chem. Soc.* p. 2061.
Nyholm, R. S. (1950d). *J. Chem. Soc.* p. 2071.
Nyholm, R. S. (1951). *J. Chem. Soc.* p. 2602.
Nyholm, R. S. (1952). *J. Chem. Soc.* p. 2906.
Nyholm, R. S., and Parish, R. V. (1956). *Chem. & Ind.* (*London*) p. 470.
Nyholm, R. S., and Ramana Rao, D. V. (1960). Unpublished work. [See Harris, C. M., Nyholm, R. S., and Phillips, D. J. (1960).]
Nyholm, R. S., and Short, L. N. (1953). *J. Chem. Soc.* p. 2670.
Nyholm, R. S., and Sutton, G. J. (1958a). *J. Chem. Soc.* p. 560.
Nyholm, R. S., and Sutton, G. J. (1958b). *J. Chem. Soc.* p. 564.
Nyholm, R. S., and Sutton, G. J. (1958c). *J. Chem. Soc.* p. 567.
Nyholm, R. S., and Sutton, G. J. (1958d). *J. Chem. Soc.* p. 572.
Nyholm, R. S., and Turco, A. (1960). *Chem. & Ind.* (*London*) p. 74.
Ooi, S., Komiyama, Y., Saito, Y., and Kuroya, H. (1959). *Bull. Chem. Soc. Japan* **32,** 263.
Orgel, L. E. (1959). *In* "International Conference on Coordination Chemistry, London," p. 93. Chemical Society, London.
Orgel, L. E. (1960). "An Introduction to Transition Metal Chemistry: Ligand-Field Theory," p. 132. Methuen, London.
Orgel, L. E. (1961). *J. Chem. Soc.* p. 3683.
Paglia, E., and Sironi, C. (1957). *Gazz. chim. ital.* **87,** 1125.
Pecsok, R. L., and Bjerrum, J. (1957). *Acta Chem. Scand.* **11,** 1419.
Perthel, R. (1959). *Z. physik. Chem.* (*Leipzig*) **211,** 74.
Peyronel, G., and Pignedoli, A. (1959). *Ricerca sci.* **29,** 1505.
Pfeiffer, P., and Glaser, H. (1938). *J. prakt. Chem.* **151,** 134.
Pfeiffer, P., Schmitz, E., and Bohm, A. (1952). *Z. anorg. u. allgem. Chem.* **270,** 287.

Phillips, J. P. (1956). *Chem. Revs.* **56,** 271.
Poulsen, J., and Bjerrum, J. (1955). *Acta Chem. Scand.* **9,** 1407.
Powell, D. B., and Sheppard, N. (1959a). *J. Chem. Soc.* p. 791.
Powell, D. B., and Sheppard, N. (1959b). *J. Chem. Soc.* p. 3089.
Prasad, S., Punna Rao, G. B., Kumar, S., Rama Reddy, V., and Kacker, K. P. (1959). *J. Indian Chem. Soc.* **36,** 129.
Rasmussen, R. S., Tunnicliff, D. D., and Brattain, R. R. (1949). *J. Am. Chem. Soc.* **71,** 1068.
Rây, P. (1961). *Chem. Revs.* **61,** 313.
Reid, J. C., and Calvin, M. (1950). *J. Am. Chem. Soc.* **72,** 2948.
Robinson, C. S., and Jones, H. O. (1912). *J. Chem. Soc.* **101,** 62.
Roof, R. B. (1956). *Acta Cryst.* **9,** 781.
Rosenheim, A., and Garfunkel, A. (1911). *Ber.* **44,** 1873.
Rosenheim, A., and Nernst, C. (1933). *Z. anorg. u. allgem. Chem.* **214,** 209.
Rosenheim, A., and Plato, W. (1925). *Ber.* **58,** 2000.
Rosenheim, A., and Sorge, O. (1920). *Ber.* **53,** 932.
Rosenheim, A., and Vermehren, H. (1924). *Ber.* **57,** 1337.
Rosenheim, A., Raibmann, B., and Schendel, G. (1931). *Z. anorg. u. allgem. Chem.* **196,** 160.
Ross, I. G. (1959). *Trans. Faraday Soc.* **55,** 1057.
Ross, I. G., and Yates, J. (1959). *Trans. Faraday Soc.* **55,** 1064.
Sacconi, L., Caroti, G., and Paoletti, P. (1958a). *J. Chem. Soc.* p. 4257.
Sacconi, L., Caroti, G., and Paoletti, P. (1958b). *J. Inorg. & Nuclear Chem.* **8,** 93.
Sacconi, L., Paoletti, P., and Cini, R. (1958c). *J. Am. Chem. Soc.* **80,** 3583.
Schilt, A. A. (1957). *J. Am. Chem. Soc.* **81,** 2966.
Schilt, A. A. (1960a). *J. Am. Chem. Soc.* **82,** 3000.
Schilt, A. A. (1960b). *J. Am. Chem. Soc.* **82,** 5779.
Schilt, A. A., and Smith, G. F. (1957). *Anal. Chim. Acta* **16,** 401.
Schilt, A. A., and Taylor, R. C. (1959). *J. Inorg. & Nuclear Chem.* **9,** 211.
Schmeltz, M. J., Miyazawa, T., Mizushima, S., Lane, T. J., and Quagliano, J. V. (1957). *Spectrochim. Acta* **9,** 59.
Schubert, M. (1932). *J. Am. Chem. Soc.* **54,** 4077.
Schwarzenbach, G., Ackermann, H., Maissen, B., and Anderegg, G. (1952). *Helv. Chim. Acta* **35,** 2337.
Shibata, S., and Sone, K. (1956). *Bull. Chem. Soc. Japan* **29,** 852.
Shibata, S., Kishita, M., and Kubo, M. (1957). *Nature* **179,** 320.
Shkolnikova, L. M., and Shugam, E. A. (1960). *Kristallografiya* **5,** 32.
Shugam, E. A. (1951). *Doklady Akad. Nauk S.S.S.R.* **81,** 853.
Shugam, E. A., and Levina, V. M. (1960). *Kristallografiya* **5,** 257.
Sidgwick, N. V. (1950a). "The Chemical Elements and Their Compounds," p. 163. Oxford Univ. Press (Clarendon), London and New York.
Sidgwick, N. V. (1950b). "The Chemical Elements and Their Compounds," p. 831. Oxford Univ. Press (Clarendon), London and New York.
Sims, P. (1959). *J. Chem. Soc.* p. 3648.
Sone, K. (1953). *J. Am. Chem. Soc.* **75,** 5207.
Stewart, F. H. C. (1958). *Chem. & Ind. (London)* p. 264.
Stewart, J. M., and Lingafelter, E. C. (1959). *Acta Cryst.* **12,** 842.
Stewart, J. M., Lingafelter, E. C., and Breazeale, J. D. (1961). *Acta Cryst.* **14,** 888.
Stosick, A. J. (1945). *J. Am. Chem. Soc.* **67,** 362.
Stouffer, R. C., and Busch, D. H. (1956). *J. Am. Chem. Soc.* **78,** 6016.

Sutton, G. J. (1956a). *Australian J. Chem.* **11,** 415.
Sutton, G. J. (1958b). *Australian J. Chem.* **11,** 420.
Sutton, G. J. (1959). *Australian J. Chem.* **12,** 122.
Sutton, G. J. (1960a). *Australian J. Chem.* **13,** 74.
Sutton, G. J. (1960b). *Australian J. Chem.* **13,** 222.
Sutton, G. J. (1960c). *Australian J. Chem.* **13,** 473.
Sutton, G. J. (1961a). *Australian J. Chem.* **14,** 37.
Sutton, G. J. (1961b). *Australian J. Chem.* **14,** 545.
Sutton, G. J. (1961c). *Australian J. Chem.* **14,** 550.
Swallow, A. G., and Truter, M. R. (1961). *Proc. Chem. Soc.* p. 166.
Tchougaeff, Z., and Fraenkel, D. (1912). *Compt. rend. acad. sci.* **154,** 33.
Tomkinson, J. C., and Williams, R. J. P. (1958). *J. Chem. Soc.* p. 2010.
Truter, M. R., and Cox, E. G. (1956). *J. Chem. Soc.* p. 948.
Tschugaeff, L. (1908). *Ber.* **41,** 2222.
Tschugaeff, L., and Kobljanski, A. (1913). *Z. anorg. u. allgem. Chem.* **83,** 8.
Tschugaeff, L., and Subbotin, W. (1910). *Ber.* **43,** 1200.
Tsuchida, R., and Yamada, S. (1955). *Nature* **176,** 1171.
Tsuchida, R., and Yamada, S. (1958). *Nature* **182,** 1230.
Tsuchida, R., Yamada, S., and Nakamura, H. (1956). *Nature* **178,** 1192.
Tsuchida, R., Yamada, S., and Nakamura, H. (1958). *Nature* **181,** 479.
Van Niekirk, J. N., and Schoening, F. R. L. (1951). *Acta Cryst.* **4,** 35.
Van Niekirk, J. H., and Schoening, F. R. L. (1952). *Acta Cryst.* **5,** 196.
Van Niekirk, J. N., and Schoening, F. R. L. (1953). *Acta Cryst.* **6,** 227.
Van Niekirk, J. N., Schoening, F. R. L., and de Wett, J. F. (1953a). *Acta Cryst.* **6,** 501.
Van Niekirk, J. N., Schoening, F. R. L., and Talbot, J. H. (1953b). *Acta Cryst.* **6,** 720.
Vasil'eva, E. V., and Nedopekin, T. K. (1959). *Tiolovye Soedinen. v Med., Ukrain. Nauch.-Issledovatel. Sanit.-Khim. Inst., Trudy Nauch. Konf., Kiev, 1957* p. 36 (publ. 1959).
Vlček, A. A. (1957a). *Nature* **180,** 753.
Vlček, A. A. (1957b). *Z. Electrochem.* **61,** 1014.
Vlček, A. A. (1961). *Nature* **189,** 393.
Wallenfels, K., and Sund, H. (1957). *Biochem. Z.* **329,** 17.
Walter, J. L., and Freiser, H. (1954). *Anal. Chem.* **26,** 217.
Watanabe, I., and Atoji, M. (1951). *Kagaku (Tokyo)* **21,** 301.
Watt, G. W., and McCarley, R. E. (1957). *J. Am. Chem. Soc.* **79,** 4585.
Weinland, R., and Scholder, R. (1923). *Z. anorg. u. allgem. Chem.* **127,** 343.
Weinland, R., and Walter, E. (1923). *Z. anorg. u. allgem. Chem.* **126,** 141.
Welsh, H. K., and Mathieson, A. Mc.L. (1953). *Acta Cryst.* **5,** 599.
Wendlandt, W., and Haschke, J. (1962). *Nature* **193,** 1174.
Werner, A. (1901). *Ber.* **34,** 2584.
Werner, A. (1920). "Neure Anschauungen auf dem Gebiete der anorganischen Chemie," 4th ed., p. 44. Vieweg, Braunschweig, Germany.
Wilkins, R. G. (1957). *J. Chem. Soc.* p. 4521.
Wilkins, R. G., and Williams, M. J. G. (1957). *J. Chem. Soc.* pp. 1763, 4514.
Wilkins, R. G., and Williams, M. J. G. (1960). *In* "Modern Coordination Chemistry" (J. Lewis and R. G. Wilkins, eds.), p. 179. Wiley (Interscience), New York.
Williams, R. J. P. (1955). *J. Chem. Soc.* p. 137.
Williams, R. J. P. (1959). *Ann. Repts. Progr. Chem. (Chem. Soc. London)* **56,** 87.
Willis, J. B., and Mellor, D. P. (1947). *J. Am. Chem. Soc.* **69,** 1237.
Wolf, L., Butter, E., and Weinelt, H. (1960). *Z. anorg. u. allgem. Chem.* **306,** 87.

Wymore, C. E., and Bailar, J. C. (1960). *J. Inorg. & Nuclear Chem.* **14**, 42.
Yamada, S., Nakamura, H., and Tsuchida, R. (1957). *Bull. Chem. Soc. Japan* **30**, No. 9, 953.
Yamada, S., Nakamura, H., and Tsuchida, R. (1958). *Bull. Chem. Soc. Japan* **31**, No. 3, 303.
Zentmyer, G. A. (1944). *Science* **100**, 294.

CHAPTER 4

Design and Stereochemistry of Multidentate Chelating Agents

HAROLD A. GOODWIN

School of Chemistry, The University of New South Wales, Kensington, N.S.W., Australia

I. Introduction

The best known and most common chelating agents are undoubtedly the bidentates. Quite early in the development of coordination chemistry, however, the existence of chelates containing an organic moiety linked to a metal atom through more than two donor atoms was established. Substances capable of being attached to a single metal atom through three, four, five, six, and even eight donor atoms are now known.

Research on multidentate chelating agents has been stimulated by a number of factors such as their interesting and very often unique stereochemical properties, their widespread occurrence in living matter, and the many practical applications found for sequestering agents such as ethylenediaminetetraacetic acid. This interest has led to the synthesis and study of a wide variety of new chelating agents and doubtless many more await discovery. The design and synthesis of such substances provide an ideal field for collaboration between organic and inorganic chemists.

The following discussion is not intended as an exhaustive survey of all known multidentate chelating agents nor is it intended as a detailed description of the chemistry of metal chelates derived from such ligands. It is planned rather to show the various structural types of multidentates which are known, examples being selected and arranged to illustrate some of the factors which can be considered in the design of new chelating agents.

II. Tridentate Chelating Agents

Examples of tridentate chelating agents are numerous and have been known for some time. However, no tridentate has been studied as extensively as the common bidentates such as ethylenediamine or 2,2′-bipyridine. Two types of tridentate may be distinguished according to their modes of attachment to an octahedral metal atom:

(*a*) those in which the donor atoms coordinate in an equatorial plane of the octahedron (planar tridentates);

(*b*) those in which the donor atoms coordinate at the vertices of one face of the octahedron (vicinal tridentates).

A. Planar Tridentates

1. *Terpyridine and Related Tridentates*

A symmetrical planar tridentate can coordinate to an octahedral metal atom and occupy the six coordination positions in only one arrangement (I),

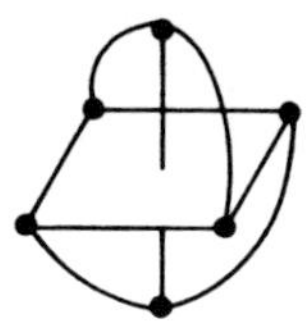

(I)

in which the molecules of the tridentate are arranged in two equatorial planes at right angles. Such a tridentate is 2,6-bis(α-pyridyl)pyridine (terpyridine) (II). Since the bonds from the nitrogen atoms to the metal

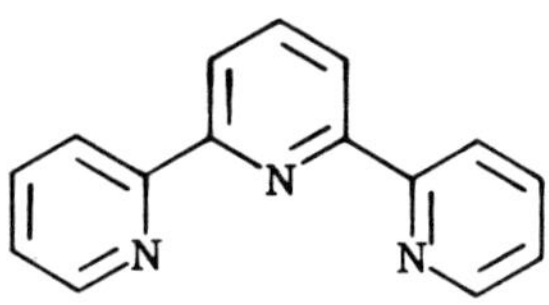

(II)

atom must be in the plane of the pyridine ring and resonance requires coplanarity of the three pyridine rings, all the bonds from terpyridine must be planar. Some strain is introduced into the terpyridine molecule when it coordinates as a tridentate, the bonds from the nitrogen atoms to the metal atom being inclined at less than 90°. Metal chelates derived from (II) are not as stable as might be expected for a tridentate such as this (Brandt *et al.*, 1954). The terpyridine molecule has been shown to remain flat when coordinated to the zinc(II) ion. Dichloroterpyridinezinc(II) has a distorted trigonal bipyramidal structure, the chlorine atoms and the central pyridine nitrogen atom occupying the equatorial plane (Corbridge and Cox, 1956).

Morgan and Burstall (1937) assigned to octahedral complexes of (II) the configuration (I) and this was supported by their inability to resolve either the stable ruthenium(II) or nickel(II) complex. Terpyridine is a symmetrical tridentate so that complexes with configuration (I) have a plane of symmetry.

The intense color and stability of many complexes of terpyridine have led to the widespread use of the base in analytical chemistry. The stability of these complexes is believed to be related to the presence of the structural arrangement

$$—N{=}C—C{=}N—C—C{=}N—$$

which gives rise to a number of resonating structures, involving π-bond formation between the metal and nitrogen atoms. This, or a closely related arrangement, occurs in a number of tridentates described recently. These have been shown, in many instances, to yield complexes qualitatively quite similar to corresponding complexes of terpyridine.

The above arrangement is conveniently obtained by condensing an aldehyde such as pyridine-2-aldehyde with an aminopyridine derivative.

The Schiff's base 8-(α-pyridylmethyleneamino)quinoline (III), obtained

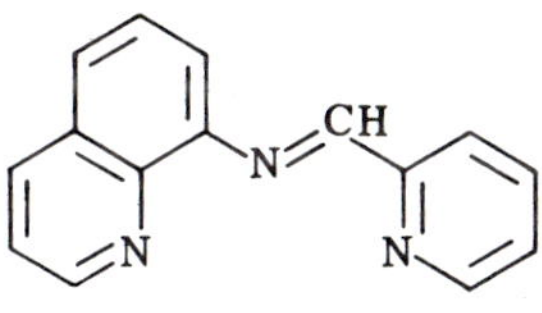

(III)

from pyridine-2-aldehyde and 8-aminoquinoline, has been shown (Dwyer *et al.*, 1953b) to function as a tridentate. Structure (III) is unsymmetrical so that octahedral complexes in which the six positions are occupied by two molecules of the tridentate exist in *d* and *l* forms. The iron(II) complex has been resolved. The closely related tridentate 2-(α-pyridylmethyleneaminomethyl)pyridine (IV) is obtained from pyridine-2-aldehyde and 2-aminomethylpyridine (Lions and Martin, 1957c). None of its complexes has been resolved.

Condensation of quinoline-8-aldehyde with 2-aminomethylpyridine

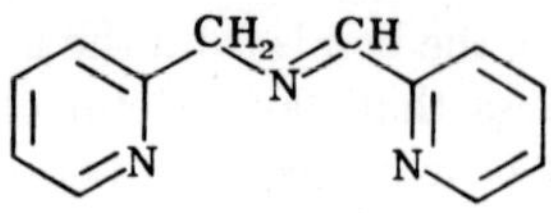

(IV)

yields a tridentate (V) which differs structurally from terpyridine, coordination with a metal atom yielding one five- and one six-membered ring. Its

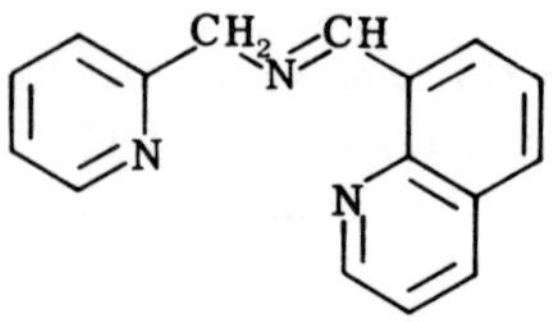

(V)

complexes are intensely colored and similar to those of terpyridine (Lions and Martin, 1957c). Substitution of diacetylmonoxime for pyridine-2-aldehyde in condensation with 2-aminomethylpyridine gives a molecule (VI) with three donor nitrogen atoms arranged as in terpyridine. The base

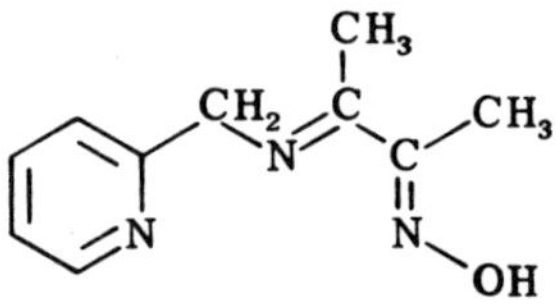

(VI)

(VI) has been shown (Lions and Martin, 1957c) to function as a tridentate. α-Pyridylhydrazine may be substituted for 2-aminomethylpyridine with little change in the nature of the resultant tridentate (Lions and Martin, 1958). The methylene group linking the pyridine ring to the azomethine nitrogen in (IV) or (V) is replaced by a secondary amino group.

Pyridine-2,6-dialdehyde may be condensed with a primary amine (Lions and Martin, 1957c) or with hydrazine (Stoufer and Busch, 1956) to yield the Schiff's bases (VII) and (VIII) respectively; both have been

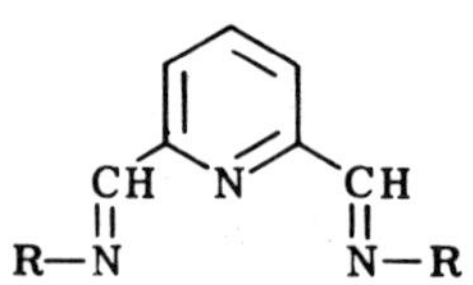

(VII)

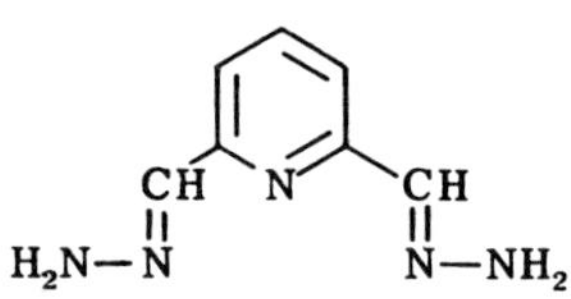

(VIII)

shown to function as tridentates.

Condensation of pyridine-2,6-dialdehyde with a diamine such as ethylenediamine results in the formation of a polymeric structure with recurring tridentate units (IX).

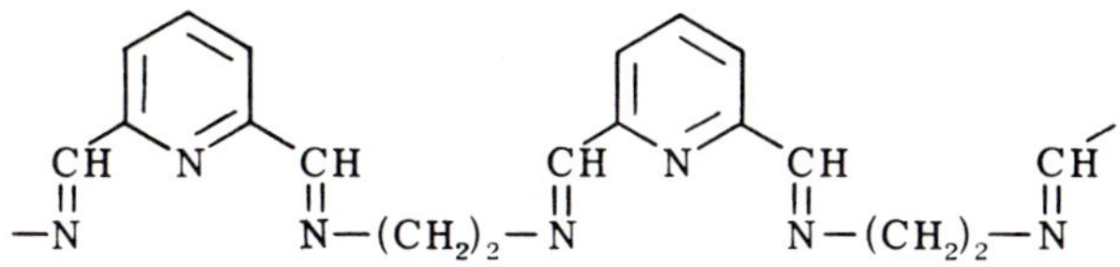

(IX)

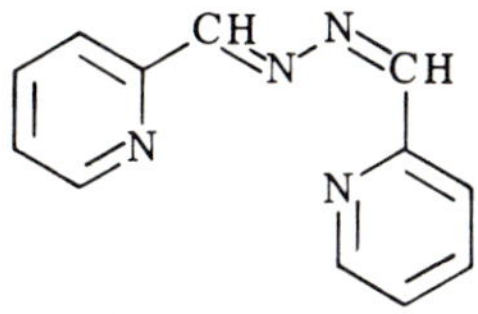

(X)

Pyridine-2-aldazine (PAA) (X) has proven to be an interesting chelating agent. It has been shown that (X) can function as a tridentate giving highly colored complexes $[M(PAA)_2]^{++}$ with iron(II) and nickel(II) salts (Stratton and Busch, 1958a). Because of the high degree of conjugation in (X) it would be attached equatorially about a metal atom and octahedral complexes have configuration (I). Such complexes should be resolvable because of the unsymmetrical nature of the ligand (after coordination). In addition to functioning as a tridentate (X) can coordinate through all four nitrogen atoms as a double bidentate (quadridentate function is sterically impossible), forming complexes of the type $[M_2(PAA)_3]^{4+}$. This formulation has been supported by conductivity, spectrophotometric, magnetic, and infrared data (Stratton and Busch, 1958b, 1960). Evidence has also been obtained for the existence, in solution, of the species $[Fe(PAA)_3]^{++}$ in which (X) is acting as a simple bidentate. Studies with the related base diacetylhydrazoneazine (XI) indicate that it functions only as a double bidentate

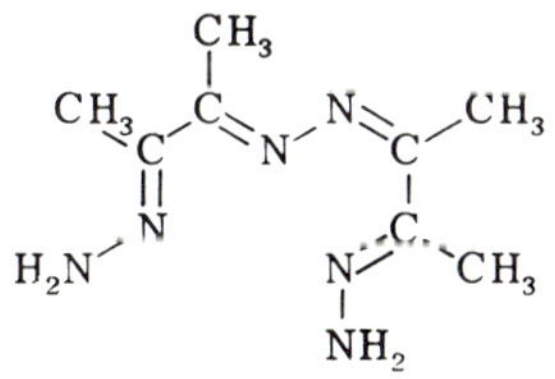

(XI)

(Stratton and Busch, 1960), yielding bridged complexes similar to those derived from (X) and octahedral metal atoms.

The base 2,3,5,6-tetrakis(α-pyridyl)pyrazine (XII) functions as a tridentate in the same manner as terpyridine (Goodwin and Lions, 1959).

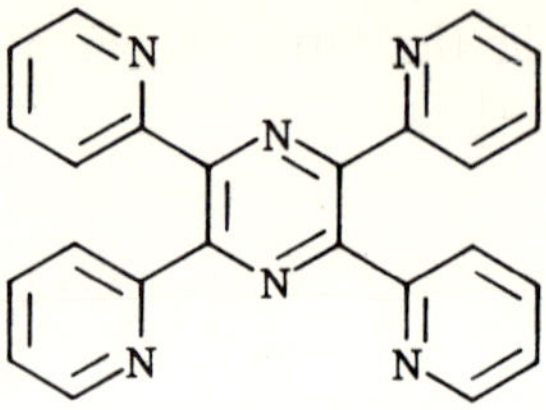

(XII)

It coordinates through three nitrogen atoms only. Coordination through all six nitrogen atoms to yield polymeric chelates does not occur for steric reasons. When the nitrogen atoms of the two pyridine rings which are attached to the pyrazine ring in the 2 and 6 positions, together with the nitrogen atom in the 1 position of the pyrazine ring, coordinate to a metal atom these two pyridine rings become coplanar with the pyrazine ring. It is then impossible for the pyridine rings attached to the 3 and 5 positions of the pyrazine ring to be simultaneously coplanar with it and to function as donors to a second metal atom. Coplanarity of these three nitrogen atoms would be essential for tridentate function. The failure of (XII) to coordinate through all six donor atoms may also be related to the reluctance of pyrazine to form bis-quaternary ammonium salts. It has recently been reported, however, that bridged complexes are formed from simple pyrazines, coordination being through both nitrogen atoms with copper(II), nickel(II), and cobalt(II) salts (Lever *et al.*, 1961).

2. *Other Planar Tridentates*

There have been prepared a large number of tridentates in which structural features within the molecules force the metal atom and the donor atoms into a planar arrangement on coordination. Although the planar arrangement in most instances has not been proven by X-ray analytical studies, models show that such an arrangement of the molecules is generally the least strained. In general, such tridentates are Schiff's base type compounds obtained by the condensation of a suitable aldehyde with a bidentate molecule containing one primary amino group. Pfeiffer and co-workers (1937) described copper(II) and nickel(II) chelates of salicylideneamino-*o*-hydroxybenzene (XIII) in which the metal atoms are apparently tricoordi-

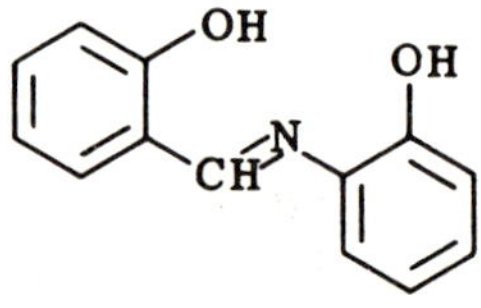

(XIII)

nated. Muto and co-workers (Kishita *et al.*, 1958; Muto, 1955, 1960) reported apparently tri-coordinated copper(II) chelates of a number of related tridentates, e.g., (XIV) and (XV). The copper(II) chelate of (XIV)

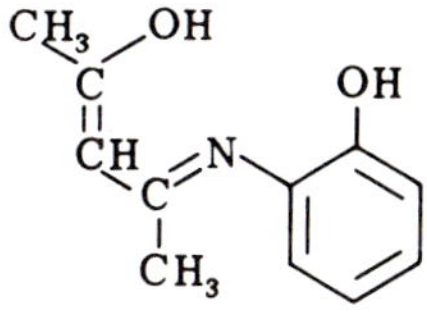

(XIV)

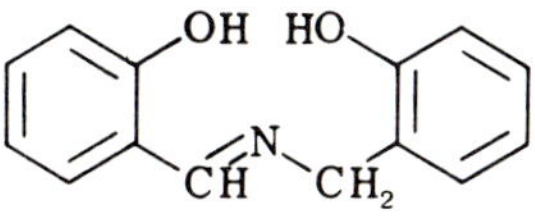

(XV)

has been shown (Barclay *et al.*, 1961a) to be dimeric. The molecule contains two essentially square coordinated copper atoms linked by oxygen bridges. The distortion from a square arrangement of the donor atoms about the copper atoms arises from the nonplanarity of the ligand molecules. The plane containing the aromatic component of the ligand molecule is at an angle of approximately 16° to the plane containing the acetylacetoneamine component. The rotation is about the bond from the nitrogen atom to the benzene ring. This nonplanarity of the ligand molecule probably arises from steric interaction between the methyl group of the acetylacetone system nearest the nitrogen atom and the hydrogen atom in the 6 position of the benzene ring. A model of the molecule shows that this interaction is considerable if the ligand remains planar (Hoskins, 1962).

Condensation of pyridine-2-aldehyde with *o*-aminophenol yields a tridentate (XVI) structurally similar to (XIII) (Muto, 1958). A structural

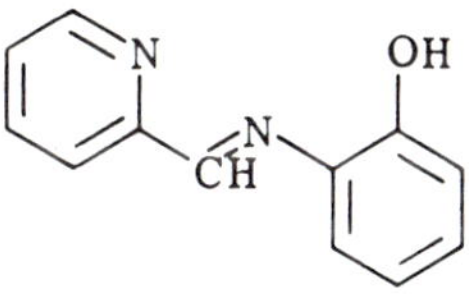

(XVI)

arrangement somewhat similar to that in (XVI) also occurs in (XVII)

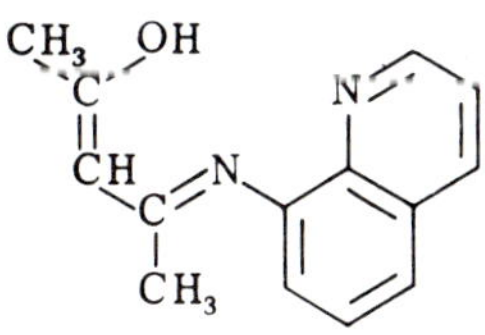

(XVII)

obtained from acetylacetone and 8-aminoquinoline (Lions and Martin, 1958). Mukherjee and Ray (1955) have prepared tridentates with general formula (XVIII), obtained by condensing salicylaldehyde with α-amino acids.

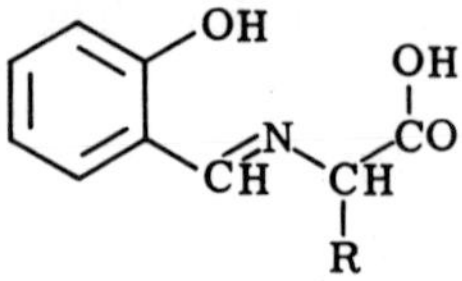

(XVIII)

Pfeiffer and Saure (1941) have considered the possible stereoisomers of octahedral metal chelates of tridentates closely related to (XIII), namely the azomethine dye (XIX) and the azo dye (XX).

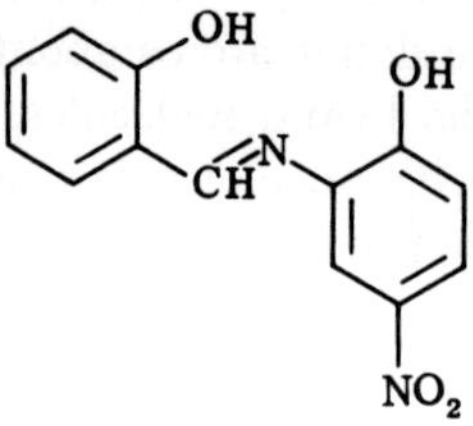

(XIX)

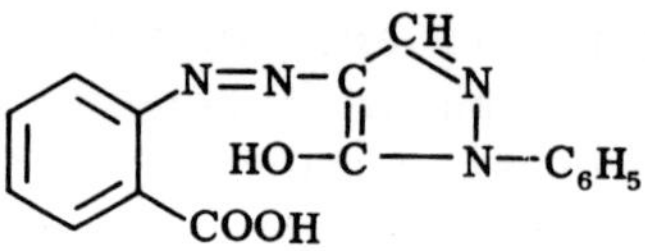

(XX)

If one assumes possible nonplanarity of the chelating agent, coordination of two molecules of unsymmetrical tridentates such as these could give rise to six isomers (three enantiomorphous pairs) in which the central donor atoms of each tridentate residue are in *cis* positions and five isomers (two enantiomorphous pairs and one inactive form) in which the central donor atoms are in *trans* positions (Martell and Calvin, 1952). Evidence was obtained for only one enantiomorphous pair, believed to have the *trans* configurations (XXIa) and (XXIb), since these are the only ones for which the bonds about the donor nitrogen atoms are planar.

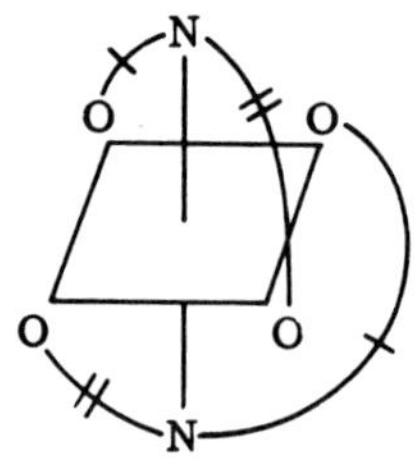

(XXI a)

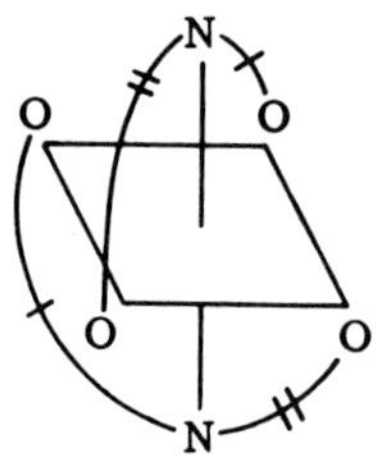

(XXI b)

Sacconi (1954a, 1954b) has prepared a number of acylhydrazones of salicylaldehyde and *o*-aminobenzaldehyde. (XXII) and (XXIII) function,

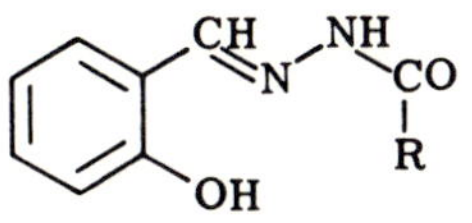

(XXII)

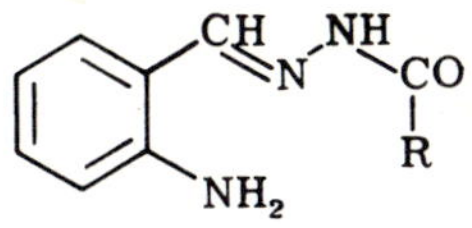

(XXIII)

in the enolic form, as tridentates. The high degree of conjugation in this form precludes a nonplanar arrangement.

B. Nonplanar Tridentates

There are many known substances whose molecules possess three donor atoms which can coordinate to a metal atom at the vertices of one face of an octahedron. Few are known, however, which can function *only* in this way. Most of the tridentates which are able to coordinate vicinally can also do so equatorially without great strain.

Vicinal attachment of a symmetrical tridentate about an octahedral metal atom can give rise to two geometrical isomers. In (XXIV) and (XXV)

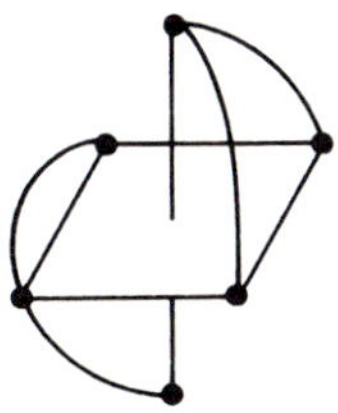

(XXIV)

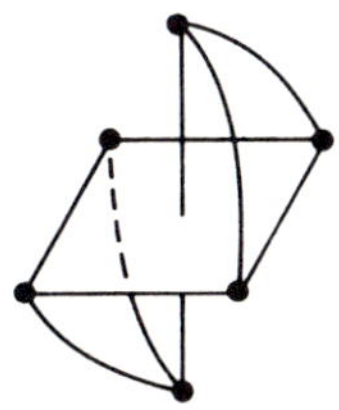

(XXV)

the central donor atoms are in *cis* and *trans* positions respectively. Complexes with configuration (XXIV) should be resolvable.

In metal chelates in which α,β,γ-triaminopropane (XXVI) functions as

$NH_2CH_2CHCH_2NH_2$
|
NH_2

(XXVI)

$NH_2CH_2CH_2NHCH_2CH_2NH_2$

(XXVII)

a tridentate the three donor atoms and the metal atom cannot be coplanar. Mann and Pope (1926b) assigned to the bis(triaminopropane)cobalt(III) ion the symmetrical configuration (XXV) but concluded that the complex also exists partly in the isomeric form (XXIV) since a trace of optical activity was observed in the complex chloride prepared from the *d*-camphor-β-sulfonate. It was noted that the equatorial attachment of (XXVI) would also give an asymmetrical octahedral structure due to the lateral displacement of the carbon atom attached to the central amino group. Such an arrangement would be most unlikely as it would involve a chain of only three carbon atoms spanning *trans* octahedral positions. (XXVI) also functions as a bidentate molecule (Mann, 1926, 1928) coordinating through the α- and β-amino nitrogen atoms (Mann, 1927). Mann considered that in the bis(triaminopropane)platinum(II) complex the platinum atom has coordination number six (1928, 1929). Although a coordination number greater than four is known for platinum(II) (Harris and Livingstone, 1962), the evidence adduced by Mann for coordination number six in this complex is not conclusive. The attachment of each molecule of the base through two nitrogen atoms only and the transfer of a proton from a coordinated amino group to the uncoordinated should yield a complex with properties similar to those observed by Mann.

Diethylenetriamine (XXVII) (Mann, 1934a) can coordinate either as a planar or a vicinal tridentate. It would seem that in no instance has any evidence been obtained for the existence of isomerism in octahedral complexes of (XXVII), nor has the configuration of complexes isolated been determined (Mann, 1934a). Demonstration of optical activity in octahedral complexes would indicate configuration (XXIV) but failure to resolve such complexes would not be indicative of any one configuration. The related base β,β'-diaminodiethylsulfide also is able to function either as a planar or vicinal tridentate. No evidence for isomerism due to the two possibilities has been obtained (Mann, 1930). The pyramidal disposition of the bonds from the sulfur atom and the relatively large size of this atom would tend to favor vicinal coordination about an octahedral metal atom. In the absence of detailed structural studies of complexes of this base, the configuration of the donor atoms about the metal atom remains uncertain.

The triarsine, bis(3-dimethylarsinylpropyl)methylarsine (triars)

$$\begin{array}{c}(CH_3)_2\text{—As—}(CH_2)_3\text{—As—}(CH_2)_3\text{—As}(CH_3)_2\\ |\\ CH_3\end{array}$$

(XXVIII)

(XXVIII) readily coordinates as a tridentate (Barclay *et al.*, 1961b). Elucidation of the structure of the five-coordinate nickel(II) complex $[Ni(triars)Br_2]$ has shown that in this complex the three arsenic atoms and the nickel atom lie in a plane (Mair *et al.*, 1960). The trimethylene groups linking the arsenic atoms in (XXVIII) allow considerable flexibility in the molecule so that the donor atoms can coordinate equatorially. Nevertheless, a twist in the chain must occur at the central arsenic atom and vicinal coordination should be possible. The formation of the copper(I) complex $[Cu(triars)I]$ (Barclay and Nyholm, 1953), in which the four donor atoms are probably arranged tetrahedrally about the copper atom, indicates that the tridentate could coordinate vicinally about an octahedral metal atom.

A number of tritertiary phosphines (XXIX, XXX, XXXI) have recently been prepared (Chatt and Watson, 1961; Hart 1960). All of these tridentates would coordinate octahedrally most readily from vicinal positions. The steric requirements of (XXXI) demand that it function in this way. Each of these phosphines was shown to displace three carbonyl groups from group VI metal carbonyls. Dipole moment and infrared data support the *cis* arrangement of the carbonyl groups in the complexes [M(triphos)$(CO)_3$], indicating that the phosphines occupy one face of the octahedron.

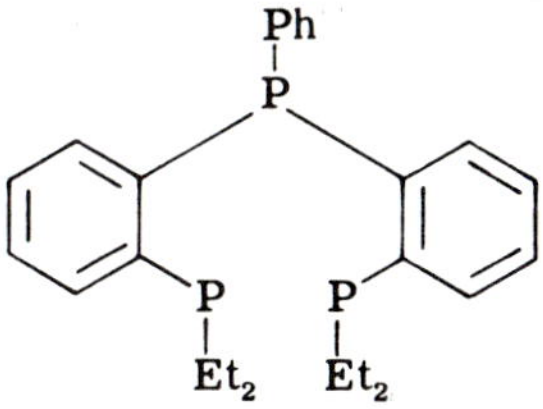

(XXIX)

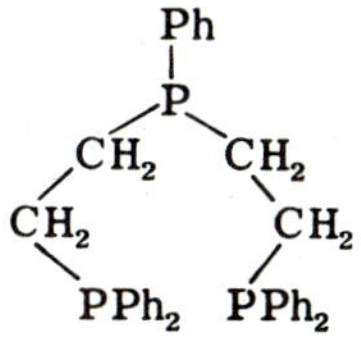

(XXX)

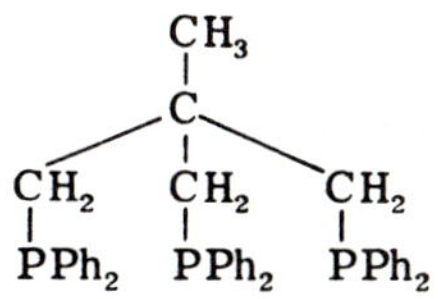

(XXXI)

Related phosphine-arsine (XXXII) and triarsine (XXXIII) tri-

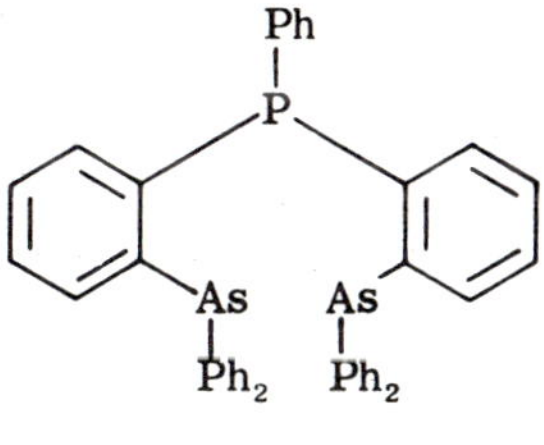

(XXXII)

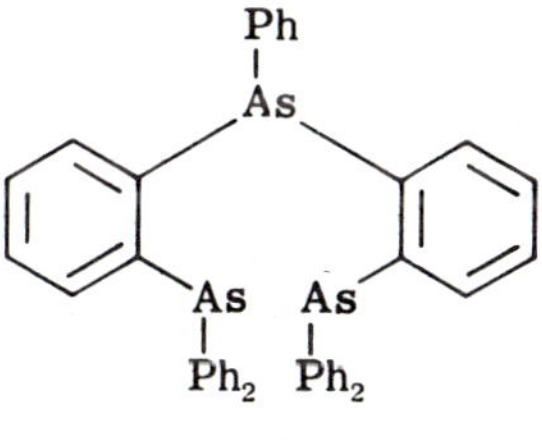

(XXXIII)

dentates are also known (Howell *et al.*, 1961). These should coordinate vicinally to octahedral metal atoms.

III. Quadridentate Chelating Agents

The first substance recognized as a quadridentate chelating agent would seem to be ethylene-bisthioglycolic acid which was shown (Ramberg and Tiberg, 1914) to yield an inner complex (XXXIV) with platinum(II). Although bis(acetylacetone)ethylenediimine and its derived copper(II)

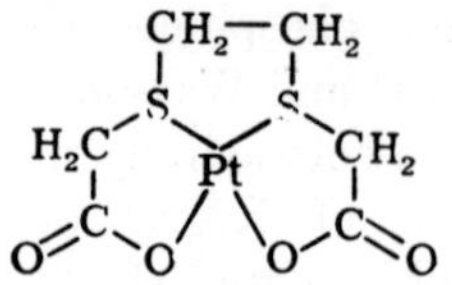

(XXXIV)

chelate were first described by Combes (1889), Morgan and Main-Smith (1925) were the first to study in detail its function as a quadridentate.

Quadridentates may be classified into three principal categories. Those in which the donor atoms are members of:

(*a*) a closed ring system;
(*b*) a continuous chain;
(*c*) a branched chain.

A. Closed Ring Quadridentates

Ring quadridentates are best illustrated by the well known and important porphyrins and phthalocyanines (Barrett *et al.*, 1936; Byrne *et al.*, 1934; Dent and Linstead, 1934; Dent *et al.*, 1934; Fischer and Friedrich, 1936; Fischer and Gleim, 1935; Fischer *et al.*, 1935; Helberger, 1937). These highly resonating molecules are necessarily planar, as are their coordination derivatives with four-covalent metal atoms (Linstead and Robertson, 1936; Robertson, 1936; Robertson and Woodward, 1940; Crute, 1959). It is noteworthy that even normally tetrahedral ions such as beryllium(II) yield phthalocyanine derivatives which are planar. Such complexes provide good examples of the phenomenon of forced stereochemistry, often associated with the coordination of multidentate chelating agents.

The reaction of tris(ethylenediamine)nickel(II) perchlorate with acetone has recently been shown to yield, among other products, a nickel(II) complex perchlorate in which the nickel atom is bound to a curious cyclic quadridentate which contains two azomethine and two secondary amino nitrogen atoms (Curtis and House, 1961). This extremely interesting reaction results in the conversion of a paramagnetic, octahedral nickel(II) complex to one which is diamagnetic and square-planar.

A number of macrocyclic tetramines have recently been reported (Stetter and Mayer, 1961). These would seem to be capable of functioning as quadridentates and their reaction with metal ions should prove interesting.

B. Open Chain Quadridentates

Most of the known quadridentates are molecules in which the four donor atoms form part of a continuous chain. These may be divided into three stereochemical types:

(*a*) those which, because of structural constraints within the molecules, can coordinate through all donor atoms to a single metal atom only in such a way that the four donor atoms lie in a plane (planar quadridentates);
(*b*) those whose molecules are so constructed that the four donor atoms cannot lie in a plane but may be orientated tetrahedrally about a metal atom (tetrahedral quadridentates);
(*c*) those whose molecules are flexible and can be orientated so that the four donor atoms can coordinate from either a planar or a nonplanar arrangement (facultative quadridentates).

1. *Planar Quadridentates*

Molecules which have four donor atoms in a plane can coordinate through all four to a square planar, four-covalent metal atom or, equatorially, to an octahedral, six-covalent metal atom. Quadridentates of this type are readily obtained by condensing a suitable aldehyde such as pyridine-2-aldehyde or salicylaldehyde with a bidentate primary amino compound such as ethylenediamine or *o*-phenylenediamine. The azomethine groups in these Schiff's bases force the donor atoms into a plane. Quadridentates derived from higher homologs of ethylenediamine may or may not function in a planar manner. Models show, for example, that substitution of $=N-(CH_2)_n-N=$ ($n > 3$) for the $=N-(CH_2)_2-N=$ sequence in quadridentates derived from ethylenediamine forces apart the carbon atoms of the methylene groups attached to the azomethine nitrogen atoms and also forces apart these nitrogen atoms. A planar arrangement of donor atoms, in this instance, is strained because of atom crowding in the central chelate loop. If n is sufficiently great (e.g., $n = 6$) the molecule becomes sufficiently flexible to present its four donor atoms at the apices of a tetrahedron circumscribing a central metal atom, and this is the preferred arrangement.

When the donor atoms are not the ultimate atoms of a chain, the terminal groups may interfere with each other spatially to such an extent that it becomes impossible to attach the whole molecule as a planar quadridentate to a metal atom. Although some distortion of the bond angles in the chelate rings of complexes derived from 1,2-bis(α-pyridylmethyleneamino)-ethane (XXXV, R = H) is necessary to bring all four donor atoms to the corners of a square, models show that it can function as a quadridentate. The corresponding Schiff's base derived from 6-methylpyridine-2-aldehyde, 1,2-bis(6′-methyl-2′-pyridylmethyleneamino)ethane (XXXV, R = CH_3),

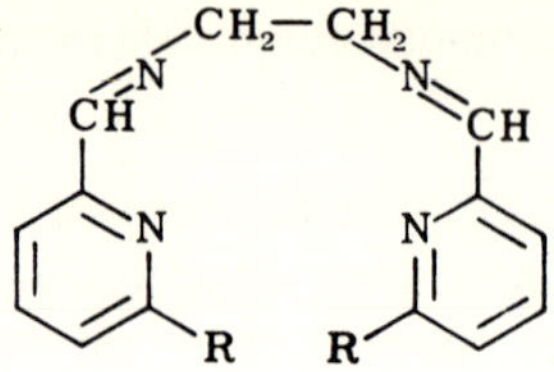

(XXXV)

cannot so function. Coordination of this with copper(II) salts results in either the formation of bridged complexes or the partial hydrolysis of the Schiff's base, leaving complexes of (XXXVI) (Goodwin and Lions, 1960).

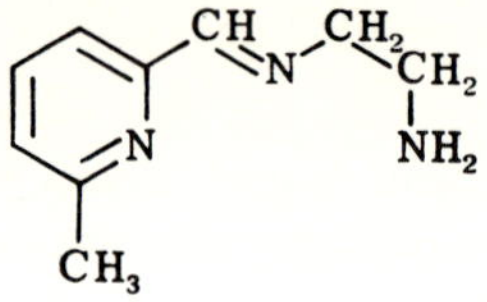

(XXXVI)

When R = CH_3, (XXXV) cannot function as a simple planar quadridentate because the methyl groups adjacent to the pyridine nitrogen atoms prevent these atoms from coming close enough to coordinate to the one metal atom.

Bis(acetylacetone)ethylenediimine (XXXVII), because of the presence

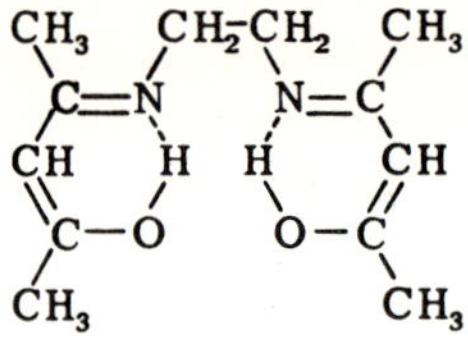

(XXXVII a)

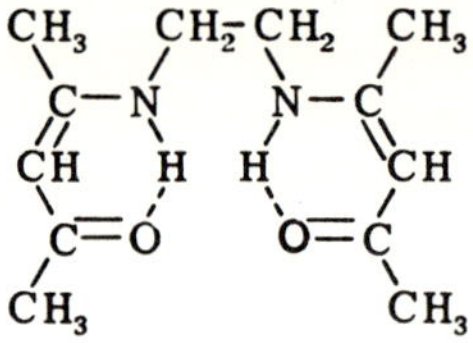

(XXXVII b)

of the azomethine groups, can function strainlessly only as a planar quadridentate. There is evidence, discussed later, to indicate that it can also coordinate in a nonplanar arrangement.

Considerable spectral evidence has been adduced to show that molecules such as bis(acetylacetone)ethylenediimine contain six-membered rings involving hydrogen bonds (Ueno and Martell, 1955, 1957; Martell *et al.*, 1958). The hydrogen-bonded structure can be formulated as an enolimine (XXXVIIa) or as a tautomeric ketamine (XXXVIIb). Proton magnetic resonance studies have shown that the ketamine tautomer is present in solution to the extent of more than 80% (Dudek and Holm, 1961). The existence of structures corresponding to (XXXVIIa) and (XXXVIIb) in metal chelates derived from amine-β-diketone condensation products is difficult to establish from infrared spectral data (Holtzclaw *et al.*, 1958). The bond lengths reported for the dimeric copper(II) complex of the tridentate (XIV) derived from acetylacetone and *o*-hydroxyaniline show that the ketamine form of the ligand is involved in coordination in this instance (Barclay *et al.*, 1961a). A recent X-ray analysis of the structure of the copper(II) chelate of (XXXVII) shows that the acetylacetone ring systems are considerably conjugated, with a possible preponderance of the structure derived from the ketamine form of the ligand. The chelate molecule, expected to be planar except for some puckering in the central chelate ring, is slightly concave towards the center (Hall *et al.*, 1962).

Perhaps the best known of the planar quadridentates is 1,2-bis(salicylideneamino)ethane (XXXVIII) (Dubsky and Sokol, 1932). The essentially

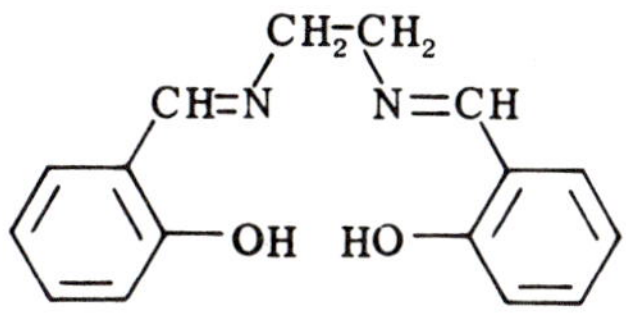

(XXXVIII)

planar nature of this molecule in derived coordination compounds has been confirmed in a number of them. Recent X-ray analysis of the copper complex of (XXXVIII) has shown it to be dimeric (Hall and Waters, 1960). Although each unit is not completely planar, the four donor atoms and the copper atom are in a plane. In the monohydrate of the zinc(II) complex of (XXXVIII), the zinc atom is five-covalent. The two nitrogen and two oxygen atoms of the ligand are coplanar but the zinc atom is 0.34 Å above this plane in the direction of the coordinated water molecule (Hall and Moore, 1960). The monohydrate of the copper(II) complex of 1,2-bis-(salicylideneamino)propane has been shown (Llewellyn and Waters, 1960) to have a similar structure, the copper atom lying 0.2 Å above the plane containing the donor atoms of the quadridentate.

The action of the cobalt(II) complex of (XXXVIII) as an oxygen carrier (Tsumaki, 1938) has been of especial interest and has led to a study of many related compounds (Diehl and Henn, 1949; Hughes *et al.*, 1946).

Condensation of salicylaldehyde with *o*-phenylenediamine yields a quadridentate (XXXIX) structurally related to, but more rigid than, (XXXVIII). It yields planar chelates with four-covalent metal atoms. (XXXIX) as well as a number of closely related quadridentates have been prepared and studied by Pfeiffer and co-workers (Pfeiffer and Glaser, 1939; Pfeiffer and Pfitzner, 1936; Pfeiffer *et al.*, 1933, 1937, 1942).

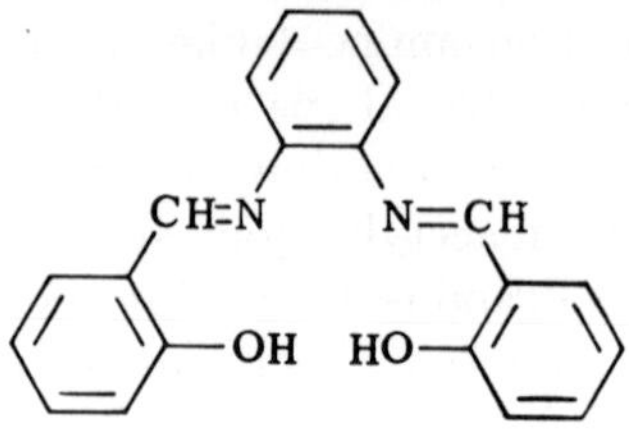

(XXXIX)

Quadridentates derived from pyridine-2-aldehyde and primary diamines have received little attention. Dwyer, Lions, and co-workers (1953b) drew attention to the function of (XXXV) (R = H) as a quadridentate and it was later briefly studied by Busch and Bailar (1956b).

Complexes derived from quadridentates such as (XXXV) differ from those derived from (XXXVIII) in that they are cationic and the chelate rings containing the terminal donor atoms are five-membered instead of six. Six-membered chelate rings containing a heterocyclic nitrogen atom can be obtained by condensing quinoline-8-aldehyde with ethylenediamine to yield 1,2-bis(8′-quinolylmethyleneamino)ethane (XL) which functions

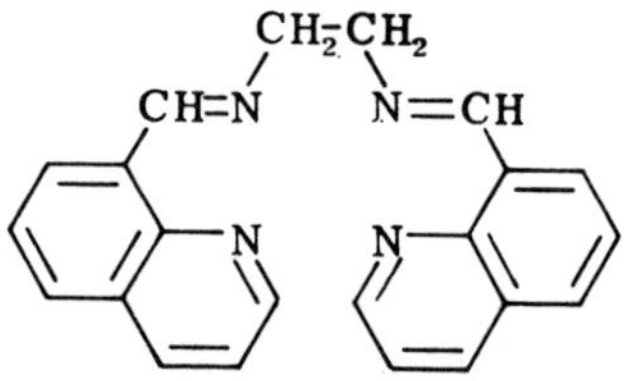

(XL)

as a quadridentate (Goodwin and Lions, 1960).

Pfeiffer and co-workers (1937) condensed *o*-aminobenzaldehyde with ethylenediamine and *o*-phenylenediamine to obtain the quadridentates (XLI) and (XLII). These are closely related structurally to (XXXVIII)

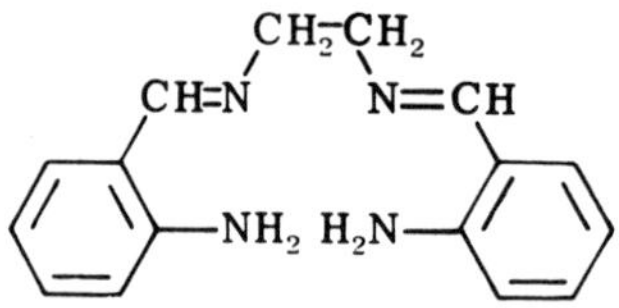

(XLI)

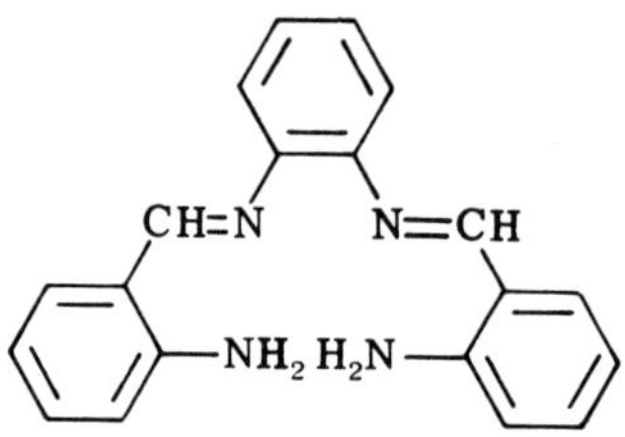

(XLII)

and (XXXIX) and coordinate similarly with loss of two protons yielding inner complexes with copper(II) and nickel(II).

Condensation of a dicarbonyl compound with a suitable molecule containing a primary amino group has been shown to yield quadridentates in a number of cases. The thiazones of Bähr (Bähr and Hess, 1952; Bähr and Schleitzer, 1955a,b; Bähr *et al.*, 1953) yield stable inner complexes (XLIII) with copper(II), nickel(II), palladium(II), and platinum(II).

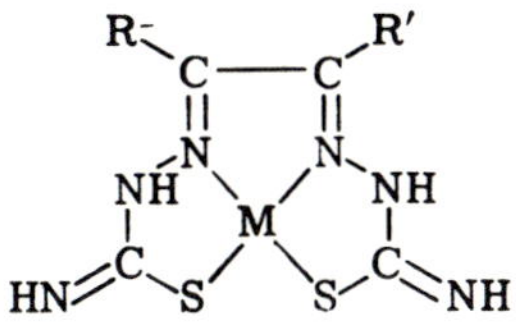

(XLIII)

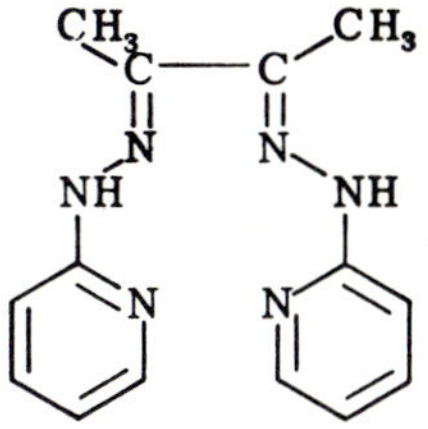

(XLIV)

The bis(α-pyridylhydrazone) of diacetyl (XLIV) (Lions and Martin, 1958) and the product of condensation of glyoxal and *o*-aminophenol (XLV) (Bayer, 1957) are similarly constituted and function as quadri-

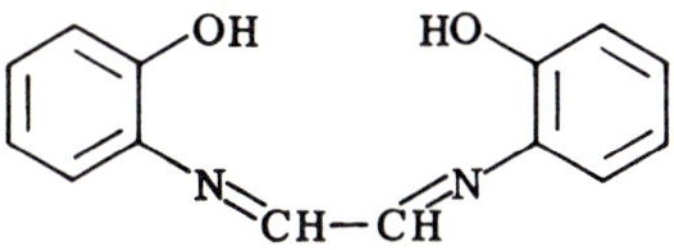

(XLV)

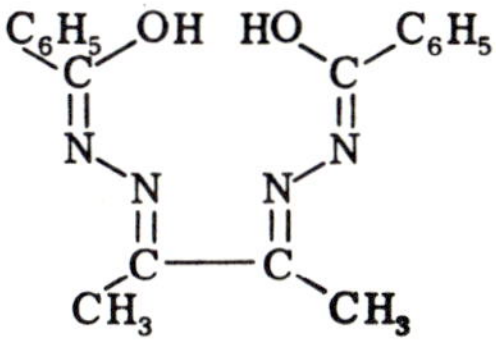

(XLVI)

dentates. Sacconi (1954b) has shown that the bis(benzoylhydrazone) of diacetyl (XLVI) loses two protons and coordinates through two nitrogen and two oxygen atoms with bivalent metal ions.

The interesting ethylene-bis(biguanide) complexes have been formulated as structure (XLVII) in which the organic moiety is functioning as a

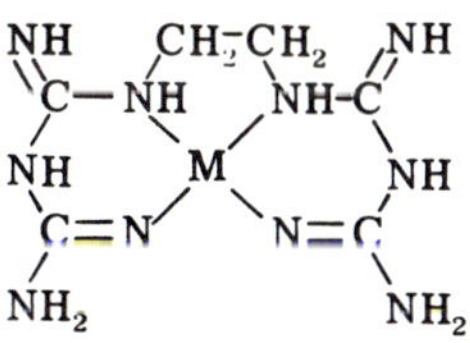

(XLVII)

planar quadridentate (Ray and Ghosh, 1943).

Morgan and Burstall (1938) prepared 2,2′,2″,2‴-tetrapyridine (XLVIII) and studied its capacity to coordinate with several metal ions.

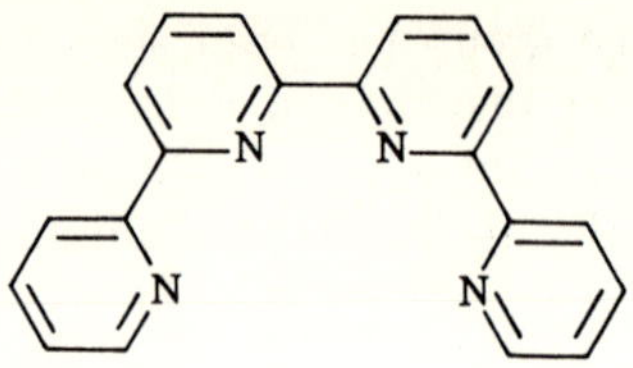

(XLVIII)

In the compounds isolated it was believed that (XLVIII) was coordinating through all four nitrogen atoms. A model of the molecule shows that if (XLVIII) does function as a quadridentate (necessarily planar) its bond angles would need to be considerably distorted in order to form a square planar arrangement about a metal atom. There is already strain in complexes derived from the tridentate terpyridine. In tetrapyridine the fourth nitrogen atom is quite far removed from the position required for square coordination (Brandt *et al.*, 1954).

2. *Tetrahedral Quadridentates*

In 2,2′-bis(salicylideneamino)-6,6′-dimethyldiphenyl (XLIX) (Lions

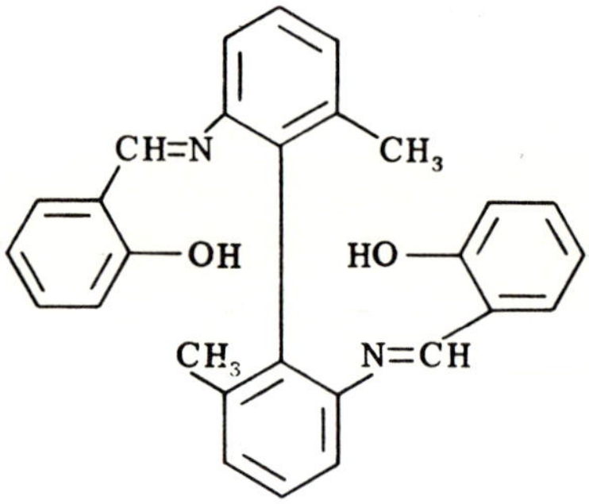

(XLIX)

and Martin, 1957a), the two oxygen and two nitrogen atoms can be brought strainlessly into a tetrahedral arrangement from which they can coordinate to a single metal atom of appropriate size. Interaction of the methyl groups attached to the benzene rings of the diphenyl system in (XLIX) prevents these rings being coplanar and thus it is sterically impossible for the four donor atoms to lie in a plane. The formation of a monomeric copper(II) complex of (XLIX) was taken to indicate that the bonds from the copper atom are tetrahedrally directed in this complex. The base 2,2′-bis(8″-quinolylmethyleneamino)biphenyl (L) similarly cannot function as a planar quadridentate (Goodwin and Lions, 1960).

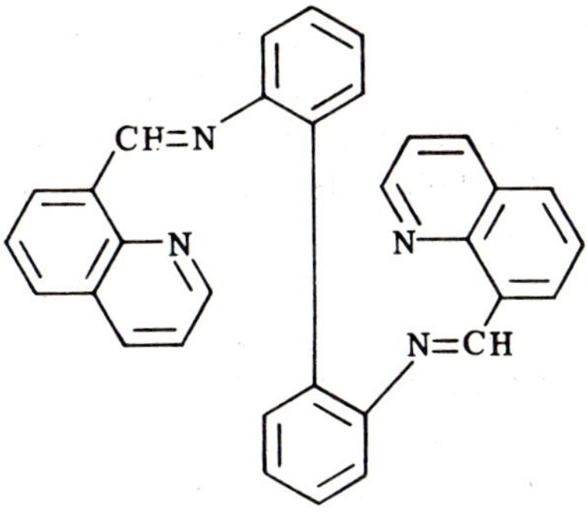

(L)

3. *Facultative Quadridentates*

The molecules of triethylenetetramine (trien) (LI) are flexible and are

$$NH_2CH_2CH_2NHCH_2CH_2NHCH_2CH_2NH_2$$

(LI)

able to coördinate to a metal atom in either a planar or a nonplanar arrangement. In the platinum(II) and palladium(II) complexes [M(trien)] [MCl_4] (Jonassen and Cull, 1949) the four donor atoms are doubtless in a plane. In the zinc complex ion $[Zn(trien)]^{++}$ the four donor atoms are probably arranged tetrahedrally about the zinc atom (Martell, 1955; Schwarzenbach, 1950).

Coordination of a quadridentate such as (LI) about a six-covalent metal atom can give three geometrical isomers (LIIa–c), in which the remaining

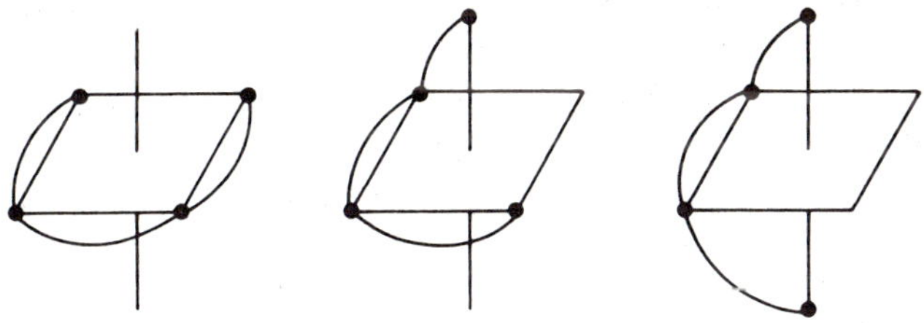

(LIIa) (LIIb) (LIIc)

octahedral positions are in *trans* (LIIa) or *cis* (LIIb and LIIc) positions. The *trans* configuration has two planes of symmetry so that complexes with this configuration would not be resolvable. Neither of the *cis* configurations has a plane, a center, or an alternating axis of symmetry and both would exist in enantiomorphous forms. Basolo (1948) prepared [Co(trien)-Cl_2]Cl and assigned to this complex a *cis* configuration on the basis of the similarity of its absorption spectrum to that of *cis*-[$Co(en)_2Cl_2$]Cl. This was confirmed by Das Sarma and Bailar (1955b) who obtained the complex in an optically active form. A corresponding *trans* complex could not be prepared. $[Co(trien)Br_2]Br$ has recently been prepared in both *cis* and *trans* forms (Selbin and Bailar, 1960). The larger size of the bromide ion apparently facilitates the formation of the *trans* complex in this instance. All three geometrical forms of $[Co(trien)(NO_2)_2]^+$ have been obtained and both *cis* isomers have been resolved (cf. Chapter 5, Section II).

1,6-Bis(α-pyridyl)-2,4-diazahexane (LIII) (Goodwin and Lions, 1960;

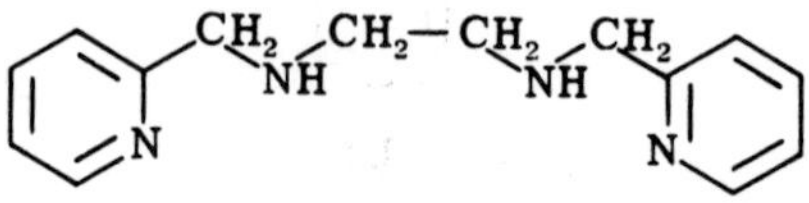

(LIII)

La Coste, 1957) and the somewhat analogous ligand, 1,8-bis(α-pyridyl)-3,6-dithiaoctane (LIV) (Goodwin and Lions, 1960) can also present their

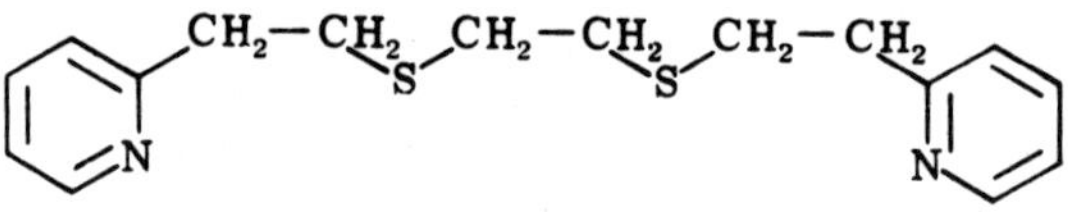

(LIV)

donor atoms to a metal atom from either a planar or a nonplanar arrangement. A planar arrangement of the donor atoms in (LIV) may be strained owing to the larger size of the sulfur atoms.

Morgan and Main-Smith (1925) claimed to have isolated three geometrical isomers of the complex ion $[Co(acacen)(NH_3)_2]^+$ [H_2(acacen)is (XXXVII)] corresponding to (LIIa–c). In addition they separated the two *cis* isomers into their enantiomorphs. It would be expected, and indeed a model shows, that (XXXVII) would function most naturally as a planar quadridentate so that octahedral complexes derived from it would have the *trans* configuration (LIIa). The azomethine groups in (XXXVII) tend to force the molecule into a plane and any nonplanar arrangement of its donor atoms in complexes must lead to severe steric strain in the ligand. Morgan and Main-Smith's inability to repeat the separation and to obtain the complex in optically active forms indicates that the *trans* form (inactive) is the most stable. It is noteworthy that nonplanarity of the donor atoms in (XXXVII) is permissible if only the ketamine form (XXXVIIb) of the ligand is involved in coordination.

C. Branched Chain Quadridentates

Branched chain quadridentates may be of three structural types:

(i)

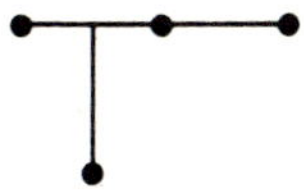

(ii)

(iii)

In type (i) the chain is bifurcated at a donor atom, in (ii) the chain is bifurcated at an atom other than a donor atom, and in (iii) a bridging group links two bidentate residues in positions between the donor atoms. Those with structure (i) are the most common and best known of the branched chain quadridentates. β,β',β''-Triaminotriethylamine (tren) (LV) cannot

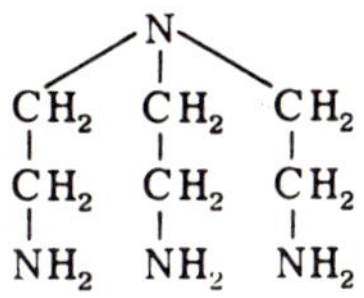

(LV)

be orientated so that the four donor atoms lie at the corners of a square but can function as a tetrahedral quadridentate (Mann and Pope, 1925, 1926a). Its platinum(II) and palladium(II) complexes present interesting stereochemical problems. The complex iodides have been formulated as $[M(tren)]I_2$ with the bonds from the metal atom presumably directed tetrahedrally. The complexes may not, however, be monomeric. Their pale color indicates that the iodine atoms are probably not coordinated to the metal atom.

In addition to a bridged octahedral complex, the anhydrous salt $Ni(tren)SO_4$ is obtained from (LV) and nickel sulfate. Molecular weight and conductivity data support the assignment of a monomeric structure to this complex in solution but do not distinguish between a tetrahedral four-covalent and an octahedral six-covalent complex with two aquo groups coordinated in *cis* positions. The thiocyanato complex $Ni(tren)(SCN)_2$ has recently been shown (Rasmussen, 1959) to have a distorted octahedral structure in which the two thiocyanato groups are coordinated in *cis* positions. Cobalt(II) gives spin-free complexes, $Co(tren)X_2$ ($X = SCN$ or I) (Barclay and Barnard, 1958), but their stereochemistry has not been determined.

Ammonia triacetic acid (LVI) is structurally related to (LV) and four-covalent complexes derived from it would be expected to be tetrahedral. In certain of its complexes (LVI) functions as a tridentate, one carboxyl

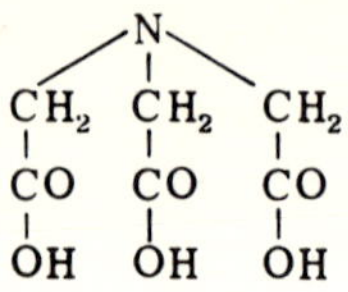

(LVI)

group remaining uncoordinated (Schwarzenbach and Biedermann, 1948; Schwarzenbach *et al.*, 1945).

Branched chain quadridentates of type (i) with phosphorus and arsenic donor atoms have been described recently. Tris(3-dimethylarsinylpropyl) arsine (LVII) (Barclay and Barnard, 1961) cannot present its four donor

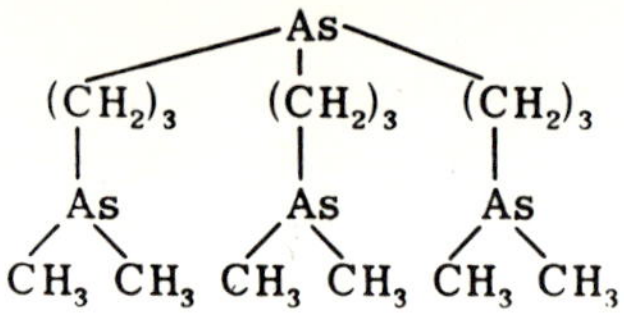

(LVII)

atoms at the corners of a square but can coordinate to either a tetrahedral metal atom or, nonequatorially, to an octahedral metal atom. Tris(*o*-diphenylarsinophenyl)arsine (LVIII) (QAS) and tris(*o*-diphenylarsinophenyl)phosphine (LIX) similarly cannot coordinate as planar quadridentates (Howell *et al.*, 1961). Coordination of (LVIII) to platinum(II) salts gives compounds with the formula $Pt(QAS)X_2$ (X = Cl, Br, I). On the basis of spectroscopic and chemical evidence, these were assigned a

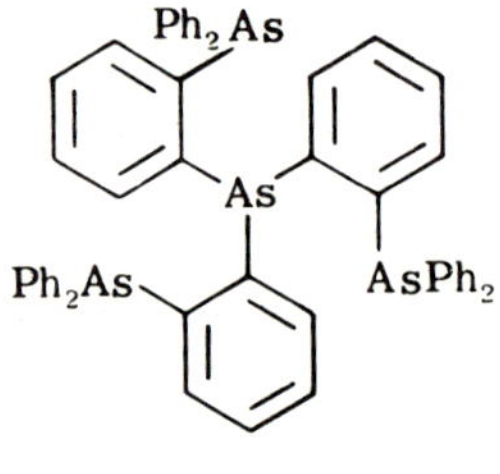

(LVIII)

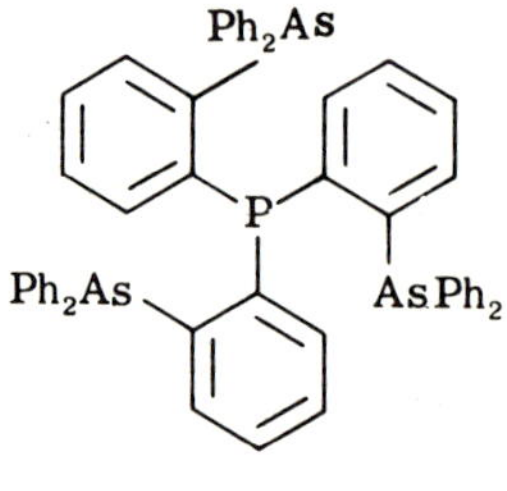

(LIX)

structure involving five-covalent platinum(II) (Brewster *et al.*, 1961). This was confirmed by X-ray crystal analysis of $[PtI(QAS)]BPh_4$ in which the complex cation was shown (Mair *et al.*, 1961) to have a trigonal bipyramidal structure (LX). An iodine atom is located at one apex of the bipyramid

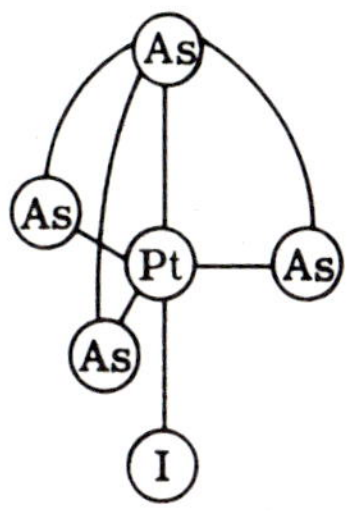

(LX)

and what is presumed to be the central arsenic atom is located at the other. The interatomic distances are approximately those expected for a regular trigonal bipyramid. The platinum atom is displaced slightly from the plane of the three equatorial arsenic atoms towards the iodine atom.

Examples of branched chain quadridentates of type (ii) appear to be lacking. Though of no special interest such quadridentates should be capable of synthesis. A tetramine such as (LXI) would be of this type. Such

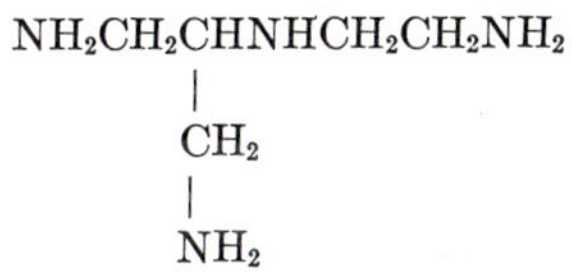

(LXI)

quadridentates would not be expected to yield planar metal chelates.

Branched chain quadridentates of type (iii) are rare. Bis(β-dicarbonyl) compounds of structure (LXII) have been shown (Kluiber and Lewis,

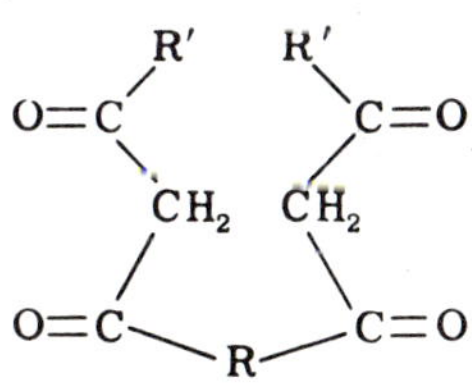

(LXII)

1960) to yield monomeric beryllium chelates if R is a polymethylene chain of six units or more. Compounds of shorter links tend to form dimers which can be polymerized on heating. A number of bis(β-dicarbonyl) compounds with structure (LXIII) have also been prepared (Martin *et al.*, 1959).

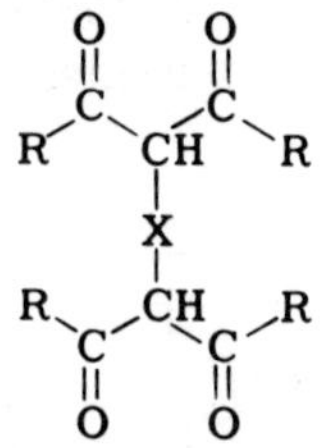

(LXIII)

Presumably these interact with metal ions to give polymeric structures as does tetraacetylethane (Charles, 1960). If the bridging group X in (LXIII) is of sufficient length monomeric chelates probably could be obtained.

Ethylenediaminetetraacetic acid (H_4EDTA) coordinates through only four donor atoms to four-covalent metal atoms. EDTA is potentially a branched chain quadridentate of type (iii). It has been shown, however, that in the four-covalent palladium(II) and platinum(II) complexes two carboxyl groups remain uncoordinated, coordination being through the two tertiary nitrogen atoms and two carboxyl groups as shown in (LXIV)

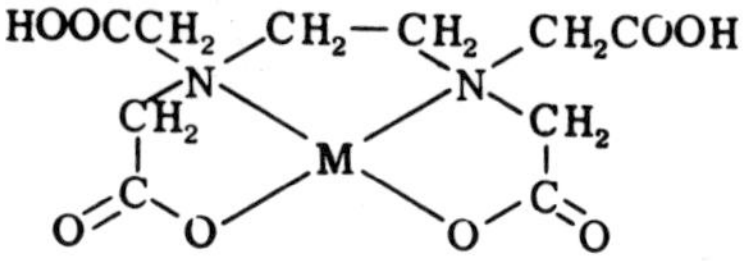

(LXIV)

(Busch and Bailar, 1956a). EDTA is here functioning as a simple linear, open chain quadridentate. It should be possible to devise molecules which are structurally related to EDTA but which contain only four donor atoms and must function as branched chain quadridentates of type (iii). A model shows, for example, that a tetramine such as (LXV) can be orientated

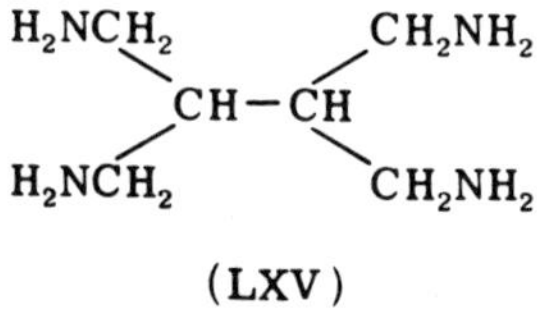

(LXV)

strainlessly in such a way that the four donor atoms can coordinate to a square planar metal atom.

IV. Quinquedentate Chelating Agents

Of the higher multidentate chelating agents, the quinquedentates are undoubtedly the least studied. There are few known examples of molecules which possess five donor atoms suitably situated for coordination to a single metal atom.

Attachment to an octahedral metal atom of a quinquedentate in which the donor atoms are members of a continuous chain of atoms can lead to one or more of the structures (LXVIa–d). In each of these structures a

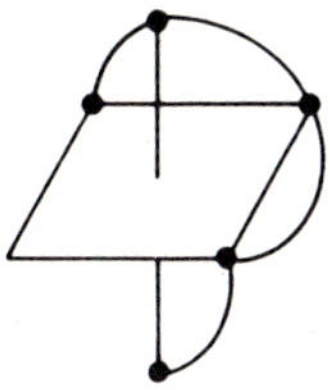

(LXVI a)

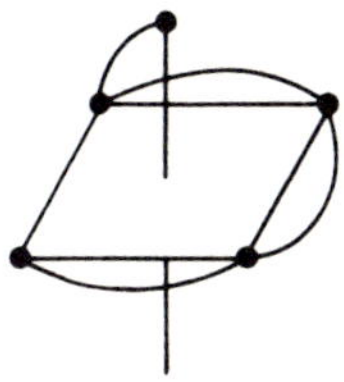

(LXVI b)

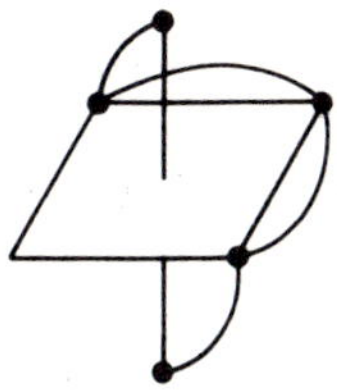

(LXVI c)

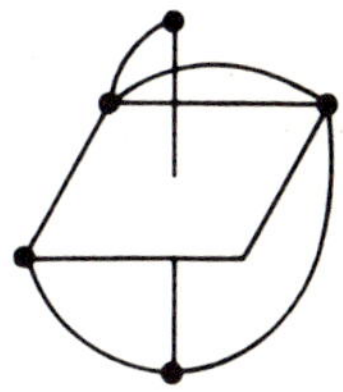

(LXVI d)

twist occurs in the coordinating molecule at at least one donor atom. Unless the molecule is so designed to allow for this twist it cannot, without excessive strain, function as a quinquedentate.

Whereas condensation of a suitable aldehyde such as pyridine-2-aldehyde or salicylaldehyde with a bidentate primary diamine will yield a quadridentate, condensation with a tridentate containing two primary amino groups will not necessarily yield a quinquedentate. Condensation of salicylaldehyde with diaminodiethylsulfide yields the Schiff's base (LXVII).

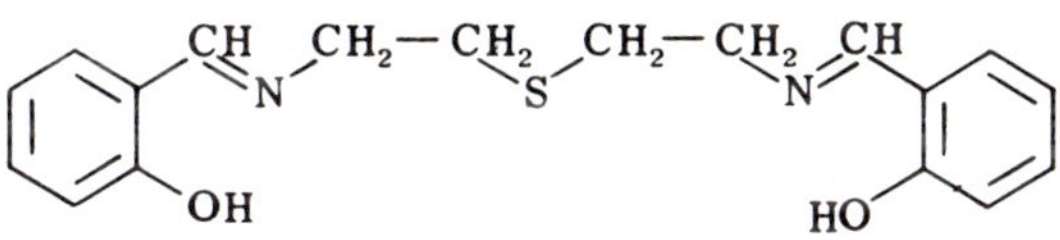

(LXVII)

Since the bonds from the azomethine nitrogen atoms must be planar the O—N—S sequence of donor atoms in (LXVII) must, after coordination, lie in an equatorial plane of the octahedron. It is impossible for the remaining oxygen and nitrogen atoms both to occupy two of the remaining octahedral sites since this second S—N—O sequence must also be planar, resulting in overlap of the terminal oxygen atoms. (LXVII) does not, in fact, function as a quinquedentate but gives rise to complexes of indefinite composition. Lengthening of the chains of carbon atoms between the sulfur and nitrogen atoms gives added flexibility to the molecule and permits nonplanarity of the O—N—S sequence (Gill, 1951). In fact, a planar arrangement of these donor atoms would not in these circumstances be favored because of atom crowding in the chelate loops containing the sulfur and nitrogen atoms.

Condensation of salicylaldehyde with diethylenetriamine gives a base (LXVIII) ($x = 2$) similarly constituted to (LXVII). Similar difficulties

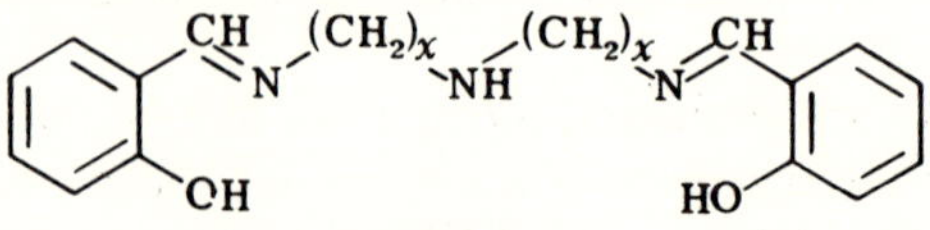

(LXVIII)

would be encountered in its coordination through all five donor atoms to a single metal atom. A cobalt(II) complex of this Schiff's base has been reported (Diehl *et al.*, 1947) but no details of its structure are known. The related base (LXVIII) ($x = 3$) gives a cobalt(II) complex which can be obtained anhydrous and which has been formulated as a five-covalent complex (Bailes and Calvin, 1947; Harle and Calvin, 1946). Although there is no direct evidence to indicate that (LXVIII) ($x = 3$) does function as a quinquedentate, a model of the molecule shows that it could possibly do so.

Reduction of the double bonds at the nitrogen atoms in (LXVII) or (LXVIII) removes the steric constraints about these atoms and the molecules become much more flexible. The base (LXIX) is able to function

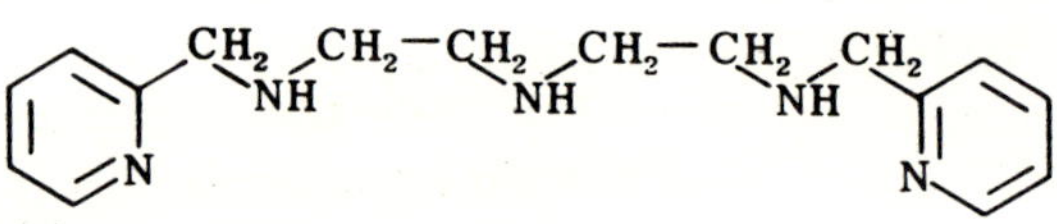

(LXIX)

readily as a quinquedentate (Hahn, 1961) and could adopt any of the configurations (LXVIa–d) with almost equal ease.

Tetraethylenepentamine (tetren) (LXX) can be attached to an octa-

$$NH_2CH_2CH_2NHCH_2CH_2NHCH_2CH_2NHCH_2CH_2NH_2$$

(LXX)

hedral metal atom and coordinate through all five nitrogen atoms. Jonassen and Frey (1953) identified from spectrophotometric studies 1:1 and 2:1 cobalt(II) complexes. Several solid complex salts of cobalt(II) and (LXX) were isolated (Jonassen and Frey, 1957) and it was assumed that in these the base coordinated through all donor atoms. Selbin (1961) has prepared the cobalt(III) complex salt $[Co(tetren)NO_2](ClO_4)_2$. Visible and infrared spectral data support the view that (LXX) functions as a quinquedentate in this instance. Octahedral complexes derived from (LXX) should be capable of adopting (though not necessarily with equal ease) any of the configurations (LXVIa–d), each of which is asymmetric. No evidence for the existence of more than one form has been obtained nor have complexes of (LXX) been resolved.

Several complexes of ethylenediaminetetraacetic acid and cobalt(III) have been reported in which the EDTA is functioning as a quinquedentate (Morris and Busch, 1956; Schwarzenbach, 1949). Infrared spectral studies of these complexes show that one carboxyl group remains uncoordinated (Busch and Bailar, 1953). Several N-substituted *N*,*N'*,*N'*-ethylenediamine-triacetic acids (LXXI) have been prepared (Bruno *et al.*, 1956). These are potentially quinquedentates and should function in this way similarly to EDTA.

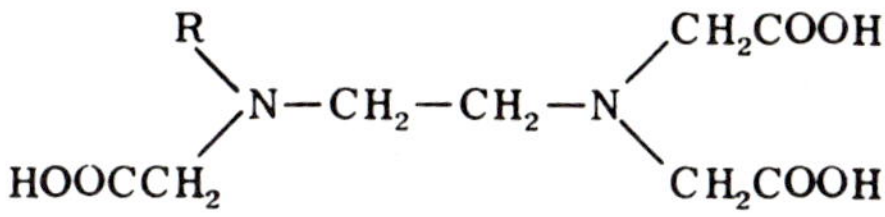

(LXXI)

The quinquedentate function of (LXXI) has not, however, been firmly established.

Examples of quinquedentates are so few that a description of the many possible structural types seems inappropriate. The known quinquedentates have either a linear, open chain structure (e.g., tetraethylenepentamine) (LXXIIa) or a branched chain structure in which the only bifurcation in the chain occurs at a donor atom (e.g., the quinquedentate polyamino-

(LXXII a)

(LXXII b)

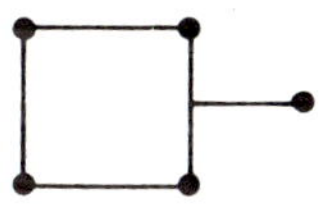

(LXXII c)

carboxylic acids) (LXXIIb).

Perhaps the most remarkable of the quinquedentates is that bound to a cobalt(III) atom in the extremely complex molecule of vitamin B_{12}. A particularly elegant X-ray structural investigation has shown that in this molecule the cobalt atom is at the center of a macrocyclic ring system which, although similar to that found in the natural porphyrins, differs strikingly in that two of the four rings present are joined directly rather than by way of an intermediate carbon atom. To this system is attached a complex side chain which contains a benzimidazole nucleus and furnishes the fifth donor atom (Hodgkin *et al.*, 1955). The quinquedentate present in vitamin B_{12} is represented by structure type (LXXIIc).

V. Sexadentate Chelating Agents

The function of ethylenediaminetetraacetic acid as a sexadentate chelating agent was postulated early in the study of its complexes. Brintzinger and co-workers (1943) prepared a cobalt(III) complex of EDTA which could be obtained anhydrous and in which the six coordination positions of the metal atom were presumably occupied by the donor atoms of the EDTA ion. The sexadentate nature of EDTA in this complex was confirmed by the infrared spectral studies of Busch and Bailar (1953).

The molecules of 1,8-bis(salicylideneamino)-3,6-dithiaoctane (LXXIII), synthesized by Dwyer and Lions (1947), can lose two protons and encompass a metal atom practically strainlessly in such a way that the six donor atoms are in positions favorable for octahedral coordination.

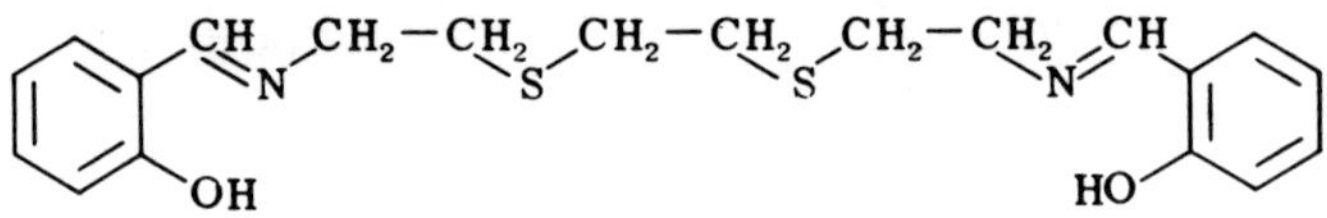

(LXXIII)

A sexadentate in which the donor atoms are members of a continuous chain of atoms may orientate itself about a metal atom in one of eight ways corresponding to the structures (LXXIVa, LXXVa, LXXVIa, and LXXVIIa) and their mirror images (LXXIVb, LXXVb, LXXVIb, and

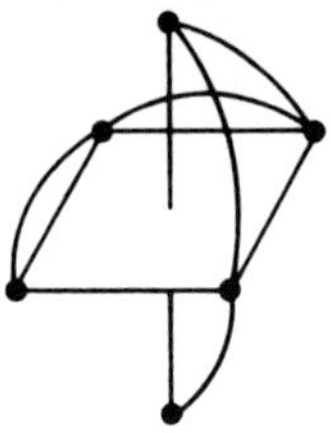

(LXXIV a)

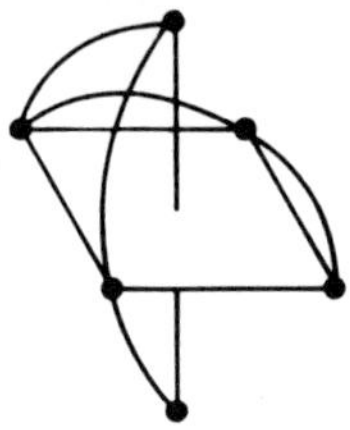

(LXXIV b)

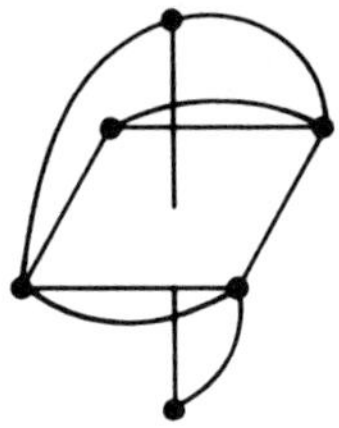

(LXXV a)

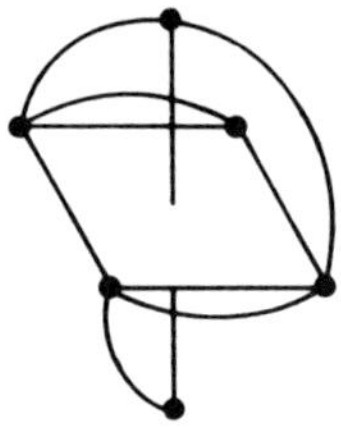

(LXXV b)

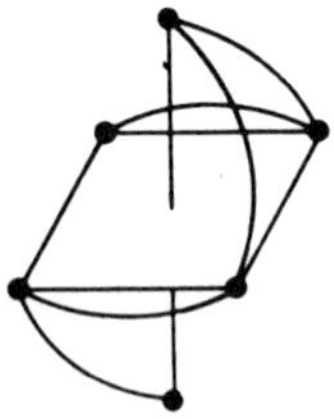

(LXXVI a)

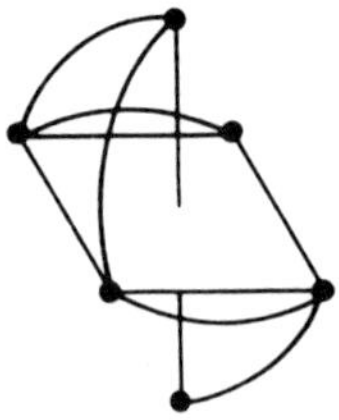

(LXXVI b)

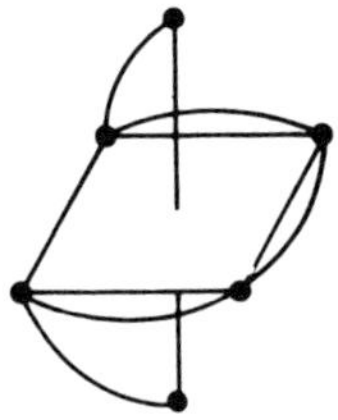

(LXXVII a)

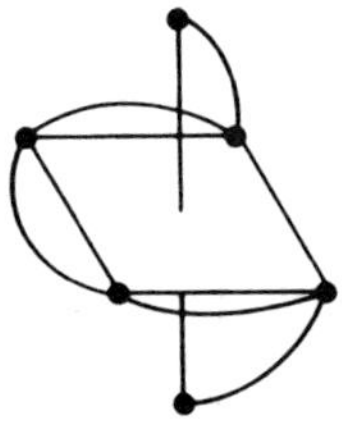

(LXXVII b)

LXXVIIb). These configurations are distinguished by the number, and relative positions in the chain, of consecutive donor atoms in an equatorial plane of the octahedron.

The number of isomers resulting from the strainless coordination of (LXXIII) about an octahedral metal atom is limited to two. The azomethine groups force each O—N—S donor atom sequence into a plane. Since the bonds from the sulfur atoms must be pyramidally directed, a twist must occur in the organic moiety at each sulfur atom. The only configurations which satisfy these demands are (LXXIVa) and its mirror image (LXXIVb). The diamagnetic cobalt(III) complex derived from (LXXIII) was resolved and the antipodes assigned these configurations (Dwyer and Lions, 1950; Dwyer *et al.*, 1950).

Lengthening of the carbon atom chains linking nitrogen and sulfur donor atoms in (LXXIII) results in some interesting stereochemical effects. For example, when the (azomethine) nitrogen atoms and the sulfur atoms are linked by a chain of three carbon atoms two series of cobalt(III) salts are obtained. These are distinguished by different absorption spectra and have both been obtained in enantiomorphous forms. One pair of enantiomorphs is similar to that derived from (LXXIII) and has been assigned configuration (LXXIV). The second is believed to have configuration (LXXVI) (Dwyer *et al.*, 1952). Salts of the two series are interconvertible. In (LXXVI) the O—N—S donor atom sequences are no longer planar but the added flexibility in chelate loops 2 and 4 (LXXVIII) allows this. The

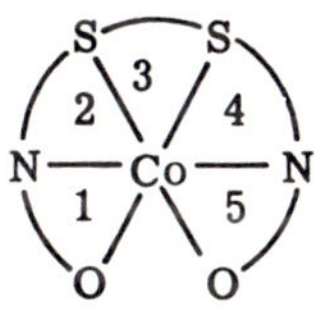

(LXXVIII)

pyramidal disposition of the bonds from the sulfur atoms still forces a twist in the organic moiety at these atoms and this can most readily be accommodated if each sulfur atom lies out of the plane containing the adjacent sulfur-metal-nitrogen ring. Consequently, configurations (LXXV) and (LXXVII) would not be favored. When, in addition, the two sulfur atoms are linked by a chain of three carbon atoms, only one series of salts is obtained. This is similar in color to those assigned the configuration (LXXIV) and is believed to have this configuration also. It would seem, however, that the molecule is sufficiently flexible for all the configurations (LXXV), (LXXVI), and (LXXVII) to be possible in addition to (LXXIV).

The polymethylene bridges between donor atoms in (LXXIII) have been replaced by one side of a benzene ring (Dwyer *et al.*, 1954). This introduces further constraints into the system since atoms attached to the benzene ring must lie in the plane of the ring. Sexadentates of this type form only one series of cobalt(III) salts to which configuration (LXXIV) was assigned.

Sexadentates analogous to (LXXIII) have been prepared in which one or both of the sulfur atoms have been replaced by an oxygen atom. The derived cobalt(III) complexes are less stable. The cobalt(III) complex of 1,8-bis(salicylideneamino)-3,6-dioxaoctane was too unstable to permit its resolution. In these complexes the usually poorly coordinating ethereal oxygen atoms are brought strainlessly into the correct position for coordination (Dwyer *et al.*, 1953a).

Substitution of a secondary amino nitrogen atom for each of the sulfur atoms in (LXXIII) gives the related sexadentate (LXXIX) obtained by

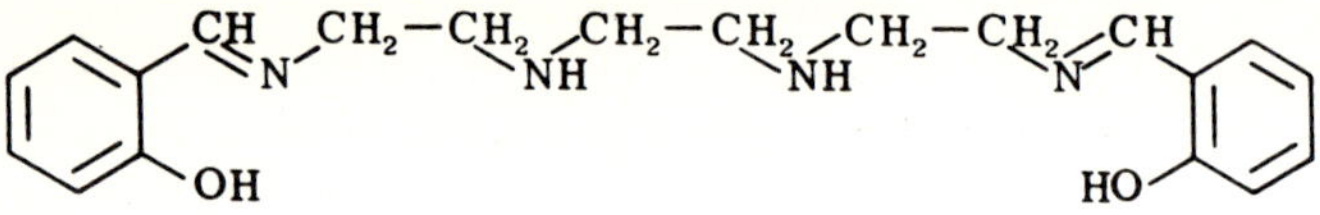

(LXXIX)

condensing salicylaldehyde with triethylenetetramine. In octahedral complexes derived from (LXXIX) the O—N—N donor atom sequence must be planar so that configuration (LXXIV) is adopted. Iron(III), aluminum(III), and cobalt(III) complexes of (LXXIX) have been obtained optically active (Das Sarma and Bailar, 1955a). The sexadentate obtained from salicylaldehyde and diglycylethylenediamine would be expected to function similarly to (LXXIX). Two apparently isomeric salts have been obtained from this sexadentate but they have not been completely characterized (Mukherjee, 1953).

The Schiff's base (LXXX), obtained from pyridine-2-aldehyde and

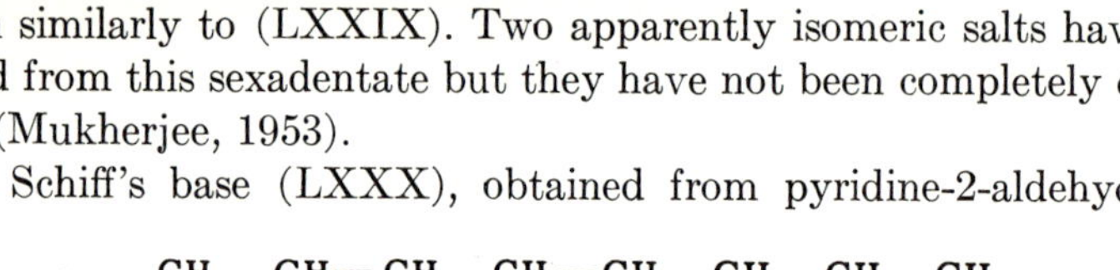

(LXXX)

1,8-diamino-3,6-dithiaoctane, functions as a sexadentate and octahedral complexes have a structure similar to that of complexes of (LXXIII) (Dwyer *et al.*, 1957).

Octahedral complexes derived from (LXXXI) must have configuration (LXXVII) in which the four central donor atoms occupy an equatorial plane. The two azomethine linkages ensure that the two nitrogen atoms of these structures and the two sulfur atoms are arranged in an equatorial plane about a metal atom while the pyramidal distribution of the bonds

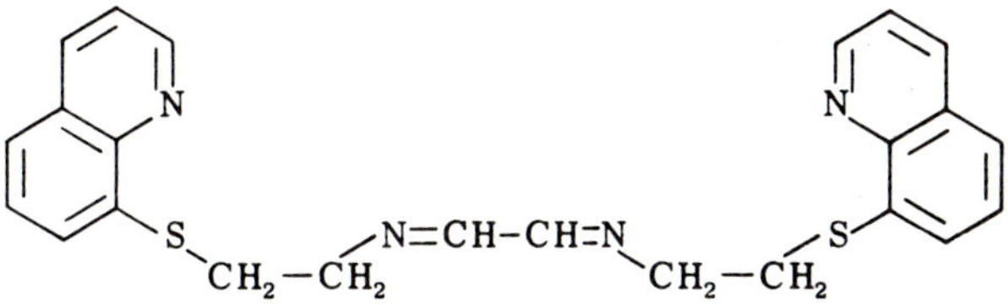

(LXXXI)

from the sulfur atoms ensures that the quinoline nitrogen atoms must lie above and below this plane (Lions, 1961).

In all the linear sexadentates so far discussed structural characteristics within the molecules have prevented complexes adopting all of the configurations (LXXIV–LXXVII). Substitution of secondary amino nitrogen atoms for the azomethine nitrogen atoms in these sexadentates would give molecules of much greater flexibility and able to yield complexes having any of these configurations, though perhaps not with equal ease. Pentaethylenehexamine (LXXXII) should be able to function in this manner.

$$NH_2CH_2CH_2NHCH_2CH_2NHCH_2CH_2NHCH_2CH_2NHCH_2CH_2NH_2$$

(LXXXII)

Although the base is known, its sexadentate function does not seem to have been established. Its coordinating properties have received little attention, apart from some solution studies (Jonassen *et al.*, 1957).

A. Structure of Sexadentates

Although the greater number of known sexadentates are of the linear, open chain type (LXXXIIIa), the most studied of all sexadentates, ethylenediaminetetraacetic acid, is not of this type but consists of a doubly

(LXXXIII a)

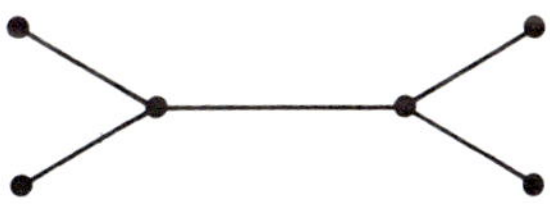

(LXXXIII b)

(LXXXIII c)

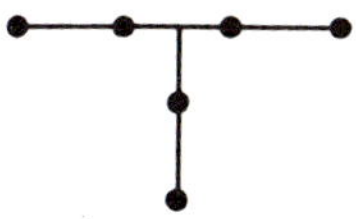

(LXXXIII d)

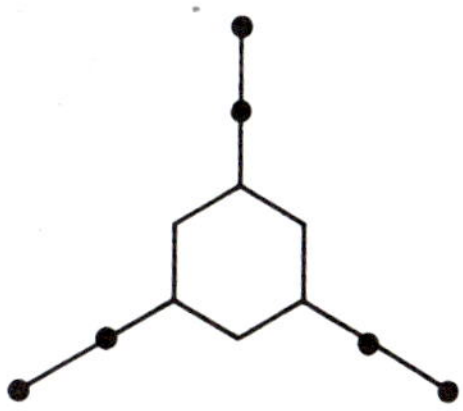

(LXXXIII e)

bifurcated chain of donor atoms, the bifurcations occurring at donor atoms (LXXXIIIb).

The number of possible arrangements of donor atoms in a sexadentate molecule is quite remarkable. Lions (1961) has indicated thirty-six different pattern types, examples of only five being known. In addition to those already discussed, sexadentates of donor atom patterns (LXXXIIIc–e) are known. In these bifurcations in the chain occur at atoms other than donor atoms.

In *N*,*N*′-ethylene-bis-2-(*o*-hydroxyphenyl)glycine (LXXXIV) (Freed-

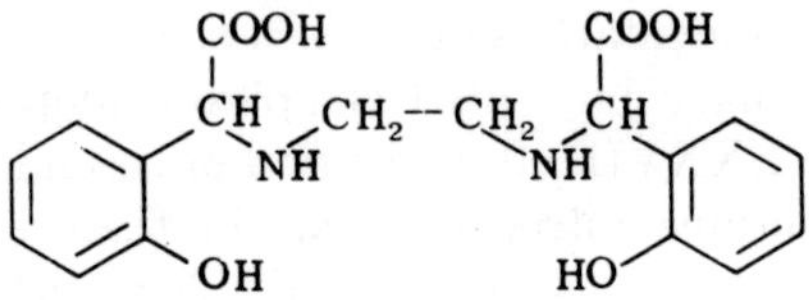

(LXXXIV)

man *et al.*, 1958), the donor atoms are arranged in pattern (LXXXIIIc). The tris(salicylidene) derivative (LXXXV) of 2-aminomethyl-1,3-diamino-

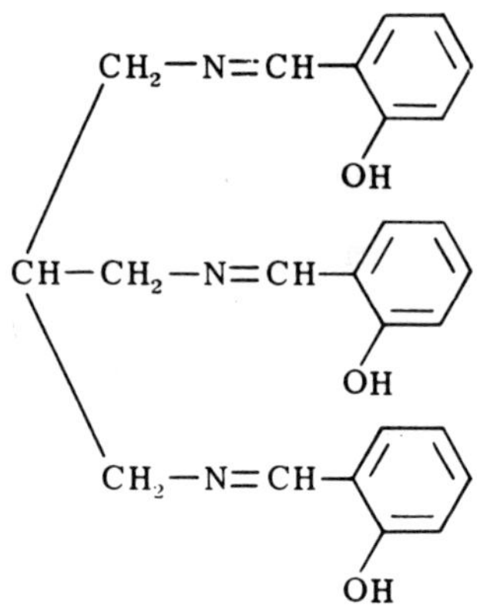

(LXXXV)

propane has a donor atom pattern as shown in (LXXXIIId). In the inner complexes derived from this and cobalt(III) and iron(III) (LXXXVI), the organic moiety must be wrapped around the metal atom so that there is a partial right-handed or left-handed spiral arrangement corresponding to enantiomorphous forms. The complexes could not be resolved (Dwyer *et al.*, 1957). Donor atom pattern (LXXXIIIe), which can be considered as a special case of (LXXXIIId), occurs in the tris(salicylidene) derivative (represented by LXXXVII) of 1,3,5-triaminocyclohexane. (LXXXVII) is able

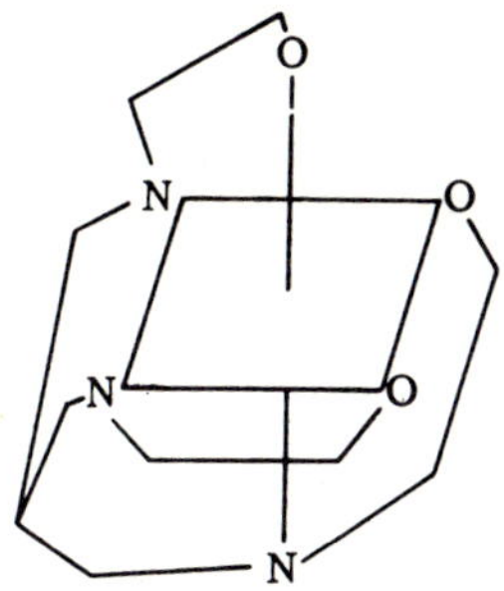

(LXXXVI)

to function as a sexadentate provided the substituents in the cyclohexane ring are in axial positions (Lions and Martin, 1957b). The arrangement of

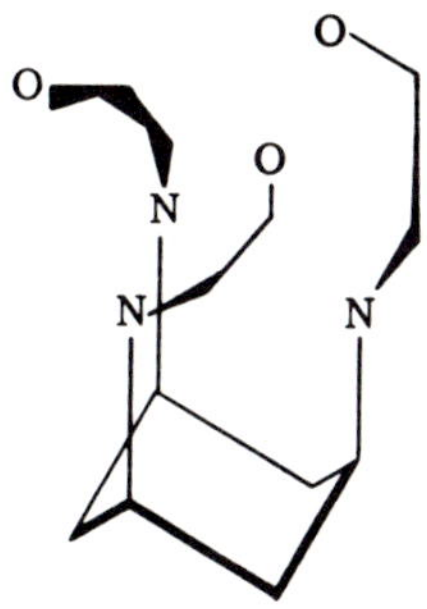

(LXXXVII)

donor atoms in octahedral complexes derived from (LXXXVII) is also a partial spiral which can be either left-handed or right-handed but none of its complexes has been resolved.

B. Design of Sexadentates

In the design of sexadentate chelating agents Lions (1961) has drawn attention to the value of the "spatial equivalence of groups." Thus, replacement of a carboxymethyl group in a useful chelating agent by an α-pyridylmethyl group provides a new molecule which can present a pyridine nitrogen donor atom to a metal atom in almost exactly the same way spatially as the oxygen of the carboxyl group was presented in the original molecule. Thus, (LXXXVIII) (La Coste, 1957) should function as a sexadentate giving complexes structurally related to those derived from EDTA. Similarly, the $—CH_2—CH_2—NH_2$ group is of approximately the same dimensions as the carboxymethyl group. It is not surprising, then, that (LXXXIX) can function as a sexadentate and that its complexes are

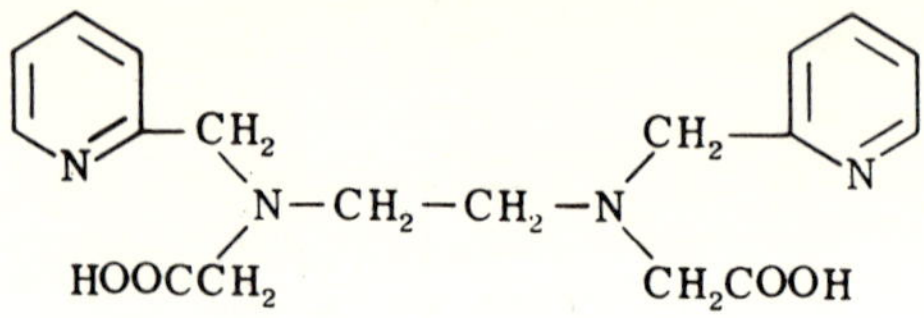

(LXXXVIII)

stereochemically similar to those derived from EDTA (Schwarzenbach and Moser, 1953).

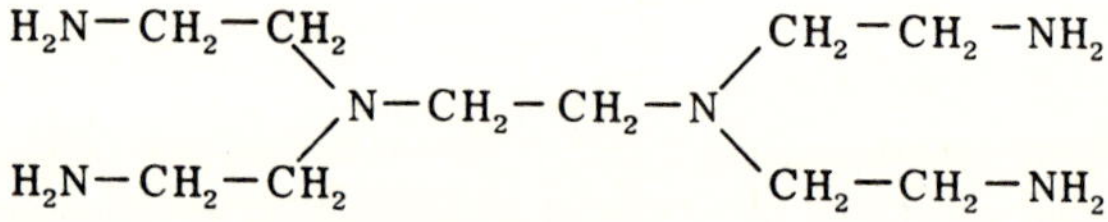

(LXXXIX)

Substitution of pyridine-2-aldehyde for salicylaldehyde in condensation reactions with primary amines in order to obtain sexadentates results in little change in the structure of the derived chelates apart from a reduction in the size of chelate loops 1 and 5 (LXXVIII). Quinoline-8-aldehyde has been condensed with 1,8-diamino-3,6-dithiaoctane (Suenaga, 1959). The octahedral complexes of the derived sexadentate must be very closely related structurally to those derived from (LXXIII). Interaction of a primary amine with a β-diketone provides a molecule which, after loss of a proton, can coordinate with a metal atom through the oxygen and nitrogen atoms with formation of a six-membered ring. Such amine–β-diketone condensation products are structurally similar to those obtained from primary amines and salicylaldehyde. Each contains an azomethine nitrogen atom and a suitably located acidic hydroxyl group. Lions and Martin (1958) prepared (XC) by condensing acetylacetone with 1,8-diamino-3,6-dithia-

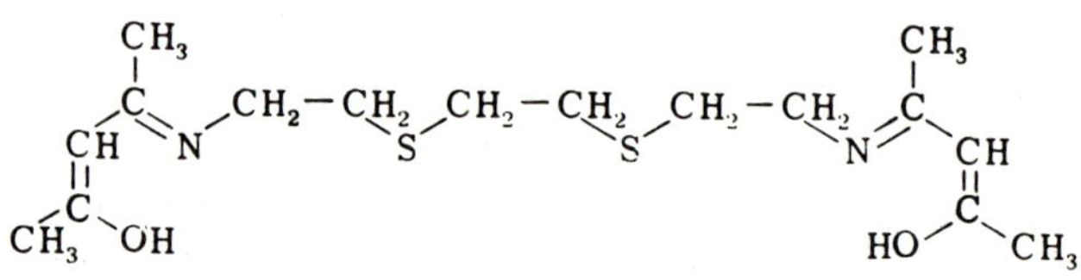

(XC)

octane. (XC) functions as a sexadentate and gives complexes with configuration (LXXIV).

VI. Higher Multidentate Chelating Agents

Although coordination numbers seven, eight, and nine have been established, greatest attention has been focused on complexes in which the metal atom has coordination number eight. In the design of molecules with eight donor atoms capable of being attached to a single metal atom, consideration must obviously be given to the stereochemistry of the metal atom. The configurations possible for an eight-covalent metal atom are the cube, the square or Archimedean antiprism, the dodecahedron and an undecahedron derived from the trigonal prism by adding atoms along the normals to two of the three rectangular faces. The dodecahedral arrangement has been shown to exist in the $[Mo(CN)_8]^{4-}$ ion (Hoard and Nordsieck, 1939) and the $[Zr(C_2O_4)_4]^{4-}$ ion (Hoard *et al.*, 1961). The $[TaF_8]^{3-}$ ion has been shown (Hoard *et al.*, 1954) to have an Archimedean antiprismatic structure.

Although potentially octadentate chelating agents are known, there does not appear to be any conclusive evidence to support octadentate function. 1,2-Bis-2-dicarboxymethylaminoethoxyethane (XCI) (Schwarzen-

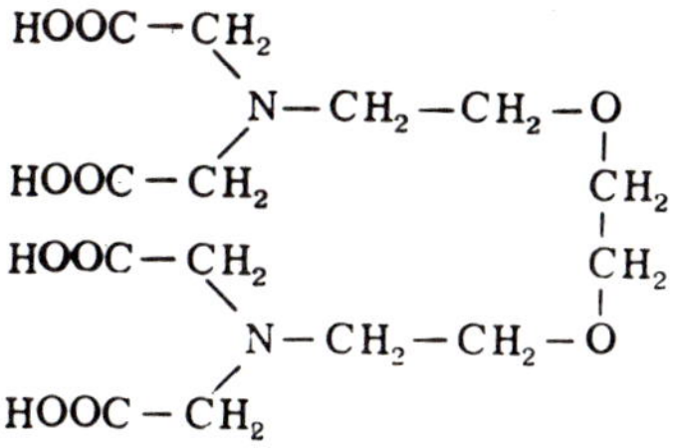

(XCI)

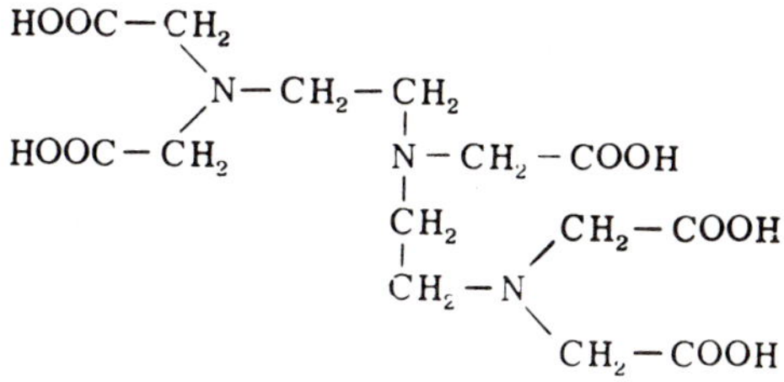

(XCII)

bach, 1955) and diethylenetriaminepentaacetic acid (XCII) (Frost, 1956) both possess eight donor atoms. These molecules possess a high degree of freedom and should be able to occupy all coordination positions about an eight-covalent metal atom. Studies on the interaction of (XCII) with zirconium(IV) indicate the formation, in solution, of a 1:1 complex in which all eight coordination positions about the metal atom are probably occupied by the eight donor atoms of the ligand (Intorre and Martell, 1960). Similarly, in the 1:1 thorium(IV) chelate, (XCII) is probably functioning as an octadentate (Bogucki and Martell, 1958). Octadentate function of (XCI) is not likely in view of the low coordinating capacity of ethereal oxygen atoms. It is significant that (XCI) fails to yield a stable zirconium(IV) chelate (Intorre and Martell, 1960).

Molecules with more than eight donor atoms are known (e.g., triethylenetetraminehexaacetic acid) but it seems unlikely that they would be able to coordinate through all donor atoms to a single metal atom.

References

Bähr, G., and Hess, E. (1952). *Z. anorg. u. allgem. Chem.* **268,** 351.
Bähr, G., and Schleitzer, E. (1955a). *Z. anorg. u. allgem. Chem.* **278,** 136.

Bähr, G., and Schleitzer, G. (1955b). *Z. anorg. u. allgem. Chem.* **280,** 161.
Bähr, G., Hess, E., and Steinkopf, E. (1953). *Z. anorg. u. allgem. Chem.* **273,** 325.
Bailes, R. H., and Calvin, M. (1947). *J. Am. Chem. Soc.* **69,** 1886.
Barclay, G. A., and Barnard, A. K. (1958). *J. Chem. Soc.* p. 2540.
Barclay, G. A., and Barnard, A. K. (1961). *J. Chem. Soc.* p. 4269.
Barclay, G. A., and Nyholm, R. S. (1953). *Chem. & Ind.* (*London*) p. 378.
Barclay, G. A., Harris, C. M., Hoskins, B. F., and Kokot, E. (1961a). *Proc. Chem. Soc.* p. 264.
Barclay, G. A., Nyholm, R. S., and Parish, R. V. (1961b). *J. Chem. Soc.* p. 4433.
Barrett, P. A., Dent, C. E., and Linstead, R. P. (1936). *J. Chem. Soc.* p. 1719.
Basolo, F. (1948). *J. Am. Chem. Soc.* **70,** 2634.
Bayer, E. (1957). *Ber.* **90,** 2325.
Bogucki, R. F., and Martell, A. E. (1958). *J. Am. Chem. Soc.* **80,** 4170.
Brandt, W. W., Dwyer, F. P., and Gyarfas, E. C. (1954). *Chem. Revs.* **54,** 959.
Brewster, J. A., Savage, C. A., and Venanzi, L. M. (1961). *J. Chem. Soc.* p. 3699.
Brintzinger, H., Thiele, H., and Müller, U. (1943). *Z. anorg. u. allgem. Chem.* **251,** 285.
Bruno, A. J., Chaberek, S., and Martell, A. E. (1956). *J. Am. Chem. Soc.* **78,** 2723.
Busch, D. H., and Bailar, J. C. (1953). *J. Am. Chem. Soc.* **75,** 4574.
Busch, D. H., and Bailar, J. C. (1956a). *J. Am. Chem. Soc.* **78,** 716.
Busch, D. H., and Bailar, J. C. (1956b). *J. Am. Chem. Soc.* **78,** 1137.
Byrne, G. T., Linstead, R. P., and Lowe, A. R. (1934). *J. Chem. Soc.* p. 1017.
Charles, R. G. (1960). *J. Phys. Chem.* **64,** 1747.
Chatt, J., and Watson, H. R. (1961). *J. Chem. Soc.* p. 4980.
Combes, A. (1889). *Compt. rend. acad. sci.* **108,** 1252.
Corbridge, D. E. C., and Cox, E. G. (1956). *J. Chem. Soc.* p. 594.
Crute, M. B. (1959). *Acta Cryst.* **12,** 24.
Curtis, N. F., and House, D. A. (1961). *Chem. & Ind.* (*London*) p. 1708.
Das Sarma, B., and Bailar, J. C. (1955a). *J. Am. Chem. Soc.* **77,** 5476.
Das Sarma, B., and Bailar, J. C. (1955b). *J. Am. Chem. Soc.* **77,** 5480.
Dent, C. E., and Linstead, R. P. (1934). *J. Chem. Soc.* p. 1027.
Dent, C. E., Linstead, R. P., and Lowe, A. R. (1934). *J. Chem. Soc.* p. 1033.
Diehl, H., and Henn, J. (1949). *Iowa State Coll. J. Sci.* **23,** 273.
Diehl, H., Liggett, L. M., Hach, C. C., Harrison, G. C., Henselmeier, L., Schwandt, R. W., and Mathews, J. (1947). *Iowa State Coll. J. Sci.* **22,** 110.
Dubsky, J. V., and Sokol, A. (1932). *Collection Czechoslov. Chem. Communs.* **3,** 548; *Chem. Abstr.* **26,** 1538 (1932).
Dudek, G. O., and Holm, R. H. (1961). *J. Am. Chem. Soc.* **83,** 2099.
Dwyer, F. P., and Lions, F. (1947). *J. Am. Chem. Soc.* **69,** 2917.
Dwyer, F. P., and Lions, F. (1950). *J. Am. Chem. Soc.* **72,** 1546.
Dwyer, F. P., Lions, F., and Mellor, D. P. (1950). *J. Am. Chem. Soc.* **72,** 5037.
Dwyer, F. P., Gill, N. S., Gyarfas, E. C., and Lions, F. (1952). *J. Am. Chem. Soc.* **74,** 4188.
Dwyer, F. P., Gill, N. S., Gyarfas, E. C., and Lions, F. (1953a). *J. Am. Chem. Soc.* **75,** 1526.
Dwyer, F. P., Gill, N. S., Gyarfas, E. C., and Lions, F. (1953b). *J. Am. Chem. Soc.* **75,** 3834.
Dwyer, F. P., Gill, N. S., Gyarfas, E. C., and Lions, F. (1954). *J. Am. Chem. Soc.* **76,** 383.
Dwyer, F. P., Gill, N. S., Gyarfas, E. C., and Lions, F. (1957). *J. Am. Chem. Soc.* **79,** 1269.
Fischer, H., and Friedrich, W. (1936). *Ann.* **523,** 154.
Fischer, H., and Gleim, W. (1935). *Ann.* **521,** 157.

Fischer, H., Haberland, H., and Müller, A. (1935). *Ann.* **521,** 122.
Freedman, H., Frost, A. E., Westerback, S., and Martell, A. E. (1958). *J. Am. Chem. Soc.* **80,** 530.
Frost, A. E. (1956). *Nature* **178,** 322.
Gill, N. S. (1951). Ph.D. Thesis, University of Sydney, Australia.
Goodwin, H. A., and Lions, F. (1959). *J. Am. Chem. Soc.* **81,** 6415.
Goodwin, H. A., and Lions, F. (1960). *J. Am. Chem. Soc.* **82,** 5013.
Hahn, K. P. (1961). M.Sc. Thesis, University of Illinois, Urbana, Illinois.
Hall, D., and Moore, F. H. (1960). *Proc. Chem. Soc.* p. 256.
Hall, D., and Waters, T. N. (1960). *J. Chem. Soc.* p. 2644.
Hall, D., Rae, A. D., and Waters, T. N. (1962). *Proc. Chem. Soc.* p. 143.
Harle, O. L., and Calvin, M. (1946). *J. Am. Chem. Soc.* **68,** 2612.
Harris, C. M., and Livingstone, S. E. (1962). *Revs. Pure and Appl. Chem.* **12,** 16.
Hart, F. A. (1960). *J. Chem. Soc.* p. 3324.
Helberger, J. H. (1937). *Ann.* **529,** 205.
Hoard, J. L., and Nordsieck, H. H. (1939). *J. Am. Chem. Soc.* **61,** 2853.
Hoard, J. L., Martin, W. J., Smith, M. E., and Whitney, J. F. (1954). *J. Am. Chem. Soc.* **76,** 3820.
Hoard, J. L., Glen, G. L., and Silverton, J. V. (1961). *J. Am. Chem. Soc.* **83,** 4293.
Hodgkin, D. C., Pickworth, J., Robertson, J. H., Trueblood, K. N., Prosen, R. J., and White, J. G. (1955). *Nature* **176,** 325.
Holtzclaw, H. F., Collman, J. P., and Alire, R. M. (1958). *J. Am. Chem. Soc.* **80,** 1100.
Hoskins, B. F. (1962). Ph.D. Thesis, University of New South Wales, Kensington, Australia.
Howell, T. E. W., Pratt, S. A. J., and Venanzi, L. M. (1961). *J. Chem. Soc.* p. 3167.
Hughes, E. W., Wilmarth, W. K., and Calvin, M. (1946). *J. Am. Chem. Soc.* **68,** 2273.
Intorre, B. I., and Martell, A. E. (1960). *J. Am. Chem. Soc.* **82,** 358.
Jonassen, H. B., and Cull, N. L. (1949). *J. Am. Chem. Soc.* **71,** 4097.
Jonassen, H. B., and Frey, F. W. (1953). *J. Am. Chem. Soc.* **75,** 1524.
Jonassen, H. B., and Frey, F. W. (1957). *J. Am. Chem. Soc.* **79,** 2454.
Jonassen, H. B., Bertrand, J. A., Groves, F. R., and Stearns, R. I. (1957). *J. Am. Chem. Soc.* **79,** 4279.
Kishita, M., Muto, Y., and Kubo, M. (1958). *Australian J. Chem.* **11,** 309.
Kluiber, R. W., and Lewis, J. W. (1960). *J. Am. Chem. Soc.* **82,** 5778.
La Coste, R. G. (1957). Doctoral Dissertation, Clark University, Worcester, Massachusetts.
Lever, A. B. P., Lewis, J., and Nyholm, R. S. (1961). *Nature* **189,** 58.
Linstead, R. P., and Robertson, J. M. (1936). *J. Chem. Soc.* p. 1736.
Lions, F. (1961). *Record Chem. Progr.* (*Kresge-Hooker Sci. Lib.*) **22,** 69.
Lions, F., and Martin, K. V. (1957a). *J. Am. Chem. Soc.* **79,** 1273.
Lions, F., and Martin, K. V. (1957b). *J. Am. Chem. Soc.* **79,** 1572.
Lions, F., and Martin, K. V. (1957c). *J. Am. Chem. Soc.* **79,** 2733.
Lions, F., and Martin, K. V. (1958). *J. Am. Chem. Soc.* **89,** 3858.
Llewellyn, F. J., and Waters, T. N. (1960). *J. Chem. Soc.* p. 2639.
Mair, G. A., Powell, H. M., and Henn, D. E. (1960). *Proc. Chem. Soc.* p. 415.
Mair, G. A., Powell, H. M., and Venanzi, L. M. (1961). *Proc. Chem. Soc.* p. 170.
Mann, F. G. (1926). *J. Chem. Soc.* p. 2681.
Mann, F. G. (1927). *J. Chem. Soc.* p. 1224.
Mann, F. G. (1928). *J. Chem. Soc.* p. 890.
Mann, F. G. (1929). *J. Chem. Soc.* p. 651.

Mann, F. G. (1930). *J. Chem. Soc.* p. 1745.
Mann, F. G. (1934a). *J. Chem. Soc.* p. 461.
Mann, F. G. (1934b). *J. Chem. Soc.* p. 466.
Mann, F. G., and Pope, W. J. (1925). *Proc. Roy. Soc.* **A109,** 444.
Mann, F. G., and Pope, W. J. (1926a). *J. Chem. Soc.* p. 482.
Mann, F. G., and Pope, W. J. (1926b). *J. Chem. Soc.* p. 2675.
Martell, A. E. (1955). *J. Phys. Chem.* **59,** 308.
Martell, A. E., and Calvin, M. (1952). "Chemistry of the Metal Chelate Compounds," p. 304. Prentice-Hall, Englewood Cliffs, New Jersey.
Martell, A. E., Linn Belford, R., and Calvin, M. (1958). *J. Inorg. & Nuclear Chem.* **5,** 170.
Martin, D. F., Shamma, M., and Fernelius, W. C. (1959). *J. Am. Chem. Soc.* **81,** 130.
Morgan, G. T., and Burstall, F. H. (1937). *J. Chem. Soc.* p. 1649.
Morgan, G. T., and Burstall, F. H. (1938). *J. Chem. Soc.* p. 1672.
Morgan, G. T., and Main-Smith, J. D. (1925). *J. Chem. Soc.* p. 2030.
Morris, M. L., and Busch, D. H. (1956). *J. Am. Chem. Soc.* **78,** 5178.
Mukherjee, A. K. (1953). *Sci. and Culture (Calcutta)* **19,** No. 2, 108.
Mukherjee, A. K., and Ray, P. (1955). *J. Indian Chem. Soc.* **32,** 505.
Muto, Y. (1955). *J. Chem. Soc. Japan* **76,** 1407.
Muto, Y. (1958). *Bull. Chem. Soc. Japan* **31,** 1017.
Muto, Y. (1960). *Bull. Chem. Soc. Japan* **33,** 1242.
Pfeiffer, P., and Glaser, H. (1939). *J. prakt. Chem.* **153,** 265.
Pfeiffer, P., and Pfitzner, H. (1936). *J. prakt. Chem.* **145,** 243.
Pfeiffer, P., and Saure, S. (1941). *Ber.* **74,** 935.
Pfeiffer, P., Breith, E., Lübbe, E., and Tsumaki, T. (1933). *Ann.* **503,** 84.
Pfeiffer, P., Hesse, Th., Pfitzner, H., Scholl, W., and Thielert, H. (1937). *J. prakt. Chem.* **149,** 217.
Pfeiffer, P., Offerman, W., and Werner, H. (1942). *J. prakt. Chem.* **159,** 313.
Ramberg, L., and Tiberg, A. (1914). *Ber.* **47,** 730.
Rasmussen, S. E. (1959). *Acta Chem. Scand.* **13,** 2009.
Ray, P., and Ghosh, S. P. (1943). *J. Indian Chem. Soc.* **20,** 291.
Robertson, J. M. (1936). *J. Chem. Soc.* p. 1195.
Robertson, J. M., and Woodward, I. (1940). *J. Chem. Soc.* p. 37.
Sacconi, L. (1954a). *J. Am. Chem. Soc.* **76,** 3400.
Sacconi, L. (1954b). *Z. anorg. u. allgem. Chem.* **275,** 249.
Schwarzenbach, G. (1949). *Helv. Chim. Acta* **32,** 839.
Schwarzenbach, G. (1950). *Helv. Chim. Acta* **33,** 974.
Schwarzenbach, G. (1955). "Die komplexometrische Titration." Enke, Stuttgart, Germany.
Schwarzenbach, G., and Biedermann, W. (1948). *Helv. Chim. Acta* **31,** 331.
Schwarzenbach, G., and Moser, P. (1953). *Helv. Chim. Acta* **36,** 581.
Schwarzenbach, G., Kampitsch, E., and Steiner, R. (1945). *Helv. Chim. Acta* **28,** 828.
Selbin, J. (1961). *J. Inorg. & Nuclear Chem.* **17,** 84.
Selbin, J., and Bailar, J. C. (1960). *J. Am. Chem. Soc.* **82,** 1524.
Stetter, H., and Mayer, K. (1961). *Ber.* **94,** 1410.
Stoufer, R. C., and Busch, D. H. (1956). *J. Am. Chem. Soc.* **78,** 6016.
Stratton, W. J., and Busch, D. H. (1958a). *J. Am. Chem. Soc.* **80,** 1286.
Stratton, W. J., and Busch, D. H. (1958b). *J. Am. Chem. Soc.* **80,** 3191.
Stratton, W. J., and Busch, D. H. (1960). *J. Am. Chem. Soc.* **82,** 4834.

Suenaga, E. (1959). *J. Chem. Soc. Japan* **80,** 500.
Tsumaki, T. (1938). *Bull. Chem. Soc. Japan* **13,** 252.
Ueno, K., and Martell, A. E. (1955). *J. Phys. Chem.* **59,** 998.
Ueno, K., and Martell, A. E. (1957). *J. Phys. Chem.* **61,** 257.

CHAPTER 5

Optical Phenomena in Metal Chelates[1]

A. M. SARGESON

Biological Inorganic Chemistry Section, The John Curtin School of Medical Research, Australian National University, Canberra, A.C.T., Australia

I. General Introduction

The rotation of the plane of polarization of plane-polarized light when it is transmitted through certain media is referred to as optical rotation

[1] The sign of the rotation of the enantiomer is denoted (+) or (−) for the Na_D line, and when the rotation is measured in another wavelength this is specified by a subscript, e.g., $(+)_{5461}$. Racemates are denoted as (±). The symbols D and L are used to describe absolute configurations relative to a standard absolute substance, e.g., D(+)$[Co(en)_3]^{3+}$ ion. The symbols D and L (small capitals) are used to denote the absolute configurations of optically active organic molecules. In the literature connected with stereospecificity the designation Dddd has been used for the (+)$[Co(+)pn_3]^{3+}$ ion. In this chapter the equivalent D[(+)(+)(+)] or (+)[(+)(+)(+)] has been used.

and the media are described as being optically active. At the molecular level, i.e., with pure liquids or true solutions, optical activity depends upon the presence of an excess, or usually the exclusive presence, of one enantiomer or optical form of a dissymmetric or asymmetric molecule or ion (cf. Section II). The optical rotatory power of a substance is expressed as its specific rotation, defined by the familiar expression $[\alpha]_\lambda^T = 100\alpha/lc$ in which α is the rotation in degrees of a solution containing c gm. of substance in 100 ml. of solution, when observed in a column l decimeters long. The molecular rotation $[M]_\lambda^T$, defined by the expression $[M]_\lambda^T = M[\alpha]_\lambda^T/100$, in which M is the molecular weight, is more significant when comparing rotatory powers. According to the theory originally proposed by Fresnel, in an optically active medium the right- and left-handed circularly polarized components of linearly polarized light travel with unequal velocities, and on emergence and recombination the plane of polarization is rotated through an angle to the original plane. Optical activity thus arises from the difference in the refractive indices $(n_l - n_r)$ of the medium towards right- and left-handed circularly polarized light. For the optical antipodes, the two values of the refractive indices are reversed, and hence the specific rotations are equal and opposite. Like the ordinary refractive index, the values of this pair of indices, and hence the specific rotation, vary with the wavelength of the light. In regions of optical transparency remote from absorption, the change in the specific rotation (rotatory dispersion) is regular though not quite linear, increasing with decreasing wavelength of the light.

In the vicinity of and within an isolated absorption band, there is also a close correspondence between ordinary refraction and optical rotation, illustrated in (Ia) and (Ib). As the absorption band is approached from the

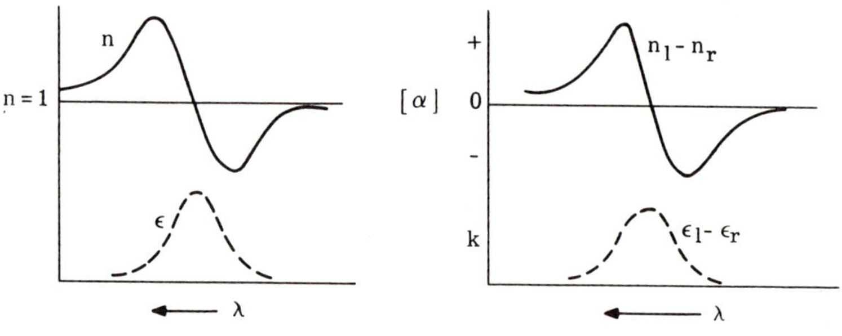

(a) (b)

(I)

long wavelength side, the difference between the two refractive indices commences to increase rapidly. The difference, and hence the specific rotation, continues to increase as the absorption band is reached, coming to a maximum on the long wavelength side of the band head. This is followed by a rapid decrease in the term $(n_l - n_r)$, which becomes zero and may reach a maximum in the opposite sense when $n_l > n_r$. The absorption band is now passed and the specific rotation which has paralleled these changes, passing through zero at the wavelength where $n_l = n_r$, increases more or less regularly, as before, but with the sign reversed. The nature of these phenomena will be evident from (Ib) and (II). A number of

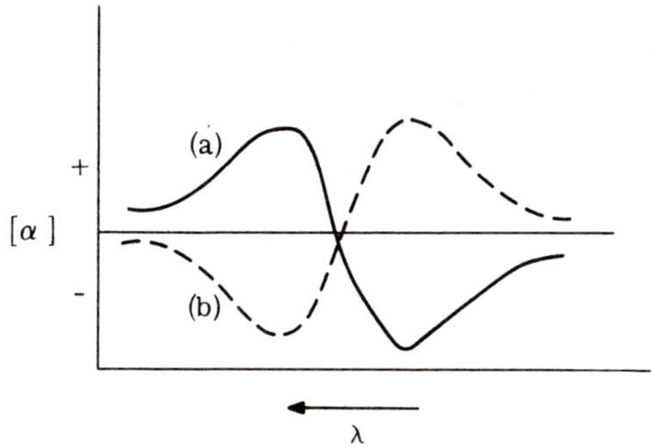

(II)

Rotatory Dispersion (R.D.) Curves are shown in Section V. It will be appreciated that the sign of the rotation in a particular wavelength (e.g., Na_D line) cannot be used to relate the configurations of different substances because of their different absorption bands.

In the vicinity of and within the absorption bands, the absorption coefficients ϵ of the substance towards the right- and left-handed circularly polarized components also become different. As a result of unequal absorption (dichroism) the emergent light beam is no longer linearly but elliptically polarized (Ib). These phenomena, circular dichroism and elliptical polarization, are known as the Cotton Effect. It should be noted, however, that not all of the absorption bands of optically active substances exhibit anomalous rotatory dispersion and the Cotton Effect. The electronic transition producing absorption must also be the cause of the optical activity.

The shape of the rotatory dispersion curve [cf. (II), curves (a) and (b)] is diagnostic of the configuration of the enantiomer being observed, since just as the enantiomers are mirror images so also are their R.D. curves. It is not possible, however, to associate the configuration of the enantiomer empirically with a particular R.D. curve; but if the sign of the rotation at a suitable wavelength could be calculated from first principles for a structure, then its absolute configuration would be established. Condon (1937) has shown that the problem, in principle, is fairly simple. The molecular rotation $[M]$ for all molecules in their lowest electronic level "a" can be expressed by the function

$$[M] = \frac{96\pi N}{hc} \cdot \frac{n^2 + 2}{3} \cdot \sum_b \frac{R_{ba}\nu^2}{\nu_{ba}^2 - \nu^2}$$

where R_{ba} is a constant, characteristic of the absorption line, called the rotational strength of the line ν_{ba}, ν_{ba} is the frequency of light in the transition $a \rightarrow b$, ν is the frequency of light at which rotation is determined, n is the refractive index, and the other constants have their usual significance. This function predicts the behavior of $[M]$ with change in refractive index and shows that the rotatory dispersion due to one absorption band has the characteristics of (Ib); i.e., the rotation passes through zero at the absorption peak and approaches zero asymptotically in the infrared and far ultraviolet. The absolute configuration of the structure is established if the sign and magnitude of the rotational strength R_{ba} for each transition can be calculated. Many workers have attempted to solve the problem but, so far, only the contributions of chromophoric groups in the presence of an asymmetric carbon atom have been computed with some success (Kuhn, 1930; Kirkwood, 1937, 1939; Kauzman *et al.*, 1940). A calculation of the absolute configuration of $(+)[Co(en)_3]Cl_3$ has been made by Kuhn (1952), but it is in disagreement with the absolute X-ray crystal analysis, which must be accepted (Saito *et al.*, 1957). A theoretical reappraisal of the problem has also been made by Moffitt (1956), but this has been criticized by Sugano (1960). Perhaps the "harmonic oscillator" model of Jones and Eyring (1961) which gives the correct configuration for $(+)[Co(en)_3]^{3+}$ ion will be more useful.

The determination of the absolute configuration of (+) and (−) tartaric acids by an X-ray method (Bijvoet *et al.*, 1951) provided organic chemists with the means of fixing the structures of all substances where configurations were known relative to tartaric acid. The absolute configuration of the $(+)[Co(en)_3]^{3+}$ ion taken in conjunction with rotatory dispersion studies allows the configuration of the enantiomers of many coordination compounds to be settled with reasonable certainty (Section V).

II. Optical Isomerism and Stereochemistry

The essential condition for optical isomerism in coordination compounds is the same as for organic substances: the structure must be nonsuperimposable on its mirror image. This does not mean, as pointed out by Jaeger (1930), that the structure need be asymmetric, for many complexes that have some elements of symmetry are resolvable. For these the description dissymmetric is more appropriate. The limiting symmetry conditions are that the molecule or ion should lack a plane or center of symmetry. The classical asymmetric carbon atom is obviously simulated by the metal atom in tetrahedral complexes of the form [M(ABCD)], which, however, are difficult to prepare and probably too labile for successful resolution. Tetrahedral chelates derived from unsymmetrical bidentate ligands satisfy the symmetry conditions for optical isomerism, and are easily prepared. The bis(benzoylpyruvato)beryllate(II) anion (III) is one of the tetrahedral

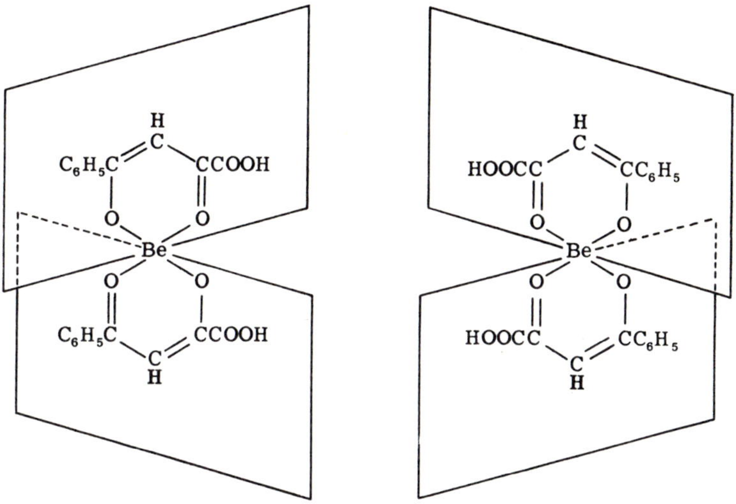

(III)

chelates that has been resolved (Mills and Gotts, 1926). The platinum atom in the octahedral complex $[Pt(py)NH_3ClBrINO_2]^0$ (IV) isolated as the

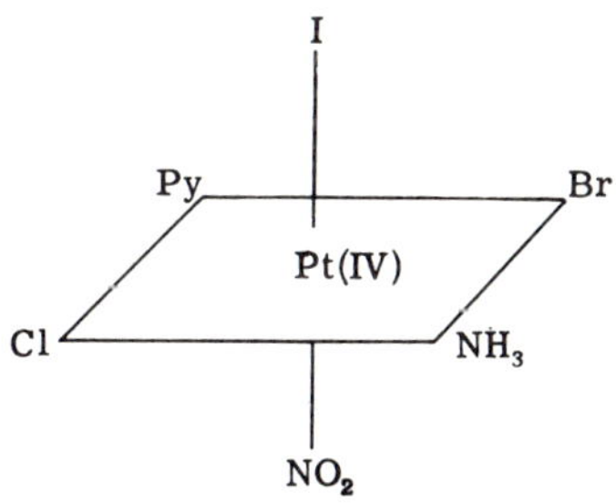

(IV)

racemic pair of one of the many geometrical isomers (Gel'man and Essen, 1950) is also an asymmetric metal atom.

In general, we are concerned with molecular asymmetry or dissymmetry in chelates (usually octahedral), and with the manner in which the ligand bridges the coordination positions. The dissymmetric tris(ethylenediamine)-

cobalt(III) ion (Va) has one threefold and three twofold axes but lacks a plane or center of symmetry. If one ethylenediamine is replaced by two

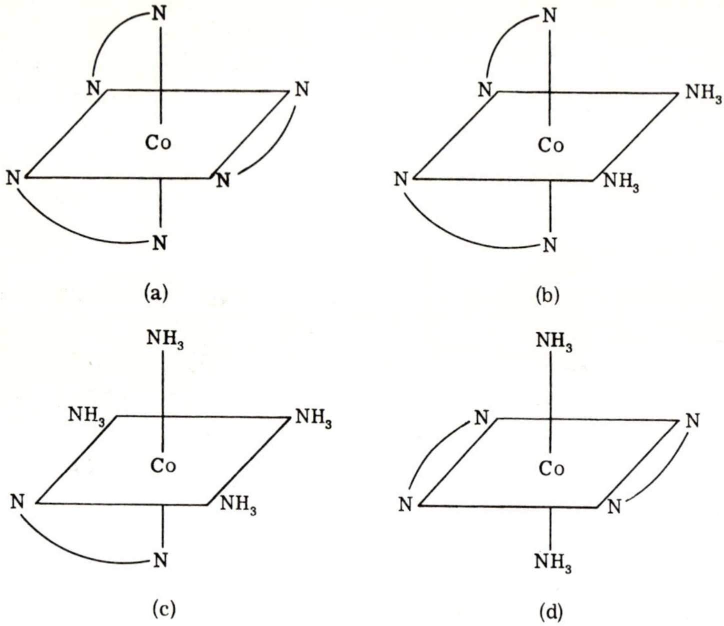

(V)

ammonia molecules (*cis*) (Vb) the complex is still dissymmetric but has lost some of its symmetry elements, namely, the threefold axis and two of the twofold axes. If another ethylenediamine is replaced by two ammonia molecules (Vc) the resulting complex has a plane of symmetry and optical isomerism is not possible. Further, if two of the ethylenediamine groups are in the same plane (*trans*) (Vd) the complex also has a plane of symmetry.

In the fixed geometry of octahedral coordination the possible existence of geometrical and optical isomers depends upon the nature and stereochemistry of the ligand molecule. Multidentate ligands may prescribe a unique geometry in conformity with the steric demands of the donor atoms, or they may be quite flexible, capable of alleviating some donor-metal angle strain, and thus permit geometrical isomerism. The bis chelates $[M(dien)_2]^{n+}$ of the tridentate ligand diethylenetriamine, $NH_2 \cdot (CH_2)_2 \cdot NH \cdot (CH_2)_2 \cdot NH_2$ (dien), can exist in three geometrical isomers (VI) of which one isomer (VIc) can exist in optical forms, though the resolution of such a chelate has not yet been performed. Because of the tetrahedral stereochemistry of 4-covalent nitrogen the puckered chelate rings in (VIb) are somewhat strained and slightly distorted from the plane containing the nitrogen atoms

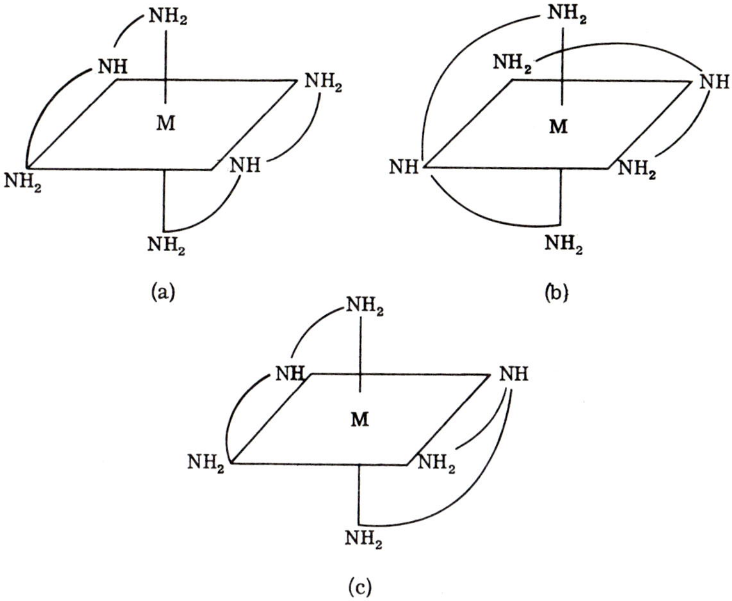

(VI)

so that this isomer might be expected to be less stable than the others. By contrast, 2,2′,2″-terpyridine has three donor nitrogen atoms and is constrained to exist in the planar form (VIb). The direction of the nitrogen-metal bonds must lie in the plane of the pyridine rings, which for maximum resonance must be coplanar, and hence the pyridine and chelate rings must also be coplanar. However, substitution in the 4 or 4″ positions (VII) would permit optical isomerism and this has been illustrated by the resolution of

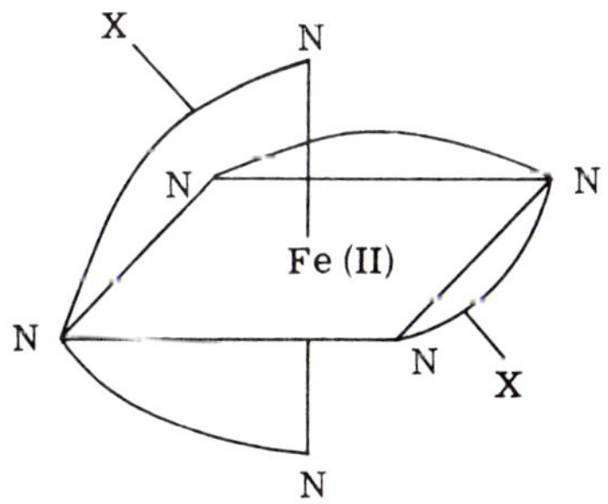

(VII)

the iron(II) chelate of the analogous 8(α-pyridylmethyleneamino)-quinoline (VIII) (Dwyer *et al.*, 1953).

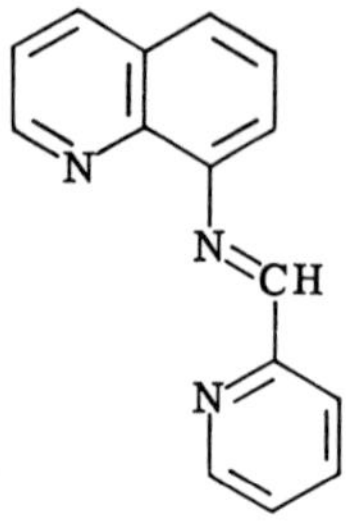

(VIII)

Two of the three possible geometrical isomers of the chelates derived from the flexible quadridentate ligand triethylenetetramine, $NH_2 \cdot (CH_2)_2 \cdot NH \cdot (CH_2)_2 \cdot NH \cdot (CH_2)_2 \cdot NH_2$ (trien), are capable of existence in optical forms (IXa,b), but the *trans* form (IXc) has a plane of symmetry. Because

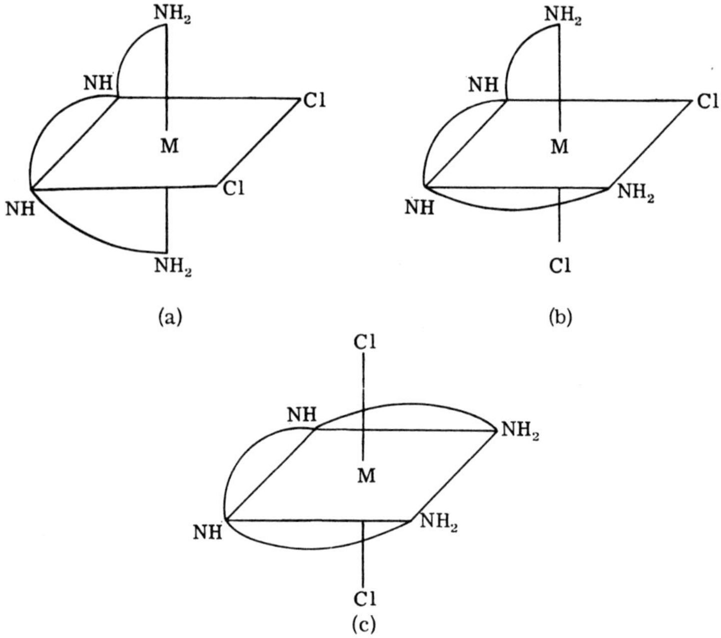

(IX)

of the tetrahedral stereochemistry of the nitrogen atoms the strain will be greatest in (IXc) and least in (IXa). Dichlorobis(triethylenetetramine)-cobalt(III) chloride has been isolated as a violet-colored substance, and by analogy with $[Co(en)_2Cl_2]Cl$, assigned the *cis* configuration (Basolo, 1948). Partial resolution by Das Sarma and Bailar (1955) and complete resolution by Sargeson and Searle (1961) have confirmed the *cis* structure.

In methanol solution, the initial violet color changes to a grey violet and this has been attributed to a partial isomerization (40%) of the *cis* isomers to *trans* (Das Sarma and Bailar, 1955). Recently all three forms of the cation $[Co(trien)Cl_2]^+$ have been isolated and both *cis* isomers have been resolved (Sargeson and Searle, 1961).

Schiff's base quadridentates derived, for example, from salicylaldehyde and ethylenediamine or *o*-phenylenediamine are usually quite inflexible: the metal-oxygen bonds must lie in the plane of the benzene ring as must the donor bonds from the doubly bonded nitrogen atoms. The resolution of one isomer of the ethylenediaminebis(acetylacetone)diamminecobalt(III) ion (Morgan and Main-Smith, 1925) is dubious, as is geometrical isomerism, since planar coordination of the quadridentate ligand is unavoidable and the ammonia molecules must consequently occupy *trans* positions.

The base 2,2′-diamino-6,6′-dimethyldiphenyl is dissymmetric because of restricted rotation of the phenyl groups and has been resolved by Meisenheimer and Horing (1927). Racemization occurs only slowly at high temperatures (Kistiakowsky and Smith, 1936), and the quadridentate bis-(salicylidene) derivative cannot under ordinary conditions adopt a planar configuration about a metal atom, since the two benzene rings of the diphenyl system cannot be coplanar. Lions and Martin (1957) have prepared the enantiomeric Cu(II) chelates (X), which are coordinated tetra-

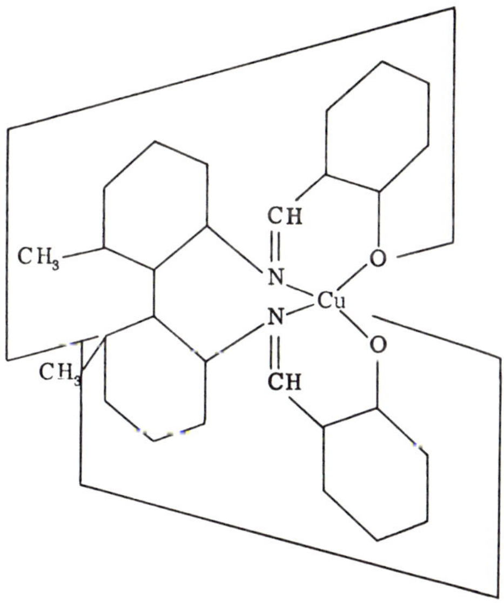

(X)

hedrally, by using the optical isomers of the ligand.

Molecular models indicate that all of the possible coordination configurations (XI) of the quinquedentate ligand tetraethylenepentamine,

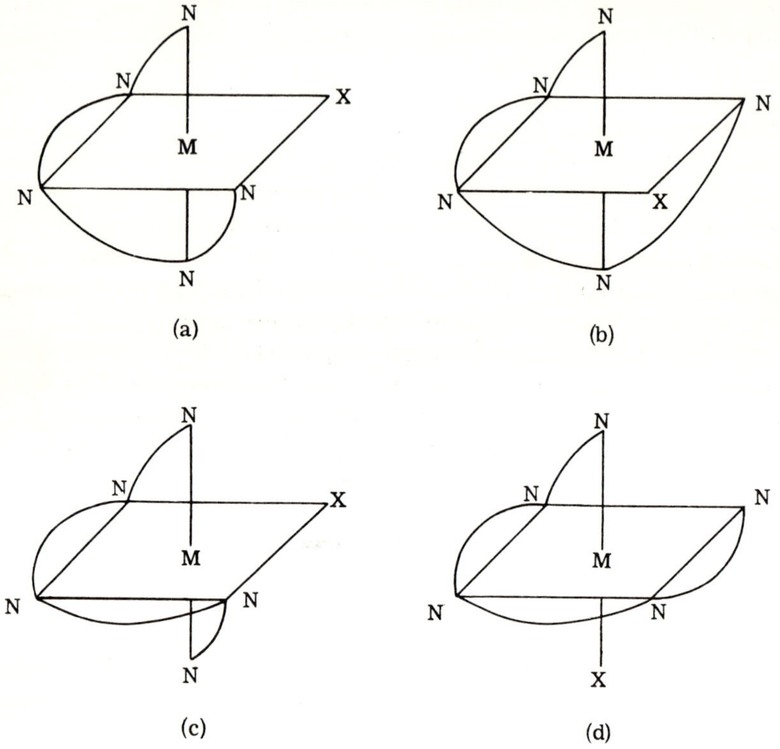

(XI)

$NH_2 \cdot (CH_2)_2 \cdot NH \cdot (CH_2)_2 \cdot NH \cdot (CH_2)_2 \cdot NH \cdot (CH_2)_2 \cdot NH_2$, are feasible, though not of the same free energy. In these chelates there are four linked-ring systems that can adopt two ring conformations (cf. Section IV). The free energies, and hence the relative stabilities of the isomers, depend not only on the cumulative effects of strain in some chelate rings, due primarily to the preferred tetrahedral bonding about the nitrogen atoms, but on the nonbonded atomic interactions peculiar to each conformation. In addition, a complete assessment of the relative stabilities of the isomers requires that the hydration or lattice energies (as relevant) be taken into account.

Similar considerations obviously apply to the diethylenetriamine and triethylenetetramine chelates containing two and three condensed rings, respectively. In the former chelate, the nonbonded atomic interactions due to two linked rings in their possible conformations may not be great, but in multicondensed ring chelates, however, they may compete with, or even outweigh, the energy effects resulting from donor-metal angle and chelate ring strain (cf. Dwyer, 1961). The interplay of these energy effects may thus limit the possibilities of isomerism; but, whereas "gross" stereochemical considerations arising from the nature of the ligand and its donor atoms are readily evaluated, the atomic interactions are not.

III. Methods of Resolution

A. Diastereoisomer Formation

Pasteur's classical "second method" is the most common and efficient method of resolving complex cations and anions, but is obviously inapplicable to nonelectrolytes. The racemic salt (±)[A]Cl can be often conveniently converted, through the silver salt of an optically active anion, Ag(−)B, to a pair of diastereoisomers. Since the diastereoisomers(+)[A]-(−)B and (−)[A](−)B are not enantiomorphs, they are theoretically separable by fractional crystallization. Practically, the ease and efficiency of the crystallization depends on the difference in the solubilities, and efficient resolution is therefore largely a problem of discovering a resolving anion or cation which gives a sufficiently large discrimination between the solubilities of the diastereoisomers. The necessity for the complete removal of the resolving agent at the end of any resolution frequently limits the choice.

Most of the early resolutions of metal complexes were carried out with organic acids or bases (or their ions): (+)tartaric acid; (+)nitrocamphor; (+)camphor- and bromcamphorsulfonic acids; (−)strychnine; (−)brucine; and (−)cinchonine. Alkaloidal salts frequently crystallize poorly, exhaustive fractional crystallization is often necessary, and traces of alkaloid are difficult to eliminate from the final product. Some resolutions which have not been substantiated by later work and some apparent slow racemizations Wahl, 1927; Burrows and Lauder, 1931; Treadwell *et al.*, 1932; Neogi and Dutt, 1938) were probably due to contamination with residual resolving agent and its slow precipitation. Strychnine, for example, is removed from diastereoisomers more efficiently as the perchlorate, than as the traditional iodide. The (+)bromcamphor sulfonate ion has proved very useful for resolving cations. Tartrates, in general, are rather too soluble in water, but diastereoisomers derived from diacetyl and dibenzoyl tartrates and the antimonyl and arsenyl tartrates are usually less soluble and crystallize well.

A recent very promising innovation has been the use of resolved metal complexes as resolving agents. The following diagram illustrates how a series of complexes can be resolved efficiently, commencing with an optically active cation first resolved through antimonyl tartrate. It is a common experience that a univalent cation is resolved most readily with a univalent anion. Such matching of charges is an obvious extension of the solubility generalization with simple inorganic compounds, i.e., AgCl, Ag_2SO_4: $BaCl_2$, $BaSO_4$.

$$(+)[CoEDTA]^- \rightarrow (+)[Co(en)_2(C_2O_4)]^+ \rightarrow (+)[Co(C_2O_4)_2gly]^{--}$$
$$\nearrow \qquad\qquad\qquad\qquad \nearrow \qquad\qquad\qquad\qquad \searrow$$
$$(+)[Co(en)_2(NO_2)_2]^+ \rightarrow (+)[Co(en)(C_2O_4)_2]^- \rightarrow (+)[Co(en)_2CO_3]^+ (+)[Co(en)_2ClH_2O]^{++}$$
$$\searrow$$
$$(-)[Co(en)(mal)_2]^-$$

Resolution can often be effected merely by dissolving the racemate salt in a minimum volume of water at room temperature and adding the solid resolving agent as one of its salts. If the diastereoisomer is too soluble to crystallize even on freezing, addition of a little methanol, ethanol, acetone, or ethyleneglycol can sometimes effect a separation (Dwyer *et al.*, 1955, 1959).

Frequently the presence of extraneous ions increases the solubility of diastereoisomers, and to achieve a separation it is necessary to remove all ions not involved in diastereoisomer formation; e.g., $(+)Co(en)_3Cl(+)$tart crystallizes much less readily in the presence of extraneous Cl^- ions (Werner, 1912a).

With complex ions of multiple charge, simple ions are sometimes included in the crystal lattice of the least soluble diastereoisomer, e.g.,

$(+)[Co(en)_3](+)_{5461}[CoEDTA]_2Cl \cdot 4H_2O$ (Dwyer *et al.*, 1955);
$(+)[Co(en)_3]Cl(+)tart \cdot 5H_2O$ (Werner, 1912a);
$(-)[Ni(phen)_3]K(+)[Co(C_2O_4)_3] \cdot H_2O$ (Dwyer and Sargeson, 1956);
$(-)[Co(en)_2(NO_2)_2]Ca(+)[Co(C_2O_2S_2)_3] \cdot H_2O$ (Dwyer and Sargeson, 1959c);
$(-)[Ni(bipy)_3]_3(+)(SbOtart)_4I_2 \cdot 18H_2O$ (Dwyer and Gyarfas, 1951a);
$(+)[Co(en)_3]Br(+)tart$ (Werner, 1912b).

Whereas the $(+)[Co(en)_3]^{3+}$ ion precipitates both the (+) and (−) forms of the $[Co(C_2O_4)_3]^{3-}$ anion as the salt $[Co(en)_3][Co(C_2O_4)_3]$, better than 90% separation can be achieved in the first precipitation of the "mixed" salt $K[Ni(phen)_3] \cdot [Co(C_2O_4)_3]$. Variation of the inactive ion can also be used to alter relative solubilities of the diastereoisomers, and so effect better separation. The diastereoisomer $(+)[Co(en)_3]Br(+)$tart (Werner, 1912b), for instance, is less soluble than $(+)[Co(en)_3]Cl(+)$tart, and $(+)[Co(en)_2(NO_2)_2]Ca(-)[Co(C_2O_2S_2)_3]$ (Dwyer and Sargeson, 1959c) is more soluble than the corresponding barium salt which precipitates both forms giving only a partial separation. The solubility of the strontium salt is intermediate, but the magnesium salt is too soluble to form crystalline diastereoisomers easily.

The splitting of the diastereoisomer to remove the resolving agent is largel. dependent on the individual solubilities of salts of the component ions. Grinding an aqueous slurry of the diastereoisomer with sodium iodide or sodium perchlorate in an ice-cold mortar is often used. The cation iodides and perchlorates are often poorly soluble in the presence of a large excess of iodide or perchlorate ion and the sodium salt of the enantiomer usually may be recovered by adding alcohol to the filtrate. A similar method involves generating tri-iodide ion in an aqueous suspension of the diastereoisomer. Complex tri-iodides are ill-defined but very insoluble, and the iodide of the enantiomer is easily recovered by bubbling sulfur dioxide through an alcoholic suspension of the tri-iodide.

B. Partial Asymmetric Synthesis

By preparing or isomerizing the complex in the presence of the resolving agent it is often possible to obtain high yields of one optical isomer at the expense of the other. The ion (+)antimonyl tartrate (1 mole) precipitates $(-)[Fe(phen)_3]^{++}$ ion (Dwyer and Gyarfas, 1949b) almost quantitatively from an aqueous racemate solution. However, if the molar ratio of complex to resolving agent is 1:2 and the solution is allowed to stand, all of the complex is converted to $(-)[Fe(phen)_3](+)(SbOtart)_2$. High yields of one optical isomer may also be obtained from the $[Ni(phen)_3]^{++}$, $[Ni(bipy)_3]^{++}$, and $[Fe(bipy)_3]^{++}$ ions, using a similar approach. It will be evident that optical lability is the essential condition for these partial asymmetric syntheses or "second order asymmetric transformations," as they are sometimes called. The more soluble enantiomer gradually racemizes and precipitates continuously as the less soluble diastereoisomer. A similar type of asymmetric synthesis was observed by Werner (1912c) when resolving the $[Cr(C_2O_4)_3]^{3-}$ ion with (−)strychninium ion. From aqueous solution he obtained the diastereoisomer $(-)(strH)_3(+)[Cr(C_2O_4)_3]$. In a solution containing *some* alcohol, however, only the diastereoisomer containing the levorotatory ion was obtained as $K(-)(strH)_2(-)[Cr(C_2O_4)_3]$. Thus, either optical form of the complex ion could be obtained, using (−)strychnine as the resolving agent. This occurred presumably because of the different solubilities of the diastereoisomers in the respective solvents, as well as rapid interconversion of the isomers.

Busch (1955) intentionally labilized the most soluble $(-)[Co(en)_3]Cl(+)$ tartrate diastereoisomer with $(\pm)[Co(en)_3]^{++}$ to improve the yield of the less soluble $(+)[Co(en)_3]Cl(+)$tart in Werner's resolution (1912a). Yields of 70% of the dextro isomer were obtained.

The ion $(+)[Co(en)_3]^{3+}$ has also been prepared by a direct second order asymmetric synthesis from the aerial oxidation of a mixture of cobalt(+)-tartrate (1 mole), ethylenediamine (3 moles), and hydrochloric acid (1 mole) in aqueous alcohol containing a little charcoal. The diastereoisomer $(+)[Co(en)_3]Cl(+)tart \cdot 5H_2O$ which separated after 30 minutes was recrystallized and converted to the chloride. A 70% yield of optically pure

$(+)[Co(en)_3]Cl_3 \cdot H_2O$ was obtained (Dwyer *et al.*, 1960). Racemization of the $(-)[Co(en)_3]^{3+}$ ion left in solution is catalyzed by the charcoal (cf. Section IV,B).

C. Adsorption Techniques

1. *Starch*

Krebs *et al.* (1954, 1956) obtained up to 65% separation of the optical isomers of organic and complex racemates, including nonelectrolytes, by passing solutions through a column of starch. Separations of 30% were common, both isomers were obtained, and large rotations were measured, 1–2° in some instances, but the substances were not purified further. The chelates resolved included: $[Co(gly)_3]^0$, $[Co(en)_3]Cl_3$, $[Cr(en)_3]Cl_3$, $K_3[Cr(C_2O_4)_3]$, $K_3[Co(C_2O_4)_3]$, $[Co(en)_2CO_3]Cl$, $[Co(en)_2(gly)]Cl$, $[Co(en)_2(NO_2)_2]Cl$, and $[Co(en)_2NO_2Cl]Cl$. A claim was also made for the resolution of $K_3[Fe(C_2O_4)_3]$ at −15°C. but only very small rotations were observed (0.03°) compared with the other separations. The column failed to resolve the labile or unstable ions $[Al(C_2O_4)_3]^{3-}$, $[Co(mal)_3]^{3-}$, $[Zn(en)_3]^{++}$, $[Cd(en)_3]^{++}$, and $[Ni(en)_3]^{++}$ at −35°C. in 60% methanol-water. Krebs *et al.* (1958) observed that there was no separation unless the compound was soluble in water or aqueous organic solvents, and that separation was better, in general, in water than in solvent mixtures. Cellulose was much inferior to starch as a resolving agent.

2. *Lactose*

A number of organic racemates have been resolved on lactose columns (Henderson and Rule, 1939; Lecoq, 1943; Prelog and Wieland, 1944). Moeller and Gulyas (1958) resolved $[Co(aca)_3]^0$ and $[Cr(aca)_3]^0$ on lactose hydrate by passing a solution of the racemate in benzene-petroleum, or benzene-hexane, through the column. The molecular rotations observed, $[M]_{5461} = 57{,}000$, were about half the values quoted by Dwyer and Gyarfas (1951b) for the resolution of $[Co(aca)_3]^0$ using a "configurational activity" method. Using a column 16 feet long and 1.5 inches in diameter, Collman and Blair (1961) repeated Moeller and Gulyas' resolution and obtained rotations as high as −1.29° for the cobalt complex, $[M]_{5461} = 445{,}000°$.

3. *Quartz*

Many partial resolutions have been claimed by the method of differential adsorption of isomers on finely powdered D or L quartz crystals (Tsuchida *et al.*, 1935, 1936; Bailar and Peppard, 1940; Busch and Bailar, 1953, 1954; Irving and Gill, 1958; Nakahara and Tsuchida, 1954; Das Sarma and Bailar, 1955; Keubler and Bailar, 1952; Schweitzer and Talbott,

1950; Karagunis and Coumoulos, 1938). They have usually been carried out by shaking or stirring an aqueous solution of the racemic complex with the active quartz powder and then filtering the solution before measuring any rotational change. Occasionally, the solution was passed down a quartz column (Karagunis and Coumoulos, 1938). Keubler and Bailar (1952) suggested that resolution is due not only to differential adsorption but also to the rate of adsorption of the isomers. Hence, in some instances it was necessary to filter the mixture quickly, before the adsorption processes had equilibrated, in order to observe any rotational change. All of these resolutions gave very low optical rotations (0.02–0.06°) and only in some instances were both (+) and (−) rotations observed. No attempt was made to separate the optically pure forms of the stable compounds.

D. Mechanical Separation

Separation of the (+) and (−) isomers by selecting the appropriate crystals requires that the complex must crystallize as a racemic mixture and the crystals must have recognizable hemihedral facets. The last condition is unnecessary if single large crystals can be grown and only one crystal is used to prepare the solution for measuring the rotation. Jaeger (1919) separated the (+) and (−) forms of $K_3[Co(C_2O_4)_3]$ by crystallizing the racemate above 13.2°C. and selecting the hemihedral crystals by hand. Similarly, single large crystals of $[Cu(en)_3]SO_4 \cdot H_2O$ grown from 80% ethylenediamine and dissolved in ethylenediamine containing a little water gave rotations of the order of ±2–3° at $\lambda = 4358$ Å. The complex racemized in 5 minutes (Gordon and Birdwhistell, 1959).

E. Configurational Activity

The concept of "configurational activity" is that the activity coefficients of the optical isomers will be different in an asymmetric environment due to the difference in interaction of the optical isomers with the same asymmetric environment. This activity difference can be reflected in two ways: (*a*) as a solubility difference and (*b*) as a shift in the equilibrium between the (+) and (−) isomers. The solubility effect has been used successfully to resolve the ion $[Co(en)_2(C_2O_4)]^+$. The complex acetate (10 gm.) and $(+)[Co(en)_3](OAc)_3$ (10 gm.) were dissolved in water (100 ml.), and $[Co(en)_2(C_2O_4)]Br$ fractionally precipitated by adding potassium bromide. The first fraction (2 gm.) contained $(+)[Co(en)_2(C_2O_4)]Br$ 90% optically pure; subsequent fractions were all *levo*. The ions (+)antimonyl tartrate, (+)bromcamphor sulfonate, and (+)camphor sulfonate gave similar results but were less effective than the $(+)[Co(en)_3]^{3+}$ ion (Broomhead, 1960). Despite the large separation (40%), the solubilities of (+) and $(-)[Co(en)_2(C_2O_4)]Br$ in the presence of $(+)[Co(en)_3]Cl_3$ are not greatly different, i.e., <5%. It would seem then that the separation depends primarily on the formation of a few crystals of the least soluble isomer, leading to a precipitation of that isomer in bulk (by seeding) until the solution is sufficiently supersaturated with the *levo* form for it to commence to separate. Werner (1914) obtained a similar separation by adding $(-)[Co(en)_2(C_2O_4)]Br$ (0.05 gm.) to a hot solution of $(+)[Co(en)_2(C_2O_4)]Br$ (2 gm. in 160 ml. H_2O). On cooling the mixture to 5–10°C. and adding alcohol (40 ml.), $(-)$ $[Co(en)_2(C_2O_4)]Br$ (0.2 gm.) of high optical purity precipitated, i.e., considerably more than the *levo* isomer added initially. Werner also obtained a similar result when he substituted $(-)[Co(en)_2$-$(NO_2)_2]Br$ for the above oxalato complex.

F. Differential Diffusion

Evidence has been obtained for different diffusion coefficients of (+) and $(-)[Co(en)_3]Cl_3$ in sucrose solutions (Carassiti, 1958). The possibility of partially separating ions, in an asymmetric medium, ionophoretically arises naturally from the concept of "configurational activity" if it is assumed that this effect is due to a difference in association of the asymmetric ions with the active medium.

G. Zone Melting

This technique for the resolution of optical isomers also depends on "configurational activity" since it assumes that the solubility of the (+) and (−) forms will be different in an asymmetric medium and that the more soluble component will move with the melted zone and the less soluble remain. However the success of the method is not only dependent on differential solubility but also on the differential rate of solution and the two factors may be either complementary or opposed. Despite these problems, Doran and Kirschner (1961) have succeeded in partially separating the diastereoisomers (+) and $(-)[Co(en)_3]Cl(+)$tart and the isomers $(+)_{5461}$ and $(-)_{5461}[Co(aca)_3]^0$ in the presence of (+)dibenzoyltartaric acid, in a column of ice.

H. "Active" Racemate Formation

When an isomorphous optically active salt $(+)a'$ is mixed with the analogous racemic salt $(\pm)a$, a racemic "mixed" racemate $(+)a'(-)a$ often separates as the less soluble material, leaving an excess of $(+)a$ in the solution (cf. Section V,A. Delèpine (1934) used this method to resolve a number of isomorphous complexes, notably the tris oxalato ions of Co(III), Rh(III), Ir(III), and Cr(III). The method is useful, however, only if the least soluble component is the "active racemate," unless $(+)a'$ and $(+)a$ crystallize as a solid solution when a partial separation of $(+)a$ and $(-)a$ is also effected.

IV. Stereospecificity

A. Introduction

When the ethylenediamine molecules in the dissymmetric chelate $[Co(en)_3]Cl_3$ are replaced by an optically active ligand such as (−)1,2-*trans*-cyclopentanediamine, an asymmetric bias is imparted to the system and the two possible optical isomers are no longer obtained synthetically in equal amounts. This preference for one optical isomer over the other is called ligand *stereospecificity*.

Early workers concluded that an optically active ligand favored the formation of one isomer to the complete exclusion of the other. Jaeger (1930) from his studies with the tris(*trans*-1,2-cyclopentanediamine)-rhodium(III) cation suggested that "the levorotary configuration of the complex ion is incompatible with the presence of three levo molecules of the base." He reached a similar conclusion with the corresponding cobalt complex, as did Smirnoff (1920) from studies with the tris(1,2-propylene-diamine)cobalt(III) and platinum(IV) cations. However, these claims for absolute stereospecificity were disproved (Lifschitz, 1925) by the isolation of both D(+)(+)(+) and L(+)(+)(+) isomers of soluble α-tris(alanine)-cobalt(III), and also by the partial separation of both optical forms of $[Co(-)(pn)_2CO_3]^+$ (Bailar and McReynolds, 1939), though the significance of these separations seems subsequently to have been largely ignored. Finally, Dwyer *et al.* (1959) have isolated D- and L-$[Co(-)(pn)_3]I_3$ and D- and L-$[Pt(-)(pn)_3]Cl_4$ (Dwyer and Garvan, 1959). Both systems had been reputed to be absolutely stereospecific. Rotatory dispersion curves of the Co(III) complexes (Section V) show that their configurations are related as are D- and L-$[Co(en)_3]Cl_3$ and are not *cis-trans* isomers, due to the relative position of the methyl group, with the same optical configuration. Both isomers were isomerized to the equilibrium mixture of D(+)(+)(+) and L(−)(−)(−) in aqueous solution using charcoal as a catalyst and the proportion of L:D was found as 7:1. This value has been revised subsequently (Dwyer *et al.*, 1961) and the corrected ratio is 15:1 (Section IV,C). No allowance had previously been made for the partial decomposition of the complexes on charcoal. Stereospecificity, in these complexes, is therefore not an absolute phenomenon but one isomer merely predominates over the other, a finding that is more in accord with results in purely organic systems. A detailed discussion of the historical background will be found in reviews by Calvin and Martell (1952) and Basolo (1956).

B. Disproportionation in Synthetic Reactions

Jaeger and Blumendal (1928) supported the theory of absolute stereospecificity with several other experiments. They found, when attempting to prepare complexes containing (+) and (−)*trans*-1,2-cyclopentanediamine by reacting *trans*-$[Co(+)(cpn)_2Cl_2]Cl$ with (−)cyclopentanediamine, that disproportionation occurred and only the (+)(+)(+) and (−)(−)(−) isomers were obtained: $3[(+)(+)(-)] \rightarrow 2[(+)(+)(+)] + [(-)(-)(-)]$. In Jaeger's own words "the expected combinations seem to be unstable from stereometric causes," and "the formation of the cation of highest possible symmetry is favored at the cost of the complexes of lower symmetry" (1930). Similarly, the reaction of *trans*-$[Co(-)(cpn)_2Cl_2]Cl$ with ethylenediamine gave both $[Co(-)(cpn)_3]^{3+}$ and $[Co(en)_3]^{3+}$ ions but the reaction of *trans*-$[Co(en)_2Cl_2]Cl$ with (−)cyclopentanediamine produced the desired cation D-$[Co(en)_2(-)(cpn)]^{3+}$. It seems unreasonable that the addition of the base in the last instance should yield but one product while disproportionation occurs in the other two essentially similar reactions. Jaeger maintained that the first dismutation was due to the inherent instability of a complex containing both *dextro* and *levo* ligands but this is at variance with the preparation of (+) $[Pt(-)(pn)_2(+)(pn)]Cl_4$ and (−)$[Pt(-)(pn)_2(+)(pn)]Cl_4$. These complexes once formed are quite stable (Dwyer and Sargeson, 1959a). Evidence has also been obtained for the existence of stable $[Co(pn)_3]^{3+}$ cations containing both (+) and (−)propylenediamine in the one ion. The salt $[Co(pn)_3]Cl_3$, prepared from racemic propylenediamine, and chromatographed on paper with butanol–H_2O–HCl (60–30–10), gave three distinct bands and a faint trace of a fourth. The first and fourth bands were identified as the racemic pairs of D[(+)(+)(+)], L[(−)(−)(−)], and L[(+)(+)(+)], D[(−)(−)(−)]. The two central bands must contain a mixture of (+) and (−)propylenediamine around the central ion though these compounds have not yet been characterized (Dwyer *et al.*, 1961).

This apparent instability of the "mixed" complexes does not arise from "inherent" factors but is due rather to the method of preparation. This has been demonstrated by reacting (−)propylenediamine with racemic *cis*- or *trans*-$[Co(en)_2Cl_2]Cl$ in water, methanol, or dimethylformamide; the resulting product contained all possible combinations of the ligand around the central ion, i.e., $[Co(en)_3]^{3+}$, $[Co(en)_2(-)(pn)]^{3+}$, $[Co(en)(-)(pn)_2]^{3+}$, and $[Co(-)(pn)_3]^{3+}$ (Dwyer *et al.*, 1961). It has been suggested that the dismutation probably occurs by electron transfer between the cobalt(III) complexes and a small amount of cobalt(II) complex in solution and, as well, by the formation of a 7- or 8-covalent "spin free" cobalt(III) intermediate. The "spin free" intermediate and the cobalt(II) ion would be labile to substitution and equilibrium between the inert cobalt(III) species and the bases would thus be established rapidly (Dwyer and Sargeson, 1959b, 1960). The reaction between cobalt(III) nitro complexes and a diamine also has been used to prepare "mixed" diamine compounds

(Werner, 1917), and this type of reaction yields a similar mixture of products, presumably by the same path (Dwyer and Sargeson, 1961b).

The composition of the products from the following reactions:

(1) $cis\text{-}[Co(en)_2Cl_2]Cl + (-)pn \xrightarrow{MeOH}$

(2) $cis\text{-}[Co(en)_2(NO_2)_2]Cl + (-)pn \xrightarrow{H_2O}$

(3) $CoCl_2 + (-)pn + 2en + HCl \xrightarrow[Charcoal]{O_2}$

(4) $CoCl_2 + 2(-)pn + en + HCl \xrightarrow[Charcoal]{O_2}$

TABLE I
PER CENT COMPOSITION OF REACTION PRODUCT

	Reaction			
Species	(1)	(2)	(3)	(4)
DL$[Co(en)_2(-)(pn)]^{3+}$	44.5	43.5	40	21
DL$[Co(en)(-)(pn)_2]^{3+}$	11	20.5	21	42
DL$[Co(-)(pn)_3]^{3+}$	13.5	5	7	32
DL$[Co(en)_3]^{3+}$	31	31	32	5

is listed in Table I. Equilibrium between the cobalt(III) complexes and the bases in reactions (3) and (4) is established rapidly in the presence of a charcoal catalyst. It is known that the rate of electron transfer between $[Co(en)_3]^{3+}$ and $[Co(en)_3]^{2+}$ on charcoal is very fast, and the Co(II) ion is labile to substitution (Dwyer and Sargeson, 1961a).

The similarity between the products of reaction (3) and those of the preparative procedures (1) and (2) leaves no doubt that, whatever the mechanism, equilibrium is reached with little retention of the intended composition. As expected, the reaction of $(+)[Co(en)_2Cl_2]Cl$ with ethylenediamine gives an almost racemic product containing only about 3% $(+)[Co(en)_3]Cl_3$ (Dwyer and Sargeson, 1959b).

Since both bases have almost identical pK_a values, the stability constants for $[Co(en)_3]^{3+}$ and $[Co(pn)_3]^{3+}$ should be very similar. Assuming this, the statistical distribution of bases among the possible compounds for a ratio of en/pn = 2/1 is, to the nearest integer: $[Co(pn)_3]^{3+}$, 4%; $[Co(pn)_2(en)]^{3+}$, 22%; $[Co(pn)(en)_2]^{3+}$, 44%; $[Co(en)_3]^{3+}$, 30%. Reaction (3) gives results surprisingly close to these values. When the en/pn ratio is reversed (1/2) as in reaction (4) the concentrations of the respective species alter to correspond to the new statistical distribution: $[Co(pn)_3]^{3+}$, 30%; $[Co(pn)_2(en)]^{3+}$, 44%; $[Co(en)_2(pn)]^{3+}$, 22%; $[Co(en)_3]^{3+}$, 4%. The agreement is additional evidence for complete equilibration in these reactions. A mixture of all of the possible products: $[Co(en)_3]^{3+}$, $[Co(en)_2(phen)]^{3+}$, $[Co(en)(phen)_2]^{3+}$, $[Co(phen)_3]^{3+}$, is obtained from the reaction of 1,10-phenanthroline (phen) and *cis*-$[Co(en)_2Cl_2]Cl$ in alcoholic solution (Buckingham *et al.*, 1961).

It seems likely that Jaeger's "mixed" complexes, as well as others prepared by this type of reaction, contain a mixture of all possible compounds, which give the correct analysis only because of the stoichiometry of the reaction mixture. Fractional crystallization, analysis, or even resolution are often not sufficient to confirm the composition of these compounds or the isomers. For instance, the product obtained by reacting (+)pn with *cis*-$[Co(en)_2Cl_2]Cl$ in methanol can be resolved as the chloride (+)tartrate and the composition of the least soluble diastereoisomer remains unchanged even after several recrystallizations. The iodide obtained from this diastereoisomer is also unaltered by recrystallization and has the apparent formula $[Co(en)_2(+)(pn)]I_3 \cdot H_2O$. The "isomer" is, in fact, a mixture of D-$[Co(en)_3]I_3 \cdot H_2O$ and D-$[Co(+)(pn)_3]I_3 \cdot H_2O$. Apparently, the two substances form a solid solution and separation by crystallization is virtually impossible (Dwyer and Sargeson, 1959b). For studies of this kind to be valid, very precise analyses are required and it is essential to recover the bases from the separate fractions to ensure that the ratio is constant. It would seem unlikely, for these reasons, that all five isomers of the much quoted example $[Co(en)(-)(pn)(NO_2)_2]Cl$ are pure (Werner, 1917). In fact, paper chromatography shows the presence of seven distinct species and the rotatory dispersion inflexions for the active isomers cross zero (530 mμ) almost 60 mμ from the active *cis*-$[Co(en)_2(NO_2)_2]^+$ and $[Co(-)(pn)_2(NO_2)_2]^+$ ions whose rotatory dispersion inflexions almost coincide (470 mμ).

A repetition of Jaeger's work with cyclopentanediamine chelates now in progress (Royer, 1961) has shown that his *trans*-diamine base is grossly impure and contains at least 10% of the *cis* form along with other impurities. Attempts to reproduce the preparation of *trans*-$[Co(cpn)_2Cl_2]Cl$, $[Co(cpn)_3]Cl_3$, and $[Co(en)_2(cpn)]Cl_3$ using the crude diamine failed in the first two instances and in the last gave a brown gelatinous mass, whose composition varied from preparation to preparation, and contained at least three components. Small amounts of the green *trans*-$[Co(cpn)_2Cl_2]Cl$ and the bright pink $[Co(cpn)_3]Cl_3$ could be prepared, however, from the crude diamine but all attempts to prepare these compounds from pure *trans*-diamine failed.

Equilibration between the complex and the bases can be avoided with the platinum(IV) complexes, since, unlike Co(II), Pt(II) is not labile to substitution and traces of Pt(II) complexes are not important. "Mixed" complexes containing ethylenediamine and/or propylenediamine can be prepared (Dwyer and Sargeson, 1959a) in dimethylformamide solution at room temperature by the reactions:

$$[Pt(en)Cl_4]^0 + 2(-)pn \rightarrow D,L\text{-}[Pt(en)(-)(pn)_2]Cl_4$$
$$[Pt(-)(pn)Cl_4]^0 + 2en \rightarrow D,L\text{-}[Pt(-)(pn)(en)_2]Cl_4$$
$$[Pt(-)(pn)Cl_4]^0 + 2(+)pn \rightarrow D,L\text{-}[Pt(+)(pn)_2(-)(pn)]Cl_4$$

The white Pt(IV) complexes form rapidly and precipitate from solution. Fractionation of $[Pt(+)(pn)_2(-)(pn)]Cl_4$ as the (+) and then as the (−)tartrates gave only the expected two isomers with (−)pn/(+)pn ratios of 1/2. In the same way D- and L-$[Pt(en)(-)(pn)_2]Cl_4$ and D- and L-$[Pt(en)_2(-)(pn)]Cl_4$ can be separated and characterized. All of these compounds are extremely stable and racemization, isomerization, or dismutation do not occur, even in boiling dilute acid or alkali solutions. The approximate amounts of D and L isomer formed (Table II) also suggest that the reaction does not occur under equilibrating conditions as it does for the cobalt complexes. The isomeric composition seems to depend only on the rate of formation of the respective isomers from the parent substance $[PtBCl_4]^0$ and the entering base.

TABLE II
ISOMER IN REACTION MIXTURE

Species	Per cent
$(+)[Pt(en)_2(-)(pn)]Cl_4$	50
$(+)[Pt(en)(-)(pn)_2]Cl_4$	55
$(+)[Pt(+)(pn)_2(-)(pn)]Cl_4$	65
$(+)[Pt(-)(pn)_3]Cl_4$	60

The ratio of (+)/(−) $[Pt(-)(pn)_3]Cl_4$—3/2 prepared this way, compared with 7/1, or greater, when equilibrium is established (Dwyer and Garvan, 1959)—shows the validity of this preparative method. Rhodium and iridium(III) should also behave in this way since their bivalent states are not easily accessible and are probably inert to substitution.

C. CONFORMATIONAL ANALYSIS

It will be clear that certain optical isomers and mixed ligand complexes in the systems examined so far are not precluded from forming on "stereometric" or other "inherent" grounds but that the experimental results are due to the methods of preparation as well as the relative stabilities of the isomers. The latter aspect has been accounted for most successfully (Corey and Bailar, 1959) by an appreciation of: "(1) the orientation of donor atoms about the central metal ion and (2) the spatial arrangements which can be assumed by the individual chelate rings and their relative stabilities"; in short by a conformational analysis of the molecule or ion. Coordinated ethylenediamine can be likened to cyclopentane, and the techniques used for stereochemical analysis of carbocyclic ring systems can be applied to chelate rings. The bond distances and angles for the Co(III)-ethylenediamine strain-free chelate ring system, when calculated by a vector analysis, agree almost exactly with those obtained from the crystal structure analysis of $[Co(en)_3]Cl_3$ (Saito *et al.*, 1957). In the strain-free system the H atoms attached to adjacent C and N atoms are completely staggered and two conformations are possible (XII).

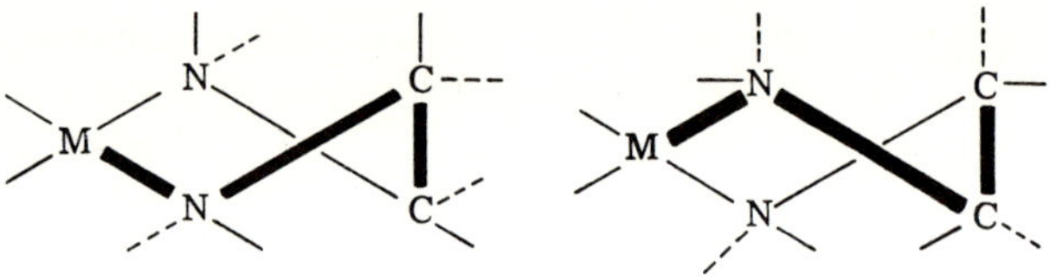

(a) *k* (b) *k'*

(XII)

The conformations of coordinated ethylenediamine rings. Bonds to the H atoms are indicated by solid and broken lines. (a) is the form found in the D-$[Co(en)_3]^{3+}$ ion.

These conformations are mirror images and are designated as *k* and *k'* forms. The four possible arrangements of the three ethylenediamine rings in $[Co(en)_3]^{3+}$, *kkk*, *k'kk*, *kk'k'*, *k'k'k'*, have different stabilities for the one complex configuration, D. The relative potential energies of the two extreme forms *kkk* and *k'k'k'* for the D-$[Co(en)_3]^{3+}$ configuration were calculated by considering the following nonbonded atomic interactions: (1) H,H; (2) H,C between the two C atoms of the ring and the axial H of the —NH_2 groups; and (3) electrostatic interactions (inverse square) between the hydrogen atoms of the donor —NH_2 groups. The energy difference between the *kkk* (stable) and *k'k'k'* (unstable) conformations was assessed as 1.8 kcal./mole (ca. 0.6 kcal./mole for each ligand.) The *kkk* form is favored mainly because of the less severe H, C interactions, and occurs to the extent of 95.5% if the energy calculation is correct. The crystal structure is in accord with the presence of the *kkk* conformation, i.e., the three C—C axes of the ethylenediamine molecules are parallel to the "short" or threefold axis of symmetry of the complex ion. The two extremes are shown in (XIII) looking down the threefold axis of symmetry. The heavy lines denote atoms and bonds at the top of the ion.

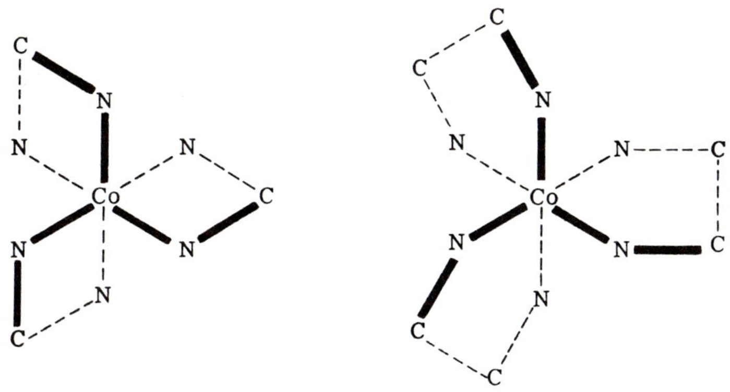

kkk *k'k'k'*

(XIII)

The C—C axes are eclipsed in the *kkk* form and staggered in the *k'k'k'* configuration. In the L-complex configuration, however, the *k'k'k'* structure is the most stable since the C—C axes are now parallel to the threefold axis of symmetry.

D(−)1,2-propylenediamine also forms a five-membered ring like ethylenediamine and theoretically can exist in the *k* and *k'* conformations. The absolute configuration (a) of D(−)pn (Reihlen *et al.*, 1932) and the two conformations of the ligand (b) and (c) for the D configuration of the complex are shown in (XIV).

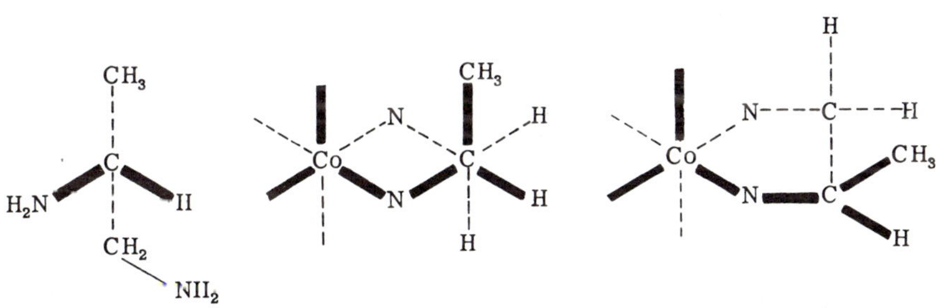

(a) *D*(-)pn (b) *k*pn (c) *k'*pn

(XIV)

In (b) the methyl group is axial and in (c) equatorial and the energy difference between these conformations has been assessed as greater than 2 kcal./mole in favor of the equatorial form. This is enough to ensure the formation of predominantly k' units for (−)pn when coordinated in the D configuration. In the L complex D(−)pn retains the k' conformation and the methyl group remains equatorial but the ring is now in the stable condition with the C—C axis parallel to the threefold symmetry axis. Thus we have two effects in opposition in the D-configuration: (*a*) the k conformation of D(−)pn is favored by 0.6 kcal./mole, but (*b*) the equatorial methyl group is much more stable than the axial (2 kcal./mole). Since the latter is the far greater effect, the ring would form the k' conformation predominantly, rather than the k conformation with an axial methyl group. Hence D- and L-complex configurations with D(−)pn ligands will not have equal stability since k' units for the propylenediamine ring will predominate in both. If we take the simplest instance, D- and L-$[Co(en)_2$D(−)$(pn)]^{3+}$, then the conformation of the individual rings would be $k'_{en}k'_{en}k'_{pn}$ for the L isomer and $k_{en}k_{en}k'_{pn}$ for the D isomer. The conformation of the ethylenediamine groups changes so as to keep the C—C axis parallel to the threefold symmetry axis of the complex ion; but the conformation of (−)pn remains unaltered in order to keep the methyl group equatorial. Thus L-$[Co(en)_2$-D(−)$(pn)]^{3+}$ should be more stable than D$[Co(en)_2$D(−)$(pn)]^{3+}$ by ca. 0.6 kcal./mole, the difference between the k and k' stabilities for one ring in one complex configuration. The equilibrium concentrations of L- and D-$[Co(en)_2$D(−)$(pn)]^{3+}$ are in the ratio of 2:1 (Dwyer *et al.*, 1961) respectively and this amounts to a free energy difference at 25°C. of ca. 0.4 kcal./mole, quite close to that calculated. The absolute configurations of these isomers have been established by comparing their rotatory dispersion curves with D- and L-$[Co(en)_3]^{3+}$ (Section V).

Cis and *trans* isomers are not possible with $[Co(en)_2(pn)]Cl_3$ but when two or three molecules of propylenediamine are coordinated the methyl groups can be either *cis* or *trans*. Whether or not some of these isomers are excluded energetically, or whether they have not been separated because their properties are so similar, is not yet known. For instance, both [Co-(en)(−)$(pn)_2]Cl_3$ and [Co(−)$(pn)_3]Cl_3$ can be resolved into two forms which are related by their rotatory dispersion curves as are D and L. The optical forms may be a mixture of *cis* and *trans*, or one may be D-*cis* and the other L-*trans*. If they are mixtures no separation has yet been detected by fractional crystallization, resolution, or chromatography and for the purposes of this article they are described as a single optical isomer, D or L.

The ratio of the equilibrium concentrations of the pairs of optical isomers as each ethylenediamine is replaced by (−)propylenediamine is given in Table III. The specificity due to each replacement is readily observed; the preponderance of the L configuration increases from 2:1 to 7:1 to 15:1 and the free energy difference between the L and D isomers increases by approximately 0.5 kcal. for each molecule of (−)propylenediamine. This effect is implicit in the conformational analysis and the agreement between the observed and calculated values is most satisfactory.

TABLE III
ISOMER RATIOS FOR COBALT(III) ETHYLENEDIAMINE AND PROPYLENEDIAMINE COMPLEXES

Isomers	Ratio	$\Delta G_{obs.}$ (kcal./mole 25°C)	$\Delta G_{calc.}$ (kcal./mole 25°C)
L[Co(en)(en)(−)(pn)]$^{3+}$/D[Co(en)(en)(−)(pn)]$^{3+}$ k' k' k' / k k k'	2/1	0.4	0.6
L[Co(en)(−)(pn)(−)(pn)]$^{3+}$/D[Co(en)(−)(pn)(−)(pn)]$^{3+}$ k' k' k' / k k' k'	7/1	1.2	1.2
L[Co(−)(pn)(−)(pn)(−)(pn)]$^{3+}$/D[Co(−)(pn)(−)(pn)(−)(pn)]$^{3+}$ k' k' k' / k' k' k'	15/1	1.6	1.8

The analysis also predicts that the stability difference between the D and L forms of α-tris(L(+)alanine)cobalt(III) would not be great. The five-membered alanine rings are essentially planar and the distinction between axial and equatorial substituents no longer applies. This accounts nicely for Lifschitz's partial isolation (1925) of both D and L α-forms, whereas the tris propylenediamine and cyclopentanediamine systems gave predominantly one isomer.

The complexes Na[CoD(−)(pn)(C_2O_4)$_2$] and [CoD(−)(pn)$_2$$C_2O_4$]Cl show little or no specificity. The concentrations observed were 60% D(−)$_{5461}$-Na[CoD(−)(pn)(C_2O_4)$_2$] and 40% L, and approximately equal amounts of the D and L(+)[Co(−)(pn)$_2$$C_2O_4$]Cl (MacDermott, 1961). They were estimated from the composition of the reaction mixture and also by isomerizing each optically pure isomer on charcoal. Like chelated alanine, the oxalate ring is planar and does not contribute to the relative stability of the isomers, hence the energy difference between the k' conformation of propylenediamine in the L and D configurations must be very small. These same effects would be expected in the [CoD(−)(pn)$_2$$CO_3$]$^+$ and *cis*-[CoD(−)-(pn)$_2$$Cl_2$]$^+$ systems.

The sexadentate D(−)propylenediamine tetraacetic acid can adopt two feasible configurations when coordinated to a cobalt(III) ion (XV), which arise from the movement of the acetate bridges when the D(−)propylenediamine ring conformation changes. The axial acetate group (O*) coordinated in the (1) position in (XVa) (k'_{pn}) becomes equatorial when the propylenediamine ring conformation changes to k in (XVb) and the acetate

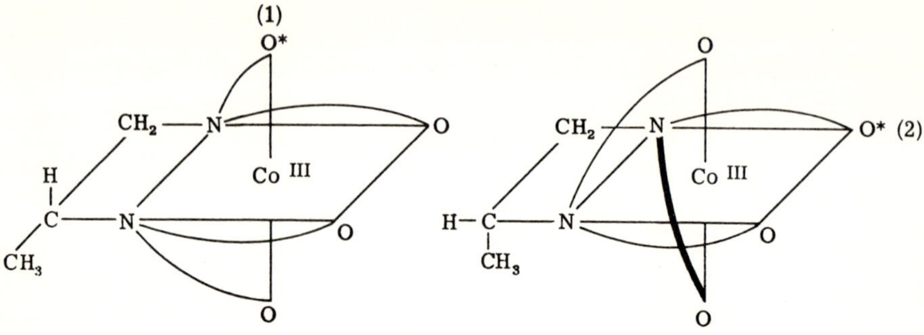

L(-) or LD (a)   D(-) or DD (b)

(XV)

now coordinates in the octahedral position (2). All the bridges must change position if the ligand is to remain sexadentate.

Optically active D(−)PDTA obtained from D(−)propylenediamine coordinates with cobalt to give $K(+)_{5461}$[CoD(−)PDTA] which has the same rotation as that obtained by resolving racemic K[CoD,LPDTA]. No evidence for the other isomer could be found (Dwyer and Garvan, 1961). This means that the free energy difference between the L(−) and D(−) isomers is of the order of 3 kcal./mole, at least. The single sexadentate ligand is, for all practical purposes, completely stereospecific and exerts a far greater effect than the bidentate (−)propylenediamine. The reason for the difference is readily apparent from (XV). In (a) the methyl group is equatorial and points away from the other atoms, but in the axial position (b) the methyl group interacts strongly with the acetate bridge in heavy type. The interaction can be demonstrated readily with molecular models of the two configurations, and really approaches a more formal steric hindrance. Bailar and Corey have estimated the energy difference between the axial and equatorial methyl conformations with bidentate propylenediamine as in excess of 2 kcal./mole. In the sexadentate PDTA chelate the interactions are even greater, and the energy difference between the D(−) and L(−) isomers will be greatly in excess of this value.

Since the absolute configuration of (−)propylenediamine is known (Reihlen *et al.*, 1932), the absolute configuration of $(+)_{5461}$K[Co(−)PDTA] (Busch and Cooke, 1961; MacDermott and Sargeson, 1961) can be deduced from the necessity for the methyl group to be in the equatorial position, and has the structure shown in (XVa). This has been designated L by comparison with $(+)_{5461}K[Co(en)(C_2O_4)_2]$ and thence $(-)[Co(en)_3]Cl_3$ (Section V).

Essentially, propylenediaminetetraacetic acid conforms with Jaeger's theory that one configuration of the complex is incompatible with one optical isomer of the ligand and thus it coordinates completely stereospecifically. The specificity is maintained provided that either of the structures shown in (XVI) remains intact.

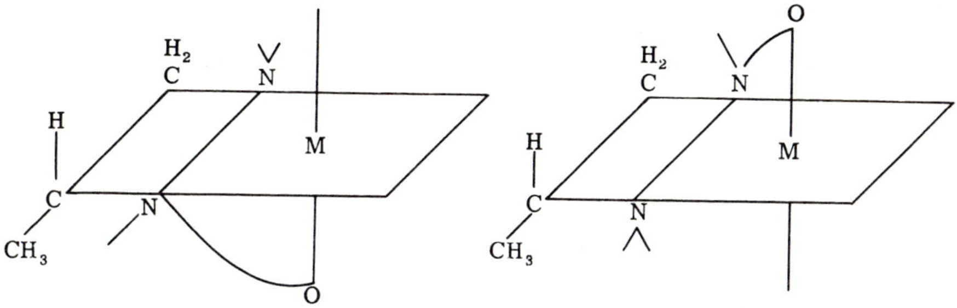

(XVI)

trans-1,2-Cyclohexanediaminetetraacetic acid (CDTA) also coordinates stereospecifically (Dwyer and Garvan, 1961) for a similar reason. *trans*-1,2-Cyclohexanediamine with the conformation (XVIIa) reacts with sodium

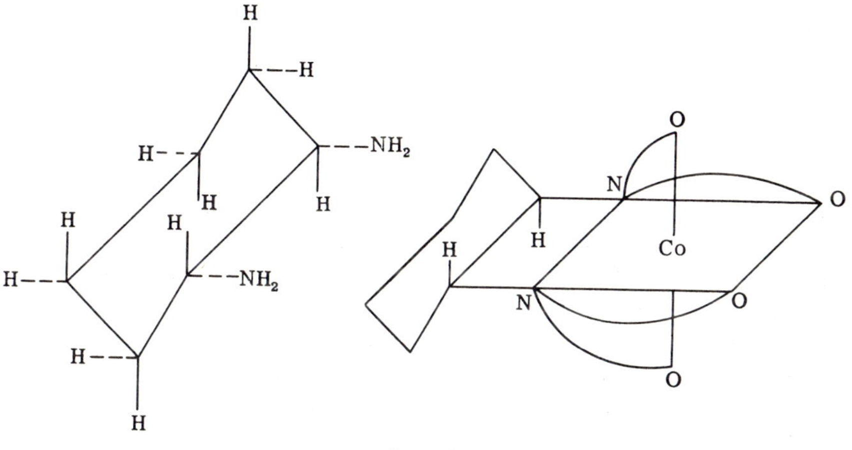

(a) (XVII) (b)

chloracetate to give optically active *trans*-1,2-cyclohexanediaminetetraacetic acid with full retention of configuration. The acid coordinates with Co(III) to give the structure (XVIIb).

In diaminocyclohexane (XVIIa) both amino groups are equatorial, since in the axial positions the ligand cannot function as a bidentate, and therefore cyclohexanediaminetetraacetic acid in this condition would not coordinate through both N atoms. If cyclohexanediaminetetraacetic acid adopts the "boat" conformation (ca. 5 kcal./mole less stable) the L structure (XVIIb) is still the only one possible, as the ligand now will not coordinate at all six positions in the D configuration.

A comparison of the rotatory dispersion curves of L$(+)_{5461}$[Co$(-)$-PDTA]$^-$ and$(+)_{5461}$[Co$(-)$CDTA]$^-$ (Section V,E) shows that these ions have the same configuration, i.e., (XVa) and (XVIIb) respectively. This means that $(-)$cyclohexanediamine has the configuration (XVIIa).

Similarly, tridentate ligands like $(-)$aspartic acid should also coordinate stereospecifically (XVIIIa). This chelate has two asymmetric centers, the

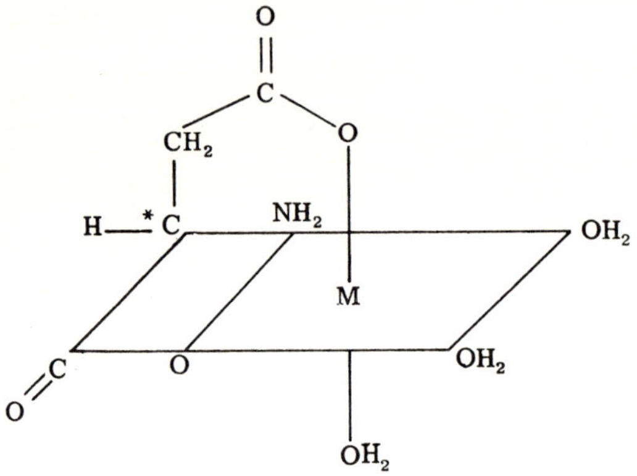

(a)

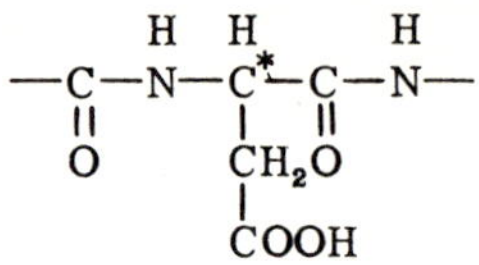

(b)

(XVIII)

metal and the C* atom, but specificity here is dependent only on the asymmetric carbon atom and not on the amino acid ring conformations. In biological systems a portion of a peptide chain such as (XVIIIb) when coordinated to a metal ion might well act as a stereospecific reaction site. The asymmetric nature of the complex is determined by the ligand and does not depend on the rate of dissociation of the unit, i.e., the complex with Mg^{++} or Ca^{++} would preserve its steric form just as easily as the inert Co(III) complex.

The influence of the conformation of substituted chelate rings on isomer stability is demonstrated by the foregoing examples and the principles enunciated can be readily applied to other systems. For instance, the structural conditions which give specificity in the propylenediaminetetraacetic acid complexes should also apply to the sexadentate 1,8-bis(salicylideneamino)-3,6-dithia-4 methyl octane (EPE),

$$HO\cdot C_6H_4\cdot CH{=}N\cdot(CH_2)_2\cdot S\cdot CH_2\cdot CHCH_3\cdot S\cdot(CH_2)_2\cdot N{=}CH\cdot C_6H_4\cdot OH.$$

The optical isomer (XIXa) should be favored because of the less severe interaction between the S—N bridge and the methyl group. Preliminary studies indicate that specificity in this system is also high (MacDermott, 1961).

The degree of stereospecificity thus varies from practically zero in the

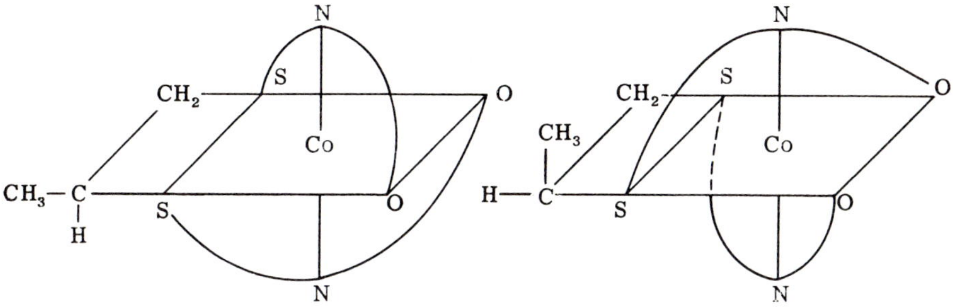

(a) (b)

(XIX)

α-trisalaninecobalt(III) chelate to the total stereospecificity of the tridentate aspartate complexes. The degree depends upon the type of interaction involved from the sum of a number of small effects as with the trispropylenediamine complexes or one large interaction as with the propylenediaminetetraacetic acid complexes to total stereospecificity imposed on the complex by the rigidity of the chelating agent, such as the asymmetric carbon atom in the aspartate ion. The last two effects are easily evaluated, the first is more difficult. However, the application of conformational analysis is obviously successful and it is very likely that Corey and Bailar's paper will stimulate much activity in this field.

V. Rotatory Dispersion and Configuration

A. Introduction

The optically active complexes of transition metal ions have one chromophoric center at least, coincident with the "center of dissymmetry"—namely the metal ion—and hence some of the absorption bands in the chromophore are optically active. For instance, the $(+)[Co(en)_3]^{3+}$ ion has optically active bands at 470 and 218 mμ* and the rotatory dispersion curve (XX) is a combination of the two individual R.D. curves which arise from these active transitions. The absorption band at 340 mμ, however, has little or no effect on the rotatory dispersion, a common feature with many cobalt(III) compounds (Kobayashi, 1943; Kuhn and Bein, 1934; Mathieu, 1936a,b; Tsuchida, 1937; Shimura, 1958). Though the R.D. curve

* *Note added in proof.* The relation between configuration and the sign of the Cotton effect and R.D. curve for the transitions in Co(III) complexes has been clarified recently by Mason (1963 *Quart. Rev.* **17**, 20).

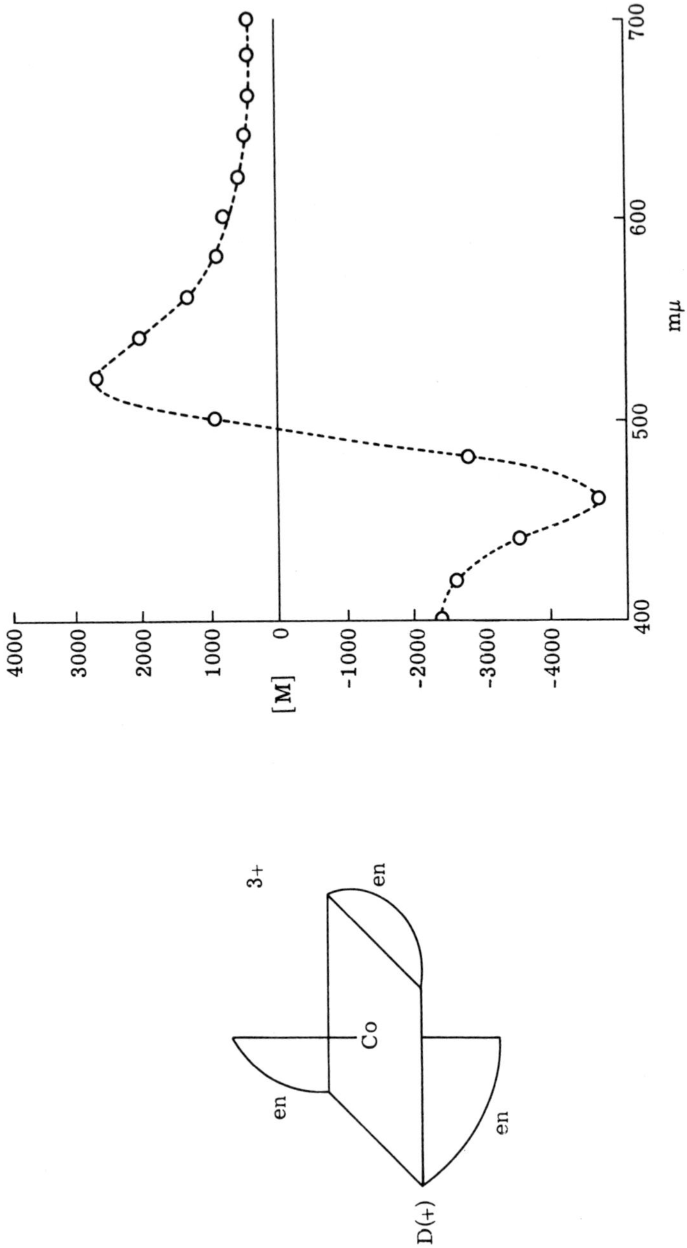
4000
3000
2000
1000
0
-1000
-2000
-3000
-4000
[M]
400
500
600
700
mμ
3+
en
Co
en
en
D(+)

(XX)

is diagnostic of configuration (Section I), it is not possible to assign the curve empirically to either the D or L structure. Some form of absolute standard is necessary and this has been provided by the absolute structural analysis of the (+)$[Co(en)_3]^{3+}$ ion (XX) (Saito *et al.*, 1957). The configurations of other cobalt complexes can be related to the D$[Co(en)_3]^{3+}$ ion but it seems unlikely that this standard can be used for complexes of other metals except perhaps Rh and Ir where the ${}^1A_{1g} \rightarrow {}^1T_{1g}$ transition is also the first optically active band. A comparison is made in Section V,B which suggests that such a relationship exists.

The Cotton Effect (elliptic polarization or circular dichroism) may also be used as a criterion for configuration since the sign of this effect, associated with the appropriate R.D. curve, is either (+) or (−) (XXI). The circular

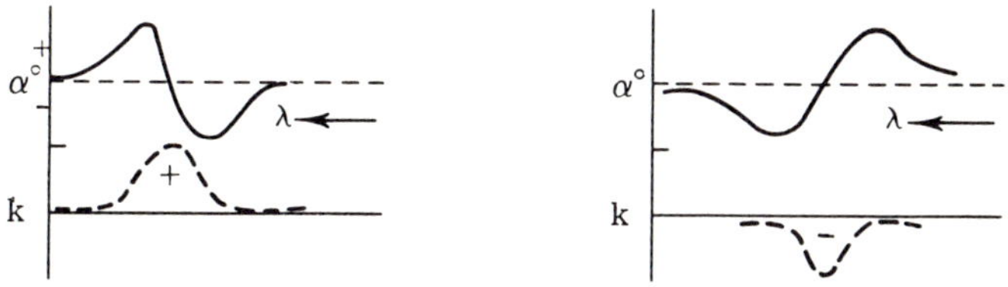

(XXI)

dichroism curve is especially valuable where the R.D. curve is complicated by a number of overlapping absorption bands since it has either a positive or negative sign for each transition and not both like the R.D. curve.

The transition metal complexes often have their first active absorption band in the visible or near ultraviolet. This often is relatively isolated and the R.D. curve can be clearly interpreted, like that of the $[Co(en)_3]^{3+}$ ion, without reference to the Cotton Effect. However, the band may be split into two or more components by departure from the "cubic" symmetry. For example, the ${}^1A_{1g} \rightarrow {}^1T_{1g}$ transition at about 470 mμ for the $[Co(en)_3]^{3+}$ ion divides into at least two levels for the ion *cis*-$[Co(en)_2Cl_2]^+$ which has one principal band at 530 mμ and a much less intense secondary band at 600 mμ. Both transitions are optically active and the R.D. curves due to each have been separated empirically, such that the "principal band" contributes the major part of the rotation in the visible region (Hidaka *et al.*, 1958) (XXII). These contributions were also recognized by Mathieu (1936a). In the optically active complex cations $[Co(en)_2Cl_2]^+$, $[Co(en)_2$-$ClBr]^+$, and $[Co(en)_2NCSCl]^+$, both "principal" and "accessory" bands appear, with the maximum rotation of the former band much greater than of the latter.

The visible absorption band of the *cis*-chloroaquobis(ethylenediamine)-cobalt(III) ion is more symmetrical than that of the *cis* dichloro ion and

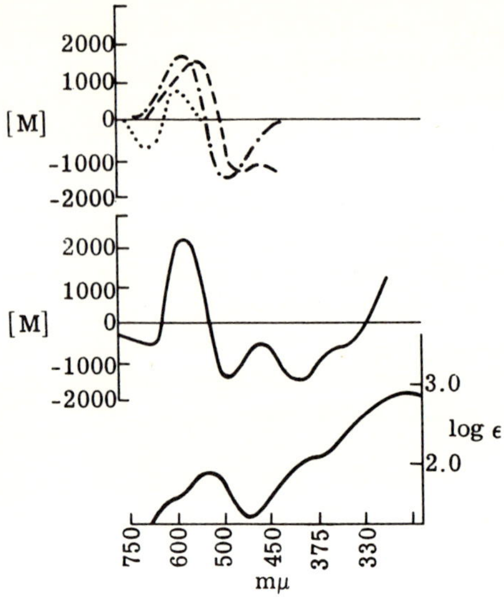

(XXII)

The rotatory dispersion curves for *cis*-dichloro- and *cis*-chloroaquo-bis(ethylenediamine)cobalt(III) ions: —— R.D. curve for $[Co(en)_2Cl_2]^+$; — · — · "principal" R.D. curve for $[Co(en)_2Cl_2]^+$; · · · · "accessory" R.D. curve for $[Co(en)_2Cl_2]^+$; – – – R.D. curve for $[Co(en)_2ClH_2O]^{++}$. [After Hidaka *et al.* (1958).]

the R.D. curve is similarly more symmetrical. The latter, in fact, is remarkably like the R.D. curve of the principal band of *cis*-$[Co(en)_2Cl_2]Cl$ (XXII), when allowance is made for the spectrochemical shift due to the replacement of a Cl^- ion by water; both have maximum molecular rotations of about 1500°. There does not seem to be any doubt, however, that the two transitions still exist in the chloroaquo ion; presumably they are now nearly superimposed and the activity of the "accessory" band is negligible.

The same situation seems likely for the other substituted cations $[Co(en)_2(H_2O)_2]^{3+}$, $[Co(en)_2(NO_2)_2]^+$, and $[Co(en)_2(NH_3)_2]^{3+}$ with the transitions gradually becoming closer, as the *Dq* difference between ethylenediamine and the substituent decreases. In fact, the visible bands of all these complexes are quite symmetrical except for a trace of asymmetry on the long wavelength side in the chloroaquo ion and a definite shoulder with the *cis*-$[Co(en)_2Cl_2]^+$ ion. It appears that the R.D. curve is, therefore, largely the contribution from the principal absorption band and this provides a means of comparing the configurations of these complexes.

Despite the early realization of the relation between rotatory dispersion and configuration, the first method suggested for correlating configurations was based on the solubility of the diastereoisomers. Werner (1912b) proposed that the least soluble diastereoisomers of similar complexes with the same resolving agent had the same configuration, e.g., $(+)[Co(en)_3]Cl(+)$-tart and $(-)[Rh(en)_3]Cl(+)$tart. This generalization was criticized by Jaeger (1930) who pointed out that "there is no plausible argument supporting his initial assumption regarding a direct connection between the configuration of the molecules and their relative or absolute solubility. This view is quite arbitrary because, in general, solubility is a so highly complicated and constituent property of matter that even where we seem to have established rules for homologous series, sometimes most unexpected and surprising exceptions spring up." Even so, the method is still retained in the literature as a means of relating configurations of optical isomers. It is an unreliable criterion when used alone, but, coupled with rotatory dispersion studies for complexes which are isomorphous, or nearly so, and especially where the diastereoisomers are identical (except for the metal), it can provide useful corroborative evidence. The limitations of the generalization will be discussed with other methods for relating configuration.

The structural relations of markedly similar complexes can also be assessed by the method of "active racemates" (Delèpine, 1921). Normal racemates crystallize in three ways: (*a*) as a racemic compound, with equal amounts of (+) and (−) isomers in the one crystal; (*b*) as a racemic mixture, where a single crystal is either wholly (+) or (−); (*c*) as a racemic solid solution where the individual crystals contain variable amounts of the (+) and (−) isomers.

If the racemate D*a*L*a* is mixed with half the amount of a molecule D*a*′ which is isomorphous or nearly so, and D*a*′L*a* crystallizes as the least soluble component, then D*a*′L*a* is referred to as the "active racemate." Delèpine used this term because the "racemate" would have an optical rotation unless D*a*′ and L*a* have equal and opposite rotations. The ion D*a*′ then has the same configuration as D*a* since it replaces the latter in the racemate phase. A number of complexes have been resolved and their configurations correlated by this method, but it is only applicable when the least soluble phase is a racemic compound.

Finally, the biological activity of complex ions can be used to relate their configurations (Dwyer *et al.*, 1956). Large complex cations inhibit a number of enzymatic reactions and marked differences in inhibitory activity have been observed between (+) and (−) isomers of the tris 1,10-phenanthroline and 2,2′-bipyridine metal ions in some systems (Chapter 9).

All four methods of correlating generic structures are now compared for several classes of complex ions.

B. Hexamine-Type Chelates

The similarity of the R.D. curves for the cations $(+)[Co(en)_3]^{3+}$, $(+)[Co(-)(pn)_3]^{3+}$, $(+)[Co(-)(pn)(en)_2]^{3+}$, $(+)[Co(-)(pn)_2(en)]^{3+}$, and

$(+)[Co(en)_2(NH_3)_2]^{3+}$ shown in (XXIII) indicates that all have the same

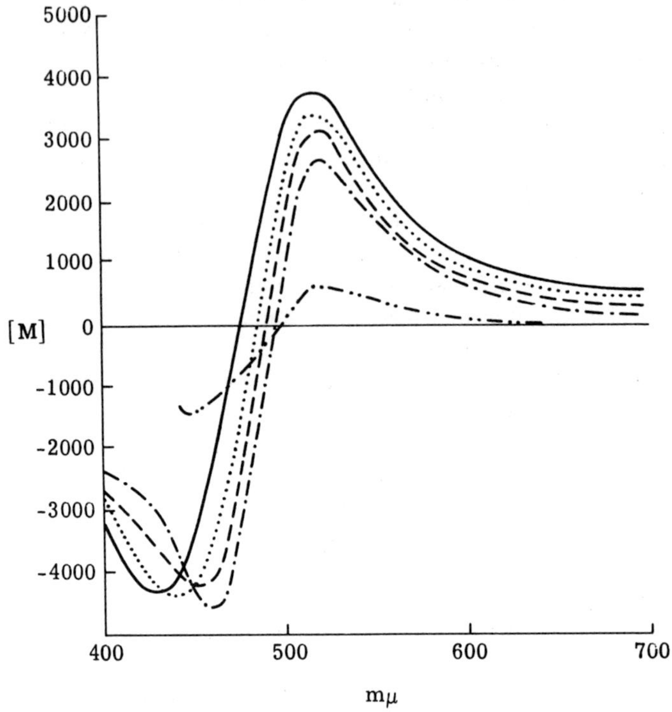

(XXIII)

The rotatory dispersion curves of the D isomers of the following ions: $[Co(en)_3]^{3+}$ —·—·; $[Co(en)_2(NH_3)_2]^{3+}$ —·······—; $[Co(en)_2(-)(pn)]^{3+}$ ———; $[Co(en)(-)(pn)_2]^{3+}$ ····; and $[Co(-)(pn)_3]^{3+}$ ——.

configuration as the $D(+)[Co(en)_3]^{3+}$ ion. The similarity is expected since they all have essentially "cubic" symmetry (in the crystal field notation), practically identical visible spectra (peak at 470 mμ), and the transition $^1A_{1g} \rightarrow {}^1T_{1g}$ (Moffitt, 1956; Jorgensen, 1956) gives rise to most of the optical rotation in the visible region. The assessment of configuration is supported by the conformational analysis of the $[Co(-)pn_3]^{3+}$ ion. From the absolute configuration of (−)propylenediamine, Corey and Bailar (1959) calculated that the $L[Co(-)(pn)_3]^{3+}$ ion was the stable isomer and this agrees with the experimental result (cf. Section IV,C). The R.D. curves of the L isomers (XXIV) are essentially mirror images of the D curves when allowance has been made for the optically active ligands.

The R.D. curves of the cations $(+)[Co(en)_3]^{3+}$, $(-)[Ir(en)_3]^{3+}$, (−)-$[Rh(en)_3]^{3+}$, and $(+)[Cr(en)_3]^{3+}$ were prepared by Mathieu (1936b), who was

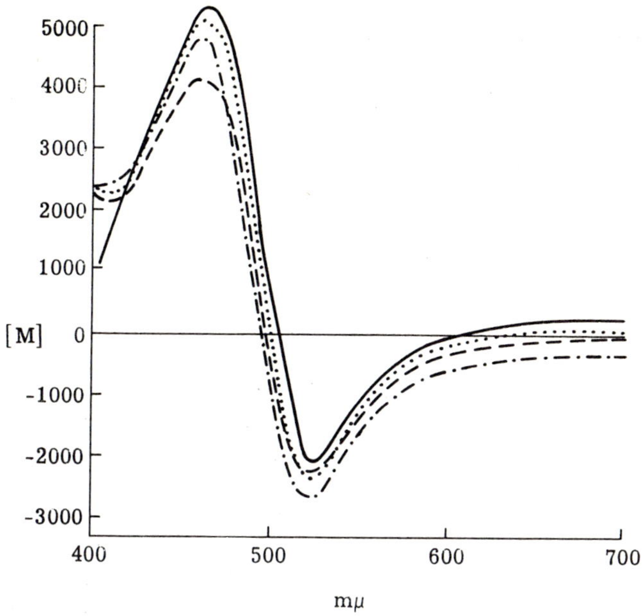

(XXIV)

The rotatory dispersion curves of the L isomers of the following ions: $[Co(en)_3]^{3+}$ – · – · –; $[Co(en)_2(-)(pn)]^{3+}$ – – –; $[Co(en)(-)(pn)_2]^{3+}$ · · · ·; and $[Co(-)(pn)_3]^{3+}$ ——.

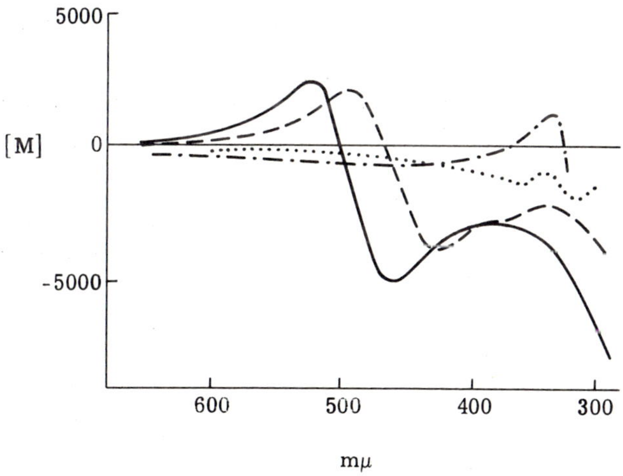

(XXV)

The rotatory dispersion curves of the following ions: $[Co(en)_3]^{3+}$ ——; $[Cr(en)_3]^{3+}$ – – –; $[Rh(en)_3]^{3+}$ – · – · –; and $[Ir(en)_3]^{3+}$ · · · ·. [After Mathieu (1936b).]

immediately impressed by their similarity (XXV). The sign of the Cotton Effect for the first optically active transition is positive in all instances, despite the total negative rotation for $[Ir(en)_3]^{3+}$ (due to a large contribution from a band further in the ultraviolet), and Mathieu concluded that these cations had the same configuration. For the d^6 systems Co(III), Rh(III), and Ir(III), the R.D. inflection coincides with the $^1A_{1g} \rightarrow {}^1T_{1g}$ transition (Jorgensen, 1956) and it seems reasonable to expect that the sign of the Cotton Effect associated with this band would be unchanged for a particular configuration.

The above assignment of configuration is also supported by Werner's solubility criterion. The diastereoisomers $(+)[Co(en)_3]Cl(+)tart{\cdot}5H_2O$ (Werner, 1912a) and $(-)[Rh(en)_3]Cl(+)tart{\cdot}4H_2O$ (Jaeger, 1930) are the least soluble. Similarly, the (+)nitrocamphor salts of the $(+)[Cr(en)_3]^{3+}$ (Werner, 1912c), $(-)[Rh(en)_3]^{3+}$ (Werner, 1912d), and $(-)[Ir(en)_3]^{3+}$ (Werner and Smirnoff, 1920) ions are the least soluble. These two sets of solubilities thus relate the configurations of the $(-)[Rh(en)_3]^{3+}$, $(-)[Ir(en)_3]^{3+}$, and $(+)[Cr(en)_3]^{3+}$ ions to the absolute D structure of the $(+)[Co(en)_3]^{3+}$ ion. Finally, Delèpine and Charonnat (1930) found that $(-)[Co(en)_3]Br_3$ and $(-)[Rh(en)_3]Br_3$ formed an "active racemate," which means that $(+)[Co(en)_3]Br_3$ and $(-)[Rh(en)_3]Br_3$ have the same configuration. Thus, all three methods agree and $(+)[Co(en)_3]^{3+}$, $(+)[Cr(en)_3]^{3+}$, $(-)[Rh(en)_3]^{3+}$, and $(-)[Ir(en)_3]^{3+}$ have the D configuration.

The tris (2,2′-bipyridine) and (1,10-phenanthroline) chelates of Fe(II), Ru(II), Os(II), and Ni(II) have been resolved by several methods and their less soluble diastereoisomers are listed in Table IV. The R.D. curves

TABLE IV
THE LEAST SOLUBLE DIASTEREOISOMERS OF THE TRIS(1,10-PHENANTHROLINE) AND (2,2′-BIPYRIDINE) CHELATES

$(+)[Ni(phen)_3](+)(SbOtart)_2{\cdot}7H_2O$	Dwyer and Gyarfas (1949a)
$(-)[Fe(phen)_3](+)(SbOtart)_2{\cdot}4H_2O$	Dwyer and Gyarfas (1949b)
$(+)[Ru(phen)_3](+)(SbOtart)_2{\cdot}2H_2O$	Dwyer and Gyarfas (1949c)
$(+)_{5461}[Os(phen)_3](+)(SbOtart)_2{\cdot}3H_2O$	Dwyer *et al.* (1950)
$(-)[Ni(bipy)_3]_3(+)(SbOtart)_4I_2{\cdot}18H_2O$	Dwyer and Gyarfas (1951a)
$(+)[Fe(bipy)_3]_3(+)(SbOtart)_4I_2{\cdot}18H_2O$	Dwyer and Gyarfas (1951a)
$(-)[Ru(bipy)_3]_3(+)(SbOtart)_4I_2{\cdot}18H_2O$	Dwyer and Gyarfas (1949d)
$(-)_{5461}[Os(bipy)_3]_3(+)(SbOtart)_4I_2{\cdot}18H_2O$	Dwyer and Gyarfas (1951a)
$(+)[Ni(bipy)_3](+)tart{\cdot}6H_2O$	Morgan and Burstall (1931)
$(-)[Fe(bipy)_3](+)tart$	Werner (1912e)
$(+)[Ru(bipy)_3](+)tart{\cdot}6H_2O$	Burstall (1936)
$(+)_{5461}[Os(bipy)_3](+)tart{\cdot}3H_2O$	Burstall *et al.* (1950)

of the tris(phenanthroline) complexes of Ni(II), Fe(II), Os(II), and Ru(II) to which we have provisionally assigned the same configuration are shown in (XXVI). A comparison with the list of less soluble diastereoisomers

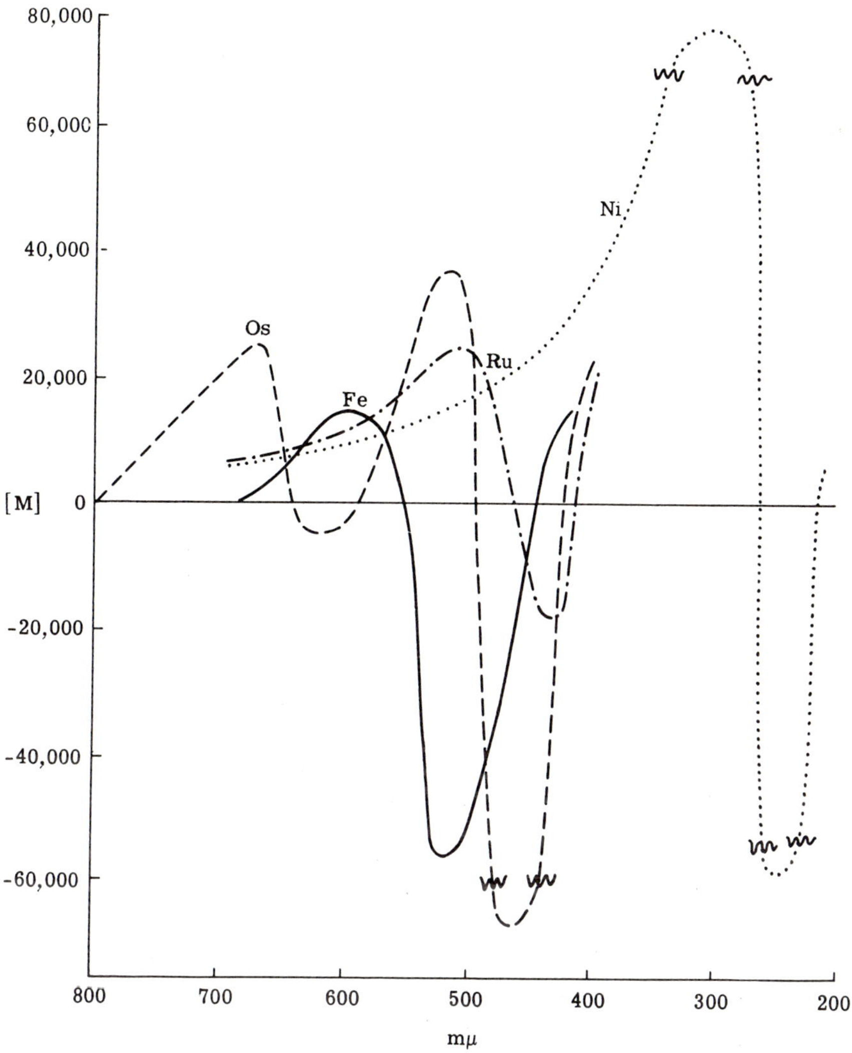

(XXVI)

The rotatory dispersion curves of the tris(1,10-phenanthroline) Ni(II), Ru(II), Fe(II), and Os(II) ions.

correlates the Ni, Os, and Ru chelates, but the (+)$[Fe(phen)_3]^{++}$ ion is anomalous. The tris(phenanthroline) chelates are highly toxic when injected intraperitoneally into mice as a result of a neuromuscular blocking action (Dwyer *et al.*, 1956). The (+)$[Ni(phen)_3]^{++}$, (+)$[Ru(phen)_3]^{++}$, and $(-)_{5461}[Os(phen)_3]^{++}$ ions are more toxic than their optical antipodes, suggesting that they have the same configuration in agreement with the solubility and R.D. correlations. The (±)tris(phenanthroline) chelates crystallize as racemic compounds at 25°C. and are suitable for the "active racemate" method of structure correlations. Unfortunately the $[Fe(phen)_3]^{++}$ cation is rather too optically labile for a definite conclusion to be reached. Similarly, no difference in toxicity is observed between the (+), (−), and (±) forms.

The corresponding tris(2,2′bipyridine) chelates have been resolved as the (+)tartrates and the iodide (+)antimonyl tartrates (Table IV). The cations $(-)[Ni(bipy)_3]^{++}$, $(-)[Ru(bipy)_3]^{++}$, and $(-)_{5461}[Os(bipy)_3]^{++}$ thus appear to have the same configuration, but the validity of the assignment is questionable since the relative solubilities of the diastereoisomers depend, in some instances, on the concentration of the resolving agent that has been used. For example either (+) or $(-)[Ni(bipy)_3]^{++}$ ion will crystallize with (+)arsenyl tartrate ion depending on how much sodium (+)arsenyl tartrate is added to the complex ion. Similarly, if the theoretical amount of (+)tartrate ion is used to resolve the $[Ni(bipy)_3]^{++}$ complex ion then $(-)[Ni(bipy)_3](+)$tartrate separates as the less soluble form instead of $(+)[Ni(bipy)_3](+)$tartrate when an excess of resolving agent is used (Jaeger and Van Dijk, 1936).

C. Tris(oxalato) Chelates

The tris(oxalato) complexes of Co(III), Ir(III), Rh(III), and Cr(III) were resolved originally as the tristrychninium salts by Werner (1912c) and subsequently with the $(+)[Ni(phen)_3]^{++}$ cation (Dwyer and Sargeson, 1956), the diastereoisomers in the latter instance being of the form $[Ni(phen)_3]K[M(C_2O_4)_3]\cdot H_2O$. The anions in both sets of less soluble diastereoisomers, $(-)_{5461}[Co(C_2O_4)_3]^{3-}$, $(+)_{5461}[Rh(C_2O_4)_3]^{3-}$, $(+)_{5461}[Ir(C_2O_4)_3]^{3-}$, and $(+)[Cr(C_2O_4)_3]^{3-}$, conform to the relative configuration assignment made by Delèpine (1934) using the isomorphous potassium salts in an "active racemate" study. The R.D. curves of the isomers listed above (XXVII) all have inflexions of the same slope for the ${}^1A_{1g} \rightarrow {}^1T_{1g}$ transition indicating the same configuration. Hence all three methods give the same result for the correlation of their relative configurations.

The same correspondence for the R.D. curves of Co, Rh, and Ir should exist here as with the ethylenediamine complexes since the same electronic transition is active in each instance (${}^1A_{1g} \rightarrow {}^1T_{1g}$) (Jorgensen, 1956). The possibility of relating the configuration of $(-)_{5461}[Co(C_2O_4)_3]^{3-}$ and D-$[Co(en)_3]^{3+}$ ions is discussed in Section V,E.

D. Bis(ethylenediamine)cobalt(III) Chelates

A most extensive study of the rotatory dispersion, absorption spectra, and circular dichroism of the series $[Co(en)_2X_2]^{n+}$ (X = Cl, Br, CNS,

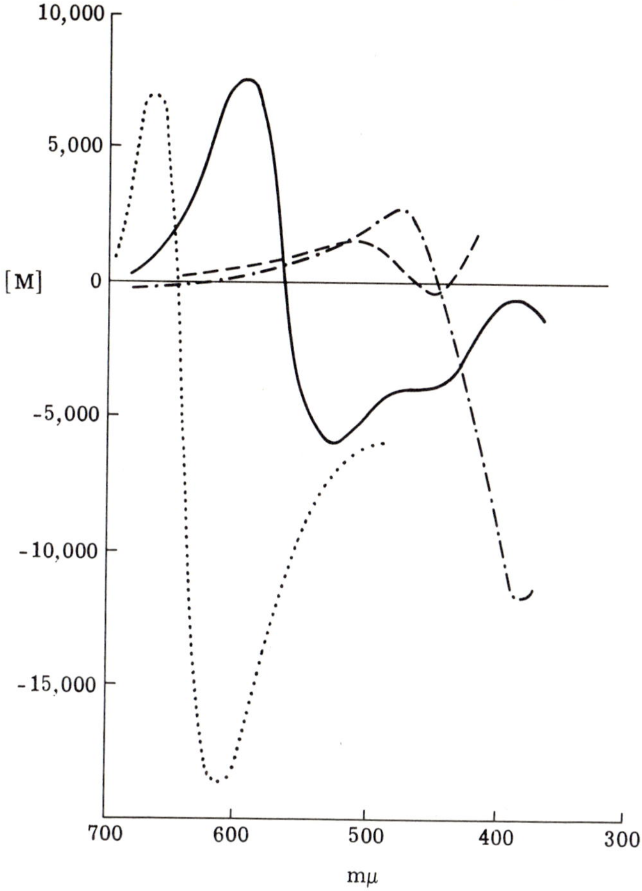

(XXVII)

The rotatory dispersion curves of the following tris(oxalato) ions with the same configuration: Co(III) · · · ·; Cr(III) ——; Rh(III) – · – · –; and Ir(III) – – –.

NH_3, NO_2, H_2O; 2X = CO_3, C_2O_4) has been carried out by Mathieu (1936a). The similarity of the rotatory dispersion and circular dichroism curves in the visible region was noted and Mathieu concluded that the configurations could be correlated by the sign of the Cotton Effect for the principal absorption band. The isomers related in this way were then compared with the solubility of the (+)bromcamphor sulfonate diastereoisomers and with several interconversions which Werner (1912b) maintained occurred without change in configuration. The excellent agreement shown in Table V is marred by only one exception: the least soluble diastereoisomer, (−)-$[Co(en)_2NH_3NO_2](+)(BCS)_2$, is in opposition to the R.D. assessment.

TABLE V

CORRELATION OF DIASTEREOISOMER SOLUBILITIES WITH THE SIGN OF THE COTTON EFFECT FOR $[Co(en)_2X_2]^{n+}$ COMPLEXES

Less soluble (+)BCS diastereoisomers or their derivatives	Isomers with a positive Cotton Effect for the principal absorption	Isomers with same configuration according to Werner (1912b)
$(+)[Co(en)_2(NH_3)_2]Cl_3$[a]	(+)	—
$(+)[Co(en)_2NH_3Cl]Cl_2$	(+)	—
$(+)[Co(en)_2NH_3Br]Cl_2$	(+)	—
$(+)[Co(en)_2NH_3NCS]Cl_2$	(+)	(+)
$(-)[Co(en)_2NH_3NO_2](NO_3)_2$	(+)	—
$(+)[Co(en)_2NH_3H_2O](NO_3)_3$[a]	(+)	—
$(+)[Co(en)_2H_2OCl]Cl_2$	(+)	—
$(+)[Co(en)_2Cl_2]Cl$	(+)	$(-)_{6563}(+)$
$(+)[Co(en)_2ClBr]Cl$	(+)	—
$(+)_{5461}[Co(en)_2ClNCS]Cl$	$(+)_{5461}$	$(+)_{5461}$
$(+)[Co(en)_2ClNO_2]Cl$	(+)	—
$(+)[Co(en)_2BrNO_2]Br$	(+)	—
$(+)[Co(en)_2NO_2H_2O]Cl_2$[b]	(+)	(+)
$(+)[Co(en)_2(NO_2)_2]Cl$	(+)	—
$(+)[Co(en)_2CO_3]Cl$[c]	(+)	(+)
$(+)[Co(en)_2C_2O_4]Cl$[c]	(+)	(+)

[a] Prepared from the less soluble diastereoisomer of the $[Co(en)_2NH_3Br]^{++}$ ion.
[b] Prepared from the less soluble diastereoisomer of the $[Co(en)_2ClNO_2]^+$ ion.
[c] Prepared from the less soluble diastereoisomer of the $[Co(en)_2Cl_2]^+$ ion.

Mathieu attempted to resolve the anomaly by transforming the (+)-$[Co(en)_2NH_3Cl]^{++}$ ion with silver nitrite (<70°C.) to the $(+)[Co(en)_2$-$NH_3NO_2]^{++}$ ion and thence by boiling with excess silver nitrite to the $(+)[Co(en)_2(NO_2)_2]^+$ ion. The configurations are now consistent with the rotatory dispersion curves, though it was realized that the result was not entirely conclusive since the possibility of an odd number of "Walden Inversions" during the chemical transformation could not be discounted.

These complexes when re-examined, relative to the absolute standard $D(+)[Co(en)_3]^{3+}$ ion and recent theoretical and kinetic developments, can be related more satisfactorily. The R.D. curves for the (+)isomers of $[Co(en)_2NH_3X]^{n+}$ where X = Cl, H_2O, NO_2, NH_3 are given in (XXVIII). All compounds have essentially the same maximum rotatory power on the long wavelength side, a positive Cotton Effect (negative slope of inflexion) which shifts towards the violet as the substituents follow the order in the spectrochemical series $Cl \rightarrow H_2O \rightarrow NH_3 \rightarrow NO_2$. Essentially the same transition is involved in each instance and, as the *Dq* difference between the substituent and the surrounding N atoms is reduced, the complexes approach the near "cubic" symmetry of the *cis*-$[Co(en)_2(NH_3)_2]^{3+}$ ion.

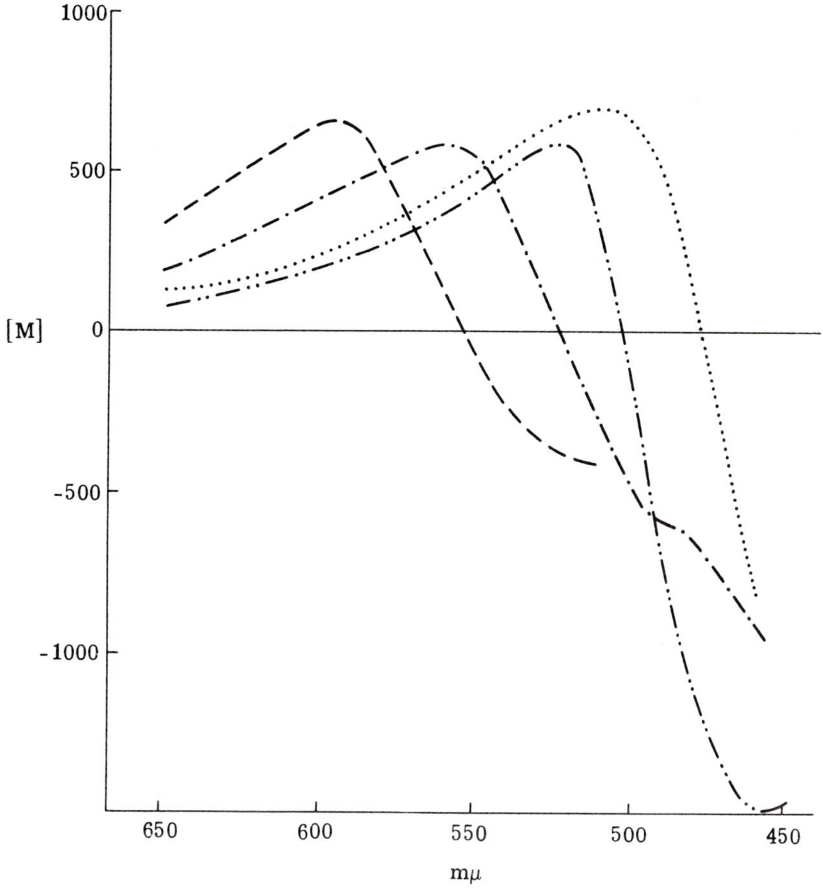

(XXVIII)

The rotatory dispersion curves of the following D *cis* ions: $[Co(en)_2NH_3Cl]^{++}$ – – –; $[Co(en)_2NH_3H_2O]^{3+}$ – · – · –; $[Co(en)_2(NH_3)_2]^{3+}$ – ·· – ·· –; and $[Co(en)_2NH_3NO_2]^{++}$ · · · ·.

This ion has already been assigned the D configuration by comparison with $D(+)[Co(en)_3]Cl_3$; hence the isomers in (XXVIII) all have the D configuration.

The following reaction sequence supports the argument above:

$$(+)[Co(en)_2NH_3Cl]^{++} \xrightarrow[H_2O]{H^+} (+)[Co(en)_2NH_3H_2O]^{3+} \xrightarrow{NO_2^-} (+)[Co(en)_2NH_3NO_2]^{+}$$

The aquation of the chloroammine to the aquoammine complex occurs with at least 75% retention of configuration (Basolo *et al.*, 1953). Nitrite ion then adds to the coordinated water molecule to form the nitrito complex, which subsequently rearranges internally to the nitro compound with retention of configuration (XXIX). Initially, the Co—O bond is not broken, and O^{18} studies show that neither of the oxygen atoms exchanges with the solvent. The reaction occurs with 93% retention of configuration (Murmann and Taube, 1956). These transformations relate the configurations conclusively and sustain Mathieu's preference for assigning generic configuration by R.D. over the diastereoisomer solubility method. They also confirm that inversion does not occur during the reaction of the chloroammino complex with silver nitrite.

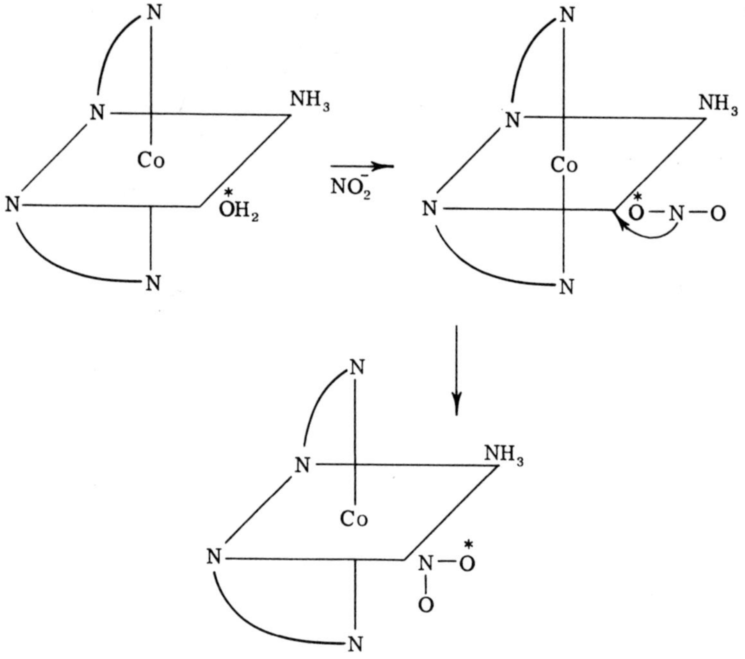

(XXIX)

Finally, it has been shown that $(+)[Co(en)_2NH_3Cl]Cl_2$ is converted to $(+)[Co(en)_2(NH_3)_2]Cl_3$ in liquid ammonia with better than 80% retention of configuration (Archer and Bailar, 1961). Amide ions accelerate the reaction enormously and most of the activity is lost, whereas ammonium ion slows it appreciably and the activity is retained. Substitution in liquid ammonia is analogous to aquation and exhibits the same stereochemical characteristics as acid and base hydrolysis. Archer and Bailar suggested that $(+)[Co(en)_2NH_3Cl]Cl_2$ and $(+)[Co(en)_2(NH_3)_2]Cl_3$ have the same configuration and the parallelism of ammonation with S_N1 aquation supports this claim.

Similarly, the cations (+)$[Co(en)_2Cl_2]^+$, (+)$[Co(en)_2ClH_2O]^{++}$, (+)-$[Co(en)_2CO_3]^+$, (+)$[Co(en)_2(H_2O)_2]^{3+}$, (+)$[Co(en)_2(NO_2)_2]^+$, and (+)-$[Co(en)_2(NH_3)_2]^{3+}$ whose R.D. curves appear in (XXX) show essentially

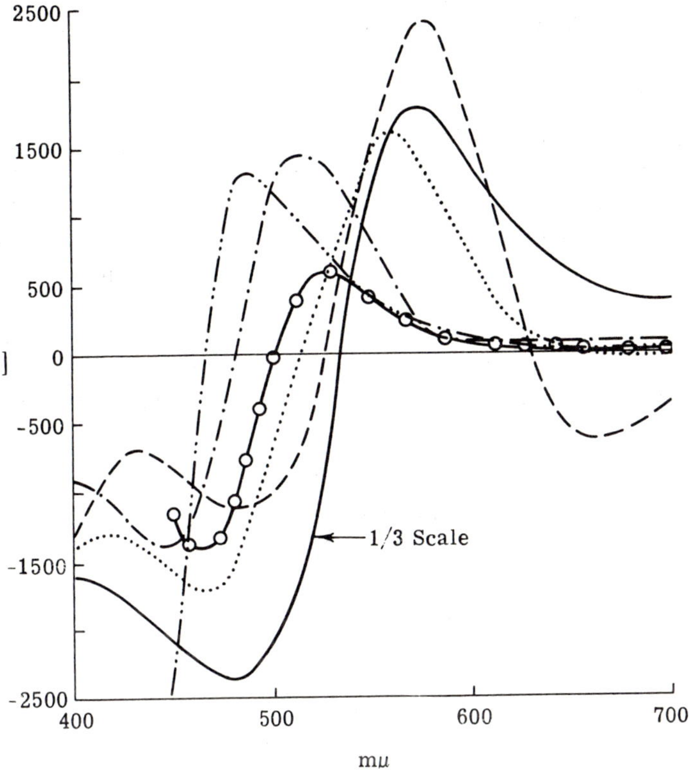

(XXX)

The rotatory dispersion curves of the following D *cis* ions: $[Co(en)_2Cl_2]^+$ – – – –; $[Co(en)_2ClH_2O]^{++}$ · · · ·; $[Co(en)_2CO_3]^+$ ——; $[Co(en)_2(H_2O)_2]^{3+}$ – · – · –; $[Co(en)_2(NO_2)_2]^+$ – ·· – ·· –; and $[Co(en)_2(NH_3)_2]^{3+}$ –O–O–O–.

the same spectrochemical gradation as the previous system. As the *Dq* difference between the substituents and remaining four N atoms diminishes, the complexes approach the essentially "cubic" symmetry of $[Co(en)_2$-$(NH_3)_2]Cl_3$ and can thus be related to the D configuration of (+)$[Co(en)_3]$-Cl_3. The maximum rotatory power varies but unlike the amminobis(ethylenediamine) complexes the symmetry also changes [e.g., $[Co(en)_2ClH_2O]^{++}$ (distorted rhombic) and $[Co(en)_2CO_3]^+$ (distorted tetragonal)], so the relatively constant maximum rotation on the long wavelength side cannot be expected. Even so, the maximum rotations of the disubstituted complexes roughly correspond and even more so if contributions from "accessory" and UV bands are discounted. These facts are taken to indicate that essentially the same active transition is involved in each instance. The configurations are also related through the reactions:

$$(+)[Co(en)_2Cl_2]^+ \xrightarrow[S_N1]{H_3O^+} (+)[Co(en)_2ClH_2O]^{++} \xrightarrow[H^+]{HCO_3^-} (+)[Co(en)_2CO_3]^+$$

$$(+)[Co(en)_2CO_3]^+ \longrightarrow (+)[Co(en)_2(H_2O)_2]^{3+} \xrightarrow{NO_2^-} (+)[Co(en)_2(NO_2)_2]^+$$

The ion $(+)[Co(en)_2Cl_2]^+$ hydrolyzes in acid solution by an S_N1 mechanism with full retention of configuration (Mathieu, 1937; Pearson *et al.*, 1956). Subsequent loss of activity is due to *cis-trans* isomerization of the $[Co(en)_2ClH_2O]^{++}$ ion (Sargeson, 1961). Optically pure $(+)[Co(en)_2ClH_2O]^{++}$ is converted to optically pure $(+)[Co(en)_2CO_3]^+$ with sodium bicarbonate (Dwyer and Sargeson, 1961b) and the rotation of the isolated product agrees with that obtained by resolution of $(\pm)[Co(en)_2CO_3]^+$ ion, hence the configuration is retained. The O^{18} studies of Hunt *et al.* (1952) and Posey and Taube (1953) suggest that the bicarbonate ion adds to the coordinated

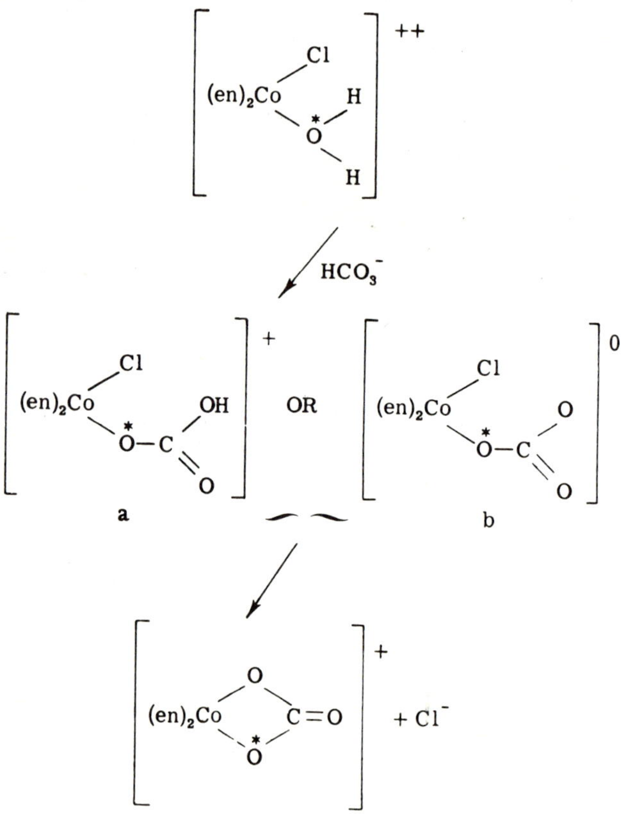

(XXXI)

water, losing one of its original oxygen atoms in the process (XXXI). The chloride ion then dissociates from the positively charged complex (XXXIa) or more likely from the nonelectrolyte (XXXIb) since the reaction is complete in about 10 minutes. An oxygen atom attached to the carbonate group then coordinates by an intramolecular mechanism to give the carbonato complex or the position is aquated from the solvent to give the aquocarbonato complex which subsequently can be isolated as the carbonato salt. Irrespective of the mechanism, the retention of the original configuration is better than 95%.

The carbonato compound dissolves in acid to form the diaquo ion also with full retention of configuration. This may be demonstrated by reconverting the diaquo cation to the carbonato when the original rotation is regained. Studies with the corresponding tetrammine complex (Posey and Taube, 1953) in acid solution indicate that the diaquo complex is obtained by the mechanism shown in (XXXII). This is consistent with the retention of optical rotatory power in the bis(ethylenediamine) complex, since aquation occurs with retention and subsequent decarboxylation without fission of the Co—O bond. (Note that only half of the coordinated water in the product contains O^{18}.)

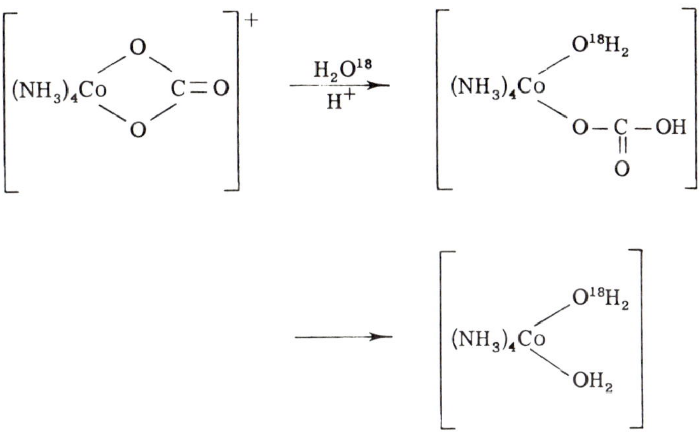

(XXXII)

Nitrite ion when added to the diaquo complex gives the *cis* dinitro complex with the same specific rotation as that obtained by resolution of the $(\pm)[Co(en)_2(NO_2)_2]^+$ ion. The mechanism of this addition has been depicted above (XXIX) and Murmann and Taube (1956) recorded 97.5% retention in the reaction. Thus, the common configuration of the $(+)[Coen_2Cl_2]^+$, $(+)[Co(en)_2ClH_2O]^{++}$, $(+)[Co(en)_2CO_3]^+$, $(+)[Co(en)_2(H_2O)_2]^{3+}$, and $(+)$-$[Co(en)_2(NO_2)_2]^+$ ions is established independently of the rotatory dispersion evidence. By analogy, $(+)[Co(en)_2C_2O_4]Cl$ will also have the same configuration as $(+)[Co(en)_2CO_3]Cl$. Unfortunately the conversion of $(+)[Co(en)_2(NO_2)_2]Cl$ to $(+)[Co(en)_2(NH_3)_2]Cl$, by reduction, or the reverse reaction, by oxidation, has not been accomplished.

E. Absolute Configuration from Stereospecificity

The absolute D configuration of (−)1,2-propylenediamine has been determined chemically with reference to D(−)alanine (Reihlen *et al.*, 1932; Schnell and Karrer, 1955) and by the absolute X-ray analysis of *trans*-$[Co(-)(pn)_2Cl_2]Cl$ (Saito and Iwasaki, 1962). The amine reacts with sodium chloroacetate, without affecting the asymmetric carbon atom, to give D(−)1,2-propylenediaminetetraacetic acid (PDTA) and the sexadentate in turn coordinates stereospecifically giving only one optical isomer of the resulting chelate (Section IV) (Dwyer and Garvan, 1961). In this way the absolute configuration of $(+)_{5461}K[Co(-)PDTA]$ is established (XXXIIIa)

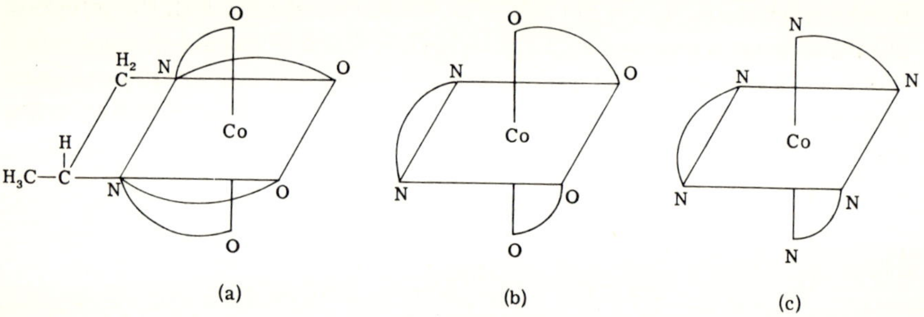

(XXXIII)

and this has been called L by comparison with L-$K[Co(en)(C_2O_4)_2]$ (XXXIIIb) and L-$[Co(en)_3]Cl_3$ (XXXIIIc) (MacDermott and Sargeson, 1961).

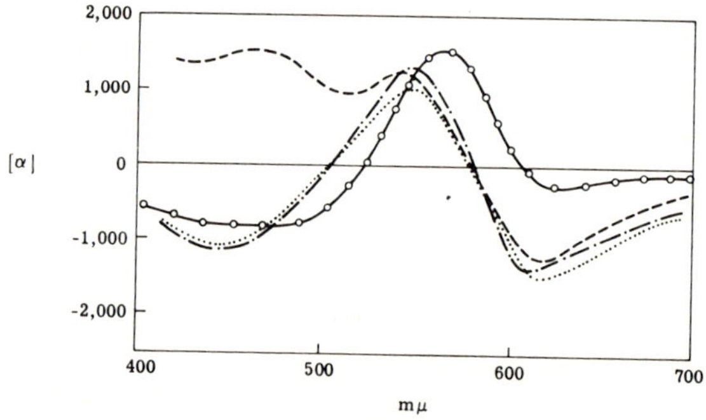

(XXXIV)

The rotatory dispersion curves of the following complexes with the same configuration: $K[Co(-)PDTA]$ – · – · –; $K[CoEDTA]$ · · · ·; $K[Co(en)(C_2O_4)_2]$ – – –; and Ba[Co-(−)CDTA]$_2$.

The R.D. curves for L$(+)_{5461}$K[Co$(-)$PDTA], $(+)_{5461}$K[CoEDTA], and $(+)_{5461}$K[Co(en)$(C_2O_4)_2$] are shown in (XXXIV). The EDTA and PDTA isomers have practically identical rotatory dispersions and the $(+)_{5461}$[CoEDTA]$^-$ ion therefore must have the L configuration, as in (XXXIIIa). Busch and Cooke (1961) have reached the same conclusion regarding the structures of the PDTA and EDTA complexes but have designated the absolute configurations as D (see Chapter 7 and the note following on the designation of absolute configurations).

The ions $(+)_{5461}$[Co(en)$(C_2O_4)_2]^-$ and $(+)_{5461}$[CoEDTA]$^-$ also have very similar spectra and rotatory dispersion curves and they are contained in the less soluble diastereoisomers when the same resolving agent is used:

$(-)$[Co(en)$_2C_2O_4$]$(+)_{5461}$[CoEDTA]$\cdot 3H_2O$,
$(-)$[Co(en)$_2C_2O_4$]$(+)_{5461}$[Co(en)$(C_2O_4)_2$],
$(-)$[Co(en)$_2(NO_2)_2$]$(+)_{5461}$[CoEDTA]$\cdot 3H_2O$,
$(-)$[Co(en)$_2(NO_2)_2$]$(+)_{5461}$[Co(en)$(C_2O_4)_2$].

For these reasons both ions have been designated as L despite their different symmetry properties. The $(+)_{5461}$[Co(en)$(C_2O_4)_2]^-$ ion has been assigned the L configuration because the R.D. curves (XXXV) in the substitution

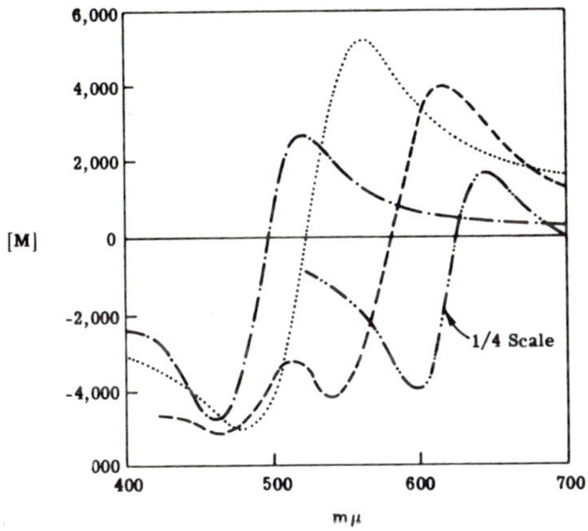

(XXXV)

The rotatory dispersion curves of the following D ions: $[Co(en)_3]^{3+}$ – · – · –; [Co-(en)$_2C_2O_4]^+$ · · · ·; [Co(en)$(C_2O_4)_2]^-$ – – –; and $[Co(C_2O_4)_3]^{3-}$ – ·· – ·· –.

sequence $(+)[Co(en)_3]^{3+}$, $(+)$[Co(en)$_2(C_2O_4)]^+$, $(-)_{5461}$[Co(en)$(C_2O_4)_2]^-$, and $(-)_{5461}[Co(C_2O_4)_3]^{3-}$ relate the configurations of the $(-)_{5461}$[Co(en)-$(C_2O_4)_2]^-$ ion to that of D$(+)[Co(en)_3]^{3+}$. The $(+)$[Co(en)$_2C_2O_4]^+$ ion has already been shown to have the D structure by analogy with the $(+)$-[Co(en)$_2CO_3]^+$ ion.

The $(+)[Co(en)_3]^{3+}$ and $(-)[Co(C_2O_4)_3]^{3-}$ ions have the same configurations and rotations of opposite sign in the Na_D line and this agrees with Kuhn's prediction (1952).

As with the previous series the R.D. inflections follow the spectrochemical shift as each ethylenediamine is replaced by an oxalate ion and the ${}^1A_{1g} \rightarrow {}^1T_{1g}$ transition is essentially the active one in each complex.

A note on the designation of absolute configurations. The ions $[CoEDTA]^-$ and $[Co(en)_3]^{3+}$ have widely differing symmetry properties and whether structure (XXXVIb) is related to (a) or (c) is largely quite arbitrary.

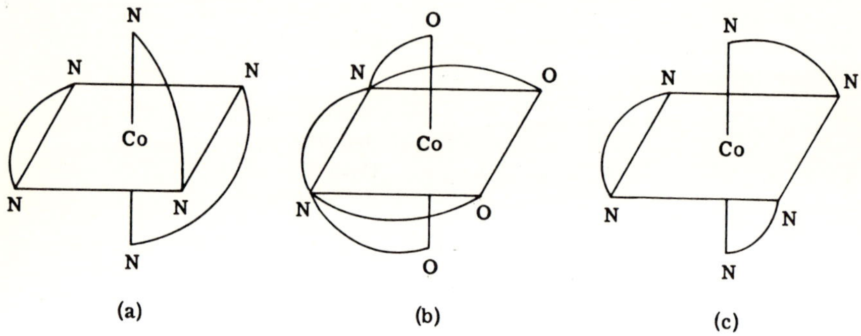

(XXXVI)

We have chosen to designate (b) as L and to relate (b) and (c) through the $(+)_{5461}[Co(en)(C_2O_4)_2]^-$ ion which has a similar spectrum, R.D. curve, and diastereoisomer solubility to $(+)_{5461}K[CoEDTA]$ and whose symmetry can be easily related to that of the $L[Co(en)_3]^{3+}$ ion. Busch *et al.* (1961) have called (b) D and the only inconsistency which arises then is that the $(+)_{5461}$-$[Co(en)(C_2O_4)_2]^-$ and $(+)_{5461}[CoEDTA]^-$ ions with very similar R.D. curves would be labeled L and D, respectively.

F. Rotatory Dispersion and Stereochemical Change

The Co(III) chelate of the sexadentate molecule 1,8-bis(salicylideneamino)-3,6-dithiaoctane, $[CoEEE]^+$, exists only in one stereoisomeric form (green), which has been resolved through the bromcamphor sulfonate (Dwyer and Lions, 1950). From the analogous molecule 1,10-bis(salicylideneamino)-4,7-dithiadecane in which the terminal ethylene chains are extended to trimethylene chains two isomers of the cation $[CoTET]^+$ can be isolated (Dwyer *et al.*, 1952). The green form of the latter chelate can be resolved through the (+)antimonyl tartrate, and the similarity between the $(+)_{5461}$ forms of the $[CoEEE]^+$ and $[CoTET]^+$ ions will be evident from the R.D. curves (XXXVII). The arrangement of the donor atoms about the metal in the $[CoEEE]^+$ ion and about the green $[CoTET]^+$ ion is shown in (XXXVIIIa).

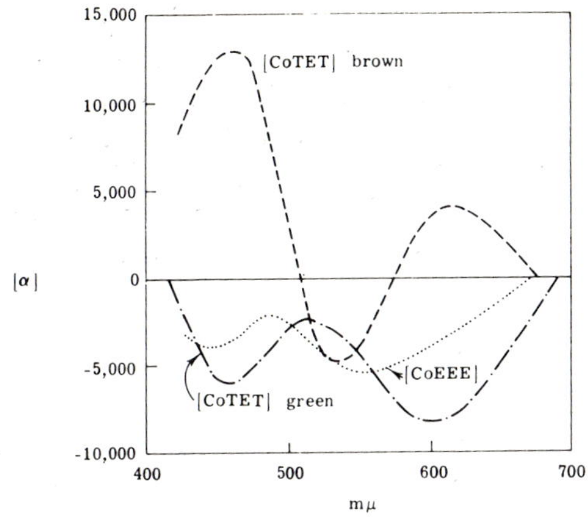

(XXXVII)

The brown isomer (XXXVIIIb) is readily obtained, with retention of optical activity but inversion of sign, by boiling the iodide of the green form in methanol solution. It was suggested (Dwyer *et al.*, 1952) that the change

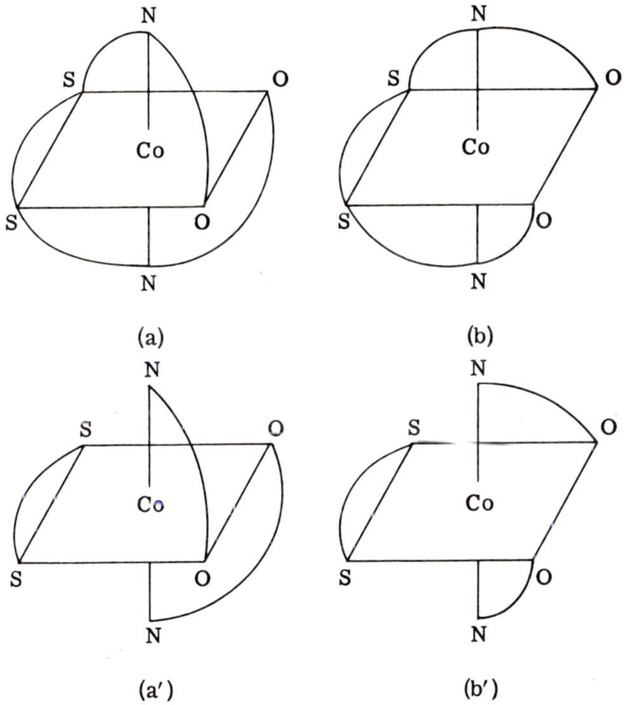

(XXXVIII)

of sign merely reflected the change in absorption spectrum on the rotatory dispersion. The retention of optical activity during isomerization can be readily explained by assuming that the metal-sulfur and metal-nitrogen bonds remain intact but the terminal oxygen atoms interchange. The rotatory dispersion curve of the brown ion (+)[CoTET]$^+$, obtained from the green ion (−)[CoTET]$^+$ (XXXVII), shows that inversion has taken place. The two structures can be related more readily if (XXXVIIIa) and (XXXVIIIb) are redrawn with the N—S links omitted (XXXVIIIa′ and b′).

A similar situation should exist with triethylenetetramine complexes since this ligand may be considered as two ethylenediamine groups joined by an ethylene bridge. The *cis*-dichlorocobalt(III) complex may be prepared in two forms, α and β, both of which have been resolved: for the α isomer $[\alpha]_D = \pm 2300°$, for the β, $[\alpha]_D = \pm 300°$. The suggested structures for the α and β isomers (XXXIX) are related as D and L if the central ethylene

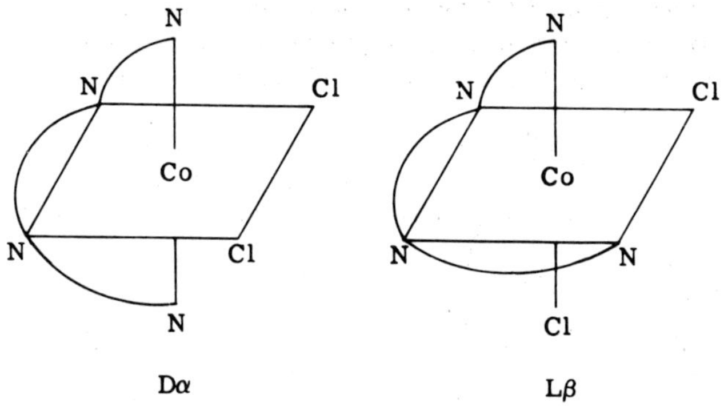

(XXXIX)

bridge is ignored. This relationship should be reflected in the rotatory dispersion curve despite the fact that they are not optical antipodes, i.e., (+)α should have the opposite configuration to (−)β. Preliminary studies on the interconversion of the α to the β system indicate that this R.D. relationship exists (Sargeson and Searle, 1961).

References

Archer, R. D., and Bailar, J. C. (1961). *J. Am. Chem. Soc.* **83,** 813.
Bailar, J. C., and McReynolds, J. P. (1939). *J. Am. Chem. Soc.* **61,** 3199.
Bailar, J. C., and Peppard, D. F. (1940). *J. Am. Chem. Soc.* **62,** 105.
Basolo, F. (1948). *J. Am. Chem. Soc.* **70,** 2634.
Basolo, F. (1956). *In* "The Chemistry of the Coordination Compounds" (J. C. Bailar, ed.), p. 313. Reinhold, New York.
Basolo, F., Stone, B. D., and Pearson, R. G. (1953). *J. Am. Chem. Soc.* **75,** 819.
Bijvoet, J. M., Peerdeman, A. F., and Van Bommel, A. J. (1951). *Nature* **168,** 271.

Broomhead, J. (1960). Thesis, Australian National University, Canberra.

Buckingham, A. D., Dwyer, F. P., and Sargeson, A. M. (1961). Unpublished work.

Burrows, G. J., and Lauder, K. H. (1931). *J. Am. Chem. Soc.* **53,** 3600.

Burstall, F. H. (1936). *J. Chem. Soc.* p. 174.

Burstall, F. H., Dwyer, F. P., and Gyarfas, E. C. (1950). *J. Chem. Soc.* p. 953.

Busch, D. H. (1955). *J. Am. Chem. Soc.* **77,** 2747.

Busch, D. H., and Bailar, J. C. (1953). *J. Am. Chem. Soc.* **75,** 4574.

Busch, D. H., and Bailar, J. C. (1954). *J. Am. Chem. Soc.* **76,** 5352.

Busch, D. H., and Cooke, D. W. (1961). *J. Inorg. & Nuclear Chem.* **23,** 145.

Busch, D. H., Cooke, D. W., Swaminathan, K., and Im, Y. A. (1961). *In* "Advances in the Chemistry of the Coordination Compounds" (S. Kirschner, ed.), p. 139. Macmillan, New York.

Calvin, M., and Martell, A. E. (1952). "Chemistry of the Metal Chelate Compounds," p. 307. Prentice-Hall, Englewood Cliffs, New Jersey.

Carassiti, V. (1958). *J. Inorg. & Nuclear Chem.* **8,** 227.

Collman, J. P., and Blair, R. P. (1961). Am. Chem. Soc. Meeting, Chicago, Illinois, September.

Condon, E. U. (1937). *Revs. Modern Phys.* **9,** 432.

Corey, E. J., and Bailar, J. C. (1959). *J. Am. Chem. Soc.* **81,** 2620.

Das Sarma, B., and Bailar, J. C. (1955). *J. Am. Chem. Soc.* **77,** 5476, 5480.

Delèpine, M. (1921). *Bull. soc. chim. France* [4]**29,** 656.

Delèpine, M. (1934). *Bull. soc. chim. France* [5]**1,** 1256.

Delèpine, M., and Charonnat, R. (1930). *Bull. soc. minéral. France* **53,** 73.

Doran, V. F., and Kirschner, S. (1961). Am. Chem. Soc. Meeting, Chicago, Illinois, September.

Dwyer, F. P. (1961). *Australian J. Sci.* **24,** 97.

Dwyer, F. P., and Garvan, F. L. (1959). *J. Am. Chem. Soc.* **81,** 1043.

Dwyer, F. P., and Garvan, F. L. (1961). *J. Am. Chem. Soc.* **83,** 2610.

Dwyer, F. P., and Gyarfas, E. C. (1949a). *J. Proc. Roy. Soc. N.S.Wales* **83,** 232.

Dwyer, F. P., and Gyarfas, E. C. (1949b). *J. Proc. Roy. Soc. N.S.Wales* **83,** 263.

Dwyer, F. P., and Gyarfas, E. C. (1949c). *J. Proc. Roy. Soc. N.S.Wales* **83,** 170.

Dwyer, F. P., and Gyarfas, E. C. (1949d). *J. Proc. Roy. Soc. N.S.Wales* **83,** 174.

Dwyer, F. P., and Gyarfas, E. C. (1951a). *J. Proc. Roy. Soc. N.S.Wales* **85,** 135.

Dwyer, F. P., and Gyarfas, E. C. (1951b). *Nature* **168,** 29.

Dwyer, F. P., and Lions, F. (1950). *J. Am. Chem. Soc.* **72,** 1545.

Dwyer, F. P., and Sargeson, A. M. (1956). *J. Phys. Chem.* **60,** 1331.

Dwyer, F. P., and Sargeson, A. M. (1959a). *J. Am. Chem. Soc.* **81,** 5272.

Dwyer, F. P., and Sargeson, A. M. (1959b). *J. Am. Chem. Soc.* **81,** 5269.

Dwyer, F. P., and Sargeson, A. M. (1959c). *J. Am. Chem. Soc.* **81,** 2335.

Dwyer, F. P., and Sargeson, A. M. (1960). *Nature* **187,** 1022.

Dwyer, F. P., and Sargeson, A. M. (1961a). *J. Phys. Chem.* **65,** 1892.

Dwyer, F. P., and Sargeson, A. M. (1961b). Unpublished work.

Dwyer, F. P., Gibson, N. A., and Gyarfas, E. C. (1950). *J. Proc. Roy. Soc. N.S.Wales* **84,** 68.

Dwyer, F. P., Gill, N. S., Gyarfas, E. C., and Lions, F. (1952). *J. Am. Chem. Soc.* **74,** 4188.

Dwyer, F. P., Gyarfas, E. C., Gill, N. S., and Lions, F. (1953). *J. Am. Chem. Soc.* **75,** 3834.

Dwyer, F. P., Gyarfas, E. C., and Mellor, D. P. (1955). *J. Phys. Chem.* **59,** 296.

Dwyer, F. P., Gyarfas, E. C., and Koch, J. H. (1956). *Australian J. Biol. Sci.* **9,** 371.

Dwyer, F. P., Garvan, F. L., and Shulman, A. (1959). *J. Am. Chem. Soc.* **81,** 290.
Dwyer, F. P., Broomhead, J. A., and Hogarth, J. W. (1960). *In* "Inorganic Syntheses" (E. Rochow, ed.), Vol. VI, p. 186. McGraw-Hill, New York.
Dwyer, F. P., Sargeson, A. M., and MacDermott, T. E. (1961). Unpublished work.
Gel'man, A. D., and Essen, L. N. (1950). *Doklady Akad. Nauk S.S.S.R.* **75,** 693.
Gordon, G., and Birdwhistell, R. K. (1959). *J. Am. Chem. Soc.* **81,** 3567.
Henderson, G. M., and Rule, H. G. (1939). *J. Chem. Soc.* p. 1568.
Hidaka, J., Yamada, S., and Tsuchida, R. (1958). *Bull. Chem. Soc. Japan* **31,** 921.
Hunt, J. P., Rutenberg, A. C., and Taube, H. (1952). *J. Am. Chem. Soc.* **74,** 268.
Irving, H., and Gill, J. B. (1958). *Proc. Chem. Soc.* p. 168.
Jaeger, F. M. (1919). *Rec. trav. chim.* **38,** 250.
Jaeger, F. M. (1930). "Optical Activity and High Temperature Measurement," pp. 92, 135–137, 155. McGraw-Hill, New York.
Jaeger, F. M., and Blumendal, H. B. (1928). *Z. anorg. u. allgem. Chem.* **175,** 161.
Jaeger, F. M., and Van Dijk, F. A. (1936). *Z. anorg. u. allgem. Chem.* **227,** 273.
Jones, L. L., and Eyring, H. (1961). *J. Chem. Educ.* **38,** 601.
Jorgensen, C. K. (1956). *Acta Chem. Scand.* **10,** 500.
Karagunis, C., and Coumoulos, B. (1938). *Nature* **142,** 162.
Kauzman, W. J., Walter, J. E., and Eyring, H. (1940). *Chem. Revs.* **26,** 339.
Keubler, J. R., and Bailar, J. C. (1952). *J. Am. Chem. Soc.* **74,** 3535.
Kirkwood, J. (1937). *J. Chem. Phys.* **5,** 479.
Kirkwood, J. (1939). *J. Chem. Phys.* **7,** 139.
Kistiakowsky, G. B., and Smith, W. R. (1936). *J. Am. Chem. Soc.* **58,** 1043.
Kobayashi, M. (1943). *J. Chem. Soc. Japan* **64,** 648.
Krebs, H., and Rasche, R. (1954). *Z. anorg. u. allgem. Chem.* **276,** 236.
Krebs, H., Diewald, J., Arlitt, H., and Wagner, J. A. (1956). *Z. anorg. u. allgem. Chem.* **287,** 98.
Krebs, H., Diewald, J., and Wagner, J. A. (1958). *Angew. Chem.* **67,** 705.
Kuhn, W. (1930). *Ber.* **8,** 284; *Trans. Faraday Soc.* **26,** 293.
Kuhn, W. (1952). *Z. Electrochem.* **56,** 506.
Kuhn, W., and Bein, K. (1934). *Z. anorg. u. allgem. Chem.* **216,** 321.
Lecoq, H. (1943). *Bull. soc. roy. sci. Liége* **12,** 316.
Lifschitz, I. (1925). *Z. physik. Chem. (Leipzig)* **114,** 485.
Lions, F., and Martin, K. V. (1957). *J. Am. Chem. Soc.* **79,** 1273.
MacDermott, T. E. (1961). Thesis, Australian National University, Canberra.
MacDermott, T. E., and Sargeson, A. M. (1961). Paper presented at ANZAAS Conference, Brisbane, June.
Mathieu, J. P. (1936a). *Bull. soc. chim. France* [5]**3,** 476.
Mathieu, J. P. (1936b). *J. chim. phys.* **33,** 78.
Mathieu, J. P. (1937). *Bull. soc. chim. France* [5]**4,** 687.
Meisenheimer, J., and Horing, M. (1927). *Ber.* **60,** 1425.
Mills, W. H., and Gotts, R. A. (1926). *J. Chem. Soc.* p. 3121.
Moeller, T., and Gulyas, E. (1958). *J. Inorg. & Nuclear Chem.* **5,** 245.
Moffitt, W. (1956). *J. Chem. Phys.* **25,** 1189.
Morgan, G. T., and Burstall, F. H. (1931). *J. Chem. Soc.* p. 2213.
Morgan, G. T., and Main-Smith, J. (1925). *J. Chem. Soc.* p. 2030.
Murmann, R. K., and Taube, H. (1956). *J. Am. Chem. Soc.* **78,** 4886.
Nakahara, A., and Tsuchida, R. (1954). *J. Am. Chem. Soc.* **76,** 3103.
Neogi, P., and Dutt, N. K. (1938). *J. Indian Chem. Soc.* **15,** 83.
Pearson, R. G., and Basolo, F. (1956). *J. Am. Chem. Soc.* **78,** 4878.

Pearson, R. G., Meeker, R. E., and Basolo, F. (1956). *J. Am. Chem. Soc.* **78,** 2673.
Posey, F. A., and Taube, H. (1953). *J. Am. Chem. Soc.* **75,** 4099.
Prelog, V., and Wieland, P. (1944). *Helv. Chim. Acta* **27,** 1127.
Reihlen, H., Weinbrenner, E., and Hessling, G. V. (1932). *Ann.* **494,** 143.
Royer, D. J. (1961). Georgia Inst. Technology, private communication.
Saito, Y., and Iwaski, H. (1962). *Bull. Chem. Soc. Japan* **35,** 1131.
Saito, Y., Nakatsu, K., Shiro, M., and Kuroya, H. (1957). *Bull. Chem. Soc. Japan* **30,** 795.
Sargeson, A. M. (1961). Paper presented at ANZAAS Conference, Brisbane, June.
Sargeson, A. M., and Searle, G. H. (1961). Unpublished work.
Schnell, S., and Karrer, P. (1955). *Helv. Chim. Acta* **38,** 2036.
Schweitzer, G. K., and Talbott, C. K. (1950). *J. Tenn. Acad. Sci.* **25,** 143.
Shimura, Y. (1958). *Bull. Chem. Soc. Japan* **31,** 315.
Smirnoff, A. P. (1920). *Helv. Chim. Acta* **3,** 177.
Sugano, S. (1960). *J. Chem. Phys.* **33,** 1883.
Treadwell, W. D., Szabados, G., and Haimann, E. (1932). *Helv. Chim. Acta* **15,** 1049.
Tsuchida, R. (1937). *J. Chem. Soc. Japan* **58,** 621.
Tsuchida, R., Kobayashi, M., and Nakamura, A. (1935). *J. Chem. Soc. Japan* **56,** 1339.
Tsuchida, R., Kobayashi, M., and Nakamura, A. (1936). *Bull. Chem. Soc. Japan* **11,** 38.
Wahl, W. (1927). *Ber.* **60,** 399.
Werner, A. (1912a). *Ber.* **45,** 121.
Werner, A. (1912b). *Bull. soc. chim. France* **11,** 1–xxiv.
Werner, A. (1912c). *Ber.* **45,** 3061.
Werner, A. (1912d). *Ber.* **45,** 1229.
Werner, A. (1912e). *Ber.* **45,** 433.
Werner, A. (1914). *Ber.* **47,** 2171.
Werner, A. (1917). *Helv. Chim. Acta* **1,** 5.
Werner, A., and Smirnoff, A. P. (1920). *Helv. Chim. Acta* **3,** 472.

CHAPTER 6

Oxidation-Reduction Potentials as Functions of Donor Atom and Ligand

D. A. BUCKINGHAM* AND A. M. SARGESON

Biological Inorganic Chemistry Section, Australian National University, Canberra, A.C.T., Australia.

I. Theoretical Aspects

For two complex ions ML_x^{n+} and $ML_x^{(n+1)+}$ of the same chemical composition, which differ only in their charge, the potential E for the reversible cell system

$$\mathrm{Pt}|\mathrm{ML}_x^{n+}, \mathrm{ML}_x^{(n+1)+}||\mathrm{H}^+_{a(\mathrm{H}^+)=1}|\mathrm{H}_2 \text{ atm (Pt)}$$

* *Present address:* Department of Chemistry, Brown University, Providence, Rhode Island.

obtained at constant temperature and pressure and without any liquid junctions, is related to the standard thermodynamic potential for the system, E°, by the expression:

$$E = E^\circ - \frac{RT}{F} \ln \frac{(a_{ML_x{}^{(n+1)+}})(a_{H_2})^{1/2}}{(a_{ML_x{}^{n+}})(a_{H^+})} \tag{1}$$

in which R is the gas constant, T the absolute temperature, F the Faraday, a_i the activity of the ion i, and E° the potential of the system when all the constituents involved in the cell reaction are at unit activity.

It is convenient to regard the measured potential E as resulting from the two separate cell reactions, the redox half cell

$$ML_x{}^{n+} \rightarrow ML_x{}^{(n+1)+} + e$$

and the hydrogen half cell

$$H^+ + e \rightarrow \tfrac{1}{2}H_2$$

Equation (1) may be rewritten:

$$E = \left(E_1{}^\circ - \frac{RT}{F} \ln \frac{a_{ML_x{}^{(n+1)+}}}{a_{ML_x{}^{n+}}}\right) - \left(E_2{}^\circ - \frac{RT}{F} \ln \frac{a_{H^+}}{(a_{H_2})^{1/2}}\right) \tag{2}$$

At unit activity of H^+ ion and one atmosphere pressure of hydrogen, the potential of the reversible hydrogen electrode is defined as zero volts at all temperatures. The second term in Eq. (2) thus disappears and the potential for the redox system becomes:

$$E = E^\circ - \frac{RT}{F} \ln \frac{a_{ML_x{}^{(n+1)+}}}{a_{ML_x{}^{n+}}} \tag{3}$$

The standard thermodynamic potential E° for the redox half cell measures the free energy change in the reaction:

$$ML_x{}^{n+} + H^+ \rightarrow ML_x{}^{(n+1)+} + \tfrac{1}{2}H_2$$

since the standard free energy change $\Delta G^\circ = -FE^\circ$ for a one electron change.

We have seen that the potential of the redox half cell: reductant $\rightarrow$ oxidant + e, is given by the expression

$$E = E^\circ - \frac{RT}{F} \ln \frac{a_{ox}}{a_{red}} \tag{4}$$

whence we obtain

$$E = E^\circ - \frac{RT}{F} \ln \frac{c_{ox}}{c_{red}} - \frac{RT}{F} \ln \frac{f_{ox}}{f_{red}}$$

in which c is the molal concentration and f is the mean activity coefficient, and

$$E = E^\circ - \frac{RT}{F} \ln \frac{f_{ox}}{f_{red}} \tag{5}$$

when $c_{ox} = c_{red}$. At low ionic strengths ($\mu < 0.1\ M$) the activity coefficient for an ion may be expressed by the limiting form of the Debye–Huckel law as

$$\log f = -0.5115\ Z_i^2 \cdot \mu^{1/2} \tag{6}$$

in water at 25°C., where Z_i is the charge on the ion and $\mu = \frac{1}{2} c_i Z_i^2$. From Eqs. (5) and (6) we obtain:

$$E = E^\circ + 0.0303(2n + 1)\mu^{1/2} \tag{7}$$

This gives the potential of the redox half cell as a function of charge n of the reductant ion and the ionic strength of the solution. Under ideal conditions at 25°C., a plot of the measured potential E against $\mu^{1/2}$ should give a straight line of slope $0.0303(2n + 1)$ and extrapolation to zero ionic strength gives the standard potential E° for the redox system, in the absence of liquid junction potentials.

Since E° is related to ΔG°, the corresponding standard entropy (ΔS°), and enthalpy (ΔH°), changes can be obtained for the cell reaction.

Partial differentiation of the thermodynamic relation: $dG = SdT + VdP$ with respect to temperature gives:

$$\left(\frac{\partial G}{\partial T}\right)_P = -\Delta S \quad \text{or} \quad F\left(\frac{\partial E}{\partial T}\right)_P = \Delta S$$

for a one electron change; whence for a not too large temperature interval:

$$\frac{F}{4.1835}\left(\frac{\Delta E}{\Delta T}\right)_P = \Delta S$$

The over-all entropy change for the cell reaction is:

$$\Delta S^\circ = \bar{S}^\circ_{ML_x{}^{(n+1)+}} - \bar{S}^\circ_{ML_x{}^{n+}} + \tfrac{1}{2} S^\circ_{H_2}(g) - \bar{S}^\circ_{H^+{}_{(aq)}}$$

At unit activity $\bar{S}^\circ_{H^+{}_{(aq)}} = 0$ by definition, and at a pressure of one atmosphere $S^\circ_{H_2}(g) = 31.2$ e.u. (Latimer, 1952). Hence on the hydrogen scale the partial molar entropy difference between oxidant and reductant is:

$$\Delta\bar{S}^\circ = \bar{S}^\circ_{ML_x{}^{(n+1)+}} - \bar{S}^\circ_{ML_x{}^{n+}} = \Delta S^\circ - 15.6 \text{ e.u.} \tag{8}$$

The entropy difference between the ions determined in this way will be used in later discussion.

Similarly the enthalpy changes ΔH° in the cell reaction can be obtained from the Gibbs–Helmholtz relation:

$$\Delta G - T\left(\frac{\partial(\Delta G)}{\partial T}\right)_P = \Delta H$$

whence $\Delta H = EF - TF(\partial E/\partial T)$ for a one electron change, and for a not too large temperature interval

$$\Delta H^\circ = \frac{E^\circ F}{4.1835} - \frac{TF}{4.1835}\left(\frac{\Delta E^\circ}{\Delta T}\right)_P$$

II. Liquid Junction Potentials

In the measurement of the redox potentials of complexes it is a common practice to use a salt bridge to connect the reference and redox half cells. The unequal rates of diffusion of the cell components at the junction of the salt bridge and half cell give rise to a small potential not accounted for in the Nernst equation and, in order to approximate to the thermodynamic potential, the "liquid junction potential" must be minimized or calculated. Reproducible "continuous mixture" junctions are difficult to set up and maintain and the potential equally difficult to calculate (Glasstone, 1942), so it is usual to minimize the junction potential. A common reference cell is the saturated calomel electrode used in conjunction with a saturated potassium chloride salt bridge. The only significant junction potential then is that of the KCl bridge with the redox half cell solution. The KCl is vastly more concentrated than the redox components and any junction potential is due largely to the diffusion of K^+ and Cl^- ions and, since the latter have nearly identical transference numbers in aqueous solution, the potential is largely eliminated. Investigations by McInnes and Yeh (1921) and Guggenheim (1930) have shown that for the saturated KCl bridge the junction potential is probably less than 1 mv. and, provided the concentrations in the redox half cell are small, such a junction is reproducible (Glasstone, 1942).

Another procedure which helps to minimize the effect of the junction potential is to standardize the calomel electrode against the hydrogen or quinhydrone electrodes. Using the salt bridge any junction potential for these systems will be included in the calibration value for the calomel.

Since the glass electrode–hydrogen electrode cell is affected negligibly by extraneous ions over a wide range of pH and ionic composition, redox potential measurements made with the glass electrode when extrapolated to zero ionic strength would give a value very close to the true E°, if not the true E° (Ogston, 1962). This method has the advantages that the glass electrode is a direct substitute for the hydrogen reference electrode, that liquid junction potentials are eliminated, and that redox potentials can be measured over a wide range in pH.

III. Reversibility

The Nernst equation is valid for reversible changes only, a condition rigorously fulfilled only if: (1) no changes take place in the cell without the passage of current; (2) every change which occurs during the passage of current may be reversed by reversing the direction of the current; (3) the net result of *all* the chemical changes which occur within the cell system is known (Harned and Owen, 1950). Condition (1) implies that the cell potential should be constant with respect to time. If this condition were rigorously enforced very few potentials could be accepted. Although no sharp distinction can be drawn between potentials which change rapidly with time and those with a slower "drift," the changes in the latter can be traced usually to: (*a*) instability of the oxidant and/or reductant towards dissociation, hydrolysis, reduction, or oxidation; (*b*) changes in the state of aggregation of a constituent, including precipitation; (*c*) the action of residual oxygen; or (*d*) the slow accommodation of the electrode. When the system is poorly poised, as it is when either reductant or oxidant is present in a large excess, or when the concentrations of both are equal but very small, the rapidity of the drift due to any one of the above causes will be accentuated.

The second condition for reversibility implies that the same potential should be found irrespective of whether the given state is reached by reducing 50% of the oxidant or oxidizing 50% of the reductant, or by preparing the appropriate mixture from the pure oxidant and reductant. It is necessary that oxidant and reductant equilibrate rapidly on the electrode surface and that the rate of electron transfer between them is fast at this surface. The most common inert electrodes are platinum, gold, and rhodium, and it seems that the potentials at the gold and rhodium are more readily reproduced than at a bright platinum surface. Finally, the net result of all chemical changes which occur in the cell reaction must be known. With complex ions it is difficult, sometimes, to know what species are present since some undergo irreversible or reversible dissociation, or hydrolysis, in aqueous solution at variable rates. The $[Fe(CN)_6]^{4-}$ ion, for instance, exchanges one cyanide ion for a water molecule but the rate is sufficiently slow to permit reproducible measurement of the redox potential of the $[Fe(CN)_6]^{4-/3-}$ system (Kolthoff and Tomsicek, 1935). However, the $[RuCl_6]^{3-/--}$ and $[OsCl_6]^{3-/--}$ systems take days to reach equilibrium because of hydrolytic reactions and it is difficult to be sure what exactly is being measured. At equilibrium in acid solution the osmium couple was shown to be a cationic system, probably $OsO^{+/++}$ (Dwyer *et al.*, 1946). A hydrolytic study of $K_2[RuCl_5H_2O]$ has demonstrated that after 70 days in 0.1 N HCl at 25°C. the solution contains 16% $[RuCl_3(H_2O)_3]^0$, 61% $[RuCl_2]_{aq}^+$, and 23% $[RuCl]_{aq}^{++}$ (Connick and Fine, 1961). It is conceivable also, in these halide systems, that the rate of electron transfer is very slow, as it is in most cobalt (II)/(III) couples, and equilibrium is probably never attained in practice. The potential obtained by Bjerrum (1941) for the $[Co(en)_3]^{++/3+}$ couple ($E° = +0.259$ volt) seems to be too positive since the $[Co(en)_3]^{3+}$ ion can be reduced slowly with silver wool in the presence of Cl^- ion ($Ag + Cl^- \rightarrow AgCl + e^-$, $E° = -0.222$ volt).

IV. Measurement of Redox Potentials

The stability of a complex couple to oxidation or reduction in aqueous solution is determined largely by its redox potential relative to the oxygen-water potential, the latter being strongly pH-dependent:

$$a_{H^+} = 1; \qquad 2H_2O \rightarrow O_2 + 4H^+ + 4e^-, \qquad E° = -1.239 \text{ volts}$$

$$a_{H^+} = 10^{-7}; \qquad 2H_2O \rightarrow O_2 + 4H^+ + 4e^-, \qquad E° = -0.815 \text{ volt}$$

$$a_{OH^-} = 1; \qquad 4OH^- \rightarrow O_2 + 2H_2O + 4e^-, \qquad E_B° = -0.401 \text{ volt}$$

Thus, if the potential of the complex couple is less negative than −0.815 volt the reductant should oxidize in moist air or neutral solution, if more negative the oxidant should reduce. Fortunately, the H_2O/O_2 couple is highly irreversible, otherwise there would be few "stable" oxidation states, and significant oxidation of water does not usually occur until the potential is more negative than about −1.5 volts in faintly acid solutions. Within the approximate potential range −0.2 to −1.5 volts usually both oxidation states can be isolated.

The experimental method of measuring redox potentials is governed largely by the position of the potential relative to the water-oxygen couple, as well as by the stability of the complex ion towards hydrolysis or dissociation of the ligands. The common methods are described in the following section.

A. Direct Measurement

(1) Equimolar amounts of oxidant and reductant are mixed in dilute (10^{-3}–10^{-4} M) aqueous solution containing two inert gold or rhodium electrodes in an inert atmosphere, usually nitrogen. The solution is allowed to equilibrate with the electrode surfaces and the potential of one measured against a saturated calomel electrode. At the null-point the electrodes are switched and the potential remeasured with the other to overcome any polarization effects produced by the coarse adjustment of the potentiometer. The ionic strength of the solution is varied by the addition of small amounts of a salt, KCl or KNO_3, or acid, and the potential versus the square

root of the ionic strength curve extrapolated to zero ionic strength to give the standard redox potential, E°. However, it is common practice to plot the potential against $\sqrt{\mu}/(1 + \sqrt{\mu})$ since the redox experiments are often carried out in moderate concentrations of electrolytes where the Debye–Huckel limiting law is not applicable. It may also be more significant to vary the ionic strength by variation in the concentration of the redox mixture, and thus determine the effect of these substances on the junction potential. Both extrapolations should give the same value of E° if the junction potential is negligible. This method, which permits a reproducibility of about 0.2 mv., can be used only when both the pure reductant and oxidant can be isolated or obtained in solution. The plot of the measured potential E against time is shown in Fig. 1(a). After equilibrium is reached, the potential should be constant if the reductant and oxidant are stable.

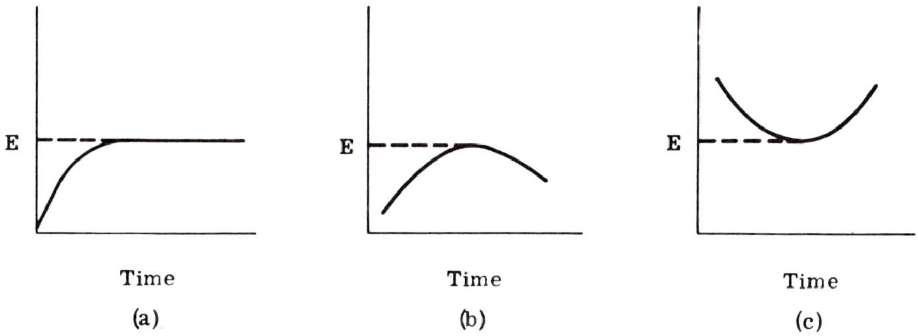

FIG. 1. The plot of the measured potential E against time.

(2) If either oxidant or reductant is unstable to reduction or oxidation, the potential may be measured with a solution of the stable component, to which is added half the equivalent amount of an oxidizing or reducing agent. The potential is then measured at intervals and when plotted against time a maximum or minimum value is obtained (Fig. 1), and this is taken as the measured potential. The curve in Fig. 1(b) corresponds to a stable reduced state, unstable oxidation state, and that in Fig. 1(c) is the reverse situation. The initial change in potential is due to establishment of equilibrium at the electrode surface and the final decay is duc to the instability of the oxidized or reduced forms, respectively. The accuracy of the potential measured depends on the relative rates of establishment of equilibrium and decay of the unstable state. It is also essential to use an oxidizing or reducing agent whose potential is considerably different from the substance to be oxidized or reduced. In favorable circumstances the potentials obtained are worth extrapolating to infinite dilution.

(3) Perhaps the most widely used method is the determination of the potential at the half equivalence point in a potentiometric titration of the stable form. The accuracy of this method is even more limited than (2) since during the titration the potential continually changes and the electrode has insufficient time to reach equilibrium. In addition, the stable form will be regenerated to some extent during the titration and, to minimize this error, the titration must be done quickly, and this in turn mitigates against equilibration. It is probably less successful than method (2).

(4) An estimate of the redox potential can be obtained by reacting the stable form of the couple with a graded series of oxidizing or reducing agents of known redox potential. A simple example is provided by the $[Ru(bipy)_2Cl_2]^{0/+}$ couple, the reduced state of which is very sparingly soluble in water. Iodine, ferric ion, and bromine fail to effect oxidation, but chlorine oxidizes the substance. The potential is therefore between -0.9 and -1.35 volts, allowing for about 0.2 volt for complete oxidation. If bromide ion is now added to the oxidized form of the couple, it is found that bromine is not produced, hence the potential must be slightly more positive than the Br^-/Br_2 couple, (-1.02 volts), i.e., approximately -0.95 volt. These semiquantitative estimates are only valid with reversible systems; failure to effect oxidation or reduction with a particular reagent may be due to mechanistic reasons, and not to the (presumed) unfavorable free-energy change.

B. Redox Potentials from Stability Constants

The redox potentials of complex couples which dissociate reversibly can be calculated from their stability constants if $E°$ for the aqueous metal ion couples is known (Perrin, 1959; Hawkins and Perrin, 1962). The theoretical basis for this will be evident if the following system is considered at equilibrium:

$$
\begin{array}{ccc}
M^{(n+1)+} + xL & \overset{\beta_x''}{\rightleftarrows} & ML_x^{(n+1)+} \\
+e \downarrow\uparrow & & \uparrow\downarrow +e \\
M^{n+} + xL & \overset{\beta_x'}{\rightleftarrows} & ML_x^{n+}
\end{array}
$$

Equation (4) gives the redox potential E of the complex couple $ML_x^{n+}/ML_x^{(n+1)+}$. At equilibrium, when $E = 0$, this becomes

$$E° = \frac{RT}{F} \ln \frac{a_{ox}}{a_{red}}$$

or

$$E° = \frac{RT}{F} \ln \frac{a_{ML_x^{(n+1)+}}}{a_{ML_x^{n+}}}$$

Now

$$\beta''_x = \frac{[a_{ML_x^{(n+1)+}}]}{[a_{M^{(n+1)+}}][a_L]^x}$$

and

$$\beta'_x = \frac{[a_{ML_x^{n+}}]}{[a_{M^{n+}}][a_L]^x}$$

hence

$$\frac{a_{ML_x^{(n+1)+}}}{a_{ML_x^{n+}}} = \frac{\beta''_x}{\beta'_x}\frac{a_{M^{(n+1)+}}}{a_{M^{n+}}}$$

and

$$E^\circ = \frac{RT}{F}\ln\frac{a_{M^{(n+1)+}}}{a_{M^{n+}}} + \frac{RT}{F}\ln\frac{\beta''_x}{\beta'_x}$$

At equilibrium also

$$E^\circ_{M^{n+}/M^{(n+1)+}} = \frac{RT}{F}\ln\frac{a_{M^{(n+1)+}}}{a_{M^{n+}}}$$

whence

$$E^\circ \text{ (complex couple)} = E^\circ \text{ (hydrated metal ion couple)} + \frac{RT}{F}\ln\frac{\beta''_x}{\beta'_x}$$

Thus the thermodynamic E° for a complex couple can be obtained from the stability constants of the complexes extrapolated to infinite dilution provided E° for the aqueous metal ions is known. However, the method is applicable only to systems which dissociate reversibly and could not be used, for example, with the $[Fe(phen)_3]^{++/3+}$ couple. Whereas the $[Fe(phen)_3]^{++}$ ion dissociates reversibly, the reaction between Fe_{aq}^{3+} ion and phenanthroline yields the dimeric cation $[Fe(phen)_2OH]_2^{4+}$ and not the $[Fe(phen)_3]^{3+}$ species. Usually, true E° (complex) values are not obtained since the thermodynamic dissociation or stability constants have not been determined, and the potential found in this way is called the "formal" redox potential, E'° (Rossotti and Rossotti, 1961). The accuracy of the method depends on the precision of both the E° (hydrated ion couple) and the stability constants. An accuracy of ± 2 mv. appears to be the limit attainable.

C. Redox Potentials from Polarography (Kolthoff and Lingane, 1952)

We assume that the electrode reaction

$$ML_x^{n+} + (y - x)L \rightarrow ML_y^{(n+1)+} + e$$

is reversible at the dropping mercury electrode, i.e., that there is no amalgamation occurring and electron transfer is rapid, that there is sufficient supporting electrolyte present to make the current entirely diffusion controlled, and that sufficient complexing agent (L) is present in solution so that its concentration on the electrode surface remains constant and independent of the current.

The potential of the dropping electrode (d.e.) at any point on either a cathodic, anodic, or composite wave is then expressed by:

$$E_{\mathrm{d.e.}} = E^\circ - \frac{RT}{F}\ln\frac{a^\circ_{\mathrm{ox}}}{a^\circ_{\mathrm{red}}} + (y - x)\frac{RT}{F}\ln a_{\mathrm{L}}$$

If we assume that the concentrations are the same as the activities, in the very dilute solutions used:

$$E_{\mathrm{d.e.}} = E^\circ - \frac{RT}{F}\ln\frac{C^\circ_{\mathrm{ox}}}{C^\circ_{\mathrm{red}}} + (y - x)\frac{RT}{F}\ln C^\circ_{\mathrm{L}}$$

where E° is the standard thermodynamic potential, C°_{ox} and C°_{red} are the concentrations at the electrode surface of the complex ion in the higher and lower oxidation states, and C°_{L} is the concentration of the uncomplexed ligand, also at the electrode surface. These concentrations are governed by the following identities:

$$C^\circ_{\mathrm{red}} = C_{\mathrm{red}} + \frac{i}{k_{\mathrm{red}}} = \frac{-(id)_a + i}{k_{\mathrm{red}}}$$

$$C^\circ_{\mathrm{ox}} = C_{\mathrm{ox}} - \frac{i}{k_{\mathrm{ox}}} = \frac{(id)_c - i}{k_{\mathrm{ox}}}$$

where (id) is the limiting diffusion current, C°_{ox} is the concentration of oxidant in the body of the solution, C°_{ox} is the concentration of oxidant at the electrode surface, and C_{red} and C°_{red} have corresponding meanings. The proportionality constant k is given by the relation

$$C^\circ_{\mathrm{red}} = \frac{i}{k_{\mathrm{red}}}$$

and

$$k_{\mathrm{red}} \propto D^{1/2}_{\mathrm{red}}$$

where D is the diffusion coefficient for the reductant. At 25°C.,

$$E_{\mathrm{d.e.}} = E_{1/2} - 0.0591\log\frac{(id)_c - i}{i - (id)_a} \tag{9}$$

where $E_{1/2}$ is the half wave potential* and

$$E_{1/2} = E^\circ - 0.0591\log\frac{k_{\mathrm{red}}}{k_{\mathrm{ox}}} + (y - x)0.0591\log C_{\mathrm{L}} \tag{10}$$

Since the curve of Eq. (9) is symmetrical about its midpoint, the half wave potential should be constant and independent of both absolute and relative concentrations of the oxidized and reduced forms. The ratio

* The sign convention for E° values adopted in this chapter leads to the opposite sign for $E_{1/2}$ values to that used in polarographic literature.

$$\frac{(id)_a}{(id)_c} = \frac{k_{\text{red}}}{k_{\text{ox}}} = \left(\frac{D_{\text{red}}}{D_{\text{ox}}}\right)^{1/2}$$

can be evaluated experimentally from the ratio of the cathodic and anodic diffusion currents in a solution containing equal concentrations of oxidized and reduced forms. This is usually sufficiently close to unity to be neglected. The shift of $E_{1/2}$ as the concentration of complexing agent varies gives the relative values of x and y, and it is obvious from Eq. (10) that if $x = y$ then $E_{1/2}$ is independent of the concentration of complexing agent and is the same as $E°$ provided $k_{\text{red}}/k_{\text{ox}} = 1$. When the complex dissociates $E°$ is equivalent to $E_{1/2}$ when C_{L} is molar, after allowance is made for the cell resistance.

Thus, redox potentials can be measured polarographically provided the rather stringent conditions of electrode reversibility, diffusion-controlled current, and constant concentration of complexing agent at the electrode surface exist. The dropping mercury electrode must behave as an inert electrode during the process. The reversibility of the electrode process can be stringently checked by determining the half wave potential anodically and cathodically. They should agree after allowance is made for the IR drop. Similarly, a mixture of equivalent amounts of oxidant and reductant should also give the same $E_{1/2}$. Deviations from this condition are frequently due to a slow electrode reaction, and the practical limit for reversibility seems to be that the standard rate constant $k_s > 0.1$ cm. sec.$^{-1}$ for the process shown. The term k_s is defined equal to k_1 and k_2 at that potential of the electrode where they are equal to each other in the process:

$$\text{oxidized form} + \text{ne} \underset{k_2}{\overset{k_1}{\rightleftarrows}} \text{reduced form (Randles, 1960)}$$

This means that the rates of electron transfer at the electrode must be rather rapid for the measurement of $E_{1/2}$ to correspond to $E°$ and this condition would probably exclude most Co(II) − Co(III) systems. However, if the solutions are dilute and reversibility is demonstrated by the anodic and cathodic waves, reasonable values of $E°$ should be obtained.

V. Factors Governing Redox Potential Values

The redox half cell may be represented by a form of Born–Haber cycle as follows:

$$\begin{array}{ccc}
\text{M}^{n+}(g) & \xrightarrow{\Delta G_I^\circ} & \text{M}^{m+}(g) \quad + (m-n)e(g) \\
\Delta G_c^{\circ\prime} \downarrow x\text{L}(g) & & x\text{L}(g) \downarrow \Delta G_c^{\circ\prime\prime} \\
\text{ML}_x{}^{p+}(g) & & \text{ML}_x{}^{q+}(g) \\
\Delta G_{\text{aq}}^{\circ\prime} \downarrow & & \downarrow \Delta G_{\text{aq}}^{\circ\prime\prime} \\
\text{ML}_x{}^{p+}(\text{aq}) & \xrightarrow[\Sigma\Delta G^\circ]{} & \text{ML}_x{}^{q+}(\text{aq}) + (m-n)e(g)
\end{array}$$

where ΔG_I° is the difference in free energy of ionization for the two gaseous metal ions, $\Delta G_c^{\circ\prime\prime}$ and $\Delta G_c^{\circ\prime}$ are the free energy changes on coordination of the ligand in the gaseous state, and $\Delta G_{aq}^{\circ\prime\prime} + \Delta G_{aq}^{\circ\prime}$ are the free energy changes for hydration of the gaseous coordinated ions, in their standard states.

A similar cycle can be presented for the $\frac{1}{2}H_2/H^+$ half cell and the overall standard free energy change for the combined cell (ΔG°) can be divided *arbitrarily* into the contributions given above; i.e.,

$$\begin{aligned}\Delta G^\circ &= \Sigma\Delta G^\circ - \Sigma\Delta G^\circ_{\frac{1}{2}H_2/H^+} = (m-n)FE^\circ \\ &= \Delta G_I^\circ + (\Delta G_c^{\circ\prime\prime} - \Delta G_c^{\circ\prime}) + (\Delta G_{aq}^{\circ\prime\prime} - \Delta G_{aq}^{\circ\prime}) - (m-n)\Sigma\Delta G^\circ_{\frac{1}{2}H_2/H^+} \\ &= \Delta G_I^\circ + \Delta(\Delta G_c^\circ) + \Delta(\Delta G_{aq}^\circ) - (m-n)\Sigma\Delta G^\circ_{\frac{1}{2}H_2/H^+}\end{aligned}$$

ΔG_I° may be replaced by I_{m-n}, the ionization potential for $M^{n+} \rightarrow M^{m+}$, since the entropy change on ionization in the gaseous state is very small (see Section VI,A). Therefore

$$\Delta G^\circ = I_{m-n} + \Delta(\Delta G_c^\circ) + \Delta(\Delta G_{aq}^\circ) - (m-n)\Sigma\Delta G^\circ_{\frac{1}{2}H_2/H^+}$$

where $\Delta(\Delta G_c^\circ)$ is the difference between the free energy of formation of the complex ions in the gaseous state and $\Delta(\Delta G_{aq}^\circ)$ is the difference between the free energy of hydration of the two complex ions. The contribution from each term in the redox half cell may now be compared with those of other systems, relative to the hydrogen half cell since the energy terms in the latter are constant (H). Thus:

$$\Delta G^\circ = I_{m-n} + \Delta(\Delta G_c^\circ) + \Delta(\Delta G_{aq}^\circ) - (m-n)H$$

Further, if redox systems of complexes with the same metal are compared, I_{m-n} is also constant, and the variables reduce to the free energy changes of coordination and hydration.

It is convenient now to divide these free energy terms into their respective enthalpy and entropy components:

$$\Delta(\Delta G_c) = \Delta(\Delta H_c) - T\Delta(\Delta S_c)$$

and

$$\Delta(\Delta G_{aq}) = \Delta(\Delta H_{aq}) - T\Delta(\Delta S_{aq})$$

The difference between the heats of the complex formation for the two ions $\Delta(\Delta H_c)$ can be further split into two terms (Orgel, 1960), the ligand field stabilization energy difference $\Delta(\Delta H_{LF})$, and $\Delta(\Delta H_{c'})$, the enthalpy change if no ligand field effect is present; i.e., $\Delta(\Delta H_c) = \Delta(\Delta H_{LF}) + \Delta(\Delta H_{c'})$. The term $\Delta(\Delta H_{c'})$ includes electrostatic, covalent, polarization, and steric effects not present in $\Delta(\Delta H_{LF})$. It is also convenient to sum all the entropy terms $\Delta(\Delta \bar{S}^\circ)$ for the half cell. Thus;

$$\Delta(\Delta\bar{S}) = [\bar{S}_{\mathrm{ML}_x{}^{q+}(g)} - \bar{S}_{\mathrm{M}^{m+}(g)} - x\bar{S}_{\mathrm{L}}(g)] - [\bar{S}_{\mathrm{ML}_x{}^{p+}(g)} - \bar{S}_{\mathrm{M}^{n+}(g)} - x\bar{S}_{\mathrm{L}}(g)] + [\bar{S}_{\mathrm{ML}_x{}^{q+}(\mathrm{aq})} - \bar{S}_{\mathrm{ML}_x{}^{q+}(g)}] - [\bar{S}_{\mathrm{ML}_x{}^{p+}(\mathrm{aq})} - \bar{S}_{\mathrm{ML}_x{}^{p+}(g)}]$$

This expression reduces to:

$$\Delta(\Delta\bar{S}) = [\bar{S}_{\mathrm{M}^{n+}(g)} - \bar{S}_{\mathrm{M}^{m+}(g)}] + [\bar{S}_{\mathrm{ML}_x{}^{q+}(\mathrm{aq})} - \bar{S}_{\mathrm{ML}_x{}^{p+}(\mathrm{aq})}]$$

The entropy of monatomic gaseous ions is given by the Sakur–Tetrode equation $S^\circ(g) = 26.03 + 1.5 \ln M$, at 25°C., where M = atomic weight. This predicts that gaseous monatomic ions of the same element have the same translational entropy. It is possible that there will be an electronic entropy term also if the ions have a multiplet ground state or low-lying excited state but the contribution from the latter term will be small (1 or 2 e.u.). The total entropy difference can be represented to a first approximation by:

$$\Delta(\Delta\bar{S}^\circ) = [\bar{S}_{\mathrm{ML}_x{}^{q+}(\mathrm{aq})} - \bar{S}^\circ{}_{\mathrm{ML}_x{}^{p+}(\mathrm{aq})}]$$

The factors contributing to the redox potentials in differing systems can now be compared by the expression:

$$\Delta G^\circ = I_{m-n} + \Delta(\Delta H^\circ_{\mathrm{aq}}) + \Delta(\Delta H^\circ_{c'}) + \Delta(\Delta H^\circ_{\mathrm{LF}}) - T(\bar{S}^\circ{}_{\mathrm{ML}_x{}^{q+}(\mathrm{aq})} - \bar{S}^\circ{}_{\mathrm{ML}_x{}^{p+}(\mathrm{aq})}) - (m - n)H(\text{const.}) \quad (11)$$

where ΔG° is the standard free-energy change for the cell reaction, I_{m-n} is the ionization potential for the parent gaseous metal ions, $\Delta(\Delta H_{\mathrm{aq}}{}^\circ)$ is the hydrational enthalpy difference between the complex ions, $\Delta(\Delta H_{c'})$ is the difference between the enthalpies of formation for the complex ions excluding the ligand field term, $\Delta(\Delta H_{\mathrm{LF}})$ is the difference between the two ligand field enthalpy terms for the complex ions and, the $\bar{S}^\circ$ terms are the partial molar entropies of the two complex ions.

VI. Evaluation of the Terms in the Born–Haber Cycle

It is interesting at this stage to examine the relative magnitude of each term in Eq. (11) for a typical redox reaction, e.g.,

$$\mathrm{Fe_{aq}}^{++} + \mathrm{H}^+ \longrightarrow \mathrm{Fe_{aq}}^{3+} + \tfrac{1}{2}\mathrm{H}_2$$

Thus:

$$\Delta G^\circ - I_3 + \Delta(\Delta H^\circ_{\mathrm{aq}}) + \Delta(\Delta H^\circ_{c'}) + \Delta(\Delta H^\circ_{\mathrm{LF}}) - T(\bar{S}^{\circ 3+}_{\mathrm{aq}} \quad \bar{S}^{\circ ++}_{\mathrm{aq}}) - H_{\mathrm{const.}} = +706 - 253 - 362 + 12 + 13 - 103 \text{ kcal.} = +13 \text{ kcal/mole}$$

E°(calc.) = −0.57 volt, E°(found) = −0.77 volt. It will be evident that the potentials arise from relatively small differences between large terms, of which the ionization potential, hydration enthalpies, and heats of formation make the largest contributions. Ligand field and entropy corrections are small terms by comparison. For the most part, the estimation of these quantities for complex couples is not possible, largely due to the lack of information for trivalent ions and their complexes. This deficiency has been noted by George and McClure (1959) and the existing information thoroughly surveyed. However, for comparative purposes it is possible to estimate trends, especially for the first transition series, and to gauge the relative magnitude of each term.

A. Ionization Potentials

The ionization potentials for simple gaseous metal ions are known quantities and can be obtained from Moore's Tables (1958). Some of these potentials are listed in Table (IV). The values are assumed to be independent of temperature and are taken as the standard enthalpy of ionization for the isolated gaseous ions at 25°C.

B. Hydration Enthalpies

Except for simple gaseous ions, these constants have not been measured experimentally, but an appreciation of their magnitude can be obtained from the Born charging energy expression:

$$\Delta G^{\circ}_{\mathrm{B.C.}} = -\frac{Z^2e^2}{2R}\left(1 - \frac{1}{D}\right)$$

This gives the free energy of formation of a spherical cavity of radius R bearing a charge Ze in a continuous medium of dielectric constant D. Differentiation of the expression with respect to temperature gives the entropy change, whence:

$$\Delta H^{\circ}_{\mathrm{B.C.}} = -\frac{Z^2e^2}{2R}\left[\left(1 - \frac{1}{D}\right) + \frac{T}{D^2}\left(\frac{\partial D}{\partial T}\right)_p\right]$$

If hydrated and complex ions are regarded as charged spheres and the solution process as an introduction of these spheres into continuous water dielectric then:

$$\Delta H^{\circ}_{\mathrm{aq}} = -\frac{167Z^2}{R}\ \mathrm{kcal./mole\ at\ 25°C}$$

The dependence of ΔG_{aq} and ΔH_{aq} (experimental) on Z^2 has been established for simple metal ions. This relationship is not unexpected since the mathematical model is relatively insensitive to water structure, and the dielectric term is small. However, the same agreement is not expected for the entropy change since

$$\Delta S^{\circ}_{\mathrm{B.C.}} = \frac{Z^2e^2}{2DR}\left(\frac{\partial \ln D}{\partial T}\right)_p$$

and is now strongly dependent on the dielectric constant. Experimentally, the agreement is poor for hydrated metal ions and the entropy change is found to vary with Z rather than Z^2 (Powell and Latimer, 1951). The Born equation is therefore probably a good estimate of ΔG°_{aq} or ΔH°_{aq} but not of ΔS°_{aq}.

C. Enthalpies of Formation, $\Delta H_{c'}$, Terms

The enthalpies of formation of di- and trivalent complexes from the aqueous metal ions and ligands may be obtained from the aqueous stability constants measured at different temperatures. For the two reactions:

$$M_{aq}{}^{3+} + L_{aq} \rightarrow ML_{aq}{}^{3+} \ldots\ldots\ldots\ldots \Delta H_{F3}$$

and

$$M_{aq}{}^{++} + L_{aq} \rightarrow ML_{aq}{}^{++} \ldots\ldots\ldots\ldots \Delta H_{F2}$$

we may write

$$\Delta(\Delta H_c) = \Delta H_{F3} - \Delta H_{F2} - \Delta(\Delta H_{aq}{}^1) + \Delta(\Delta H_{aq}{}^2)$$

in which $\Delta(\Delta H_{aq}{}^1)$ and $\Delta(\Delta H_{aq}{}^2)$ refer to the ions ML^{3+}, ML^{++} and M^{3+}, M^{++}, respectively. Since the hydration differences can be estimated from the Born equation, $\Delta(\Delta H_c)$ can be determined once ΔH_{F3} and ΔH_{F2} are known. For ions with d°, d^5, or d^{10} configurations like Ca^{++}, Mn^{++}, Zn^{++} and Sc^{3+}, Fe^{3+}, Ga^{3+}, where there is no ligand field contribution, this gives $\Delta(\Delta H_{c'})$ immediately, which permits an estimate of its value for other ions of the same transition series. The $\Delta H_{c'}$ term should increase more sharply for the trivalent ions than the divalent as the atomic number increases; i.e., it should parallel the trend for the over-all heats of hydration of these ions. Unfortunately, there is little information available on the formation of trivalent or high valency complexes but if we assume the electrostatic picture of complex formation the $\Delta(\Delta H_{c'})$ term should parallel the difference between

$$\left(\frac{3}{R^2_{(M^{3+})}} - \frac{2}{R^2_{(M^{++})}}\right)$$

for the pairs of ions Sc^{3+}, Ca^{++}; Fe^{3+}, Mn^{++}; and Ga^{3+}, Zn^{++}. Their respective ionic radii (Å) are Ca^{++}, 0.99; Mn^{++}, 0.80; Zn^{++}, 0.74; Sc^{3+}, 0.81; Fe^{3+}, 0.64; and Ga^{3+}, 0.62. Hence the trivalent state is favored over the divalent and the higher atomic numbers over the lower. The difference is partly diminished, however, by the higher polarizabilities of the lower valence states of metals with the higher atomic numbers. From the first to the third transition series, the $\Delta(\Delta H_{c'})$ term for the smaller ions Fe^{3+}, Fe^{++} will be larger than for Os^{3+}, Os^{++} and the larger polarizability of the Os(II) state would enhance this difference. The polarizability difference between Fe^{3+} and Os^{3+} would not be as large as for the divalent states.

The $\Delta(\Delta H_{c'})$ terms can be obtained for the aqueous metal di- and trivalent ions from the experimental hydration enthalpies (Brewer *et al.* 1950; see Section VII) by subtracting $\Delta(\Delta H_{LF})$, and $\Delta(\Delta H_{c'})$ increases as the atomic number increases.

D. Entropy Terms

For complexes of different metals with the same ligand $\Delta S°$ for the reaction:

$$M_{aq}^{++} + L_{aq} \rightarrow ML_{aq}^{++}$$

is roughly constant (George and McClure, 1959) and the same could be expected for the trivalent ions:

$$M_{aq}^{3+} + L_{aq} \rightarrow ML_{aq}^{3+}$$

The entropy difference between these two reactions

$$\Delta(\Delta S°) = \bar{S}_{ML_{aq}^{3+}} - \bar{S}_{ML_{aq}^{++}} - \bar{S}_{M_{aq}^{3+}} + \bar{S}_{M_{aq}^{++}}$$

should also be constant since for the aqueous metal ions $(\bar{S}°_{M_{aq}^{3+}} - \bar{S}°_{M_{aq}^{++}})$ is roughly constant, ~40 e.u. (Table I).

TABLE I
Entropies $\bar{S}°$ of Ions in Aqueous Solution at 25°C.[a]

Ion	$\bar{S}°$ (cal. deg.$^{-1}$ mole^{-1})		Ion	$\bar{S}°$ (ca. deg.$^{-1}$ mole^{-1})	
H^+	0.00				
Fe^{++}	−25.9 ± 1.0 (−27.1)		Fe^{3+}	−61 ± 5 (−70.1)	
Mn^{++}	−19.1	(−20)	Mn^{3+}	−59	
			Cr^{3+}	−65	(−73.5)
Co^{++}	−27 (−24)	(−22 exptl.)			
Ni^{++}	−31 (−25)	(−23 exptl.)			
			Rh^{3+}	−60	
			V^{3+}	−65	
Cu^{++}	−26.5 ± 1 (−23.6)				
Zn^{++}	−25.7 ± 1 (−25.45)				

[a] Entropies obtained from Powell and Latimer (1951), Staveley and Randall (1958), Conway (1952).

This generalization seems to apply to metal ions from all transition series and the entropy term for metal ion couples with the same ligand may be considered constant provided the ionic charges are constant. An increase in charge on the metal chelate involves an unfavorable entropy increase due to the more ordered arrangement of solvent molecules in the vicinity of the charged ion. The $T\Delta\bar{S}°$ term then favors the lower charged ion. For example the ferrous aquo ion is stabilized by 0.56 volt by the entropy term whereas the oxidant in the $Fe(CN)_6^{4-/3-}$ couple is stabilized by 0.62 volt by the same term. It follows that the ionic entropy should be partly a surface charge density effect and an increase in the ionic size should rapidly diminish the entropy contribution. This is probably demonstrated by the much smaller $\Delta\bar{S}°$ term for the $[Fe(phen)_3]^{++/3+}$ couple (+5 e.u.) compared with the $Fe_{aq}^{++/3+}$ system (−40 e.u.). The same pair of couples shows the possible significance of the contribution of the entropy terms to the redox potentials of different complexes of the same metal ions; the redox potential difference here is only about 0.4 volt. Fortunately the entropy terms are often small for the larger complexes.

TABLE II

THE PARTIAL MOLAR ENTROPY DIFFERENCES FOR SOME REDOX COUPLES

Redox couple	Entropy differences (cal. mole⁻¹ deg.⁻¹)
$\bar{S}°[Fe(CN)_4(bipy)]^- - \bar{S}°[Fe(CN)_4(bipy)]^{--}$ [a]	+34.6 ± 1.6
$\bar{S}°[Fe(CN)_6]^{3-} - \bar{S}°[Fe(CN)_6]^{4-}$ [a]	+48
$\bar{S}°[Fe(phen)_3]^{3+} - \bar{S}°[Fe(phen)_3]^{++}$ [a]	+5.2 ± 2
$\bar{S}°[Fe(bipy)_3]^{3+} - \bar{S}°[Fe(bipy)_3]^{++}$ [a]	+8.2 ± 2
$\bar{S}°[Ru(bipy)_3]^{3+} - \bar{S}°[Ru(bipy)_3]^{++}$ [a]	+0.2 ± 2
$\bar{S}°[Os(bipy)_3]^{3+} - \bar{S}°[Os(bipy)_3]^{++}$ [a]	{−0.8 ± 1; −0.4 ± 1
$\bar{S}°[Os(bipy)_2PyCl]^{++} - \bar{S}°[Os(bipy)_2PyCl]^+$	+5.1 ± 1
$\bar{S}°[IrCl_6]^{3-} - \bar{S}°[IrCl_6]^{4-}$ [a]	+19.6
$S°_{(H_2)} = 31.2$	

[a] George *et al.* (1959).

E. LIGAND FIELD EFFECTS AND ELECTRON-PAIRING ENERGIES

It is possible to estimate the ligand field term $\Delta(\Delta H_{LF})$ from a knowledge of the degree of splitting of the d electron levels in the oxidized and reduced complexes. ΔH_{LF} itself is equal to the sum of the energy changes to which the electrons of the complex are subjected during the process $M^{n+} \rightarrow ML_x^{n+}$. It is assumed that the only electrons affected by the ligand field are those which are nonbonding and/or π-bonding in the complex. ΔH_{LF} thus follows from a knowledge of the relevant ligand field stabilization energies and d electron repulsion energies. The latter are important when these electrons are paired, but are otherwise ignored, although this cannot be theoretically justified (Griffith, 1961). The ligand field stabilization energy may be obtained spectroscopically as -0.6Δ per e_g electron plus 0.4Δ per t_{2g} electron (Δ = 10 Dq).* Pairing energies are also obtained from spectroscopic data (see Table III).

* *Dq* is defined in Chapter (2).

TABLE III
MEAN ELECTRON-PAIRING ENERGIES FOR GASEOUS IONS (KCAL./MOLE)[a]

Metal	2+ state	3+ state
Cr	67 (70[b])	—
Mn	73 (74)	80 (81)
Fe	50 (55[b])	85 (74[b])
Co	64 (69[b])	60 (61[b])
Ni	—	77 (86)

[a] Values in brackets are calculated from electrostatic interaction parameters B,C for *d* electrons (George and McClure, 1959).
[b] Estimated from typical B,C values for divalent and trivalent ions.

Some idea of the relative magnitudes of these terms may be gained from Tables IV, V, VI, and VII.

The literature contains numerous papers on this subject, together with a certain amount of controversy, but a useful summary is that of George and McClure (1959). Generally Δ tends to be about 50% greater for trivalent than for divalent transition metal ions of the same series, and there are similar differences between ions of the same valency belonging to different transition series. Ions of the same valency within a particular transition series display much smaller changes in Δ and in pairing energies, but the changes are by no means negligible (Orgel, 1960; Griffith and Orgel, 1957).

The mean electron-pairing energies for some ions in the gaseous state are given in Table III. In the complexes the spin-pairing energies are considerably reduced, probably by as much as 20–30% (George and McClure, 1959).

VII. Aquo Couples

We have already seen that the change in free energy for a redox couple relative to the hydrogen electrode is given by

$$\Delta G^\circ = I_{m-n} + \Delta(\Delta H_{\text{aq}}) + \Delta(\Delta H_{c'}) + \Delta(\Delta H_{\text{LF}}) - T\Delta\bar{S}^\circ - (m-n)H$$

and the experimental hydration enthalpy difference $\Delta(\Delta H_{\text{aq}})_{\text{exptl.}}$ for the gaseous metal ions of the couple is given by

$$\Delta(\Delta H_{\text{aq}})_{\text{exptl.}} = \Delta(\Delta H_{\text{aq}}) + \Delta(\Delta H_{c'}) + \Delta(\Delta H_{\text{LF}})$$

It follows that

$$\Delta G^\circ = I_{m-n} + \Delta(\Delta H_{\text{aq}})_{\text{exptl.}} - T\Delta\bar{S}^\circ - (m-n)H$$

and if we assume the absolute potential of the hydrogen electrode (H) is -4.5 volts then E° for the aqueous metal ion couples $M^{++/3+}$ can be calculated from the third ionization potentials, I_3, $\Delta(\Delta H_{aq})_{exptl.}$, and $\Delta\bar{S}°$. These quantities are readily available for the di- and trivalent ions of the first transition series (Table IV) and the calculated and experimental $E°$ values are plotted in Fig. 2. Close agreement is obtained for all potentials

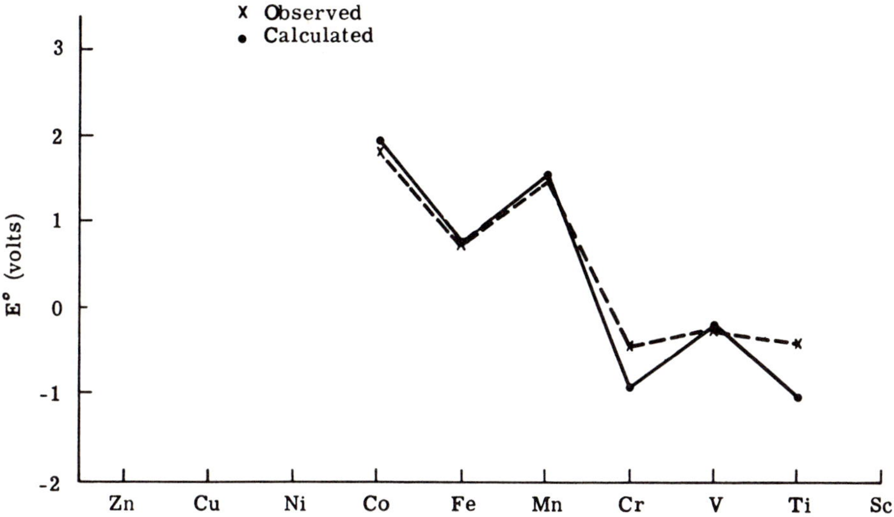

FIG. 2. The observed and calculated $E°$'s for $M_{aq}^{++/3+}$ ion couples.

except those of titanium and chromium, the former of which is too positive by about 0.6 volt. Latimer (1952) suggested that the measured potential is too negative so that good agreement could be expected here also.

It is evident that there is no regular trend for the plot of $E°$ against the atomic number and this has been ascribed to the ligand field effect (Jorgensen, 1956b; George and McClure, 1959). After allowance had been made for the $\Delta(\Delta H_{LF})$ term Jorgensen found that the potentials increased by approximately 1.5 volts per unit of atomic number with a "hump" between manganese and iron of about 3 volts. The "hump" was ascribed to the special stability of half filled shells with a maximum value of the total spin. By extrapolation Jorgensen predicted the following redox potentials: $Sc^{++/3+}$, +2.7 volts; $Ti^{++/3+}$, +1.6 volts; $Ni^{++/3+}$, −3.1 volts; $Cu^{++/3+}$, −2.6 volts; and $Zn^{++/3+}$, −4.8 volts.

In the absence of ligand field effects (i.e., assuming for all complexes a spherically symmetrical d electron cloud) we expect to find a simple correlation between redox potential and ionization potential. In practice, of course, this situation is only approached by d^0, d^5, and d^{10} ions, but it is possible to observe the correlation by making allowance for $\Delta(\Delta H_{LF})$ in a "corrected" ΔG° value (George and McClure, 1959). In Fig. 3 is shown the superimposition of the "corrected" ΔG° values and the I_3 curve when plotted against atomic number. The agreement is excellent for V, Cr,

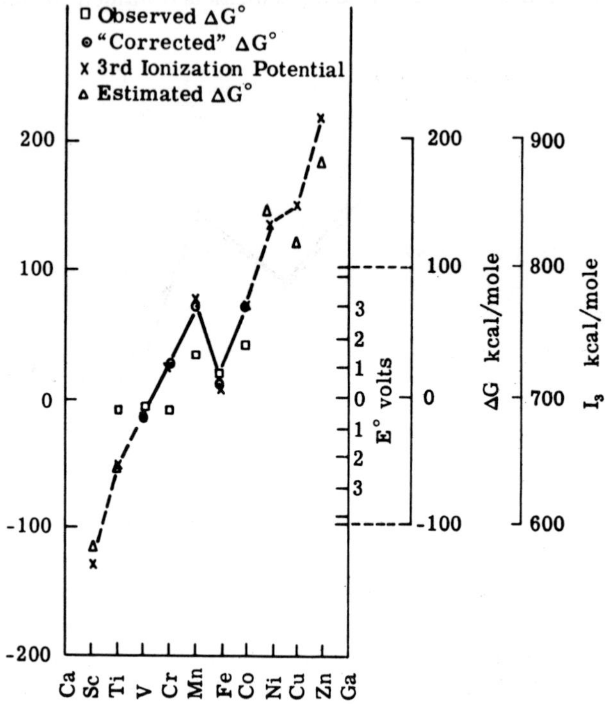

FIG. 3. The potentials of $M_{aq}^{++/3+}$ couples corrected for ligand field effects and fitted to the ionization potential curve for the gaseous ions (after George and McClure, 1959).

Mn, Fe, and Co. The value of the $Ti^{++/3+}$ potential, however, seems far too negative since the $\Delta(\Delta H_{LF})$ term is negligible if Dq for Ti_{aq}^{++} $\sim$1000 cm.$^{-1}$. This agrees with Latimer's contention that the measured potential is far too negative and the curve predicts an increase of about +2 volts for this potential. Similarly, the following potential values are implied by extrapolation: $Ni^{++/3+}$, -5.5 volts; $Cu^{++/3+}$, -4.5 volts; $Zn^{++/3+}$, -6.5 volts; and $Sc^{++/3+}$, $+4.5$ volts. They are considerably different from the values predicted by Jorgensen (1956b) and this difference is largely due to the method of assessment. Whereas Jorgensen has extrapolated his values from the known "corrected" redox potentials by adding 1.5 volts for each unit of atomic number and then making the ligand field correction, George and McClure have made the ligand field addition to the "corrected" ΔG° assuming that the latter coincides with the ionization potential curve.

To calculate the ligand field term Jorgensen assumes $Dq^{++} = 1200$ cm.$^{-1}$ and $Dq^{3+} = 2000$ cm.$^{-1}$ whereas George and McClure assume that $Dq^{3+}/Dq^{++} = 1.5$ and adopt the known Dq values for divalent ions. Another significant difference is the total ligand field contribution for Co(III). Jorgensen has assessed the stabilization of the trivalent ion as 22 kcal./mole (24 Dq = 127 kcal./mole less the pairing energy 105 kcal./mole) whereas George and McClure's estimate, 45 kcal./mole, probably arises from the difference between the experimental heat of hydration and the interpolated heat of hydration less the ligand field term. If the latter is correct, the pairing energy for Co(III) aquo ion is considerably less than supposed.

The hydration enthalpies for aquated ions are not known experimentally but may be estimated by the Born equation using the Goldschmidt radius for the ion plus the diameter of a water molecule (2.76 Å) as the effective radius of the aquated ions. The difference between the experimental hydration enthalpy of the gaseous ion, the Born hydration enthalpy of the aqueous ion, the latent heat of evaporation for six molecules of water, and the ligand field enthalpy term gives the enthalpy of formation less the ligand field effect, i.e.,

$$\Delta H_{c'} = (\Delta H_{\text{aq}})_{\text{exptl.}} - \Delta H_{\text{aq}} - \Delta H_{\text{LF}} - 62.4$$

Values of $\Delta H_{c'}$ and $\Delta(\Delta H_{c'})$ calculated in this way are given in Table IV. Both are large compared with the ligand field terms and while $\Delta(\Delta H_{\text{aq}})$ is virtually constant from Ti to Co, the $\Delta(\Delta H_{c'})$ term increases slightly; i.e., $\Delta H_{c'}$ is larger for trivalent ions than divalent and the difference increases with atomic number. This result is not unexpected since a small difference in the radius of the hydrated gaseous ion would not affect the hydration enthalpy greatly.

The simple aquo ions of the second and third transition series are less common and, as they are usually in high valence states, extensively hydrolyzed, so that it is difficult to determine what species are present in solution. Consequently a comparison of the contributions to their redox potentials is not easy to make. One system which probably does not suffer from these disadvantages is the Ag^+/Ag^{++} couple ($E^\circ = -1.98$ volts) and it is interesting to compare it with the Cu^+/Cu^{++} couple (-0.167 volt) The second ionization potentials for Cu (468 kcal./mole) and Ag (495 kcal./mole), respectively, stabilize the Cu^{++} ion relative to Ag^{++}. In addition, stabilization of the Cu(II) state arises from the $\Delta(\Delta H_{\text{aq}})_{\text{exptl.}}$ term (-368 kcal./mole for the copper couple and -295 kcal./mole for the silver system). The $\Delta(\Delta H_{\text{LF}})$ term stabilizes Cu(II) by 22 kcal./mole and Ag(II) probably by about 30 kcal./mole and tends to make the silver potential more positive than the copper. It is evident, however, that the ionization potential and hydration enthalpies (exptl.) are the significant

TABLE IV

THE BORN–HABER CYCLE TERMS FOR $M_{aq}^{++/3+}$ IONS

Ion	I_3	$(\Delta H_{aq})_{exptl.}$	$\Delta(\Delta H_{aq})_{exptl.}$	ΔH_{aq}(Born)	$\Delta(\Delta H_{aq})$(Born)	ΔH_{LF}	$\Delta(\Delta H_{LF})$	$\Delta H_{c'}$	$\Delta(\Delta H_{c'})$
				kcals./mole.					
Ti^{++}	648	−446	−581	−186	−257	27	−4	−295	−328
Ti^{3+}		−1027		−443		23		−623	
V^{++}	685	−453	−600	−187	−249	43	−3	−285	−354
V^{3+}		−1053		−436		40		−639	
Cr^{++}	713	−460	−645	−188	−254	24	+36	−310	−355
Cr^{3+}		−1105		−442		60		−665	
Mn^{++}	776	−445	−653	−183	−258	0	+36	−324	−359
Mn^{3+}		−1098		−441		36		−683	
Fe^{++}	706	−468	−604	−187	−253	12	−12	−331	−363
Fe^{3+}		−1072		−440		0		−694	
Co^{++}	772	−497	−629	−187	−255	21	+1	−351	−373
Co^{3+}		−1126		−442		22		−724	
Ni^{++}	834	−507	—	−189	—	—	—	−351	—
Ni^{3+}		—		—		—		—	
Cu^{++}	849	−507	—	—	—	—	—	—	—
Cu^{3+}		—		—		—		—	
Zn^{++}	915	−492	—	—	—	—	—	—	—
Zn^{3+}		—		—		—		—	

factors in making the copper potential so much more positive than the silver value.

VIII. The Potentials of Fe(II)/(III) Complex Couples

Two classes of redox systems are found with iron complexes, "high spin" and "low spin," both reversible. The relative magnitudes of each of the term contributions to the potential are given in Table (V) for a representative from each class, the aquo and the phenanthroline complex couples.

TABLE V
THE REDOX POTENTIALS OF SOME OCTAHEDRAL Fe(II)/(III) COMPLEXES

Ligand	$E°$ (volts)	$\Delta(\Delta H_{aq})$ (kcal./mole)	$\Delta(\Delta H_{c'})$ (kcal./mole)	$\Delta(\Delta H_{LF})$ (kcal./mole)	$T\Delta\bar{S}$ (kcal./mole)
H_2O	−0.77	−250	−362	+12	+13
8-OH-quinoline	+0.15	+25	│	—	—
EDTA	−0.12	+70	inc.	approx. +12	—
C_2O_4	+0.01	+230	↓	—	—
CN	−0.36	+240	↑	↑	−14
Cyclopentadiene	−0.41	−24	inc.	slight increase favoring Fe (II)	—
1,10-Phenanthroline	−1.06	−130	≫ −362	>+12	−1.5

The values of the terms are known reasonably well for the aquo couple: $\Delta(\Delta H_{c'})$ is obtained by difference from the experimental hydration enthalpies of the gaseous ions, the Born hydration enthalpies, and the ligand field effects. For the tris(phenanthroline) chelate system some of the quantities are unknown and estimates must be made. If we assume that the phenanthroline Dq value is three times that of water, then the ferric state is stabilized by about 20 kcal./mole. However, the pairing energies 50 kcal./mole for Fe(II) and 85 kcal./mole for Fe(III) strongly stabilize the ferrous state. It also seems likely that $Dq^{3+}/Dq^{++} < 1.5$ since the d^6 "spin-paired" Fe(II) system is capable of forming three metal-to-ligand π bonds (Orgel, 1960). Thus the divalent state is probably quite strongly stabilized and the $\Delta(\Delta H_{LF})$ factor tends to make the potential of the tris-(phenanthroline) chelate system more negative than the aquo system. The diminution in hydration enthalpy for the large $[Fe(phen)_3]^{n+}$ ions relative to the smaller aquo ions is compensated by the increase in the formation enthalpy term $\Delta(\Delta H_{c'})$ due to the highly polarizable phenanthroline ligand. The entropy terms stabilize the $[Fe(phen)_3]^{3+}$ ion relative to the Fe^{3+} aquo ion; hence the combined ligand field and $\Delta(\Delta H_{c'})$ factors must be the decisive terms in establishing the potential order. The values of the redox potentials and of the Born–Haber cycle terms for some "high spin" and "low spin" types are given in Table (V).

The 8-hydroxyquinoline, EDTA, and oxalate chelates form a series in which the charges, and thus the hydration enthalpies, increase sharply and favor the divalent state. The formation enthalpy differences, however, tend to stabilize the Fe(III) state in the order oxalate > EDTA > 8-OH-quinoline when viewed purely as a charge effect. Ligand polarizabilities diminish this effect but not enough to alter the order; thus hydration and enthalpy effects tend to cancel. The ligand field energies are probably of the same order as for the aquo system since the bonding atoms are predominantly oxygen. Estimates made by George and McClure (1959) are: 8-OH-quinoline, 20 kcal./mole; EDTA, 17 kcal./mole; oxalate, 22 kcal./mole. On these grounds the redox potentials might be expected to be fairly close. Comparing the tris(oxalato) and hexa(aquo) systems, the hydration enthalpies are about equal but of opposite sign, but the $\Delta(\Delta H_c')$ term for the oxalate (an interaction between positive central ions with six negative charges) will be much greater than for the aquo system. If the $\Delta(\Delta H_c')$ term is about -900 kcal./mole, favoring the Fe(III) state, then the difference between this term, the ionization potential, and hydration energy terms is about $+30$ kcal./mole. This is approximately 50 kcal./mole less positive than the value for the aquo ions, and the Fe(III) state is stabilized. The value of $\Delta(\Delta H_{c'})$ for oxalate ion is of the right order, since the energy per bond for $CrCl_2$ is -272 kcal./mole, and for $CrCl_3$ it is -413 kcal./mole. It is probably the large enthalpy of formation which decides the potential order in this instance; unfortunately the entropy terms are unknown.

The principal difference between the "high spin" and "low spin" complexes probably lies in the pairing energy and larger ligand field contribution. The order for the latter is probably CN > cyclopentadiene > phen. The difference between 24 Dq for Fe(II) and 20 Dq for Fe(III) is small. However, the mean pairing energy difference of 70 kcal./mole in favor of Fe(II) should tend to stabilize all the potentials in this state. In this context Fe bis(cyclopentadiene) complexes can be viewed as slightly flattened octahedral structures (Orgel, 1960).

The hydration enthalpies differ widely from the large positive value for the cyano complexes to the negative contributions in the phenanthroline chelates. Cyclopentadiene is assumed to coordinate as a singly negatively charged ligand. The hydration energy term thus predicts a negative potential order of phen > cyclopentadiene > CN. However, this trend is opposed by the large $\Delta(\Delta H_{c'})$ term which is greatest for cyanide and least for phenanthroline, and which strongly favors the oxidized state. It is probably this term or the entropy term contribution (CN, −14 kcal./mole; phen, −1.5 kcal./mole) which decides the potential order, since the ligand field pairing energy factor is opposed to the observed order.

IX. Hexacyano Metal Complex Couples

The redox potentials of the Cr, Mn, Fe, and Co hexacyano complexes in the +2 and +3 oxidation states are given in Table VI along with other relevant data.

In these systems the hydration enthalpy and over-all entropy differences will be almost constant and the $\Delta(\Delta H_{c'})$ term probably increases slightly with atomic number. The major contributions in determining the redox potential order for these spin-paired complexes are made by the terms associated with the ionization potentials, the ligand field effects, and the electron-pairing energies. The relative order of the redox potentials (as they become more negative), predicted from the ionization potentials, is Fe > Cr > Co > Mn. However, after the electron-pairing energy term is accounted for, the order alters to Cr $\gtrsim$ Mn > Fe > Co and this is the order observed except that the cobalt potential should lie between Cr and Mn. The pairing energies are not as large in the complexes as in the free gaseous ions, but it is assumed that they scale down proportionately in these very similar compounds.

If we assume, for the moment, that the ratio of $Dq^{3+}/Dq^{++} = 1.5$ and that the *Dq* value is roughly constant for the cyanide ligand then, despite the larger stabilization of the bivalent states of Cr, Mn, and Fe in *Dq* units, the trivalent states are slightly stabilized energetically in the order Cr < Mn < Fe. On the other hand, the Co(III) state is strongly stabilized by 24 Co(III)–18 Co(II) *Dq* units, making the potential much more positive than predicted by ionization potentials and pairing energy terms. Taking the value of *Dq*(III) to be 3400 cm.$^{-1}$ for the CN^- ligand and of *Dq*(II) to be 2200 cm.$^{-1}$, then the Cr(III) state is stabilized by approximately 13 kcal./mole, the Mn(III) state by 30, the Fe(III) state by 42, and the Co(III) state by the very large contribution of 120 kcal./mole. Though the ratio Dq^{3+}/Dq^{++} may well be roughly constant, for cyanide complexes, the actual values for CN^- ion are not: $[Co(CN)_6]^{3-}$ ion, $Dq =$ 3400 cm.$^{-1}$; $[Cr(CN)_6]^{3-}$ ion, $Dq = 2630$ cm.$^{-1}$. Unfortunately, pairs of *Dq* values for the two oxidation states of these metals are not known, and for this reason the above calculations are speculative. Irrespective of these considerations, however, the ligand field term stabilizes the Co(III) state sufficiently to make the Co(II) complex a powerful reducing agent and the potential of the cobalt complex couple far more positive than the manganese couple.

TABLE VI

THE REDOX POTENTIALS AND BORN–HABER CYCLE TERMS FOR HEXACYANO METAL ION COUPLES

	$E°$ (volts)	$E°_{M^{++/3+}aq}$ (volts)	I_3 (kcal./mole)	Pairing energy for gaseous ions (kcal./mole)			LFSE (Dq units)	
				M^{3+}	M^{++}	Difference	M^{3+}	M^{++}
$[Cr(CN)_6]^{4-/3-}$	+1.28	+0.41	712	0	66	−66	12	16
$[Mn(CN_6)^{4-/3-}$	+0.22	−1.51	776	80	152	−72	16	20
$[Fe(CN_6)]^{4-/3-}$	−0.36	−0.77	706	170	100	+70	20	24
$[Co(CN_6)]^{4-/3-}$	+0.83	−1.82	772	120	64	+56	24	18
$[V(CN)_6]^{4-/3-}$	+0.8	+0.25	685	0	0	0	8	12
$[Ni(CN)_6]^{4-/3-}$	−2	−3	834	77	—	+77	18	12

	$E°$ (volts)	I_3 (kcal./mole)	Pairing energy difference	LFSE		Difference
				M^{3+}	M^{++}	
$[Fe(CN_6)]^{4-/3-}$	−0.36	706	↓ Decreases and favors the divalent state	↓ Increases	↓ Increases	↑ Increases
$[Ru(CN)_6]^{4-/3-}$	−0.86	655				
$[Os(CN)_6]^{4-/3-}$	−0.75	576				

It is interesting now to compare the potentials of the hexacyano nickel(II)/(III) and vanadium (II)/(III) couples relative to the above systems. The $[Ni(CN)_6]^{4-}$ ion, of course, is unstable with respect to dissociation to the planar $[Ni(CN)_4]^{--}$ ion, but exists in the presence of a large excess of CN^- ion. We assume that all other terms are approximately equal except the third ionization potentials and the pairing energies. The first of these terms is 58 kcal./mole higher than for cobalt, and the pairing energy of the gaseous Ni^{3+} ion is about 77 kcal./mole, compared with 100 kcal./mole for the gaseous Co^{3+} ion. Consequently, the Ni(II) state is stabilized enormously and the potential of the hexacyano Ni(II)/(III) couple is probably more negative than -2.0 volts. Ligand field effects have been neglected, but the two large terms above should greatly outweigh the LFSE difference between the two ions. On the other hand, the third ionization potential of vanadium is 20 kcal./mole smaller than that of iron, with the result that the V(III) state is stabilized relative to the Fe(III) state. There are no pairing energies to be considered here, and for this reason the potential of the $[V(CN)_6]^{4-/3-}$ system would be expected to lie between the values for the corresponding chromium and manganese complex system, provided the Dq values are comparable. Both the V(II) and V(III) hexacyano complexes can be prepared. The former is a powerful reducing agent, though neither anion is very stable to dissociation in aqueous solution. A comparison of the aquo and hexacyano couples shows that the bivalent state is considerably destabilized by coordination with cyanide. This is largely due to the $\Delta(\Delta H_{c'})$ term which, by virtue of the interaction of the positive metal ion and negative ligand, strongly favors the trivalent state and is responsible for diminishing the effect of the ionization potential. It would be expected then that the $[V(CN)_6]^{4-}$ ion would be a better reducing agent than the V_{aq}^{++} ion.

In the $[M(CN)_6]^{4-/3-}$ couples of the iron triad, the $\Delta(\Delta H_{aq})$ term is approximately constant since the ionic sizes probably differ only by a few hundredths of an Angstrom unit, and the entropy term will also be constant. The ionization potentials predict an order for the redox potentials of Os > Ru > Fe (with Fe the most negative), compared to the order Ru > Os > Fe actually observed. The pairing energies, greatest for the trivalent states, decrease rapidly in the order Fe > Ru > Os. The difference also decreases and hence acts to stabilize the reduced state of Fe more than of Os. This effect is in the same direction as the ionization potential term. The ligand field effect stabilizes the bivalent states in the order Os > Ru > Fe, since metal to ligand double bonding will be a maximum for the d^6 system, and for osmium and ruthenium relative to iron. This term opposes the order directed by the ionization potentials. The enthalpy of formation less the ligand field contribution should be larger for trivalent than for bivalent ions on electrostatic grounds, and greater for osmium and ruthenium than iron on a polarizability basis. The difference between these contributions for the bi- and trivalent states should be larger for Fe than Os mainly because of the larger polarizability of the osmium(II) state. For this reason the trivalent states are stabilized and the Fe(III) state more than the Os(III) state. This enhances the ligand field effect and the two combine to partly displace the order set by the ionization potentials and pairing energy terms.

X. Ethylenediaminetetraacetato Chelate Couples

The vanadium and iron chelates in the +2 and +3 states are "spin-free," whereas the cobalt chelate is "spin-free" in the +2 state, and "spin-paired" in the +3 state. The redox potentials are shown in Table VII. The hydration enthalpy differences and the entropy terms will be relatively constant for the ions $[M(EDTA)]^{--}$ and $[M(EDTA)]^{-}$, but a slight stabilization of the +3 states will be imposed by the $\Delta(\Delta H_{c'})$ term which becomes slightly larger in the order V < Fe < Co. The $\Delta(\Delta H_{c'})$ term acts in opposition to the ionization potential term which makes the vanadium couple the most positive, followed by iron and cobalt in that order.

The ligand field Dq values are close to those for the aquo ions and since we are interested in the difference between the bi- and trivalent states the $\Delta(\Delta H_{LF})$ terms for the aquo and EDTA systems should be similar. Therefore, the $\Delta(\Delta H_{LF})$ term for vanadium is about zero, the Fe(II) state is stabilized by about 12 kcal./mole, and the $\Delta(\Delta H_{LF})$ term for cobalt is practically zero if the pairing energy for the Co_{aq}^{3+} ion is used (105 kcal./mole). It would seem that the ligand field terms are not very significant and that ionization potentials dominate the redox potential order.

Since the third ionization potentials for manganese and cobalt are similar, 776 and 772 respectively, the $\Delta(\Delta H_{LF})$ term of about 36 kcal./mole for the manganese couple stabilizes the manganese(III) state and requires the manganese potential to be more positive than that of the cobalt couple. The manganese potential has not been measured but the Mn(III) complex is obtained from the divalent chelate by oxidation with lead dioxide, sodium bismuthate, or dichromate ion (Pribil, 1949). Dichromate ion at pH 3 ($E° = -1.21$ volts) places an upper limit on the potential of about -1.0 volt and, since bromine ($E° = -1.07$ volts) does not oxidize the Mn(II) state and Fe^{++} ion ($E° = -0.77$ volt) reduces the Mn(III) chelate, the value -1.0 volt seems to be of the right order. The disagreement between the predicted (> -0.6 volt) and approximate experimental value could be due to the estimate of the ligand field stabilization of the Co_{aq}^{3+} ion. If the value 45–50 kcal./mole for the Co_{aq}^{3+} ion, obtained by George and

TABLE VII

THE REDOX POTENTIALS AND BORN–HABER CYCLE TERMS FOR SOME EDTA CHELATE COUPLES

Chelate couple	E° complex (volts)	$E^{c}{}_{M^{++/3+}{}_{aq}}$ (volts)	I_3 (kcal./mole)	$\Delta(\Delta H_{c'})$ (kcal./mole)	$\Delta(\Delta H_{aq})$ (kcal./mole)	$\Delta(\Delta H_{LF})$ (kcal./mole)	Pairing energies (kcal./mole)
$[V(EDTA)]^{--/-}$	+1.03	+0.25	685	↑	↑	0	—
$[Fe(EDTA)]^{--/-}$	−0.12	−0.77	706			+12	—
$[Co(EDTA)]^{--/-}$	−0.60	−1.82	772	Slight inc. with inc. in at. no.	Relatively constant	−103	100
$[Mn(EDTA)]^{--/-}$	−1.0	−1.50	776			−36	—
$[Cr(EDTA)]^{--/-}$	~+1	+0.41	713	↓	↓	−35	—

McClure (1959), is used the $\Delta(\Delta H_{LF})$ terms for Co and Mn chelates are now similar and their redox potentials should also be similar.

Chromium on the other hand has a third ionization potential of 713 kcal./mole, close to that of iron, 706 kcal./mole, but the $\Delta(\Delta H_{LF})$ term favors the Cr(III) state by about 35 kcal./mole. The stabilization of the Fe(II) state is 13 kcal./mole and this would make the potential of the Cr chelate couple considerably more positive than the iron chelate couple. The $[Cr(II)EDTA]^{--}$ ion, if it exists, should be a powerful reducing agent.

Relative to the aquo ions the EDTA chelate couples are all much more positive (Table VII); i.e., the trivalent states are much more stabilized. This is largely due to the difference in the free energy of formation of the gaseous complexes relative to the formation of the hexaaquo ions, which, incidentally, also favors the trivalent state. Ligand field effects are approximately equal, and this is reflected in the roughly constant difference (0.5–0.7 volt) between the potentials of the aquo and EDTA couples for V, Mn, and Fe. The difference for the cobalt couples is 1.22 volts but both values are uncertain since the cobalt systems are irreversible.

The potentials (Table VII) refer to the electrode reactions:

$$[Fe(EDTA)]^{--} \rightarrow [Fe(EDTA)]^{-} + e$$

$$[V(EDTA)]^{--} \rightarrow [V(EDTA)]^{-} + e$$

$$[Co(H_2O)(EDTA)]^{--} \rightarrow [Co(EDTA)]^{-} + e$$

(Schwarzenbach *et al.*, 1949, 1951, 1953). It is probable that the Fe(II) and V(II) complexes have a coordinated water molecule also, and we have assumed the potentials to be for the same electrode process in each instance. The assumption is not quite valid since the $[Fe(III)EDTA]^{-}$ ion has been shown to be seven-coordinate (Hoard *et al.*, 1961) in the solid state, whereas the Co(III) compound is six-coordinate in the same state (Weakliem and Hoard, 1959). These coordination numbers are probably preserved in solution and it is difficult to gauge the effect of the additional water molecule in the Fe(III) state but it should be secondary, compared with the large contributions made by the ionization potential term.

XI. Metal Phenanthroline-Type Chelate Couples

The ease of preparation of Fe(II) and Fe(III) tris-1,10-phenanthroline (phen), 2,2′-bipyridine (bipy), and bis-2,2′,2″-terpyridine (trpy) complexes, their use as redox titration indicators, the ease of substitution of the aromatic ring systems, and the reversible nature of the couples have contributed greatly to their popularity for redox potential studies. The similarity between all three systems is well known and is exemplified by a

TABLE VIII

OXIDATION-REDUCTION POTENTIALS OF PHENANTHROLINE-TYPE COUPLES

Couple	I (kcal./mole)	Aquo potential	Phenanthroline (volts)	Bipyridine (volts)	Terpyridine (volts)
Co(II)/(III)	772	−1.84	−0.42[a]	−0.37[a]	—
Fe(II)/(III)	706	−0.771	−1.120[b]	−1.096[b]	−1.09[c]
Ru(II)/(III)	655	−0.083	−1.314[d]	−1.304[d]	−1.30[c]
Os(II)/(III)	576	+0.25[e]	−0.877[f]	−0.878[f]	−0.987[g]
Cu(I)/(II)	468	−0.167	−0.174[h]	−0.120[h]	+0.080[h]
Ag(I)/(II)	495	−1.98	−2.22[i]	−2.15[j]	

[a] From Paglia and Sironi (1957).
[b] From Dwyer and McKenzie (1947).
[c] Estimated from data by Dwyer and Gyarfas (1954).
[d] Estimated from data by Dwyer (1949).
[e] Estimate by Dwyer (1951).
[f] From Dwyer *et al.* (1951).
[g] From Dwyer and Gyarfas (1954).
[h] From James and Williams (1961).
[i] From Scrocco and Ragazzini (1954).
[j] From Scrocco and Marmani (1954).

comparison of their redox potentials in Table VIII. In fact it would seem that only the conjugated system —N=C—C=N— is necessary to reproduce the bonding properties of these aromatic ligands (Krumholtz, 1953).

For metal complex couples containing the same ligand, the hydration and entropy terms will be constant. The $\Delta(\Delta H_{c'})$ term probably increases only slightly with atomic number and the potential order should be decided largely by the ionization potentials, electron-pairing energies, and ligand field effects.

The principal difference between the tris(phenanthroline) Fe(II)/(III) and Co(II)/(III) couples probably lies in the ligand field term which is strongly in favor of the Co(III) state instead of the Co(II) state as with the Fe(II)/(III) couple. This makes the cobalt potential (−0.42 volt) more positive than the latter (−1.1 volts) (Paglia and Sironi, 1957). The difference is smaller than expected, however, because of the higher ionization potential (I_3) for cobalt.

The Fe(II)/(III) tris(phenanthroline) system has already been discussed relative to the aquo couple (Section VIII) and the most significant differences appeared in the hydration and formation enthalpies, the former smaller than the aquo couple and the latter larger so that they offset each other. The entropy, ligand field, and pairing energy terms largely cancelled, leaving the phenanthroline potential (−1.1 volts) not greatly different from that of the aquo couple (−0.77 volt). On the other hand the ligand field and entropy terms for the $[Co(phen)_3]^{++/3+}$ couple both stabilize the trivalent ion relative to the hexaaquo system, to the extent of +1.4 volts.

Reduction of phenanthroline by Cr^{++} ion precludes the measurement of the $[Cr(phen)_3]^{++/3+}$ potential (Dwyer and Wooldridge, 1949). But both $[Cr(bipy)_3]Br_2{\cdot}6H_2O$ and $[Cr(bipy)_3](ClO_4)_3$ have been isolated (Barbieri and Tettamanzi, 1932) and the magnetic moment of the bivalent state (3.27 B.M.) indicates that it is spin-paired. Unfortunately, the dark brown chromous salt, although stable in the solid state, rapidly lightens in color in solution, due to oxidation, and the potential would be difficult to measure. However, it would seem from this behavior that the potential value is not greatly different from the aquo couple (+0.41 volt). The third ionization potentials for iron and chromium are similar hence the difference in redox potential must arise largely from the ligand field effects and electron-pairing energy differences. The ferrous state is stabilized by a ligand field >12 kcal./mole while the chromium(III) state is stabilized by 12 Dq(III) − (16 Dq(II) − π). The large pairing energy term for Cr(II) (π = 67 kcal./mole) destabilizes this state sufficiently to make it a strong reducing agent.

Similarly, the tris(bipyridine) Mn(II)/(III) and V(II)/(III) potentials have not been measured. This may be due to the poor coordinating ability of the trivalent state with these bases, as suggested by King and Garner (1952) for vanadium, and by the fact that Mn(II) solutions containing excess bipyridine when oxidized by ammonium peroxydisulfate give $[MnO_2(bipy)_2]S_2O_8$ (Burstall) and not the tris Mn(II) complex.

The phenanthroline-type chelates for the iron triad have potentials in a narrow range (Os $\geq$ Fe > Ru) (Table VIII) and the terms which decide the potential order are difficult to assess. The decrease in ionization potential (I_3) from the first to the third transition series predicts a potential order of Os > Ru > Fe (Os the most positive) and the pairing energy differences between tri- and bivalent states give the same result with an over-all stabilization of the bivalent states. The ligand field differences also stabilize the bivalent states, but of osmium more than of iron. The $\Delta(\Delta H_{c'})$ term, however, favors the trivalent ions but is largest for Fe > Ru > Os. The last two terms then diminish the effect of the ionization potentials and pairing energies and partly displace the order predicted by I_3, probably because of the smaller I_3 difference between Fe and Ru than between Ru and Os.

The effect of ionization potential on redox potential is more readily observed for the $[Cu(phen)_2]^{+/++}$ and $[Ag(phen)_2]^{+/++}$ system. Here I_2 for silver is larger than for Cu and the silver redox potential is correspondingly much more negative. The $\Delta(\Delta H_{c'})$ term probably also stabilizes the Cu(II) state relative to Ag(II), but the ligand field effect favors Ag(II) over Cu(II).

In both instances the univalent state is slightly stabilized over the bivalent state when compared with the corresponding aquo couples (Table VIII).

XII. Ruthenium Chelate Couples

The complexes of ruthenium and osmium are most suitable for redox studies since they are known to exist in several oxidation states and in general they are well defined and inert to dissociation of their ligands. The potentials for the ruthenium (II)/(III) couples in Table IX (Dwyer

TABLE IX
THE OXIDATION-REDUCTION POTENTIALS OF Ru(II)/(III) AND (III)/(IV) COMPLEXES IN IM·H_2SO_4

Oxidation states	Complex ions	$E°$ (volts)
Ru(II)/(III)	$[Ru(bipy)_3]^{++/3+}$	−1.26
	$[Ru(bipy)(py)_3Cl]^{+/++}$	−0.89
	$[Ru(bipy)(py)_3H_2O]^{++/3+}$	−1.04
	$[Ru(bipy)(py)_2(H_2O)_2]^{++/3+}$	−0.78
	$[Ru(bipy)_2(py)_2]^{++/3+}$	−1.25
	$[Ru(bipy)_2(NH_3)_2]^{++/3+}$	−0.88
	$[Ru(bipy)_2(en)]^{++/3+}$	−0.74
	$[Ru(bipy)(py)_4]^{++/3+}$	−1.25
	$[Ru(bipy)(py)_2(aca)]^{+/++}$	−0.62
	$[Ru(bipy)(aca)_2]^{\circ/+}$	−0.5[a]
	$[Ru(bipy)Cl_4]^{--/-}$	−0.4[a]
	$[Ru(bipy)(py)Cl_3]^{-/\circ}$	−0.4[a]
Ru(III)/(IV)	$[Ru(bipy)Cl_4]^{-/\circ}$	−1.0[a]

[a] Potentials estimated.

and Goodwin, 1957) are less precisely known than for the corresponding osmium couples since they are more negative and the oxidized state correspondingly less stable. They were measured by potentiometric titration in acid solution, or by adding half the equivalent amount of oxidizing agent to obtain equal concentrations of oxidant and reductant *in situ*. For these reasons only the gross effects of the substituents can be evaluated. Substitution of a negatively charged ligand in the $[Ru(bipy)_2(py)_2]^{++/3+}$ system (−1.25 volts) moves the potential to a more positive value by about 0.3–0.4 volt, e.g., $[Ru(bipy)_2(py)Cl]^{+/++}$, $E° = -0.89$ volt. Two large factors contributing to the potential change are the differences in the hydration and formation enthalpies ($\Delta(\Delta H_{aq})$ and $\Delta(\Delta H_{c'})$. Assuming the complexes to be spherically symmetrical, and that the effective radius of the Ru(II) complex reduces from 5.6 to 5.3 Å, and the Ru(III) complex from 5.5 to 5.2 Å, when the pyridine is substituted by Cl^- ion, the difference in the ΔH_{aq} terms is about 60 kcal./mole favoring the oxidized state with the higher charge. The hydration term tends to make the potential more negative for the substituted couple. Opposed to the hydration term is the $\Delta(\Delta H_{c'})$ term, which favors the oxidized state. On electrostatic grounds the $\Delta(\Delta H_{c'})$ term would be larger for anionic substituents than for neutral pyridine. The difference in ligand field effects is probably also small, for though $LF_{py} > LF_{Cl^-}$ the remaining five equivalent substituents contribute most of the field.

It seems then that the $\Delta(\Delta H_{c'})$ term is slightly larger than the hydration term and decides the potential order, since the entropy difference between the two sets of ions is of the order 4–5 e.u. (0.06 volt) and contributes little to the potential shift (cf. Os complexes).

The potential shifts may also be viewed in another manner, since the complexes are stable to dissociation and hydrolysis, and may be considered as separate entities. The ionization potential for the lower charged complex would be less than for the higher and this would stabilize the trivalent state with the lower charge relatively. Opposed to this consideration is the hydration term, and from the observed potential we deduce the former term must be predominant. A similar situation arises for the $[Ru(bipy)(py)_4]^{++/3+}$ ($E^\circ = -1.25$ volts), $[Ru(bipy)_2(aca)]^{+/++}$ ($E^\circ = -0.62$ volt), and $[Ru(bipy)(aca)_2]^{\circ/+}$ ($E^\circ = -0.5$ volt) couples. Again the $\Delta(\Delta H_{aq})$ term favors, relatively, the Ru(III) state and the higher charge more than the lower. On the other hand, the $\Delta(\Delta H_{c'})$ term increases as the number of negatively charged ligands increases and this stabilizes the oxidized state with the potentials becoming more positive as the charge decreases. The entropy differences are probably also small in these couples.

The ligand field for pyridine is greater than for acetylacetone and the $\Delta(\Delta H_{LF})$ term should slightly favor the Ru(III) state; i.e., Dq Ru(III) > Dq Ru(II). As the pyridine is replaced by acetylacetone the effects of the $\Delta(\Delta H_{LF})$ term should increase as the π-bonding contribution due to acetylacetone will be less than for pyridine. The Ru(III) state is thus stabilized in the order of decreasing complex charge and the potential becomes more positive.

The effect of the ionization potential term on the availability of oxidation states is also readily observed with this element. The Ru(III)/(IV) couple $[Ru(bipy)Cl_4]^{-/\circ}$ has a potential of -1.0 volt whereas the potential of the Ru(II)/(III) system $[Ru(bipy)Cl_4]^{--/-}$ is only about -0.4 volt. The reduction in the negative value of the potential is expected for the decrease in the ionization potential term. The hydration enthalpy differences favor the highest charge, in this instance the lowest oxidation state, tending to diminish the effect of the ionization potential term. Similarly, the $\Delta(\Delta H_{c'})$ term stabilizes the Ru(IV) state, making the (III)/(IV) potential more positive, relative to the (II)/(III) couple, and diminishing the effect of the ionization potential term. The pairing energies and ligand field terms stabilize the Ru(IV) over the Ru(III) state, and similarly the Ru(III) over the Ru(II) state, but it is difficult to assess whether the difference for these terms between Ru(IV) and Ru(III) is greater or less than between Ru(III) and Ru(II). However, in view of the favorable metal to ligand double bonding for the d^6[Ru(II)] system, and the lower pairing energy for the Ru(IV) state, it is probable that $\Delta(\Delta H_{LF})$ for Ru(IV) − Ru(III) is greater than $\Delta(\Delta H_{LF})$ for Ru(III) − Ru(II). Hence the $\Delta(\Delta H_{LF})$ term probably also reduces the effect of the ionization potential term.

The redox potentials of the couples $[Ru(bipy)_2(py)_2]^{++/3+}$, $[Ru(bipy)_2(NH_3)_2]^{++/3+}$, and $[Ru(bipy)_2(en)]^{++/3+}$ follow the order of the base strength of the substituent. This result is not unexpected, since the $\Delta(\Delta H_{aq})$ and $\Delta(\Delta H_{LF})$ terms would be approximately equal in all three instances, and the $\Delta(\Delta H_{c'})$ term should roughly parallel the base strength of the substituent. This type of effect is more complicated, however, than a linear relationship with the base strength of the substituted ligand (cf. Section XIV). The change in the nature of the π bonding from the ruthenium to the ligand has also been neglected, and for Ru(II) and pyridine may be quite appreciable. As a result, the potential of the $[Ru(bipy)_2(py)_2]^{++/3+}$ couple is probably more negative than would be anticipated simply by considering the base strength of pyridine compared with ammonia and ethylenediamine.

XIII. Osmium Chelate Couples

The osmium(II)/(III) phenanthroline and bipyridine complexes have redox potentials in the vicinity of −0.8 volt which means both forms can be isolated and are stable in a redox cell. Moreover the complexes are inert to dissociation of their ligands so that the cell constituents are only the complex ions and their attendant anions. The potentials measured at varying ionic strength and extrapolated to infinite dilution give $E°$ with a maximum deviation of ±0.2 mv. The $E°$'s of a series of complexes are listed in Table X and the effect of change in charge, degree of chelation, and halogen substitution can be readily observed (Buckingham *et al.*, 1961) For such a large series of chelates all of the same metal, the changes in the various contributions are smooth and regular, and are best ascribed to the obvious characteristics of the ligands.

A. Charge Effect

Two of the largest contributions to the $[Os(bipy)_2(py)_2]^{++/3+}$ potential are the hydration energy difference and the $\Delta(\Delta H_{c'})$ term. If the charge of the ions is now reduced by one unit while keeping the ions essentially spheri-

TABLE X
REDOX POTENTIALS OF SOME OSMIUM COMPLEXES

Complex	$E°$ (volts)
$[Os(py)(bipy)(trpy)]^{++/3+}$	−0.8700
$[OsCl(bipy)(trpy)]^{+/++}$	−0.5622
$[OsBr(bipy)(trpy)]^{+/++}$	−0.5670
$[OsI(bipy)(trpy)]^{+/++}$	−0.5660
$[Os(py)_2(bipy)_2]^{++/3+}$	−0.8339
$[OsCl(py)(bipy)_2]^{+/++}$	−0.4823
$[OsBr(py)(bipy)_2]^{+/++}$	−0.4861
$[OsI(py)(bipy)_2]^{+/++}$	−0.4875
$[Os(py)_4(bipy)]^{++/3+}$	−0.8033
$[OsCl(py)_3(bipy)]^{+/++}$	−0.4247
$[OsBr(py)_3(bipy)]^{+/++}$	−0.4434
$[OsI(py)_3(bipy)]^{+/++}$	−0.4506
$[Os(bipy)_3]^{++/3+}$	−0.8836
$[Os(trpy)_2]^{++/3+}$	−0.9866
$[Os(trpy)(py)_3]^{++/3+}$	−0.7986
$[Os(aca)(bipy)_2]^{+/++}$	−0.1539

cally symmetrical as in the $[Os(bipy)_2(py)Cl]^{+/++}$ couple, then both these factors should be affected appreciably. The $\Delta(\Delta H_{aq})$ term decreases since it is dependent on the square of the charges on the ions and the difference decreases sharply as the charges decrease. The $\Delta(\Delta H_{c'})$ term, however, increases due to the greater interaction of the Cl^- ion over pyridine with the osmium(III) center. The former effect stabilizes the osmium(II) state and the latter the osmium(III) and, since the potential becomes more positive by about 0.4 volt, the latter is the greater effect.

The ligand field term is probably quite small even though the Cl^- ion has an appreciably different ligand field from pyridine. If this term contributes at all it probably does so through a slight destabilization of the osmium(II) state with the substitution of pyridine by Cl^- ion; also the entropy term difference between the two couples is small, about 5 e.u. (0.07 volt).

A similar reduction in potential is observed for Br^- and I^- substitution and also by the replacement of two pyridine molecules with one acetylacetone ion. The potential in the last instance becomes more positive by 0.7 volt.

B. CHELATE EFFECT

The di- and trivalent osmium complexes containing bipyridine, terpyridine, and pyridine have identical sizes and hence constant contributions from the $\Delta(\Delta H_{aq})$ term. Similarly, the hydration term should be constant for the $[Os(bipy)(py)_3Cl]^{+/++}$, $[Os(bipy)_2(py)Cl]^{+/++}$, and $[Os(bipy)(trpy)Cl]^{+/++}$ couples. The entropy contributions within both sets of compounds will also be constant and changes in redox potential which arise from different ligand arrangements must be due largely to $\Delta(\Delta H_{c'})$ and $\Delta(\Delta H_{LF})$ terms.

The stability constants of pyridine, bipyridine, and terpyridine complexes cannot be correlated with the basic strength of the ligands and hence on the availability of σ electrons at the donor N atoms and it is almost certainly the π-accepting properties of these ligands which account for the stability of their complexes. Further, the electron involved in oxidation and reduction comes ultimately from the t_{2g} orbitals of the metal, those in fact which are used to form π bonds to the ligand. Therefore, the potentials of these osmium complexes should depend strongly on the degree of stabilization of their $t_{2g}\pi$ orbitals and the effect should be greater for the osmium(II) than for the osmium(III) ions since the latter have one less electron. The decrease in conjugation, terpyridine > bipyridine + pyridine > three pyridines is correlated with a decreasing ability to form metal – ligand π bonds and consequent destabilization of the t_{2g} level. This results in a relative stabilization of osmium(III) as terpyridine is successively substituted by bipyridine and by pyridine in the series $[Os(trpy)_2]^{++/3+}$ (−0.9866 volt), $[Os(py)(bipy)(trpy)]^{++/3+}$ (−0.8700 volt), and $[Os(py)_3(trpy)]^{++/3+}$ (−0.7986 volt); $[Os(bipy)_3]^{++/3+}$ (−0.8836 volt), $[Os(py)_2(bipy)_2]^{++/3+}$ (−0.8339 volt), and $[Os(py)_4(bipy)]^{++/3+}$ (−0.8033 volt). The halo-substituted complexes show a similar gradation in potential $[Os(bipy)(trpy)Cl]^{+/++}$ (−0.5622 volt), $[Os(py)(bipy)_2Cl]^{+/++}$ (−0.4823 volt) and $[Os(py)_3(bipy)Cl]^{+/++}$ (−0.4247 volt), as the conjugation is reduced.

The enthalpy of formation terms $\Delta(\Delta H_{c'})$ are not easy to assess but if we assume that the cumulative polarizability of three pyridines is smaller than terpyridine then the osmium(III) state would be stabilized for terpyridine and its potential would tend to be the most positive. This is opposed to the observed order, hence the ligand field is probably even larger than formerly supposed.

C. Halide Effect

The three sets of complexes containing coordinated chloride, bromide, or iodide ion in Table X show a variation in the potential order which depends on the remaining ligands. The system $[Os(bipy)(trpy)X]^{+/++}$ has a potential order of Br < I < Cl, $[Os(bipy)_2(py)X]^{+/++}$ of Cl > Br > I, and $[Os(py)_3(bipy)X]^{+/++}$ of Cl > Br > I in the sense that the greatest potential is the least negative.

There are probably two effects acting here in opposition, the polarizabilities of the halogens tending to stabilize the osmium(III) state for I >

Br > Cl, and the donation of π electrons from metal to ligand which is greatest for osmium(II) and for I > Br > Cl. The π effect then makes the iodo potential the most negative. Another factor which must be considered is the π-accepting character of the nitrogen-containing ligands, greatest for terpyridine and least for pyridine. It would seem from the observed potentials that the π effect for the halogen predominates except for the $[Os(trpy)(bipy)X]^{+/++}$ system where the large π-accepting ability of terpyridine and bipyridine reduces the donation to the halogen and allows the polarizabilities to partially reverse the expected order. It must be realized however that the potential differences considered here are only of the order of 0.01 volt = 0.2 kcal. and that small entropy or hydration effects could account for the observed results.

XIV. The Effect of Substitution in the Ligand

Ligand substituents may affect the redox potential by: (*a*) influencing the basicity of the donor atoms; (*b*) affecting the π-bonding ability of the ligand; (*c*) changing the entropy and enthalpy of hydration of the complex ions; or (*d*) by purely steric effects preventing the ligand from acquiring the most favorable orientation about the central metal ion.

The effect of substitution in the ligand on the basicity of the donor atoms and hence on the stability of the coordination compounds has been extensively studied by observing changes in the potentials of the

TABLE XI

POTENTIALS OF TRIS-(SUBSTITUTED 1,10-PHENANTHROLINE) IRON(II)/(III) COMPLEXES[a]

Substituent	$E°$ (0.1 *M* acid) (volts)	$E'°$ (1 *M* acid) (volts)	Substituent	$E°$ (0.1 *M* acid) (volts)	$E°$ (1 *M* acid) (volts)
H	−1.10	−1.06	3,6,7-Me_3	−0.99	—
3-Me	−1.07	−1.03	3,4,6,7-Me_4	−0.84	—
5-Me	−1.06	—	3,4,6,8-Me_4	−0.89	—
3,4-Me_2	−0.97	−0.93	—	—	—
3,8-Me_2	−1.03	—	3,4,7,8-Me_4	−0.85	—
4,5-Me_2	−0.95	—			
4,6-Me_2	−0.95	—			
4,7-Me_2	−0.88	−0.87	5-NO_2	—	−1.25
5,6-Me_2	−1.00	−0.97	5-NO_2, 6-Me	—	−1.22
3,4,6-Me_3	−0.92	—	5-Br	—	−1.12
3,4,7-Me_3	−0.88	—	5-Cl	—	−1.12
3,5,7-Me_3	−0.93	−0.89	5-Phenyl	—	−1.08

[a] Data from: Brandt and Smith (1949), Smith and Richter (1944), Brandt and Gullstrom (1952).

couples $[M(phen)_3]^{n+/(n+1)+}$ and $[M(bipy)_3]^{n+/(n+1)+}$ where $M = Fe^{++/3+}$, $Ru^{++/3+}$, and $Cu^{+/++}$. For example, the potentials obtained by Brandt and Smith (1949) with a number of methyl-substituted phenanthroline Fe(II)/(III) complexes show a regular effect (Table XI). Methyl groups in the 3 and 4,6 positions increase the potential by 0.03 and 0.15 volts, respectively, and 3,4,6 substitution stabilizes the oxidant also by 0.18 volt.

Several authors (Schwarzenbach *et al.*, 1955; Martell and Calvin, 1953; Bjerrum, 1950) have suggested that a linear relationship exists between the logarithm of the stability constant β_3 for a series of complex ions of the same metal and the acid dissociation constant pK_a of the ligand, of the form

$$\log \beta_3 = a(pK_a) + b \tag{12}$$

The significance of "a" and "b" has been discussed by Irving and Rossotti (1954, 1956) who maintain that only markedly similar ligands would be expected to follow the empirical relation closely, where the substituents do not hinder complex formation and no drastic change occurs in the bonding properties of the ligand.

The standard redox potential $E°$ for a metal chelate couple may be related to the aquo potential, $E°_{aq}$, by the expression

$$E° = E°_{aq} - 2.303\,\frac{RT}{nF}\,[\log \beta_{ox} - \log \beta_{red}] \tag{13}$$

where β_{ox} and β_{red} are the appropriate stability constants for the complex ions. Now if the empirical relation in Eq. (12) holds then

$$E° = E°_{aq} - 2.303\,\frac{RT}{nF}\,[pK_a(a_{ox} - a_{red}) + (b_{ox} - b_{red})]$$

and a linear relationship should exist between $E°$ and the pK_a of the attached ligand. For the tris-5-chloro-, 5-nitro-, 5-bromo-, and 5-methyl-1,10-phenanthroline Fe(II) complexes, Brandt and Gullstrom (1952) found a linear correlation between pK_a and $\log \beta_3$ and almost a linear correlation between $E(1M\,H_2SO_4)$ and pK_a (Fig. 4). The agreement between the two sets of results is not unexpected since the latter depends on the former. However, more recent studies show that the correlation is obtained only for a series of ligands which are structurally closely related, form similar bonds, and have the same number and size of chelate rings (Jones *et al.*, 1958; Schwarzenbach *et al.*, 1955; Uusitalo, 1957). Thus, deviations from linearity in the plot of $E°$ against pK_a were observed by Tomkinson and Williams (1958) for a series of tris-(substituted 8-hydroxyquinoline) Fe(II)/(III) couples (Fig. 5). The effect of the substituent was related to the change in the ligand-metal bond and the marked deviations of the 5-formyl and 5-cyano couples to the π acceptor properties of these groups.

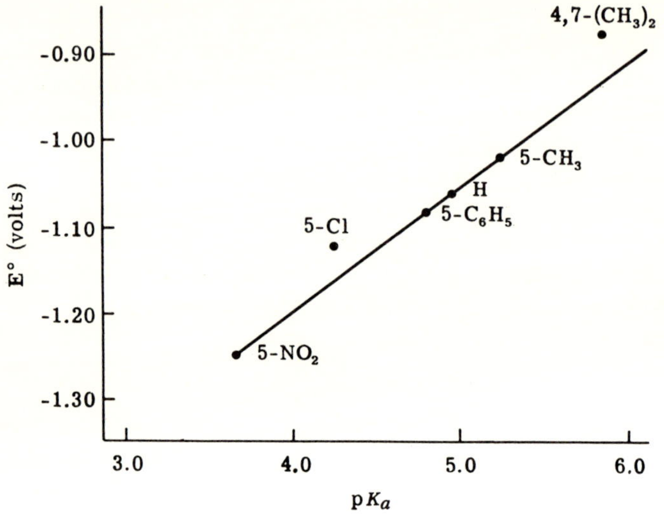

FIG. 4. The relation between $E°$ and basicity for substituted phenanthroline complexes of iron(II)/(III) (Brandt and Gullstrom, 1952).

The remaining ligands all contain electron donor substituents and fall on a regular curve.

It is conceivable that the deviations, or part thereof, could be ascribed to hydration and entropy effects. For example, bulky substituents might be expected to increase the structure-breaking effect of the complex on the solvent and so increase the entropy of solvation. This would be different for the two oxidation states of the metal and would contribute a small change to $\Delta(\bar{S}°)$. The substituents will also alter $\Delta(\Delta H_{aq})$ but if these effects were of primary importance in determining the deviations from linearity in the plot of $E°$ against pK_a then a regular trend could be expected as the size of the substituent increased. Such a correlation has not been observed. For example, the redox potentials of a series of osmium(II)/(III) couples $[Os(pyX)(bipy)(trpy)]^{++/3+}$, when X = H, 3-$CH_3$, 4-$C_3H_7$, 4-$C_2H_5$, and 4-$CH_3$, show (Fig. 6) that not only is there no linear relationship between pK_a and $E°$ but also that there is no regular trend with the size of the substituent (Buckingham *et al.*, 1961).

It can be inferred that variations in $\Delta(\bar{S}°)$ and $\Delta(\Delta H_{aq})$ are only of minor importance when the position of substitution is far removed from the coordination center. It is evident then that there is no exact relation between the pK_a of the ligand and $E°$ or even the stability constant β. The enthalpy changes on the formation of the conjugate acid of the ligand mainly reflect differences in the σ electron density of the donor atoms,

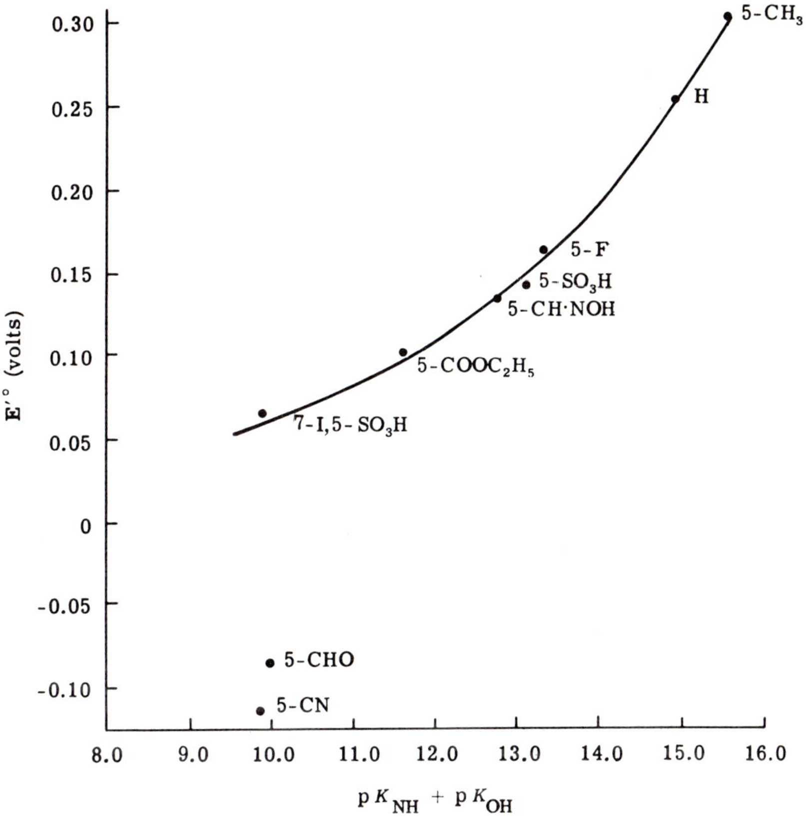

FIG. 5. The relation between E'° and basicity for substituted 8-hydroxyquinoline complexes of iron(II)/(III) (Tomkinson and Williams, 1958).

whereas ΔH changes for the formation of metal complexes will be affected by both σ- and π-bonding capabilities of *both* ligand and metal ion. The greater sensitivity of E° to π effects can arise because E° (complex) is a function of two stability constants ($\log \beta_{ox} - \log \beta_{red}$) both of which change in the same sense with respect to changes in σ effects (i.e., pK_a) but which may change in opposite senses because of π effects (Tomkinson and Williams, 1958).

The redox potentials of substituted phenanthroline and bipyridine complexes of Cu(I)/(II) have been measured in 50% dioxane-water (James and Williams, 1961) and the results are given in Figs. 7 and 8. For both series Cu(II) is stabilized relative to Cu(I) as the pK_a of the ligand increases. However, the aquo couple in 50% dioxane-water is -0.25 volt so that the unsubstituted phenanthroline stabilizes the Cu(I) state slightly,

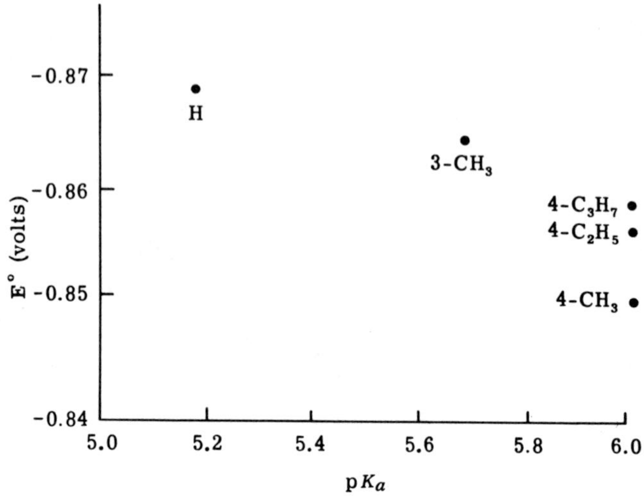

FIG. 6. Plot of $E°$ against pK_a of the substituted pyridine in the complexes [Os(pyX)-(bipy)(trpy)]$^{++/3+}$.

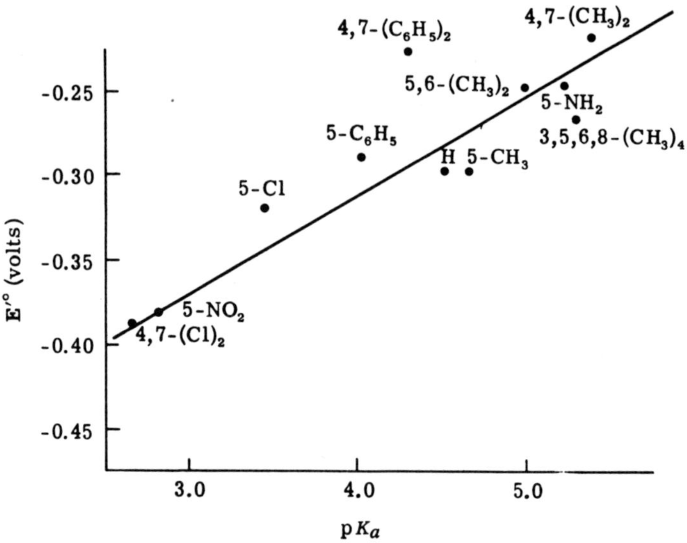

FIG. 7. The relation between $E'°$ and basicity for substituted phenanthroline complexes of copper(I)/(II) (James and Williams, 1961).

relative to water. James and Williams take this to mean that "the unsaturated nature of phenanthrolines stabilises cuprous with respect to cupric ion, but that the greater donor power of the phenanthroline nitrogen atoms

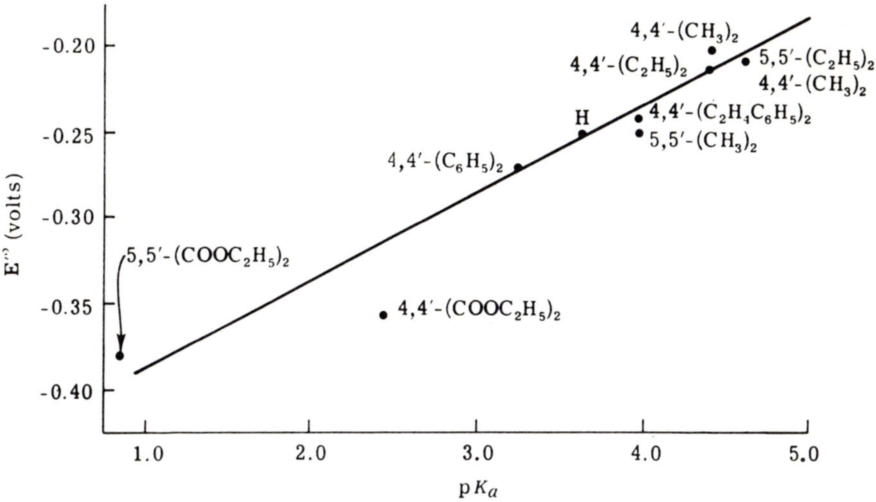

FIG. 8. The relation between E'° and basicity for substituted bipyridine complexes of copper(I)/(II) (James and Williams, 1961).

than of water stabilises cupric with respect to cuprous." It would seem then that an increase in the σ donor strength stabilizes the higher valent state and an increase in π acceptor strength stabilizes low valency states. The low potential of the 4,4′-diphenyl-2,2′-bipyridine couple (−0.269 volt) is in accord with this finding. The high potentials of the 4- and 5-phenyl-substituted phenanthrolines were attributed to steric strain in the ligand itself.

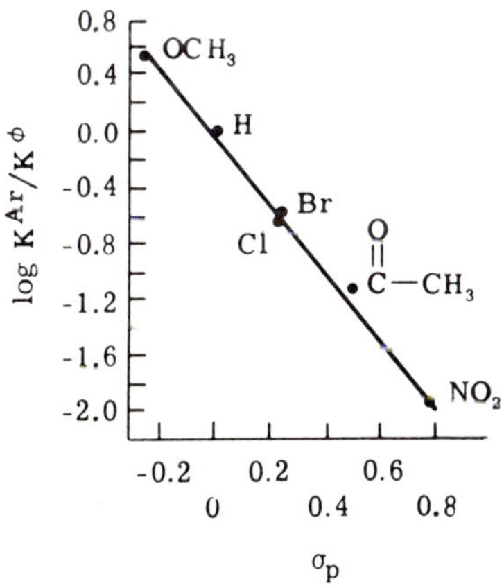

FIG. 9. The redox potential difference between phenylferrocene and *p*-substituted phenylferrocenes plotted against the Hammett σ constant (after Mason and Rosenblum, 1960).

An interesting example of the effect of ligand substitution on redox potential is shown by Mason and Rosenblum (1960) for a series of para-substituted phenylferrocene complexes. The potentials were obtained by a potentiometric titration of the ferrous complex in 75% acetic acid–25% dilute perchloric acid mixtures, using a Cr(VI) solution.

Since it is not possible to measure the basic strength of the ligands, the differences between the formal oxidation potentials of phenylferrocene (ϕ) and the substituted ferrocenes (Ar) are plotted against the Hammett substituent constants (Fig. 9). $(E^\circ_{Ar} - E^\circ_{\phi})/0.0591 = \log (K^{Ar}/K^{\phi})$. The linear relationship between the two is not unexpected since the ligands have very similar structures and the Hammett constant is a measure of both σ- and π-bonding effects of the substituent. The electron donating groups on the phenyl ring stabilize the oxidized state while electron-withdrawing substituents lower the potential more than phenylferrocene itself, stabilizing the reduced form.

ACKNOWLEDGMENT

The authors would like to acknowledge the help and advice of Professor A. N. Hambly and Dr. E. Magnusson.

References

Barbieri, G. A., and Tettamanzi, A. (1932). *Atti reale. accad. nazle. Lincei* **15,** 877.

Bjerrum, J. (1941). "Metal Ammine Formation in Aqueous Solution," p. 227. Haase Copenhagen.

Bjerrum, J. (1950). *Chem. Revs.* **46,** 381.

Brandt, W. W., and Gullstrom, D. K. (1952). *J. Am. Chem. Soc.* **74,** 3532.

Brandt, W. W., and Smith, G. F. (1949). *Anal. Chem.* **21,** 1313.

Brewer, L., Bromley, L. A., Gilles, P., and Lofgren, N. L. (1950). *In* "Chemistry and Metallurgy of Miscellaneous Materials" (L. L. Quill, ed), p. 76. McGraw-Hill, New York.

Buckingham, D. A., Dwyer, F. P., and Sargeson, A. M. (1961). Unpublished work.

Burstall, R. W. Private communication to R. S. Nyholm.

Connick, R. E., and Fine, D. A. (1961). *J. Am. Chem. Soc.* **83,** 3414.

Conway, B. E. (1952). "Electrochemical Data," p. 25. Am. Elsevier, New York.

Dwyer, F. P. (1949). *J. Proc. Roy. Soc. N.S.Wales* **83,** 134.

Dwyer, F. P. (1951). *Revs. Pure and Appl. Chem. (Australia)* **1,** 77.

Dwyer, F. P., and Goodwin, H. A. (1957). Unpublished work.

Dwyer, F. P., and Gyarfas, E. C. (1954). *J. Am. Chem. Soc.* **76,** 6320.

Dwyer, F. P., and McKenzie, H. A. (1947). *J. Proc. Roy. Soc. N.S.Wales* **81,** 93.

Dwyer, F. P., and Wooldridge, H. (1949). *J. Proc. Roy. Soc. N.S.Wales* **83,** 235.

Dwyer, F. P., Humpoletz, J. E., and Nyholm, R. S. (1946). *J. Proc. Roy. Soc. N.S.Wales* **80,** 242.

Dwyer, F. P., Gibson, N. A., and Gyarfas, E. C. (1951). *J. Proc. Roy. Soc. N.S.Wales* **84,** 80.

George, P., and McClure, D. S. (1959). *In* "Progress in Inorganic Chemistry" (F. A. Cotton, ed.), Vol. I, p. 381. Wiley (Interscience), New York.

George, P., Hanania, G. I., and Irvine, D. H. (1959). *J. Chem. Soc.* p. 2548.
Glasstone, S. (1942). "An Introduction to Electrochemistry," p. 212. Van Nostrand, Princeton, New Jersey.
Griffith, J. S. (1961). "The Theory of Transition Metal Ions," pp. 315–319. Cambridge Univ. Press, London and New York.
Griffith, J. S., and Orgel, L. E. (1957). *Quart. Revs. (London)* **11,** 381.
Guggenheim, E. A. (1930). *J. Am. Chem. Soc.* **52,** 1315.
Harned, H. S., and Owen, B. B. (1950). *In* "The Physical Chemistry of Electrolyte Solutions," p. 297. Reinhold, New York.
Hawkins, C. J., and Perrin, D. D. (1962). *J. Chem. Soc.* p. 1351.
Hoard, J. L., Smith, G. S., and Lind, M. (1961). *In* "Advances in the Chemistry of Coordination Compounds". (S. Kirschner, ed.), p. 296. Macmillan, New York.
Irving, H. M., and Rossotti, H. S. (1954). *J. Chem. Soc.* pp. 2904, 2910.
Irving, H. M., and Rossotti, H. S. (1956). *Acta. Chem. Scand.* **10,** 72.
James, B. R., and Williams, R. J. P. (1961). *J. Chem. Soc.* p. 2007.
Jones, J. G., Poole, J. B., Tomkinson, J. C., and Williams, R. J. P. (1958). *J. Chem. Soc.* p. 2001.
Jorgensen, C. K. (1956a). *Proc. 10th Solvay Congr. in Chem., Brussels,* p. 355.
Jorgensen, C. K. (1956b). *Acta Chem. Scand.* **10,** 1505.
King, W. R., and Garner, C. S. (1952). *J. Am. Chem. Soc.* **74,** 3709.
Kolthoff, I. M., and Lingane, J. J. (1952). "Polarography," 2nd ed., Vol. I, p. 217. Wiley (Interscience), New York.
Kolthoff, I. M., and Tomsicek, W. J. (1935). *J. Phys. Chem.* **39,** 945.
Krumholtz, P. (1953). *J. Am. Chem. Soc.* **75,** 2163.
Latimer, W. M. (1952). "Oxidation Potentials," 2nd ed., p. 30. Prentice-Hall, Englewood Cliffs, New Jersey.
McInnes, D. A., and Yeh, Y. L. (1921). *J. Am. Chem. Soc.* **43,** 2563.
Martell, A. E., and Calvin, M. (1953). "The Chemistry of the Metal Chelates," p. 76. Prentice-Hall, Englewood Cliffs, New Jersey.
Mason, J. G., and Rosenblum, M. (1960). *J. Am. Chem. Soc.* **82,** 4206.
Moore, C. E. (1958). *Natl. Bur. Standards (U.S.) Circ.* 467, Table 34, Vol. III.
Ogston, A. G. (1962). Private communication.
Orgel, L. E. (1960). "An Introduction to Transition Metal Chemistry," p. 167. Methuen, London.
Paglia, E., and Sironi, C. (1957). *Gazz. chim. ital.* **87,** 1125.
Perrin, D. D. (1959). *J. Chem. Soc.* p. 290.
Powell, R., and Latimer, W. M. (1951). *J. Chem. Phys.* **19,** 1139.
Pribil, R. (1949). *Czechoslov. Chem. Communs.* **14,** 320, 626.
Randles, J. E. B. (1960). *Ric. sci.* **5,** 3.
Rossotti, F. J. C., and Rosotti, H. (1961). "The Determination of Stability Constants," p. 128. McGraw-Hill, New York.
Schwarzenbach, G. (1949). *Helv. Chim. Acta* **32,** 839.
Schwarzenbach, G., and Heller, H. (1951). *Helv. Chim. Acta* **34,** 576.
Schwarzenbach, G., and Sandera, J. (1953). *Helv. Chim. Acta* **36,** 1089.
Schwarzenbach, G., Anderegg, G., Schneider, W., and Senn, H. (1955). *Helv. Chim. Acta* **38,** 1147.
Scrocco, E., and Marmani, G. (1954). *Atti accad. nazl. Lincei, nazl. Rend. Classe sci. fis. mat. e nat.* **16,** 637.
Scrocco, E., and Ragazzini, M. (1954). *Atti accad. nazl. Lincei, nazl. Rend. Classe sci. fis. mat. e nat.* **16,** 489.

Smith, G. F., and Richter, F. P. (1944). *Ind. Eng. Chem.* **16,** 580.
Staveley, L. A. K., and Randall, T. (1958). *Discussions Faraday Soc.* **No. 26,** 157.
Tomkinson, J. C., and Williams, R. J. P. (1958). *J. Chem. Soc.* p. 2010.
Unsitalo, E. (1957). *Ann. Acad. Sci. Fennicae, Ser. A* **No. 87.**
Weakliem, H. A., and Hoard, J. L. (1959). *J. Am. Chem. Soc.* **81,** 549.

CHAPTER 7

Metal Chelates of Ethylenediaminetetraacetic Acid and Related Substances

F. L. GARVAN*

Biological Inorganic Chemistry Section, Australian National University, Canberra, A.C.T., Australia

I. Ethylenediaminetetraacetic Acid

A. GENERAL PROPERTIES

Among the many polyaminocarboxylic acids that have been synthesized over the past 30 years, ethylenediaminetetraacetic acid (H_4EDTA) still

* *Present address:* Christian Brothers' Training College, Strathfield, N.S.W., Australia.

seems to hold pride of place. Recently other such acids have been made that have advantages over H_4EDTA in specific cases, but the ease with which the acid is prepared, certainly at the present, outweighs the slight advantages of other similar chelate compounds (Martell, 1961).

Because of its almost innumerable commercial and analytical uses, ethylenediaminetetraacetic acid has come to be called by several names: Calsol, Chelaton, Complexone(II), Complexone(III) (disodium salt), Titra Ver, Trilon B, Versene, Iminol D, Nervanaid, Nullapon, Sequestrene, Sequestrol, and Sequestric Acid.

Ethylenediaminetetraacetic acid is a white anhydrous crystalline solid melting at 240°C. with decomposition. It is almost insoluble in water and has a solubility minimum of 3.05×10^{-4} mole/liter at pH 1.6 (Klygin *et al.*, 1959). Common organic solvents such as ethanol, acetone, ether, and benzene will not dissolve the acid but it is soluble in a hot solution of formamide from which it will crystallize on cooling. It is readily soluble in alkaline solutions. As the abbreviation H_4EDTA suggests,[1] the acid is tetrabasic. The ionization equilibria are characterized by the following values for the equilibrium constants expressed as their pK values (Schwarzenbach and Ackermann, 1947): $pK_1 = 1.996$; $pK_2 = 2.672$; $pK_3 = 6.161$; $pK_4 = 10.262$. The acid forms a series of mono-, di-, tri-, and tetrasodium salts, which increase in water solubility with increasing degree of neutralization. The disodium salt, $Na_2H_2EDTA{\cdot}2H_2O$, can be readily obtained in a pure state so that it is used as a primary standard in analysis. Blaedel and Knight (1954) have reported a detailed study of the solubility of the salt.

Information on the organic chemistry of H_4EDTA is scanty. This is due in part to the tendency of workers to concentrate on the inorganic aspects of the acid, viz., complex formation with metal ions, and the industrial and analytical applications which stem from this. Nevertheless, the chemical nature of the acid itself is such that to react it with the ordinary modifying reagents usually requires conditions so drastic that the compound is broken down. It is not known to form an anhydride or acid chloride, and cannot be esterified by the usual procedures. The acid is

[1] H_4EDTA will be used as the abbreviation for the solid acid, while $(H_3EDTA)^-$, $(H{\cdot}EDTA)^{3-}$, will be used when a precise statement about the species involved is required. EDTA (without any indication of charge) will be used as a general abbreviation for any of the anionic species of ethylenediaminetetraacetic acid, when the charge on the species is not known or when what is meant may be easily understood from the context; e.g., we may speak of the metal chelates of EDTA. In metal chelates, if one or more carboxyl groups are not coordinated to the central metal ion, this will be indicated in the formula; e.g., $[Co(H{\cdot}EDTA)Cl]^-$ means that one of the carboxyl groups is uncoordinated and protonated; the sixth octahedral position is occupied by chlorine. Again, in general, the structure of the metal chelate will be understood from the text.

broken down by strong oxidizing agents but no definite products have been isolated. There is no report of its reduction to any known compounds. The ampholytic nature of the acid is believed to be the cause of its solubility in strong mineral acids from which crystalline compounds of the type $H_4EDTA \cdot H_2SO_4$ can be obtained (Aiken, 1956; Beck and Gorog, 1959). This kind of compound is the only suitable starting material for the synthesis of esters (U.S. Patent, 1947). With organic amines such as triethanolamine or cyclohexylamine, H_4EDTA forms salts which readily crystallize and are soluble in organic solvents (Aiken, 1956).

B. Synthesis of Polyaminocarboxylic Acids

Only the most important syntheses that are being used at the present time on a large scale will be discussed.

1. *Condensation of an Amine with a Monohalogenated Carboxylic Acid (Usually Monochloroacetic Acid) as the Sodium Salt*

The reaction shown in Eq. (1) is straightforward and is readily carried out in mildly alkaline conditions. The tendency for the halogen to hydrolyze

$$NH_2CH_2CH_2NH_2 + 4\,ClCH_2COONa \xrightarrow{4\,NaOH} [-CH_2N(CH_2COONa)_2]_2 + 4\,NaCl + 4\,H_2O \tag{1}$$

to hydroxyl, accompanied by undesirable side reactions, is probably the cause of yields no higher than 80–85% being attained. The method is versatile and can be used to synthesize nitrilotriacetic acid, (U.S. Patent, 1938), *trans*-1,2-cyclohexanediaminetetraacetic acid (Schwarzenbach and Ackermann, 1949), and 1,2-propylenediaminetetraacetic acid (Dwyer and Garvan, 1959). Aiken (1956) has obtained fully N-substituted polyacetic acids from polyamines, and, by reducing the amount of chloroacetic acid, intermediate compounds such as ethylenediaminediacetic acid can be made. The reaction has been extended to include preparation of the acids $(-CH_2NRCH_2COOH)_2$ where R is a long chain alkyl or substituted phenyl group (British Patent, 1954).

2. *Reaction of an Amine with Hydrocyanic Acid (Strecker Synthesis) or Sodium Cyanide ("Carboxymethylation of Amines") and an Aldehyde, Usually Formaldehyde*

The Carboxymethylation of Amines, Eqs. (5) and (6), is the most generally applied synthesis. The Strecker Synthesis, Eqs. (2)–(4), is less preferable because of the large quantities of hydrogen cyanide that have to

$$CH_2O + HCN \longrightarrow HOCH_2CN \tag{2}$$

$$4\,HOCH_2CN + NH_2CH_2CH_2NH_2 \longrightarrow [-CH_2N(CH_2CN)_2]_2 + 4\,H_2O \tag{3}$$

$$[-CH_2N(CH_2CN)_2]_2 + 8\,H_2O + 4\,HCl \longrightarrow [-CH_2N(CH_2COOH)_2]_2 + 4\,NH_4Cl \tag{4}$$

Strecker synthesis

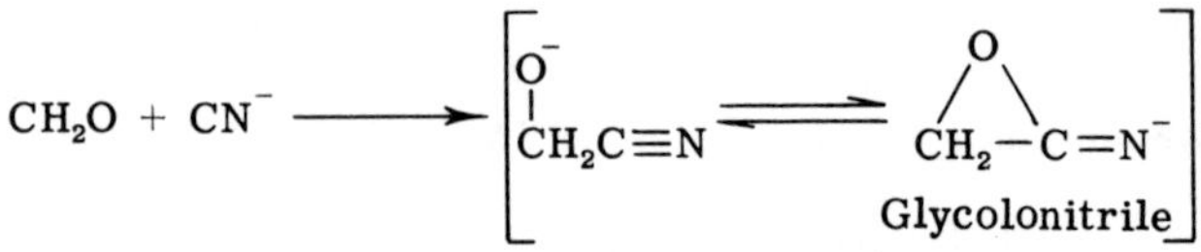

(5)

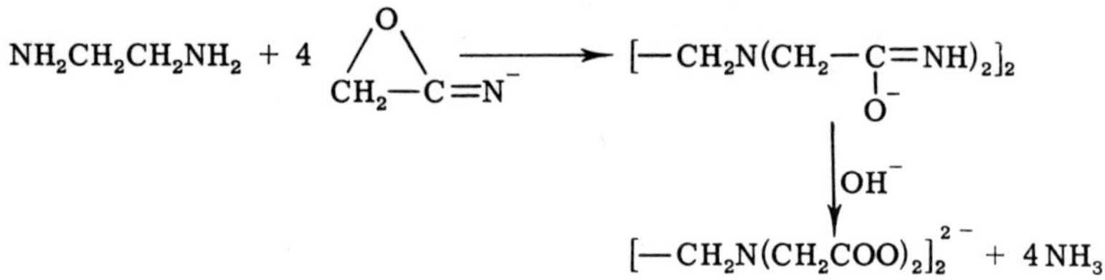

(6)

Carboxymethylation of amines

be handled. The Carboxymethylation of Amines has the disadvantage that ammonia is formed which can also condense with the reactants to give by-products. It is removed therefore by employing reduced pressures at pH 11. A study of the mechanism of the reaction has shown that a glycolonitrile is formed. By controlling the amount of glycolonitrile formed and its tendency to hydrolyze, yields for the preparation of H_4EDTA as high as 96% have been effected in the laboratory (Smith *et al.*, 1949; Martell and Bersworth, 1950; Ziemlak *et al.*, 1950).

Theoretically, by the use of higher aldehydes poly-α-alkyl acetic acids can be made, although steric hindrance may prevent their formation (Irving *et al.*, 1958). This all-embracing method has also been used to make dialkyl and diaryl diacetic acids (British Patent, 1955a,b), substituted glycines, polyacetic acids of *trans*-1,2-cyclohexanediamine, and *N*-hydroxyethylethylenediaminetriacetic acid and its homologs (British Patent, 1955c). Very recently it has been utilized to prepare *cis*-1,2-cyclohexanediaminetetraacetic acid which seems to have defied all other methods of preparation (Kroll and Gordon, 1960).

3. *Condensation of an Alkyl Halide with an Amino Acid*

The chief disadvantage of this type of reaction is the general immiscibility of the reactants. This causes long reaction times of the order of 20 hours. In combination with the "Carboxymethylation of Amines" the

$$ClCH_2CH_2Cl + 2\,NH(CH_2COONa)_2 \xrightarrow{2\,NaOH} [-CH_2N(CH_2COONa)_2]_2 + 2\,NaCl + 2\,H_2O \qquad (7)$$

method is useful for the synthesis of di-C-substituted ethylenediaminetetra-acetic acids (Aiken, 1956).

Another interesting application of this reaction is the production of polymeric analogs of H_4EDTA. Compound (I) can be synthesized by

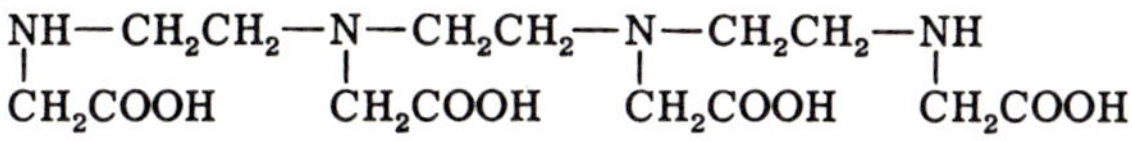

(I)

condensation of 2 moles of ethylenediaminediacetic acid with ethylene dichloride. The two terminal N-hydrogen atoms can be further condensed to give, theoretically at least, compounds of type (II).

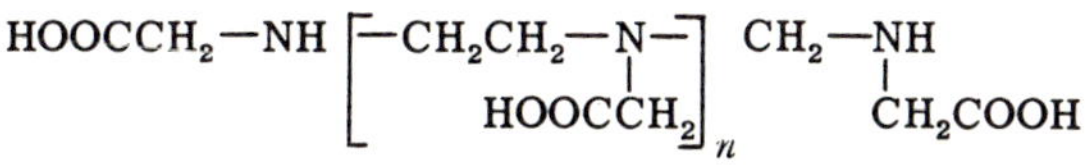

(II)

4. *Oxidation of Polyethanolamines*

The main difficulty in the reaction in Eq. (8) is in preventing the oxidation of the strongly reactive amino groups which are susceptible to attack

$$[-CH_2N(CH_2CH_2OH)_2]_2 + 4\,KOH \longrightarrow [-CH_2N(CH_2COOK)_2]_2 + 8\,H_2 \qquad (8)$$

by alkalis and oxidizing reagents. By limiting the amount of water present so that hydrolysis of the amino group is kept at a minimum, and by the application of pressure, side reactions are cut down (U.S. Patent, 1945a). Cadmium oxide is used as a catalyst, but lower yields result than with reactions described by Eqs. (1)–(6). The required ethanolamine is prepared from ethylene oxide and the appropriate amount of amine (Dumas and Stas, 1840).

5. *Condensation of an Amine with 2 or More Moles of an Amino Acid*

The reaction in Eq. (9) of ethylenediamine and the sodium salt of glycine takes place in aqueous solution, preferably when air is excluded by

$$NH_2CH_2CH_2NH_2 + 4\,NH_2CH_2COONa \longrightarrow [-CH_2N(CH_2COONa)_2]_2 + 4\,NH_3 \qquad (9)$$

bubbling nitrogen through the reaction mixture, removing the unwanted ammonia at the same time. Other α-amino acids can be condensed to give various polycarboxylic acids (U.S. Patent, 1945b, 1946).

6. *Condensation of an Aromatic Compound with Formaldehyde and Iminodiacetic Acid*

This reaction is readily accomplished (1–4 hours at 60–70°C.) and yields are high with *para*-substituted phenols (Schwarzenbach *et al.*, 1952).

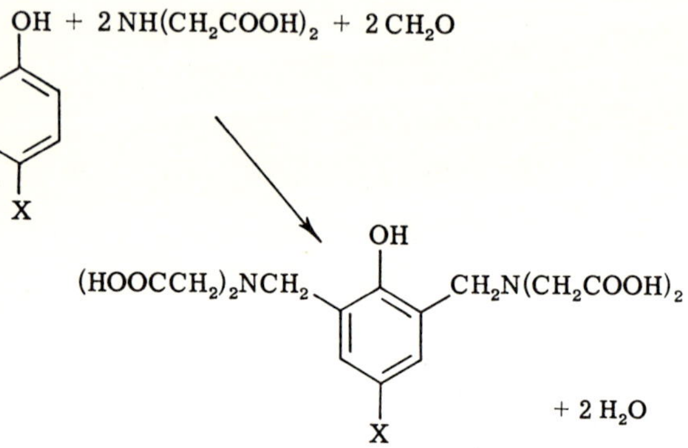

(10)

$(X = CH_3,\ Cl,\ SO_3^-)$

The synthesis seems versatile for the preparation of aromatic compounds (Anderegg *et al.*, 1954).

7. *Isolation of the Polyaminocarboxylic Acid from the Reaction Mixture*

In most of the reactions described above sodium salts of the acids have been prepared. The isolation of the actual acid from the reaction mixture requires comment. H_4EDTA and allied symmetrical compounds are practically insoluble in water at their isoelectric points. Hence, if they are reasonably pure they can be precipitated by adjusting the pH. However the ampholytic nature of these compounds leads to anomalous solubility, especially in the presence of inorganic compounds and reaction by-products. Furthermore, relatively unsymmetrical or substituted acids (e.g., *N*-hydroxyethylethylenediaminetriacetic acid) have considerable water solubility and techniques such as the use of ion-exchange resins have to be employed. Dwyer and Garvan (1959, 1961) have shown that the optical forms of 1,2-propylenediaminetetraacetic acid and *trans*-1,2-cyclohexanediaminetetraacetic acid are far more soluble in water than the racemic forms and their isolation could be effected only by using ion-exchange resins.

C. Mode of Ionization

From a determination of the heat content changes for the ionization of the first two protons from H_4EDTA, Tillotson and Staveley (1958) showed that ions of the type $(H_5EDTA)^+$ and $(H_6EDTA)^{++}$ can exist in acid solution. Beck and Gorog (1959) and Klygin and his associates (1959) have also supported this. Tillotson and Staveley have discussed some of the possible structures for the species in the ionization process from $(H_5EDTA)^+$ to $(EDTA)^{4-}$. Their proposals for the most likely structures are consistent with the infrared spectroscopic work of Chapman (1955) and the pK values of homologs of H_4EDTA as determined by Schwarzenbach and Ackermann (1948). Although Schwarzenbach and Ackermann (1947), Martell and his co-workers (1949), and Charles (1956) have proposed other structures, the work of Tillotson and Staveley is the most comprehensive to date.

D. Complex Formation

Ethylenediaminetetraacetic acid has six atoms (four oxygen and two nitrogen atoms) through which, potentially at least, it can bond to a metal ion. If the acid is sexadentate it can be seen from (III) that five five-membered rings are formed and the charge on the central metal ion is reduced by four. EDTA forms very stable metal chelates with practically every metal in the periodic table and almost invariably the ratio of metal ion chelated to the chelating molecule is 1:1. The various factors responsible

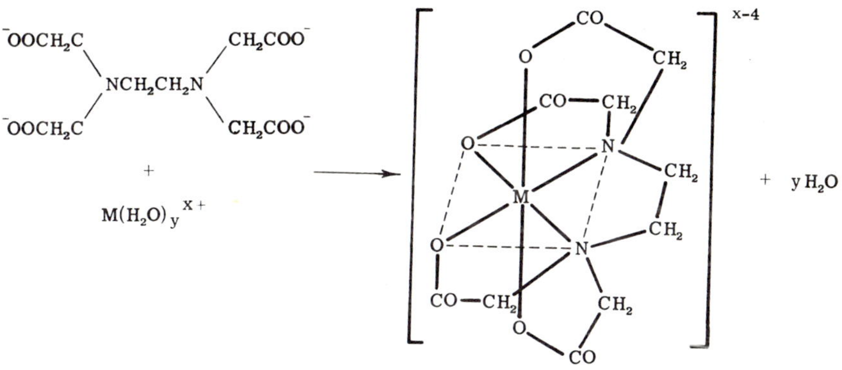

(III)

for the high stability of these metal chelates will now be considered.

1. *Formation of Stable Five-Membered Rings*

Schwarzenbach (1952) has proved conclusively that coordination compounds containing chelates are more stable than structurally similar complexes containing simple ligands. There is also abundant evidence in the literature to show that when the heterocyclic ring in a metal chelate contains five atoms the resulting complex is generally most stable. Mann (1927, 1928) has admirably demonstrated this with 1,2,3-propanetriaminetetrachloroplatinum(IV). With metal chelates of EDTA these two effects are combined to give very stable complexes.

However, it has been doubted whether the nitrogen atoms of EDTA actually take part in ring formation with the alkaline earth ions, since these metals have little or no tendency to coordinate with amines. Calcium was believed to be bound only through the four oxygen atoms. This would require the formation of one eleven-membered ring and two eight-membered rings. Such big rings are unstable especially in aqueous solutions where the competition between water molecules and the chelate for the metal ion favors completely the hydrated form of the ion. Busch and Bailar (1956) have isolated platinum(II) complexes containing EDTA as a bidentate chelate coordinated through the two nitrogen atoms, as well as a quadridentate chelate coordinated through the two nitrogen atoms and two of the four oxygen atoms. The crystal structure determinations of the Ni(II), Cu(II), and Co(III) complexes of EDTA (see p. 38) show that five-membered rings are formed (Weakliem and Hoard, 1959; Smith and Hoard, 1959).

An interesting investigation by Schwarzenbach and Ackermann (1948) on the calcium complexes of EDTA and its homologs further emphasizes the importance of the five-membered ring in these compounds. Stability constants (log β) for the calcium chelates (IV) decrease considerably as n increases. Whether these homologs have the sexadentate (as illustrated) or quadridentate form has no bearing on the effect that increasing n would have on the stability constants, provided the same structure is formed throughout.

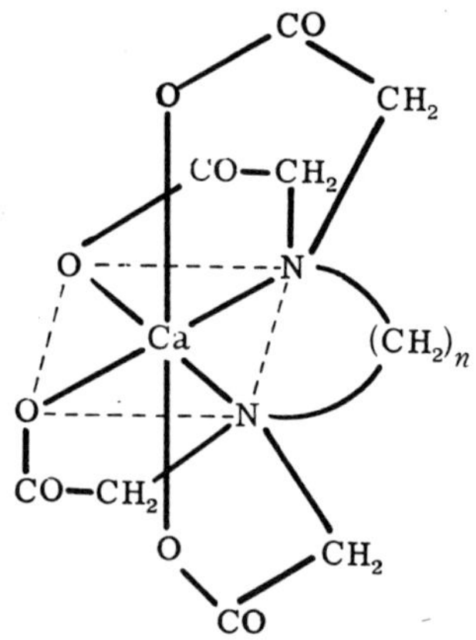

(IV)

2. *Multiple Rings Giving Increased Complex Stability*

During recent years quantitative data have been obtained to prove that the stability of coordination compounds containing chelating agents increases as the number of rings formed per chelating agent is increased. Table I illustrates this fact with metal chelates of EDTA. In each case it has been assumed that the complexes compared have similar octahedral structures (Martell and Calvin, 1952).

TABLE I
COMPARISON OF THE STABILITIES OF COMPLEXES CONTAINING DIFFERENT MULTIDENTATE CHELATES

No. of rings per chelate	Chelate	Complex	Log β	Per cent dissociation of 0.001 M complex
2	Iminodiacetic acid	$[Zn(IDA)_2]^{--}$	13.5	2×10^{-1}
5	H_4EDTA	$[Zn(EDTA)]^{--}$	16.6	1.6×10^{-5}
2	Methyliminodiacetic acid	$[Ca(MIDA)_2]^{--}$	7.5	3
5	H_4EDTA	$[Ca(EDTA)]^{--}$	10.6	1.6×10^{-2}

3. *Thermodynamic Quantities Associated with Complex Formation*

Although coordination compounds of the alkaline earth ions are not unknown (Sidgwick, 1950), the comparatively high stability of their complexes with H_4EDTA is unexpected. The factors in the two sections above could be applied to other multidentate chelates, e.g., polyamines, yet no such chelates form complexes of comparable stability with these metals. Recently the thermodynamics of these EDTA complexes have been studied in an effort to account for their stability. Calvin and Bailes (1946) first proposed that the entropy increase associated with chelate ring formation in general makes a major contribution to the stability as measured by the equilibrium formation constant. This theory has been extended to EDTA complexes, many of which have been studied thoroughly. Table II, which comes from the investigations of Carini and Martell (1954), Charles (1954), Care and Staveley (1956), Staveley and Randall (1958), and Betts and Dahlinger (1959), summarizes the results of this work.

The most significant conclusion to be drawn from the thermodynamic data is that the stability of the metal chelates is due primarily to the ΔS values which are large and positive. An entropy increase would be expected for a system such as this where the charge of the ions is decreased, where the ion formed is larger than the ions from which it is produced allowing a more favorable distribution of charge, and where an increase in the number of particles occurs. If EDTA were sexadentate in all the ions examined, it would be expected that the complexes would have essentially the same shape, differ little in size, and interact with the solvent in much the same way. The entropy of the complex ion in solution would then be

TABLE II

THERMODYNAMIC QUANTITIES OF METAL CHELATES OF EDTA[a]

Cation	ΔH	ΔF	ΔS	$S°$	$\Delta S + S°$	r
Mg^{++}	+ 3.14	−11.65	50.5	−28.2	22.3	0.66
Ca^{++}	− 6.45	−14.34	26.9	−13.2	13.7	0.99
Sr^{++}	− 4.11	−11.57	25.4	− 9.4	16.0	1.12
Ba^{++}	− 4.83	−10.40	19.0	+ 3	22	1.34
Mn^{++}	− 5.45	−18.6	44.5	−18	26.5	0.80
Co^{++}	− 4.4	−21.9	59.7	−22	38	0.72
Ni^{++}	− 8.35	−24.96	56.7	−23	34	0.69
Cu^{++}	− 8.67	−25.20	56.4	−23.6	32.8	0.72
Zn^{++}	− 5.61	−22.12	56.3	−25.45	30.8	0.74
Cd^{++}	−10.08	−22.07	40.9	−14.6	26.3	0.97
Pb^{++}	−14.08	−24.18	34.5	+ 5.1	39.6	1.20
Al^{3+}	+12.58	−21.60	116.6	−74.9	41.7	0.51
Y^{3+}	+ 0.32	−24.26	83.8	—	—	0.92
In^{3+}	− 7.23	−33.46	89.5	−62	27.5	0.81
La^{3+}	− 0.8	−20.72	66.8	−39.2	27.6	1.14
Ce^{3+}	− 0.47	−21.07	69.1	−41.3	27.8	1.07
Pr^{3+}	− 0.8	−21.49	69.4	−43.0	26.4	1.06
Nd^{3+}	− 0.8	−21.89	70.7	−44.5	26.2	1.04
Sm^{3+}	− 0.8	−22.54	72.9	−47.1	25.8	1.00
Eu^{3+}	− 0.16	−22.72	75.7	−48.4	27.3	0.98
Gd^{3+}	+ 0.43	−22.94	78.4	−49.5	28.9	0.97
Tb^{3+}	+ 1.5	−23.62	84.3	−50.9	33.4	0.93
Dy^{3+}	+ 1.5	−24.25	86.3	−52.3	34.0	0.92
Ho^{3+}	+ 1.25	−24.61	86.7	−53.8	32.9	0.91
Er^{3+}	+ 1.5	−25.06	89.1	−55.2	33.9	0.89
Tm^{3+}	+ 1.58	−25.44	90.6	−56.5	34.1	0.87
Yb^{3+}	+ 1.32	−25.91	91.3	−57.7	33.6	0.86
Lu^{3+}	+ 0.6	−26.11	89.7	−58.8	30.9	0.85

[a] The units of ΔH and ΔF are kcal./mole, of ΔS, cal./mole degree, and of the ionic radius, angstrom units. The ionic radii are taken from Ahrens' table (1952) for six-fold coordination.

almost constant or show a gradual change with the radius of the metal ion. This would also apply to $\Delta S + S°$ which differs from the standard entropy of the complex only by the constant entropy of the $(EDTA)^{4-}$ ion ($S°$ is the standard entropy of the metal ion). Obviously this trend is not found.

It can be seen that for ions of a given charge, ΔS is much bigger for a triply charged cation than for a doubly charged ion. From considerations of the interactions between the ion and the water molecules which are displaced when the complex is formed, and the interactions with the more remote molecules, Staveley and Randall (1958) have shown that for an ion of given charge ΔS is controlled to a first approximation by the radius of the metal ion. They suggest that the EDTA chelates of the alkaline earth ions and divalent metal ions without closed electronic configurations have different structures. Smith and Hoard (1959) have shown that both the Cu(II) and Ni(II) chelates of EDTA contain one coordinated water molecule with the chelate attached at five positions through two nitrogen and three oxygen atoms. It might then be inferred that EDTA is sexadentate in the alkaline earth metal chelates. However, it will be seen in Section II that this inference does not agree entirely with other experimental evidence. To explain the anomalous behavior of the Mn(II) chelate Staveley and Randall (1958) have suggested that the EDTA is quadridentate. These authors are at pains to point out that it is probably an oversimplification to think that in metal chelates of EDTA, in solution at least, the EDTA offers a definite number of symmetrically placed points of attachment to the metal ion, and that if this is less than six the deficiency is made up by a number of water molecules.

Wheelwright *et al.* (1953) originally proposed that the break that occurs at gadolinium in the sequence of the stability constants of the lanthanon chelates of EDTA could be explained by the fact that with the earlier, larger ions the EDTA acts as a sexadentate chelate, but at gadolinium it changes to a quinquedentate chelate. It was thought that at gadolinium a critical size of the cation is reached that causes an abrupt change of coordination number. This proposal was later withdrawn by Schwarzenbach and Gut (1956) when it was found that a similar "gadolinium break" occurs in the sequence of stability constants with *trans*-1,2-cyclohexanediaminetetraacetic acid and nitrilotriacetic acid. Staveley and Randall (1958) have suggested that EDTA is probably quinquedentate in the lanthanon chelates and that the change in the stability constants is due to the varying degrees of stabilization from the interaction of the 4*f*-orbitals with the ligand field.

The data in Table II for the lanthanons are from the work of Betts and Dahlinger (1959). It is obvious that the $\Delta S + S°$ values divide the lanthanons into two groups. Those from La to Gd have $\Delta S + S° = 27.1 \pm 0.9$ cal./mole °C. and those from Tb to Lu have $\Delta S + S° = 33.3 \pm 0.7$ cal./mole °C. (only Sm, Gd, and Lu are outside the range quoted, by a small margin). The authors stressed that the validity of the ensuing arguments must depend on the correctness of their assumptions for calculating these thermodynamic quantities and they are somewhat diffident about their results. In fact, they obtained two series of values for $\Delta S + S°$ depending on the method of calculation. The consoling thing, however, is that, no matter which method is used, the same kind of division appears in the $\Delta S + S°$ values. With this preface they suggest that the stereochemistry of the metal chelates changes after Gd (cf. Wheelwright *et al.*,

1953) but they propose that those from La to Gd have the EDTA attached at five positions with a water molecule in the sixth, and those from Tb to Lu have the EDTA coordinated at four positions with two water molecules in the remaining positions. However, they make the strange suggestion that when the EDTA is quadridentate it is coordinated through three oxygen atoms and one nitrogen atom. This would require the formation of at least one chelate ring containing eight atoms. Such a ring would be extremely unstable and most unlikely to occur.

It is to be hoped that very soon someone will investigate the X-ray crystal structures of suitable metal chelates from each of these two Ln-EDTA groups, so that their stereochemistry will be definitely decided. It will be seen in Section II that the radius of the central metal ion does indeed appear to influence the stereochemistry of EDTA in its metal chelates. The nature of the bonding must also be an important factor. Rossotti (1960) has discussed these factors from the thermodynamic point of view.

4. *pH of the Reaction Mixture*

The pH is important in complex formation with H_4EDTA because the species of the acid present in solution is greatly pH-dependent (Pecsok, 1952).

$$H_4EDTA \underset{H^+}{\overset{OH^-}{\rightleftharpoons}} \underset{\text{pH } 2.3}{\overset{50\%}{(H_3EDTA)^-}} \underset{H^+}{\overset{OH^-}{\rightleftharpoons}} \underset{4.5}{\overset{95\%}{(H_2EDTA)^{--}}} \underset{H^+}{\overset{OH^-}{\rightleftharpoons}} \underset{8.1}{\overset{100\%}{(H{\cdot}EDTA)^{3-}}} \underset{H^+}{\overset{OH^-}{\rightleftharpoons}} \underset{12.5}{\overset{100\%}{(EDTA)^{4-}}}$$

Martell (1952) has shown from the conductance titration of the acid with calcium hydroxide that chelation is at a minimum when the pH is low. Whether complex formation will take place at low pH depends on the stability of the complex. The pH of the solution can also affect the species formed. It will be seen in Section II that many complexes of EDTA have a water molecule coordinated to the central ion and the following equilibrium is pH-dependent:

$$[MH_2O(EDTA)]^{--} \underset{H^+}{\overset{OH^-}{\rightleftharpoons}} [M(EDTA)OH]^{3-}$$

It is also possible that at high pH the OH^- can actually displace a coordinated carboxyl group from the central metal ion and thereby change the multidentate function of the EDTA, while at intermediate values of pH certain quinquedentate aquo complexes change to sexadentate complexes.

II. The Metal Chelates of Ethylenediaminetetraacetic Acid

In the following discussion stability constants of the metal chelates are not listed unless they are relevant. They are available from the work of

Bjerrum *et al.* (1957). [An excellent review of the biological aspects of polyaminocarboxylic acids is available in *Federation Proceedings* **20,** 1–263 (1961).]

A. Group IA, Lithium, Sodium, and Potassium

The low values of the stability constants show that very weak complexes are formed with lithium and sodium. Brintzinger and Munkelt (1948) first prepared the lithium chelate which they incorrectly formulated as a simple salt, $Li_4EDTA \cdot 4H_2O$. It was originally thought that potassium may be complexed, and tetramethylammonium hydroxide was used for titration against aminocarboxylic acids (Schwarzenbach *et al.*, 1945). Subsequently Schwarzenbach and Ackermann (1947) showed that the potassium ion did not result in any appreciable error in acid-base titrations. The infrared spectra of the solid tetra-alkali salts show slight evidence for sodium and potassium complexes (Sawyer and Paulsen, 1958).

B. Group IB

1. *Copper*

Although Smith and Hoard (1959) have determined the crystal structure of the Cu(II) chelate of EDTA, it is of great interest to follow the development of the conclusions of various workers from indirect evidence.

Brintzinger and Hesse (1942) first prepared the greenish blue complex formulated as $[Cu(H_2EDTA)]$. It is sparingly soluble in water and pH titrations showed it to be a dibasic acid. Qualitative tests indicated that a complex is formed and Brintzinger and Hesse thought the EDTA acted as a quadridentate chelate but they did not prove it. Pfeiffer and Offermann (1942) isolated various solid salts and Hill-Cottingham (1955) has shown that the complex is not affected by sunlight.

On the basis of chemical evidence three schools of thought emerged concerning the stereochemistry of the copper chelate. Firstly, some workers considered the complex $[Cu(H_2EDTA)]$ to be square-planar with the EDTA acting as a quadridentate chelate (Brintzinger and Hesse, 1942; Chaberek and Martell, 1952; Bennett and Schmidt, 1955; Kirschner, 1956). The only evidence to support this conclusion came from the infrared spectrum of $[Cu(H_2EDTA)] \cdot H_2O$. Kirschner (1956) reported two strong bands of comparable intensity in the carbonyl region corresponding to two free carboxylic acid groups (frequency peak, 1718 cm.$^{-1}$) and two coordinated carboxyl groups (frequency peak, 1615 cm.$^{-1}$). Evidence was also found for Cu—N bonds.

Other workers believed that a water molecule is coordinated to the copper, with the remaining five octahedral positions occupied by EDTA which has one free carboxyl group (Klixbull-Jorgensen, 1955; Care and Staveley, 1956; Garvan, 1959). Klixbull-Jorgensen interpreted, according to the crystal field theory, the absorption spectrum of the Cu(II)-EDTA complex in solution at about pH 8; he also compared the complex with the square-planar complex bis(glycinato)copper(II). Both investigations showed that the EDTA complex is octahedral. However, as the evidence was not absolutely conclusive, it was thought that the EDTA is quinquedentate with a water molecule coordinated in the sixth octahedral position. It was observed that this water molecule can occupy two positions with respect to the two nitrogen atoms giving rise to two[2] isomers, (V) and (VI). This possibility of isomerism rendered the interpretation of the spectrum difficult. However, (V) and/or (VI) were supported by comparison with Ni(II) complexes.

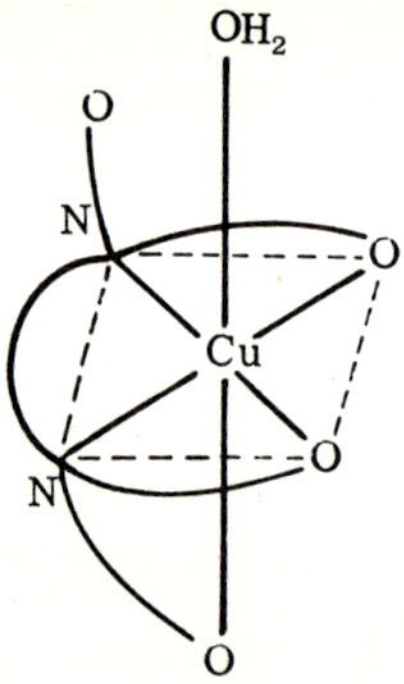

(V)

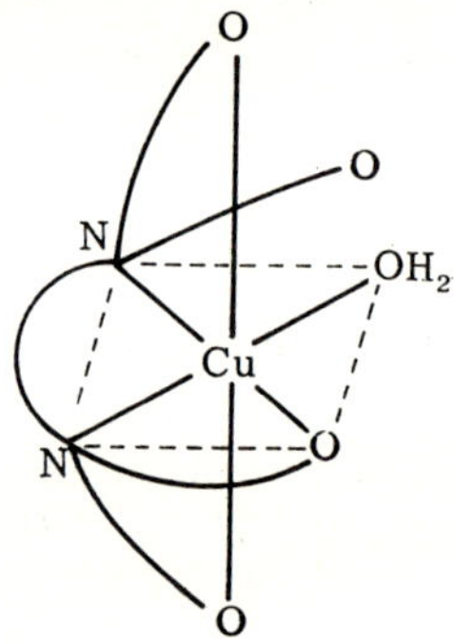

(VI)

Care and Staveley (1956) found evidence for the formation of a hydroxo complex in a study of the thermodynamics of the Cu(II)-EDTA chelate. A structure of the type (V) or (VI) satisfied the data obtained, since the hydroxo species could result from the aquo complex by the loss of a proton. However, Rossotti (1958) has pointed out that the hydroxo complexes of EDTA metal chelates could be equally well formed by replacement of a coordinated carboxyl group by a hydroxyl group at high pH.

Garvan (1959) prepared the complexes formulated as $H[CuH_2O$-

[2] Actually, as Smith and Hoard (1959) have pointed out, there are four possible isomers, but the other two (not illustrated) arise from the different possible arrangements of the carbon atoms between the two nitrogen atoms, and from coordination in polar positions of two carboxyl groups attached to one nitrogen atom. These last two isomers are much more unstable than those illustrated in the text and will not be considered further in this chapter. The carbon atoms between the two nitrogen atoms will be assumed to adopt always the *gauche* configuration.

(H·EDTA)] and Ba[CuH_2O(EDTA)]·$4H_2O$ and showed them to be isomorphous with the corresponding nickel complexes by X-ray powder photographs. Infrared spectra, magnetic moments, and dehydration studies showed that the nickel chelates contain quinquedentate EDTA with a water molecule coordinated in the sixth octahedral position. Hence the copper chelates must also contain quinquedentate EDTA. Neither Smith and Hoard (1959) nor Garvan (1959) could agree that the peaks at 1718 $cm.^{-1}$ and 1615 $cm.^{-1}$ reported for H[CuH_2O(H·EDTA)] by Kirschner (1956) have comparable intensities. The higher frequency is less intense than the lower, indicating fewer uncoordinated than coordinated carboxyl groups. There is also a sharp band at 3606 $cm.^{-1}$ due to the OH stretching vibration and this was ascribed by Garvan to a coordinated water molecule (cf. Ni(II) chelates, Section II,M,5).

Finally, Kirschner (1956) thought the complex K_2[Cu(EDTA)]·$4H_2O$ contains sexadentate EDTA. His infrared studies indicated that in the acid copper complex the chelate is quadridentate and the proposal made was that as the acid complex is deprotonated by the addition of potassium hydroxide the two free carboxyl groups coordinate to the copper. The potassium salt of the copper chelate shows only one sharp peak in the carbonyl region at 1615 $cm.^{-1}$. The disappearance of the higher frequency peak, at 1718 $cm.^{-1}$ in the acid salt of the copper chelate, was believed to be due to the equivalence of all the carboxyl groups which were regarded as being coordinated to the copper. Sawyer and Paulsen (1958) have criticized this conclusion on the grounds that it would be difficult to resolve uncoordinated carboxylate groups in the presence of the 1615 $cm.^{-1}$ peak. Although Kirschner (1956) obtained optically active solutions of the copper chelate no definite conclusion can be drawn from this concerning its stereochemistry.

Smith and Hoard (1959) have solved most of these problems by a crystal structure study. They showed that H[CuH_2O(H·EDTA)] is isomorphous with the corresponding nickel chelate which contains quinquedentate EDTA with the water molecule coordinated in the equatorial position where it is *cis* to one nitrogen atom and *trans* to the other (VI). The two carbon atoms between the two nitrogen atoms are above and below the plane containing the nickel ion and the two nitrogen atoms, in the *gauche* configuration (see p. 39).

The potential for the reaction

$$[Cu(EDTA)]^{3-} \rightleftharpoons [Cu(EDTA)]^{--} + e^-$$

is −0.13 volt showing that the higher oxidation state is just slightly more stabilized with respect to the lower oxidation state than in the Cu(I)/Cu(II) system (Belcher *et al.*, 1955). The Cu(I)-EDTA complex could not be isolated, but from the above redox potential a high stability constant was calculated (log β = 18). This is an example of a labile complex having a high stability constant.

2. *Silver and Gold*

A weak complex is formed with Ag(I) (log β = 7.32). However, Ringbom and Linko (1953) have shown that it is sufficiently stable to cause dissolution of silver chloride by EDTA in alkaline solution. No gold chelates have been reported, but when one drop of 1% gold solution is treated with a solution of EDTA a purple ring develops that can be used as a spot test. The color was thought to be caused by finely divided gold (Hynes *et al.*, 1950).

C. Group IIA

1. *Beryllium*

Brintzinger and Munkelt (1948) and Pribil (1950) reported that beryllium does not complex with EDTA. This conclusion is unreliable as it was based on the precipitation of beryllium hydroxide with ammonia. Some solid complexes have been isolated that are extremely soluble in water giving a pale yellow color in concentrated solution (Brintzinger and Munkelt, 1948). The small size of the beryllium ion would certainly make the complex unstable (Care and Staveley, 1956).

2. *Calcium*

The high stability of the Ca(II) chelate with EDTA has stimulated research on this complex, especially with regard to its stereochemistry. The white crystalline compounds $K_2[Ca(EDTA)]$ and $Na_2[Ca(EDTA)]\cdot 6H_2O$ are readily soluble in water (Pfeiffer and Offermann, 1942; Pfeiffer and Simons, 1943). Pfeiffer and Simons studied the stereochemistry of $Na_2[Ca(EDTA)]\cdot 6H_2O$ and their work was the first serious attempt to unravel the structure of an EDTA metal chelate. It was thought at first that the EDTA is quadridentate with a square-planar distribution of bonds through two nitrogen and two oxygen atoms. To test this theory an attempt was made to prepare the calcium chelate of $HOOCCH_2(CH_3)NCH_2CH_2N(CH_3)CH_2COOH$. A simple salt (*sic*) resulted as evidenced by precipitation of calcium oxalate, on the addition of oxalate ion. It was then concluded that the stability of $[Ca(EDTA)]^{--}$ must arise from coordination of all the carboxyl groups, so that the chelate was believed to be sexadentate. The only way to demonstrate this, it was decided, was to resolve the complex. However, if the chelate were quinquedentate or quadridentate the complex could also be resolved, so that even a successful resolution would not prove the sexadentate function of the EDTA. The only symmetrical structure is the *meso* form of the square-planar configuration. Although the brucine salt could be easily crystallized no resolution was effected. This was to be expected, as whatever the structure of the calcium chelate, the interactions between the alkaline earth ions and the nitrogen and oxygen atoms of the $(EDTA)^{4-}$ anion would be expected to be essentially electrostatic (Keller and Parry, 1956; Care and Staveley, 1956; Sawyer and Paulsen, 1958). The fact that the EDTA complexes of calcium, strontium, and barium exchange almost instantaneously with radioactive calcium ion makes the chances of resolution very remote (Astakhov and Fomenko, 1957).

Since this early work by Pfeiffer and Simons, opinion has oscillated between a square-planar and an octahedral structure for the calcium chelate. Schwarzenbach and Ackermann (1947) have also proposed that a water molecule occupies one of the octahedral positions since the complex shows a slight tendency to take up hydroxyl ion in solution. A proton would then be lost according to the equation

$$[CaH_2O(EDTA)]^{--} + OH^- \rightarrow H_2O + [Ca(EDTA)OH]^{3-}$$

The pK value (11.39) for

$$[CaH_2O(EDTA)]^{--} \rightleftharpoons H^+ + [Ca(EDTA)OH]^{3-}$$

shows that the aquo complex is an extremely weak acid. This slight uptake of OH^- might also be explained if one carboxyl group of the sexadentate EDTA detaches at high pH. Hydroxyl ion could then coordinate to the calcium in the unoccupied octahedral position.

It is a very difficult matter to ascribe sexadentate function to EDTA in its calcium chelate from chemical evidence alone. Since the complex is colorless, different species such as $[Ca(EDTA)OH]^{3-}$ are not as easily detected as with highly colored complexes, e.g., $[Co(EDTA)]^-$ and [Co-$(EDTA)OH]^{--}$, and nothing short of a complete crystal structure can unequivocally solve the problem. Even then it is not impossible that a modified species exists in solution. Nevertheless the evidence summarized below suggests strongly that $[Ca(EDTA)]^{--}$ is octahedral containing sexadentate EDTA at least in the solid state and in solution at pH 5–10. (i) It has been shown above that the nitrogen atoms of EDTA are coordinated to the calcium (Section I,D,1). (ii) The failure of the compound $HOOCCH_2$-$(CH_3)NCH_2CH_2N(CH_3)CH_2COOH$ to form a stable complex with calcium suggests an octahedral configuration. (iii) The thermodynamic data discussed in Section I,D,3 point to the conclusion that EDTA is sexadentate in its calcium chelate. (iv) pH values in the titration of Ca^{++} with H_4EDTA

are increased considerably when proton acceptors in the form of acetate ions are added. If two of the carboxyl groups of EDTA were not coordinated, little change would be expected (Johnson and Callis, 1956).

3. *Magnesium, Strontium, and Barium*

Several metal chelates of these elements have been isolated (Pfeiffer and Offermann, 1942; Pfeiffer and Simons, 1943; Brintzinger and Munkelt, 1948; Vorisek, 1959). Brunisholz (1957b) has described a convenient method for preparing $H_2[Mg(EDTA)]\cdot 6H_2O$ and has recommended it instead of $Na_2[Mg(EDTA)]\cdot xH_2O$ in analysis. The acid salts, $H_2[M(EDTA)]\cdot yH_2O$ (M = Mg, Sr, Ba), are comparatively insoluble in water, whereas the dialkali salts are quite soluble (0.1–0.2 *M* at 30°C.).

The infrared frequency of the carboxyl groups in the solids, Na_2[M-(EDTA)], falls uniformly with increasing ionic radius suggesting the order of increasing stability as Ba < Sr < Ca < Mg (Sawyer and Paulsen, 1958). Stability constants measured in solution agree with this sequence except that Mg and Ca are reversed (Bjerrum *et al.*, 1957). The tendency for Mg(II) to be more strongly hydrated in solution than Ca(II) would be absent in the solid state, thereby possibly accounting for this reversal. However, Sawyer and Paulsen stressed the unreliability of these infrared measurements for making such fine distinctions. The uncertainty of resolving free carboxylate ions from coordinated carboxyl groups in Na_2 M-(EDTA)] prevented any definite conclusions about stereochemistry from being made. Hoard *et al.* (1961a) report that the infrared spectra of $H_2[Mg(EDTA)]\cdot 6H_2O$ and $H_2[Ba(EDTA)]\cdot 4H_2O$ show that probably there is at least one uncomplexed carboxylic acid group. It can then be inferred that at least the EDTA is not sexadentate in these acid salts.

The stereochemistry of the complexes of Mg, Sr, and Ba has not been studied extensively. Schwarzenbach and Ackermann (1947) found that the magnesium chelate is a weak hydroxyl ion acceptor and an aquo complex was proposed, $[MgH_2O(EDTA)]^{--}$. The unreliability of such conclusions has already been stressed. Care and Staveley (1956) and Martell (1956) have ascribed the decrease in stability of the strontium and barium chelates with respect to the calcium chelate to the increased size of the metal ions which changes the sexadentate function to quinquedentate, a molecule of water occupying the sixth position.

Finally, unstable complexes of the type $[M(H\cdot EDTA)]^-$ have been reported by Schwarzenbach and Ackermann (1947). These complexes may be considered as intermediates in the formation of $[M(EDTA)]^{--}$. Their concentration, even at low pH, is always less than the concentration of $[M(EDTA)]^{--}$. The tridentate attachment of $(H\cdot EDTA)^{3-}$ proposed by Geigy (1955) is without experimental proof.

D. Group IIB, Zinc, Cadmium, and Mercury

Brintzinger and Munkelt (1948) first isolated the compounds Zn[Zn(EDTA)], Cd[Cd(EDTA)]·$4H_2O$, and H_2[Hg(EDTA)]·H_2O and concluded that the cadmium compound is a simple salt. Stability constant determinations prove that quite stable metal chelates are formed with all three metals (Bjerrum *et al.*, 1957). Both cadmium and mercury complexes have been thoroughly studied in polarographic investigations (e.g., see Pecsok, 1952; Goffart *et al.*, 1953; Matyska *et al.*, 1955; Watters *et al.*, 1956; Schmid and Reilley, 1958).

Because of the comparable sizes of Ca^{++} and Cd^{++}, Care and Staveley (1956) originally interpreted the thermodynamic data for the cadmium chelate as indicating that the EDTA is sexadentate. This interpretation does not agree with a later explanation given by Staveley and Randall (1958) and discussed in Section I,D,3.

The mercury chelate behaves anomalously. Its high stability probably arises from strong covalent Hg—N bonds. Table III summarizes the

TABLE III
Stability Constants of Metal Chelates of $(HOOCCH_2)_2N—(CH_2)_n—N(CH_2COOH)_2$

Metal	Log β				
	$n = 2$	$n = 3$	$n = 4$	$n = 6$	$n = 8$
Ca	10.70	7.12	5.05	4.40	4.51
Cd	16.62	13.45	11.87	11.70	11.99
Hg	21.80	19.70	20.81	21.38	21.83

stability constants (log β) for the Ca(II), Cd(II), and Hg(II) chelates of $(HOOCCH_2)_2N—(CH_2)_n—N(CH_2COOH)_2$ as n increases from 2 to 8 (Schwarzenbach and Anderegg, 1955). The Ca(II) and Cd(II) chelates behave as expected: as the size of the chelate ring increases the complex becomes less stable. However, the mercury chelate with $n = 8$ is even more stable than with $n = 2$. It is probable, therefore, that the carboxyl groups are free in higher homologs, and the stability throughout is due to strong N—Hg—N bonds. Since the latter tend to be linear, larger rings are favored (Sidgwick, 1950; Schwarzenbach and Anderegg, 1955). The initial drop in log β for $n = 3, 4$ may indicate coordination of some carboxyl groups in $[Hg(EDTA)]^{--}$.

E. Group IIIA and the Lanthanons

1. *Scandium and Yttrium*

Brintzinger and Munkelt (1948) prepared the compound $Y[YH_2O$-$(EDTA)]_3 \cdot 21H_2O$ and regarded it as a simple salt. However, it has quite a high stability constant ($\log \beta = 18.0$) which lies between those for dysprosium and holmium and is in agreement with the usual group trend and comparable ionic sizes. Its stereochemistry will be discussed below. The high stability of the scandium chelate ($\log \beta = 23.1$) enables it to be easily separated from the lanthanons on ion-exchange resins (Iya and Loriers, 1953).

2. *The Lanthanons*

The gradual variation in stability of the lanthanon chelates of EDTA has provided an excellent basis for the separation of the lanthanons by ion-exchange techniques (e.g., see Vickery, 1952, 1954; Holleck and Hartinger, 1954; Achard, 1955; Topp, 1956; Fuger, 1957; Meinhold and Kremers, 1957). Separations have also been effected through solubility differences of acid, potassium, and ammonium salts (e.g., see Brunisholz, 1955; 1957a; Marsh, 1955). Spectra of and term splitting in these metal chelates have been studied by Moeller and Brantley (1950) and Holleck and Eckardt (1954). By comparison with the alkaline earth EDTA chelates Jones (1955) has found that the bonding in the lanthanon chelates is essentially electrostatic.

A discussion of the stereochemistry of the lanthanon chelates has already been given from the thermodynamic data obtained for them. Other evidence will now be discussed. The complex $[NdH_2O(H \cdot EDTA)]$ is a moderately strong acid with $pK = 3.7$ (Moeller *et al.*, 1955). These authors have shown from infrared measurements that the complexes $[NdH_2O(H \cdot EDTA)]$ and $[YH_2O(H \cdot EDTA)] \cdot xH_2O$ contain an uncomplexed carboxylic acid. Furthermore, the complexes $Na[LnH_2O(EDTA)] \cdot zH_2O$ are isomorphous for Ln = Pr, Nd, Sm, Gd, Y, and on the basis of differential thermal analyses a water molecule was believed to occupy the sixth position in the octahedral configuration; the complexes decomposed before the last amounts of water were removed (Moeller *et al.*, 1955; Moeller and Horwitz, 1959). Rossotti (1958) is not convinced by these dehydration studies. However, linked with the other evidence given above the conclusions of Moeller and his associates seem quite reasonable. Moeller, Moss, and Marshall attempted to resolve the yttrium chelate with alkaloids and by adsorption on optically active quartz but without success. Their deductions made about the ionic character of these complexes in view of this negative evidence are not convincing.

Wald (1961) has obtained the infrared spectra of the acid salts of the chelates of Nd, Sm, Ho, and Er. He confirms the work of Moeller *et al.* (1955) for the Nd chelate but the other chelates reveal only one broad absorption band in the carbonyl region that suggests the EDTA is sexadentate. Wald does not believe this possible, and attributes the failure to resolve the carboxylic acid group to association. He tentatively suggests that the structures proposed by Betts and Dahlinger (1959) (cf. Section I,D,3) obtain for the lanthanons after gadolinium. Hoard *et al.* (1961a) have suggested that the lanthanon chelates are sexadentate and seven-coordinate, $[LnH_2O(EDTA)]^-$. This probably applies to sodium or potassium salts and not to acid salts since Moeller *et al.* (1955) have shown the presence of an uncomplexed carboxylic acid group in some of the acid salts of lanthanon chelates.

F. Group IIIB, Aluminum, Gallium, Indium, and Thallium

Saito and Terrey (1956) have isolated acid, sodium, potassium, and ammonium salts of the chelates of Al, Ga, and In. The presence of a water molecule in the octahedral sphere is indicated from potentiometric titrations and by the difficulty with which the last molecule of water is removed. However, Hoard *et al.* (1961a) have shown that $NH_4[Al(EDTA)]\cdot 2H_2O$ is isomorphous with $NH_4[Co(EDTA)]\cdot 2H_2O$ and the cobalt chelate has been shown to contain sexadentate EDTA (Weakliem and Hoard, 1959). Hoard and his co-workers (1961a) also report that the infrared spectrum of the acid salt of the aluminum chelate shows there is probably an uncomplexed carboxylic acid group present, and this confirms the work of Saito and Terrey above, at least for the acid salts. It seems that at some intermediate range of pH the quinquedentate aquo complex $[AlH_2O(H\cdot EDTA)]$ is converted into the sexadentate complex $[Al(EDTA)]^-$.

The Tl(III) chelate was predicted to have a much lower stability than the In(III) chelate as it is easily reduced to Tl(I) and is very unstable to heat. This anomalous behavior is ascribed to the greater ionization potential of Tl(III) over In(III), for generally the ionization potential decreases with atomic number in a group of the periodic table (Saito and Terrey, 1956).

Saito and Terrey (1956) have replaced the coordinated water molecule in the In(III) chelate to give the compounds $Na_2[In(EDTA)Br]\cdot 4H_2O$, $(NH_4)_2[In(EDTA)Cl]\cdot 2H_2O$, and $NH_4[InNH_3(EDTA)]\cdot H_2O$. These substituted complexes are easily soluble in water but rapidly hydrolyze to the aquo complex. Dissociation constants of the coordinated water show that it is more firmly held in the Tl(III) complex than in the In(III) complex.

Moeller and Graham (1957) have attempted the resolution of $[GaH_2O(EDTA)]^-$ using brucine but without success.

G. Group IVA

1. *Titanium*

From polarographic evidence the complexes [Ti(IV)(EDTA)], [Ti(IV)O-(EDTA)]$^{--}$, and [Ti(III)(EDTA)]$^{-}$ have been identified (Blumer and Kolthoff, 1952; Pecsok and Maverick, 1954). Below pH 2 [Ti(IV)(EDTA)] is stable for at least 17 days. At higher pH a yellow color develops after 60 hours. The complex is in equilibrium with Ti(IV)O^{++} ion. Above pH 2.5 [Ti(IV)O(EDTA)]$^{--}$ is the predominant species which can be polarographically reduced to [Ti(III)(EDTA)]$^{-}$. Likewise, below pH 2 [Ti-(IV)(EDTA)] can be reduced to [Ti(III)(EDTA)]$^{-}$, but solutions of the latter reduce water with evolution of hydrogen. The absorption spectrum of [Ti(III)(EDTA)]$^{-}$ has been examined with reference to the crystal field theory but its stereochemistry has not been discussed (Klixbull-Jorgensen, 1955).

2. *Zirconium and Hafnium*

Morgan and Justus (1956) report that the Zr(IV) and Hf(IV) chelates are polymeric structures. At both extremes of the pH range (3–9) of stability, precipitates of Zr(IV) and Hf(IV) hydrous oxides were obtained. The Zr(IV) chelate isolated from aqueous solution at pH 6 contains 1.1 EDTA molecules per Zr atom. The stability constant of the 1:1 Zr(IV) chelate has been obtained by spectrophotometric comparison with the Cu(II) chelate. Relative ease of hydrolysis shows that the Hf(IV) chelate is half as stable as the Zr(IV) chelate which is 12.5 times as stable as the Cu(II) chelate. Intorre and Martell (1960) report the identification of the complex [Zr(IV)$(H_2O)_2$(EDTA)] in which the EDTA is sexadentate with the two water molecules also coordinated to the zirconium in an Archimedean antiprism structure.

H. Group IVB, Tin and Lead

Smith (1961) has reported that Sn(II) forms a complex with EDTA but he offers no suggestions as to its stereochemistry.

Brintzinger and his co-workers (1943) regarded the complex Pb[Pb-(EDTA)]·H_2O as a salt since it gave the usual qualitative tests for Pb^{++}. Obviously, these tests were afforded by the uncomplexed lead ion. The acid salt of the lead chelate is difficultly soluble in water. This insolubility of the acid salts of lead chelates of polyaminocarboxylic acids is sometimes used to isolate the acid from reaction mixtures after removal of the lead. By contrast, the disodium salts of these lead chelates are very soluble. While lead(II) acetate is readily chelated by EDTA, lead(II) nitrate gives no complex unless hydrogen peroxide is added. The function of the hydrogen peroxide is not understood. Because the lead chelate is about one hundred million times as stable as the calcium chelate the latter can be used in the treatment of lead poisoning.

The thermodynamics of the formation of $[Pb(EDTA)]^{--}$ show the usual large, positive entropy increase, while the heat evolved is the largest of the complexes studied (see Table II). This suggests the possibility of sexadentate EDTA. Comparison with the stabilities of the Pb(II) chelates of ethylenediaminediacetic acid and ethylenediaminedipropionic acid suggests that $[Pb(EDTA)]^{--}$ contains sexadentate EDTA (Chaberek and Martell, 1952).

I. Group VA

1. *Vanadium*

Schwarzenbach and Sandera (1953) have isolated the complexes $Na[V(III)(EDTA)]\cdot 4H_2O$ (yellow), $Na_2[V(IV)OH_2O(EDTA)]\cdot 2H_2O$ (deep blue), and $Ba[V(IV)OH_2O(EDTA)]\cdot 5H_2O$ (blue). $Na[V(III)(EDTA)]\cdot 4H_2O$ can be completely dehydrated forming a red-brown powder which dissolves in water to give the original color of the hydrated complex. The V(III) and Fe(III) chelates have similar acidic properties. The yellow color of $[V(III)(EDTA)]^-$ changes to red with alkali due to the formation of $[V(III)(EDTA)OH]^{--}$. Above pH 12 a brown precipitate forms. Below pH 2 $[V(III)(EDTA)]^-$ does not take up protons, indicating the EDTA is sexadentate. It seems probable that the V(III) chelate has the same structure as the Fe(III) chelate (see Section II,M,1). For the reaction:

$$[V(II)(EDTA)]^{--} \rightleftharpoons [V(III)(EDTA)]^- + e^-$$

$E°$ = +1.026 volts, showing the +3 oxidation state is powerfully stabilized over the +2 state. (For the couple:

$$V^{++} \rightleftharpoons V^{3+} + e^-$$

$E°$ = +0.255 volt, Schwarzenbach and Sandera, 1953.)

$[V(IV)OH_2O(EDTA)]^{--}$ takes up protons indicating a free carboxyl group, while $[V(IV)O(EDTA)OH]^{3-}$ forms slowly on the addition of alkali. When $Ba[V(IV)OH_2O(EDTA)]\cdot 4H_2O$ is dehydrated only four water molecules can be removed. These facts indicate that EDTA is quinquedentate, a water molecule occupying one of the octahedral positions. For the reaction:

$$[V(III)(EDTA)]^- + 2H_2O \rightleftharpoons [V(IV)OH_2O(EDTA)]^{--} + 2H^+ + e^-$$

$E°$ is −0.802 volt. Hence the +3 state is stabilized with respect to the +4 state on the addition of EDTA to V^{3+}.

Ringbom *et al.* (1957) have found evidence for the formation of an EDTA complex of the vanadyl ion, $[VO_2(EDTA)]^{3-}$ (greenish yellow). Below pH 3 a proton is taken up to form $[VO_2(H{\cdot}EDTA)]^{--}$, accompanied by a slight color change. The pK value (3.60) for the dissociation of this proton suggests that one carboxyl group of EDTA is not coordinated, as the formula above indicates. It is difficult to understand how EDTA could be quinquedentate, as two of the octahedral positions must be occupied by the two oxide ions of VO_2^+. It is possible that the complex is seven-coordinate like the structure proposed for the Fe(III) chelate. The weak complex $[V(V)(EDTA)]^+$ has been detected in solution by Sajo (1958). In the pH range 1–3.5 the stability constant (7.07) remains constant, but above pH 3.5 it varies, indicating formation of different species.

2. *Niobium and Tantalum*

Niobium(V) and tantalum(V) have recently been shown to form complexes with EDTA in solution (Kirby and Freiser, 1961a,b). They can be reduced to the +4 oxidation state and the Ta(IV) chelate is claimed to be the first definite instance of Ta(IV) in aqueous solution. Their stereochemistry has not been discussed.

J. Group VB, Antimony and Bismuth

Jardin (1959) has reported the formation of an antimony chelate without referring to its stereochemistry. The bismuth(III) chelate, H[Bi-(EDTA)] has been isolated as a white soluble powder by Brintzinger and Munkelt (1948). A solution of the complex gives precipitates with hydrogen sulfide, sodium hydroxide, and sodium stannite (Na_2SnO_2), but not with ammonia nor on dilution with water. These reactions point to the formation of a weak complex. The analysis indicated no water, but this could be unreliable. Although octahedral Bi(III) complexes are not unknown, it is doubtful if EDTA is sexadentate as the above formula suggests (Sidgwick, 1950).

K. Group VIA

1. *Chromium*

The chromium(III) chelate has been extensively studied. The first complex isolated was formulated as H[Cr(EDTA)] on analytical evidence (Brintzinger *et al.*, 1943). This suggests the EDTA is sexadentate. Schwarzenbach and Biedermann (1948) prepared the complex again and showed that its correct formula is $[CrH_2O(H{\cdot}EDTA)]$ in which the water molecule is so firmly coordinated that it cannot be removed at 100°C. under reduced pressure. The infrared spectrum shows two bands in the carbonyl region: the higher frequency peak at 1742 cm.$^{-1}$ is less intense than the one at 1650 cm.$^{-1}$ and indicates the presence of an uncomplexed carboxylic acid group (Dwyer and Garvan, 1960b). Solutions of the complex have absorption maxima at 396 and 538 mμ (Klixbull-Jorgensen, 1955; Cellini and Valiente, 1955). The complex is not very soluble in water and its magnetic moment (3.84 B.M.) is the normal value for Cr(III) complexes (Klemm, 1944). Hamm (1953) was unable to explain the slight change in absorption he observed on the addition of the first equivalent of alkali to $[CrH_2O(H \cdot EDTA)]$ since according to the crystal field theory there should be no change, as perturbations from the ligands decrease with the sixth power of the distance. The recent work of Hoard and his associates (1961a) enables us to explain this problem. Hoard has shown that at an intermediate range of pH the Cr(III) chelate contains sexadentate EDTA and this complex has been isolated in the solid state. No doubt the change in absorption observed by Hamm is due to at least partial formation of sexadentate $[Cr(EDTA)]^-$. Schwarzenbach (1949) has shown that $[CoH_2O(EDTA)]^-$ and $[Co(EDTA)]^-$ have almost identical absorption spectra. So the change in the absorption spectra of the Cr(III) chelates would be expected to be similarly only slight. Hence when one equivalent of alkali is added to the quinquedentate aquo complex $[CrH_2O(H \cdot EDTA)]$ the equation might be best written:

$$2[CrH_2O(H \cdot EDTA)] + 2OH^- \rightarrow [Cr(EDTA)]^- + [CrH_2O(EDTA)]^- + 3H_2O$$

There is probably an equilibrium mixture of the quinquedentate aquo complex and the sexadentate complex. The second equivalent of alkali causes the formation of the blue hydroxo complex through deprotonation of the coordinated water molecule or substitution of the newly coordinated carboxyl group. The environment of the Cr(III) is now changed profoundly resulting in a change of spectrum (absorption maxima now at 390, 590 mμ). The pK for

$$[CrH_2O(H \cdot EDTA)] \rightleftharpoons [CrH_2O(EDTA)]^- + H^+$$

is 3.1, showing that $[CrH_2O(H \cdot EDTA)]$ is a moderately strong acid, and for

$$[CrH_2O(EDTA)]^- \rightleftharpoons [Cr(EDTA)OH]^{--} + H^+$$

the pK is 7.52 (Schwarzenbach and Biedermann, 1948; Hamm, 1953). Furlani *et al.* (1960) have determined these pK values as 2.27 and 7.41, respectively. They also detected in solution the green complex $[Cr(EDTA)(OH)_2]^{3-}$ in which the EDTA is quadridentate. No salts of this green complex could be isolated. For the reaction

$$[Cr(EDTA)OH]^{--} + H_2O \rightleftharpoons [Cr(EDTA)(OH)_2]^{3-} + H^+$$

they report a pK of 12.25.

As pointed out earlier two isomers can exist for the ion $[CrH_2O(EDTA)]^-$ depending on whether the water molecule is *cis* to both nitrogen atoms or *cis* to one and *trans* to the other. Klixbull-Jorgensen (1955) has suggested that the intermediate form of $[CrH_2O(EDTA)]^-$ observed by Hamm (1953) and discussed below is perhaps another geometric form. It is possible that the properties of both isomers, if both isomers are formed, are not sufficiently dissimilar to be detected easily.

The kinetics[3] of the formation of $[CrH_2O(EDTA)]^-$ from a Cr(III) salt and Na_2H_2EDTA show that at room temperature it takes 50 hours for complete chelation, but a boiled solution reacts almost immediately (Hamm, 1953; Cellini and Valiente, 1955). Hamm has shown from spectrophotometric studies that the formation of $[CrH_2O(EDTA)]^-$ is accomplished by several slow steps. Towards 90% completion of the reaction there is good agreement between experimental and calculated values. After this, considerable divergence indicates slow-step processes until finally $[CrH_2O(EDTA)]^-$ is formed. It is possible that during this last stage both geometric isomers are formed and one of these slowly transforms into the more stable isomer.[4] Another possibility that must now be considered is that the sexadentate $[Cr(EDTA)]^-$ is also formed in the equilibrium mixture. The small differences in absorption did not permit rate studies to be carried out during this last phase of the complex formation.

When chromium(III) nitrate and Na_2H_2EDTA are first mixed in solution a pale green color develops that is almost identical with the color developed when oxalate and chromium(III) ions are mixed. Hamm (1953) has shown that a detailed kinetic study reveals an initial mechanism in the EDTA-chromium system similar to that in the oxalate-chromium system (Hamm and Davis, 1953).

Hoard and his co-workers (1961a) from unit cell, density, and space group data have shown that $NH_4[Cr(EDTA)]\cdot 2H_2O$ and $Rb[Cr(EDTA)]\cdot 2H_2O$ are isomorphous with $NH_4[Co(EDTA)]\cdot 2H_2O$ and $Rb[Co(EDTA)]\cdot 2H_2O$. Since the latter have been shown to contain sexadentate EDTA (Weakliem and Hoard, 1959), it must be concluded that the EDTA is also sexadentate in the chromium complexes in which there is no proton to fix the uncoordinated carboxyl group that is definitely present in $[CrH_2O(H\cdot EDTA)]$. This requires the postulate that in an intermediate pH range

[3] Margerum (1959) has reviewed the coordination kinetics of EDTA complexes, in which he discusses the sexadentate and quinquedentate functions of EDTA as revealed by ultraviolet spectroscopy during the formation of metal chelates. Apparently there are several forms of 1:1 complexes formed during chelation. Some of these could be the geometric isomers of the quinquedentate species.

[4] Similarly, Morris and Busch (1959) have accounted for the deviations observed from the first order rate law at long reaction times in the hydrolysis of quinquedentate Co(III) chelates in terms of simultaneous reactions of different geometric isomers.

(5–9), both $[CrH_2O(EDTA)]^-$ and $[Cr(EDTA)OH]^{--}$ are converted, though perhaps slowly, to sexadentate $[Cr(EDTA)]^-$.

Dwyer and Garvan (1960b) were unable to resolve the chromium chelate although they were able to resolve the structurally identical PDTA complex (Dwyer and Garvan, 1961). In this PDTA complex they were able to demonstrate that the optical stability arises from the stereospecificity forced on the chromium by the optically active chelating agent. The fact that the structurally similar anion, ethylenediaminebis(oxalato)chromate(III), racemizes by an intramolecular mechanism with a half-life of less than 4 minutes at 20.6°C. (Bushra and Johnson, 1939) very strongly suggests that a similar rapid rearrangement in the Cr(III)-EDTA complex is responsible for the failure to observe any optical activity. The mechanism for this rearrangement will be discussed in Section II,M,4.

2. *Molybdenum and Tungsten*

Sajo (1958) has shown that molybdates and tungstates form EDTA metal chelates similar to those of vanadium(V), although the molybdenum(V) chelate is slightly less stable than the vanadium(V) chelate.

L. Group VIIA, Manganese

The colorless Mn(II) chelate is oxidized by lead dioxide or sodium bismuthate to the ruby red Mn(III) chelate in faintly acid solution (Pribil and Hornychova 1949, 1950). Although the Mn(III) chelate is easily reduced to the Mn(II) chelate, it affords a colorimetric method for analyzing manganese. Yoshino and his co-workers (1961) have isolated red crystals of the Mn(III) chelate formulated as $K[Mn(H_2O)_2(EDTA)]\cdot H_2O$ in which they believe the EDTA is quadridentate. They suggest an attachment of the EDTA similar to that which Betts and Dahlinger (1959) proposed for some of the lanthanon chelates, but which seems very unlikely. The complex is light-sensitive and easily decomposed by heat giving off carbon dioxide. It is worth noting that Staveley and Randall (1958) assigned a quadridentate attachment of EDTA in the Mn(II) chelate to explain its anomalous thermodynamic quantities. Martell (1956) assigned a sexadentate attachment in the Mn(II) chelate but without any good experimental proof. Hoard *et al.* (1961a,c), from the early stages of a crystal structure determination of a complex thought to be $Mn[Mn(H\cdot EDTA)]_2\cdot 10H_2O$, maintain that the EDTA is sexadentate and a water molecule is also coordinated to the manganese in a seven-coordinate structure (VII). The above formula as written by Hoard is difficult to reconcile with their proposals as (H·EDTA) suggests that one of the carboxyl groups is protonated and therefore not coordinated to the manganese. The final outcome of this work will be most interesting.

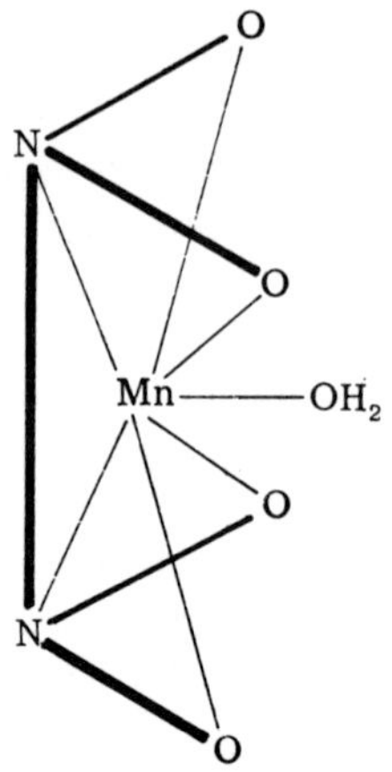

(VII)

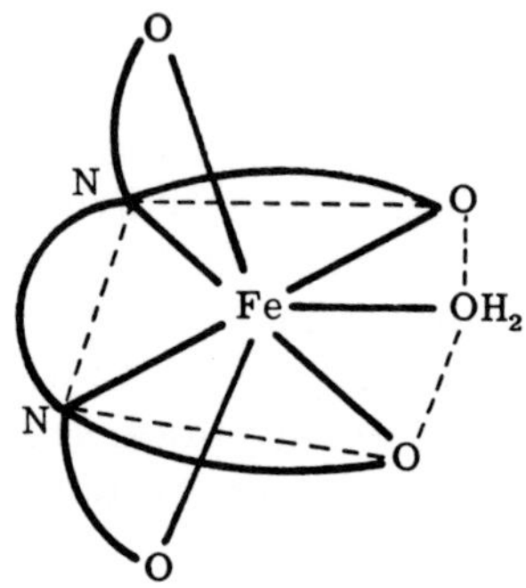

(VIII)

M. Group VIII

1. *Iron*

The colorless Fe(II) chelate has not been isolated but is easily oxidized to the pale yellow Fe(III) chelate even by dissolved oxygen when the pH is only moderately high (Long *et al.*, 1948; Jones and Long, 1952). Qualitative tests show that the Fe(II) chelate is moderately stable provided oxidizing conditions are excluded. The complex absorbs strongly below 300 mμ, while pH titrations show that $H_2[Fe(EDTA)]$ is a strong dibasic acid. $E°$ for the reaction

$$[Fe(EDTA)]^{--} \rightleftharpoons [Fe(EDTA)]^{-} + e^{-}$$

is greatly pH-dependent indicating formation of different species (Schwarzenbach and Heller, 1951). At pH 4–6, $E° = -0.1172$ volt showing that the Fe(III) chelate is stabilized over the Fe(II) chelate with respect to the Fe(II)/Fe(III) couple. Jones and Long (1952) have shown that the exchange between Fe^{++} and $[Fe(EDTA)]^{--}$ at pH $\leqslant 4.7$ is essentially instantaneous, as is the charge transfer process between $[Fe(EDTA)]^{--}$ and $[Fe(EDTA)]^{-}$ which occurs at pH 2 between the actual metal chelates and not between any decomposition products of these complexes (Reynolds *et al.*, 1961). Exchange between Fe^{*++} and $[Fe(EDTA)]^{-}$ is slow. Jones and Long (1952) studied the exchange of Fe^{*3+} with $[Fe(EDTA)]^{-}$ and found a complicated expression for the rate equation. However, a reasonable interpretation has been given for each of the terms in the equation (Jones and Long, 1952; Basolo and Pearson, 1958). The half-life of exchange varies from 0.5 to 400 hours at high and low pH, respectively.

Brintzinger *et al.* (1943) isolated the metal chelates formulated as

H[Fe(EDTA)] and NH_4[Fe(EDTA)]·H_2O. From an analytical study of H[Fe(EDTA)], Jones and Long (1952) confirmed the absence of water in their effort to demonstrate that the EDTA is sexadentate. In the light of the work to be presented below it seems that this analytical study is suspect. H[Fe(EDTA)] is a very strong acid indicating that there is no free carboxylic acid group and that the EDTA is sexadentate. When two equivalents of alkali are added to H[Fe(EDTA)] the deep orange $[Fe(EDTA)OH]^{--}$ is formed (Schwarzenbach and Heller, 1951). Because of this reaction it was believed that a water molecule is coordinated to the iron which immediately suggests that one of the carboxyl groups is not coordinated. This explanation would not then agree with the strong acid character of the acid salt of the Fe(III) chelate. One explanation of this was thought to be that at high pH a carboxyl group detaches and a hydroxyl ion immediately coordinates to the metal. Opportunely, Hoard and his associates (1961a,b) have solved this dilemma by an X-ray crystal structure determination. They have shown that the Fe(III) chelate contains sexadentate EDTA with a water molecule coordinated to the metal in a structure "loosely describable as pentagonal bipyramidal" (VIII). Hoard believes that these sexadentate seven-coordinate structures will prove to be quite common in the metal chelates of EDTA.

The magnetic moment of the Fe(III) chelates indicates five unpaired electrons (Klemm, 1944), showing that there is not sufficient energy from crystal field stabilization to pair the electrons in the 3*d*-orbitals. Garvan (1959) was unable to resolve the Fe(III) chelate with active $[Co(en)_2(NO_2)_2]^+$ even though easily crystallizable salts were obtained. So far no unequivocal evidence has been produced to confirm the resolution of an Fe(III) complex having a weak crystal field. Although Thomas (1922) reported the resolution of $[Fe(C_2O_4)_3]^{3-}$, Long (1941) and Basolo (1953) were unable to repeat the work.

Solutions of the Fe(III) chelate are indefinitely stable in the dark, but on exposure to sunlight a light-induced reduction to the Fe(II) chelate occurs, similar to that observed with $[Fe(C_2O_4)_3]^{3-}$ (Jones and Long, 1952; Hill-Cottingham, 1955).

2. *Ruthenium and Osmium*

Indirect evidence for the formation of Ru(III) and Os(IV) chelates is afforded by their interference in the spectrophotometric determination of Ir(IV) with EDTA (MacNevin and Kriege, 1956). Their stereochemistry is unknown.

3. *Cobalt*

Cobalt(II) chelates have been isolated as the pink solids, $H[CoH_2O(H{\cdot}EDTA)]{\cdot}2H_2O$, $Na_2[CoH_2O(EDTA)]{\cdot}H_2O$, and $Co[CoH_2O(EDTA)]{\cdot}$

$2H_2O$ (German Patent, 1952; Astakhov and Verenikin, 1955; Shimi and Higginson, 1958). The complex has an absorption spectrum similar to that of the aquated Co(II) ion (Long *et al.*, 1948; Ichimonji, 1955). Oxygen does not oxidize it, but hydrogen peroxide, sodium hypochlorite, and lead dioxide easily effect the oxidation. Yalman (1961) has examined the kinetics of the oxidation by hydrogen peroxide and shown the intermediate formation of a bridged peroxo complex. The difficulty of dehydration indicates the presence of a coordinated water molecule in the Co(II) chelate. Thiocyanate ion produces a deep violet color, due presumably to $[Co(EDTA)CNS]^{3-}$. Some indication has been found for the coordination of NH_3 and CN^- (Schwarzenbach, 1949).

The Co(III) chelate has an intense violet color with an absorption maximum at 536 mμ. It is stable in the pH range 1–11. Compounds of the type $M[Co(EDTA)]\cdot xH_2O$ have been isolated where M = H, Li, Na, K, Rb, Cs, NH_4, *cis*- and *trans*-$[Co(en)_2(NO_2)_2]$, and *trans*-$[Co(en)_2Cl_2]$. The complexes $Me[Co(EDTA)]_2\cdot yH_2O$ have also been synthesized for Me = Mg, Ca, Ba, Pb, Co (Schwarzenbach, 1949; Dwyer *et al.*, 1955; Mori *et al.*, 1956).

Many workers have attempted to show that $[Co(EDTA)]^-$ contains sexadentate EDTA. Evidence from dehydration and magnetic moments (Brintzinger *et al.*, 1943; Dwyer *et al.*, 1955), infrared spectra (Busch and Bailar, 1953; Mori *et al.*, 1956; Morris and Busch, 1956), and solution studies (Schwarzenbach, 1949) point to the fact that the EDTA is indeed sexadentate. Weakliem and Hoard (1959) have definitely established from a crystal structure determination that the complexes $Rb[Co(EDTA)]\cdot 2H_2O$ and $NH_4[Co(EDTA)]\cdot 2H_2O$ contain sexadentate EDTA in an octahedral configuration.

The $[Co(EDTA)]^-$ ion was the first metal chelate of EDTA to be resolved (Busch and Bailar, 1953). Since then more efficient procedures have been devised of which the most useful seems to be that of Dwyer and Garvan (1960a). The active isomers exhibit different rotations and an increase in $[\alpha]_D$ with increasing concentration (Douglas and Erdman, 1957). Active solutions are stable at room temperature for several months but above 75°C. the racemization is measurable. It has been established that the complex undergoes a very slow pH-independent (in acid solution) racemization, as well as a more rapid base-catalyzed racemization (Dwyer *et al.*, 1955; Busch and Im, 1958; Busch *et al.*, 1961). The pH independence of the rate of racemization in acid solution excludes the formation of five-coordinate intermediates which would be stabilized by increasing the hydrogen ion concentration. Since the activation energy for the racemization is unusually large, some kind of intramolecular rearrangement of the kind proposed by Bailar (1958) seems most probable, (IX) $\rightleftharpoons$ (X) $\rightleftharpoons$ (XI).

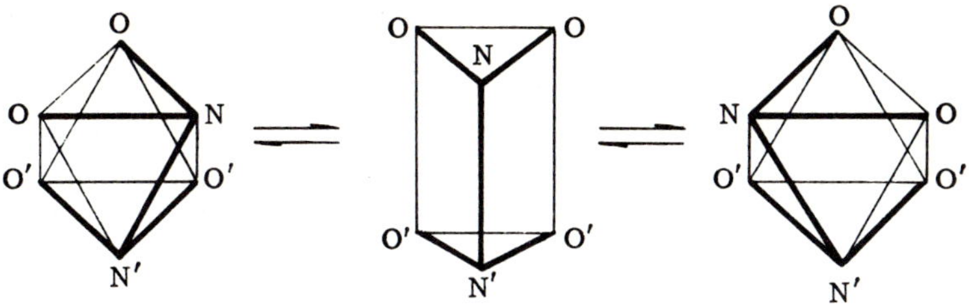

(IX) (X) (XI)

However, the process envisaged by Ray and Dutt (1941, 1943) for the racemization of the tris(biguanide)cobalt(III) ion is a similar mechanism and may occur also (XII) ⇌ (XIII) ⇌ (XIV).

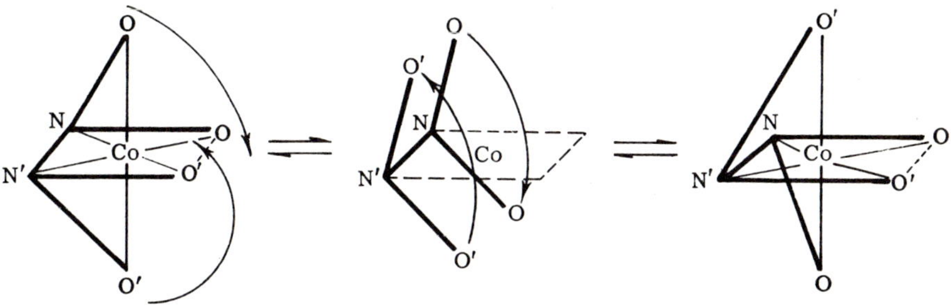

(XII) (XIII) (XIV)

Busch *et al.* (1961) have offered the following explanation as to why base should catalyze the racemization of $[Co(EDTA)]^-$. The structure of $[Co(EDTA)]^-$ as determined by Weakliem and Hoard (1959) would allow the approach of an OH^- on the side of the octahedron occupied by the four oxygen atoms, thereby promoting the transformation into a trigonal biprism with the OH^- projecting into the rectangular face. This symmetrical intermediate is the kind of structure determined for the seven-coordinate Mn(II) chelate of EDTA (VII) (Hoard *et al.*, 1961c).

Busch and his associates (1961) have studied the base hydrolysis of $[Co(EDTA)]^-$ in detail. They are able to explain all the features of this hydrolysis by an S_N2 mechanism in which bond forming is dominant in the transition state. It will be seen below that in other reactions of this type dissociation processes appear to occur.

The electron transfer racemization of $[Co(EDTA)]^-$ has been investigated by Im and Busch (1961a). The racemization occurs through a thermal process:

$$(+)[Co(EDTA)]^- \rightleftharpoons (-)[Co(EDTA)]^-$$

and an electron transfer process:

$$(+)[Co(EDTA)]^{-} + [Co(EDTA)]^{--} \rightleftharpoons [Co(EDTA)]^{--} + (-)[Co(EDTA)]^{-}$$

The most reasonable mechanism for the electron transfer reaction appears to involve direct electron transfer from the reductant to the oxidant, with the original coordination sphere intact in the transition state. The rate constants obtained by Im and Busch by the use of optical activity are compatible with those obtained by the use of isotopic tracers (Adamson and Vorres, 1956).

The optical stability of $[Co(EDTA)]^{-}$ at once suggests its use as a resolving agent. Dwyer and Garvan (1960a) obtained the active isomers of $[Co(EDTA)]^{-}$ using active $[Co(en)_2(NO_2)_2]^{+}$. By reversing the process, optically active $[Co(en)_2(NO_2)_2]^{+}$ can be obtained in good yield (Dwyer *et al.*, 1961). The active $[Co(EDTA)]^{-}$ can also be used to resolve $[Co(en)_2(C_2O_4)]^{+}$ which formerly has been only partially resolved (Werner, 1912; Werner and Bosshart, 1914). The active isomers of $[Co(en)_2(NO_2)_2]^{+}$ and $[Co(en)_2(C_2O_4)]^{+}$ have already been used to resolve over thirty compounds (Dwyer *et al.*, 1961).

A most significant reaction of $[Co(EDTA)]^{-}$ is that with ethylenediamine, in which $[Co(en)_3]^{3+}$ is formed. If active $[Co(EDTA)]^{-}$ is used there is some retention of configuration (Dwyer *et al.*, 1955; Kirschner *et al.*, 1957; Dwyer and Garvan, 1958). Busch and his co-workers (1961) have given a reasonable interpretation of the experimental results. In 50% ethylenediamine $(-)_{5461}[Co(EDTA)]^{-}$ produces 56.5% of $(-)[Co(en)_3]^{3+}$, while the same reaction in anhydrous ethylenediamine produces 63% of the same isomer. Busch assumed that the replacement of the EDTA by ethylenediamine proceeds with no rearrangement and with only terminal groups being replaced, in a purely statistical distribution. Such a process predicts that there should be 67% of $(-)[Co(en)_3]^{3+}$ formed. When allowances have been made for the base-catalyzed racemization of active $[Co(EDTA)]^{-}$ under the conditions of the experiment, Busch calculated that $(-)[Co(en)_3]^{3+}$ should be present to the extent of 57.6% at equilibrium. This value compares much better with the experimental value of 56.5% than would be expected. The apparent success of this work led Busch on to deduce the absolute configuration of the active isomers of $[Co(EDTA)]^{-}$ since the absolute configuration of $[Co(en)_3]^{3+}$ has been definitely established by Saito and his associates (1955, 1957). The absolute configurations of the $[Co(EDTA)]^{-}$ and $[Co(PDTA)]^{-}$ ions are discussed in Chapter 5, Section IV,C.

Kirschner *et al.* (1957) have been able to effect a partial resolution of 1,2-propylenediamine by reacting the racemic base with $(+)_{5461}[Co(EDTA)]^{-}$. Since the isomer $(+)[Co(-)(pn)_3]^{3+}$ is thermodynamically less stable than $(+)[Co(+)(pn)_3]^{3+}$, the $(+)$pn replaces the EDTA more quickly than does the (−)pn. By reversing the procedure using (−)pn and racemic $[Co(EDTA)]^-$ they brought about a partial resolution of $[Co(EDTA)]^-$.

As well as being sexadentate, EDTA is quinquedentate in the complexes $[Co(H{\cdot}EDTA)X]^-$ and $[Co(EDTA)X]^{--}$ in which X = NO_2, Br, Cl, H_2O, OH (Schwarzenbach, 1949; Morris and Busch, 1956; Mori *et al.*, 1956; Shimi and Higginson, 1958; Dwyer and Garvan, 1958). The pK of the acid salts is about 3 and is little affected by the nature of the foreign substituent. This is in excellent agreement with the pK values of the acid salts of other quinquedentate chelates of EDTA. The nitro complexes are stable in solution though boiling produces decomposition. The blue-green chloro and bromo complexes readily lose the halogen ligand in aqueous solution to form violet $[Co(EDTA)]^-$. Ag^+ and Hg^{++} make this conversion almost instantaneous.

The nitro, bromo, and chloro complexes are prepared by oxidation of the Co(II) chelate with nitrous acid, bromine, and chlorine, respectively. The substitution and oxidation are probably effected by the respective attack of $[CoH_2O(EDTA)]^{--}$ with NO_2^+, Br^+, and Cl^+, the mutual attraction aiding the process. The intermediate seven-coordinate Co(II) complex formed would be oxidized by losing an electron to the attacking species, followed by dissociation of the water molecule, leaving the incoming group bonded to the cobalt. The chloro complex can be prepared in better yield by heating $[Co(EDTA)]^-$ with hydrochloric acid. Evidently one carboxyl group is strained and can be detached and protonated in the presence of the strong acid.

The aquo and hydroxo complexes, neither of which has been isolated pure, are in the following equilibria:

$$\underset{\text{blue}}{[Co(EDTA)OH]^{--}} \underset{pH>8}{\overset{pH<8}{\rightleftharpoons}} \underset{\text{violet}}{[CoH_2O(EDTA)]^-} \underset{pH>3}{\overset{pH<3}{\rightleftharpoons}} \underset{\text{violet}}{[CoH_2O(H{\cdot}EDTA)]}$$

$[CoH_2O(EDTA)]^-$ reverts to $[Co(EDTA)]^-$ rapidly on boiling and more slowly on standing. The hydroxo complex is prepared by oxidation of the Co(II) chelate at pH 9 with hydrogen peroxide. It has been suggested that this blue complex is not a hydroxo complex but an isomer of $[Co(EDTA)]^-$ in which the two nitrogen atoms occupy *trans* positions in a distorted octahedron (Mori *et al.*, 1956). The existence of such a complex is very doubtful from stereochemical considerations.

The kinetic studies of Shimi and Higginson (1958) show that $[Co(EDTA)]^-$ forms from $[CoH_2O(H{\cdot}EDTA)]$ and $[Co(EDTA)OH]^{--}$ by an S_N1-type process with the formation of a five-coordinate intermediate. The unbound carboxylate ion in $[CoH_2O(EDTA)]^-$ displaces the ligand water molecule in an S_N2-type reaction. An investigation on the rates of the reactions in Eqs. (11), (12), and (13) has led to the conclusion that the mechanism is most simply described as S_N1 in the rate-determining steps. The observation that metal ions with great affinities for halogens (Ag^+, Hg^{++}, Cd^{++}, Pb^{++}) greatly accelerate the removal of Br^- and Cl^- agrees with the proposal that the breaking of the Co(III)—X bond is rate-determining. Furthermore, the rate is unchanged over the pH range 1–7, indicating that the free carboxylic acid group or carboxylate ion takes no part in the rate-determining reaction (Shimi and Higginson, 1958; Morris and Busch, 1959; Dyke and Higginson, 1960).

$$[Co(EDTA)OH]^{--} \rightleftharpoons [Co(EDTA)]^- + OH^- \quad (11)$$

$$[Co(EDTA)Br]^{--} \rightleftharpoons [Co(EDTA)]^- + Br^- \quad (12)$$

$$[Co(EDTA)Cl]^{--} \rightleftharpoons [Co(EDTA)]^- + Cl^- \quad (13)$$

Busch and Bailar (1953) partially resolved the bromo complex by adsorption on active quartz. Trituration with silver oxide yielded active $[Co(EDTA)]^-$ but the retention of configuration could not be determined. Dwyer and Garvan (1958) resolved the quinquedentate series $[Co(EDTA)X]^{--}$ for X = NO_2, Br, Cl, using active $[Co(en)_2(NO_2)_2]^+$. They showed that the removal of the halogen with Ag^+ or Hg^{++} proceeds with complete retention of configuration. The reaction involved in preparing active $[Co(H{\cdot}EDTA)Cl]^-$ from active $[Co(EDTA)]^-$ and hot hydrochloric acid was shown to occur with at least 75% retention of configuration. The exact figure is sure to be much greater as the reaction is accompanied by decomposition to Co(II). Likewise, the elimination of the nitro group from $[Co(H{\cdot}EDTA)NO_2]^-$ occurs with at least 80% retention of configuration.

Like the sexadentate complex, the chloro and bromo complexes react with ethylenediamine to form $[Co(en)_3]^{3+}$. When the active isomers are used there is 55% retention of configuration, in good agreement with the value for the sexadentate complex and with Busch's theoretical treatment given above. However, the active nitro complex forms $[Co(en)_3]^{3+}$ but with complete loss of configuration. Busch and Cooke (1961) maintain that the absence of rotation in this product strongly supports the mechanism proposed for the replacement of EDTA by ethylenediamine in $[Co(EDTA)]^-$, $[Co(EDTA)Cl]^{--}$, and $[Co(EDTA)Br]^{--}$ (Busch *et al.*, 1961). They believe that the nitro group would block one of the two positions required by the particular isomer of the intermediate, $[Co(en)(EDTA)]^-$, which leads to retention of configuration. This blockage would prevent any retention of configuration from taking place.

These quinquedentate Co(III) chelates offer an excellent opportunity for studying the difference between coordinated and unattached carboxyl groups in the one entity by infrared techniques. Busch and Bailar (1953) first realized this and their work has stimulated similar investigations (Kirschner, 1956; Morris and Busch, 1956; Mori *et al.*, 1956; Sawyer and Paulsen, 1958). In the complexes $[Co(H{\cdot}EDTA)X]^-$ ($X = NO_2$, Br, Cl) the carboxylic acid peak is around 1730 cm.$^{-1}$ and the coordinated carboxyl peak about 1640 cm.$^{-1}$. $[Co(EDTA)]^-$ has only one peak at 1638 cm.$^{-1}$ showing the equivalence of all four carboxyl groups coordinated to the cobalt. The neutral salts of the chloro and nitro complexes, $[Co(EDTA)X]^{--}$, reveal two bands at 1648, 1650 cm.$^{-1}$ (coordinated carboxyl groups) and 1600, 1604 cm.$^{-1}$ (free carboxylate ion), respectively, showing that complexed carboxyl groups and free carboxylate ions in the one coordination compound can be distinguished provided the metal forms strong covalent bonds. However, the limit of resolution of the spectrophotometer may prevent this distinction from being made, especially with metals such as the alkaline earths, copper, and nickel, since with these the carboxyl-metal link is essentially electrostatic with a frequency almost identical with the carboxylate ion frequency (Sen *et al.*, 1955).

The visible absorption spectra of $[Co(H{\cdot}EDTA)X]^-$ ($X = NO_2$, Cl, Br) support the quinquedentate function of $(H{\cdot}EDTA)^{3-}$ and give bands due to the coordination of the nitro, chloro, and bromo groups to cobalt (Mori *et al.*, 1956; Shimura and Tsuchida, 1956).

4. *Rhodium and Iridium*

MacNevin and his associates (1958) were able to show that rhodium(III) chloride is complexed by EDTA but they could not isolate pure compounds. Dwyer and Garvan (1960b) prepared the acid salt $[RhH_2O(H{\cdot}EDTA)]$ from rhodium(III) hydroxide and H_4EDTA in water at 145°C. Debye–Scherrer powder photographs showed that this complex is isomorphous with the quinquedentate complex $[CrH_2O(H{\cdot}EDTA)]$. The infrared spectrum and pH titrations also confirmed the quinquedentate attachment of the EDTA and the coordination of the water molecule.

For the reactions

$$[RhH_2O(H{\cdot}EDTA)] \rightleftharpoons [RhH_2O(EDTA)]^- + H^+$$

and

$$[RhH_2O(EDTA)]^- \rightleftharpoons [Rh(EDTA)OH]^{--} + H^+$$

the pK values are 2.32 and 9.12. However, Dwyer and Garvan (1960b) showed that $K[Cr(EDTA)]{\cdot}2H_2O$ is isomorphous with $K[Rh(EDTA)]{\cdot}2H_2O$, and, since Hoard *et al.* (1961a) have shown $NH_4[Cr(EDTA)]{\cdot}2H_2O$ contains sexadentate EDTA, it can be reasonably assumed that the EDTA becomes sexadentate in $K[Rh(EDTA)]{\cdot}2H_2O$ (cf. Section II,K,1).

The rhodium chelate has been resolved into its optical isomers with active $[Co(en)_2(NO_2)_2]^+$. The active isomers racemize at 96°C. with a half-life of 163 minutes. Absorption spectra of various samples extracted during the racemization showed that apparently little actual decomposition takes place. When one equivalent of alkali is added to optically active K[Rh(EDTA)]·$2H_2O$ the absorption spectrum changes, indicating formation of the hydroxo complex [Rh(EDTA)OH]$^{--}$, and the specific rotation falls to approximately half the original value. [Rh(EDTA)]$^{-}$ is sensitive to ultraviolet light and active solutions can be rendered inactive after 2 hours' exposure to a mercury ultraviolet lamp, but no loss in activity occurs when the crystalline material is irradiated for 6 hours. Although probable, it has not been established that the photo and thermal racemizations have the same mechanism. If the potassium salt of the rhodium chelate is sexadentate it would probably racemize by the mechanism proposed by Bailar (1958) for the cobalt chelate. If the complex exists in solution as the equatorial aquo complex (cf. VI), Bailar's mechanism would convert the (+) equatorial isomer into the (−) polar isomer. Hence the loss of activity by this mechanism would be strictly isomerization and not racemization. Another possibility is that the detached carboxyl group coordinates to form a seven-coördinate structure (cf. VII) which can be symmetrical, causing racemization. Finally, it is also possible that first one of the coordinated carboxyl groups detaches to form a diaquo species which then racemizes by an intramolecular mechanism similar to that proposed by Bailar. Four

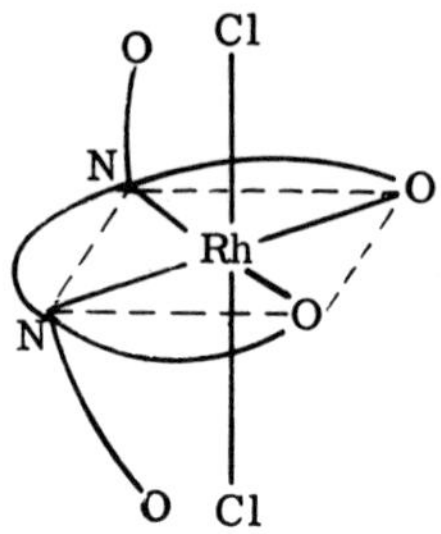

(XV)

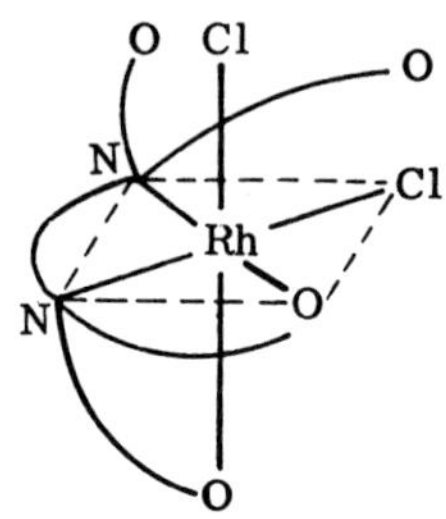

(XVI)

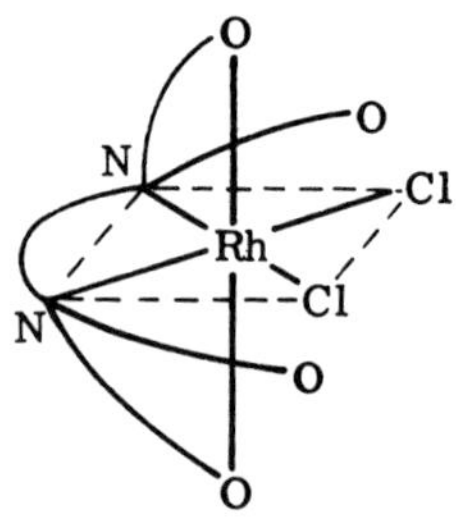

(XVII)

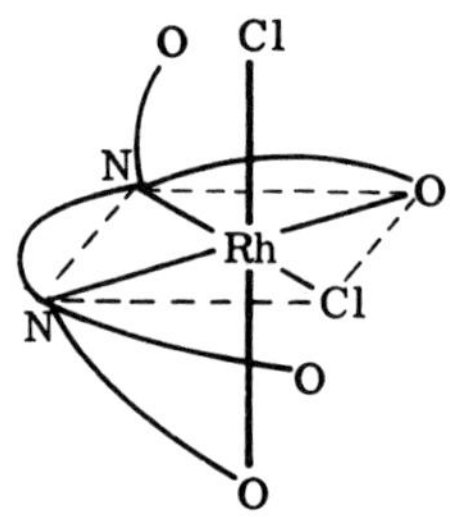

(XVIII)

such disubstituted geometric isomers can be formed of which only types (XVI) and (XVIII) can give a symmetrical intermediate in the proposed intramolecular mechanism. The other disubstituted geometric isomers simply change from one to the other, but with inversion, in the same way as the polar and equatorial isomers mentioned above.

The dichloro and dibromo complexes $K[Rh(H_2EDTA)Cl_2]$ and $K[Rh(H_2EDTA)Br_2]$ have been made from $K[Rh(EDTA)]\cdot 2H_2O$ with concentrated hydrochloric and hydrobromic acids, respectively, at 90°C. The infrared spectra of these complexes show two peaks of comparable intensity in the carbonyl region, corresponding to two free carboxylic acid groups and two coordinated carboxyl groups. The active complexes prepared from active K[Rh(EDTA)] do not undergo photoracemization in aqueous solution, but, in the dark, the halogen atoms slowly dissociate to give the original complex with complete retention of configuration. The four geometric isomers possible for these disubstituted complexes are illustrated in (XV)–(XVIII). All can exist in (+) and (−) forms and (XV) can also exist in a *meso* form (Dwyer and Garvan, 1960b).

MacNevin and Kriege (1954, 1956) have detected the formation of an Ir(IV) chelate from the appearance of a strong absorption band at 313 mμ.

5. *Nickel*

The Ni(II) chelate has been isolated as the acid, sodium, barium, and nickel salts (Brintzinger and Hesse, 1942; Astakhov and Verenikin, 1955; Dwyer and Garvan, 1961). Smith and Hoard (1959) have determined the crystal structure of $H[NiH_2O(H\cdot EDTA)]$ and shown that it contains quinquedentate EDTA with the water molecule coordinated in the octahedral sphere in the equatorial position in which it is *cis* to one nitrogen atom and *trans* to the other. Magnetic moment data (Klemm and Raddatz, 1942), pK values (Cook and Long, 1951, 1958), visible absorption spectra (Klixbull-Jorgensen, 1955), and infrared spectra (Smith and Hoard, 1959; Garvan, 1959) also indicated the above attachment.

The exchange of Ni^{*++} ion with $[NiH_2O(EDTA)]^{--}$ reveals a complicated mechanism that is greatly pH-dependent. No fewer than nine steps were proposed to account for the rate law observed. The mechanism visualized is a series of rapidly established preliminary equilibria between H^+ and $[NiH_2O(EDTA)]^{--}$, followed by a competition between either unimolecular decomposition of the protonated complex ions or bimolecular collision and exchange with Ni^{++} ion (Cook and Long, 1958).

Table IV summarizes the exchange between the complexes listed and $[NiH_2O(EDTA)]^{--}$ at pH 9. The small variation in the rate compared with the ionic charge difference was interpreted by a mechanism involving a direct transfer of nickel atoms resulting from bimolecular collision (Calkins and Hall, 1958).

TABLE IV
THE RATE OF EXCHANGE OF VARIOUS COMPLEXES WITH $[NiH_2O(EDTA)]^{--}$

Complex	Ionic charge difference	Rate (moles liter^{-1} sec.$^{-1}$ $\times 10^5$)
$[Ni(glycinate)_2]^{\circ}$	2	3.8
$[Ni(glutamate)_2]^{--}$	0	1.7
$[Ni(lysinate)_2]^{+?}$	>2	9.9
$[Ni(NH_3)_4]^{++}$	4	8.4
$[Ni(en)_2]^{++}$	4	5.3

The infrared spectrum of $H[NiH_2O(H{\cdot}EDTA)]$ reveals an extremely sharp peak at 3610 cm.$^{-1}$ which must be due to the OH stretch of the coordinated water molecule. The deuterated complex $D[NiD_2O(D{\cdot}EDTA)]$ also has this sharp peak at the expected frequency of 2660 cm.$^{-1}$. The same peak is observed in $H[CuH_2O(H{\cdot}EDTA)]$. Unfortunately, this peak does not always appear even in cases in which the presence of the coordinated water molecule has been definitely established, e.g., $[CrH_2O(H{\cdot}EDTA)]$ and $[RhH_2O(H{\cdot}EDTA)]$. Evidently the different crystal lattices cause hydrogen-bonding in these last two complexes which prevents detection of this sharp peak (Garvan, 1959).

Dwyer and Garvan (1961) made the barium salt of the nickel chelate and showed it to be isomorphous with the corresponding copper chelate. The nickel chelate is paramagnetic (2.98 B.M.) and on heating at 150°C. under reduced pressure only four of the five water molecules could be removed. This suggests, but does not prove, that the EDTA is still quinquedentate in $Ba[NiH_2O(EDTA)]{\cdot}4H_2O$ and $Ba[CuH_2O(EDTA)]{\cdot}4H_2O$. Even though well crystallized salts were obtained with active $[Co(en)_2(NO_2)_2]^+$ the nickel chelate could not be resolved (Dwyer and Garvan, 1961).

6. *Palladium and Platinum*

MacNevin and Kriege (1954, 1955a,b, 1956) demonstrated that Pd(II) forms a stable complex with EDTA which can be used for the spectrophotometric and volumetric determination of the metal. Busch and Bailar (1956) have isolated the following complexes: $[Pt(H_2EDTA)]{\cdot}3H_2O$, $[Pd(H_2EDTA)]{\cdot}H_2O$, $[Pt(H_4EDTA)Cl_2]{\cdot}5H_2O$, and $[Pd(H_4EDTA)Cl_2]{\cdot}5H_2O$. The first two complexes contain quadridentate EDTA and the latter two bidentate EDTA (through two nitrogen atoms). This was deduced from infrared spectra, analyses, equivalent weights, dehydration of the complexes, and pK values of the uncomplexed carboxylic acid groups. Models show that the *meso* form of the quadridentate chelate is strained. The racemic form could not be resolved using strychnine, quinine, and active $[Co(en)_3]^{3+}$.

N. The Actinons

$[Th(H_2O)_2(EDTA)]$ has been isolated as a white, difficultly soluble, crystalline compound (Brintzinger *et al.*, 1943). Both water molecules are believed to be bound to the metal, and the EDTA coordinated through all six positions in a square Archimedean antiprism structure (Bogucki and Martell, 1958). At high pH the complex hydrolyzes to form a binuclear diolate complex, $[Th(EDTA)OH]_2^-$.

The complex $[U(H_2O)_2(EDTA)]$ has been isolated as pale green leaflets (Brintzinger *et al.*, 1943). Its structure is probably similar to the Th(IV) chelate above. The uranyl complex $[UO_2(H_2EDTA)]\cdot H_2O$ has been obtained as a yellow precipitate.

Potentiometric pH titrations and ion-exchange and spectrophotometric studies show that stable 1:1 metal chelates are formed with Pu(III), Pu(IV), Pu(V), and Pu(VI)O_2 (Foreman and Smith, 1957; Moskvin, 1959). Complexes of the types $[Pu(III)_2(EDTA)]^{++}$ and $[Pu(IV)_2(EDTA)]^{4+}$ have also been identified in solution. The plutonyl complex has a higher stability than the uranyl complex. Hydrolysis and polymerization of the uranyl ion at low pH to give ions of charge density too low for stable chelate formation may account for this difference (Foreman and Smith, 1957). Chelate formation has also been reported for Np(V) and Am(III) (Zolotov and Novikov, 1959; Moskvin, 1959).

III. 1,2-Propylenediaminetetraacetic Acid and Its Metal Chelates

A. General Discussion

1,2-Propylenediaminetetraacetic acid (H_4PDTA) offers excellent opportunities for elucidating reactions of metal complexes with polyaminocarboxylic acids because of the presence of the asymmetric carbon atom. Since the acid can be labeled by its optical activity it is extremely useful in carrying out ligand exchange studies with metal complexes and in investigating the principle of stereospecific limitation in complexes containing a single optically active organic molecule. The similarity to H_4EDTA also allows the comparison of rates and mechanisms with those already determined for EDTA metal chelates by isotopic tracer techniques. No doubt the active acid will also find use in biological systems.

In view of the extraordinary potential of H_4PDTA it is surprising to find that in the literature, until very recently, the only references to the acid were in patents (see Dwyer and Garvan, 1959, for references), in which the preparation of the acid and some of the esters and sodium salts are quoted in relation to a number of general preparative reactions for polyaminocarboxylic acids. Dwyer and Garvan (1959) found that the detailed method of Smith *et al.* (1949) for the preparation of H_4EDTA when applied to H_4PDTA gave only a viscous sirup which could not be induced to crystallize. However, the racemic acid can be easily prepared by condensation of racemic 1,2-propylenediamine with sodium chloroacetate at 20°C. The inactive acid is in every way similar to H_4EDTA. Its infrared spectrum shows peaks at 1700 cm.$^{-1}$ (strong) and 1620 cm.$^{-1}$ (weak). The 1700 cm.$^{-1}$ peak is due to carboxylic acid groups. Evidently hydrogen-bonding has decreased the frequency as double bond character is lost for the COOH group. Broad absorption in the higher frequency region (2500–3100 cm.$^{-1}$) can also be ascribed to association of carboxyl groups (Dwyer and Garvan, 1959).

Unlike the racemic acid, the optical forms of H_4PDTA are quite soluble in water and do not crystallize when the reaction mixture is acidified. Special procedures using ion-exchange resins had to be used to obtain the pure optical forms which crystallize from water as the monohydrate with $[\alpha]_D = \pm 47°$. The rotation changes to ±39°, ±37°, ±30°, ±22°, and ±42° in the presence of 1, 2, 3, 4, and 15 equivalents of alkali. The optically active acids (monohydrates) show bands in their infrared spectra at 1727 cm.$^{-1}$ (strong), 1630 cm.$^{-1}$ (weak), and 1570 cm.$^{-1}$ (strong). The carboxylic acid groups show the normal frequency, revealing less effective hydrogen-bonding than in the racemic acid. The 1570 cm.$^{-1}$ band can be assigned to OH bending from the water of crystallization. The OH stretching frequency is also shown by a strong 3470 cm.$^{-1}$ band. The differences in hydrogen-bonding are possibly due to different crystal lattices in the racemic and active acids (Dwyer and Garvan, 1959).

B. Metal Chelates of 1,2-Propylenediaminetetraacetic Acid

1. *Cobalt*

The Co(III) chelate containing PDTA is almost identical in every way with its EDTA analog except that the PDTA complex is more soluble. There seems no room for doubt that PDTA is sexadentate in its Co(III) chelate. The absolute stereospecificity and absolute configuration of $[Co(PDTA)]^-$ are fully discussed in Chapter 5, Section IV,C. For the present it is sufficient to point out that, since there is absolute stereospecificity, resolution of the complex should also effect the resolution of the chelating agent. Dwyer and Garvan (1959) resolved $[Co(PDTA)]^-$ with active $[Co(en)_2(NO_2)_2]^+$ and recovered the optically active acid from the complexes according to Eqs. (14)–(16).

$$[Co(PDTA)]^- + 6CN^- \rightarrow [Co(CN)_6]^{3-} + (PDTA)^{4-} \quad (14)$$

$$2[Co(CN)_6]^{3-} + 2(PDTA)^{4-} + 5Cu^{++} \rightarrow Cu_3[Co(CN)_6]_2 \downarrow + 2[Cu(PDTA)]^{--} \quad (15)$$

$$[Cu(PDTA)]^{--} + H_2S \rightarrow CuS \downarrow + (H_2PDTA)^{--} \quad (16)$$

After filtration of the precipitate in Eq. (15) all extraneous cations were exchanged for H^+ with an ion-exchange resin so that in Eq. (16) H_4PDTA crystallized after removal of the insoluble Cu(II)S. It must be noted here that, in itself, recovery of the pure active acid from the complexes cannot be used as an argument for the stereospecificity found. It is now only too evident that in the past such reasoning has led to many false conclusions.

The optically active ion $[Co(PDTA)]^-$ does not undergo thermal racemization at 100°C. although a slow decomposition takes place (Dwyer and Garvan, 1961; Im and Busch, 1961b). Because of the stereospecificity of the ligand no racemization occurs when equimolar solutions of $(+)_{5461}$-$[Co(-)(PDTA)]^-$ and the Co(II) chelate of $(-)$(PDTA) are mixed. However, complete loss of activity occurs when equivalent amounts of $(+)_{5461}$-$[Co(-)(PDTA)]^-$ and the Co(II) chelate of $(+)$(PDTA) are mixed. The rate law established by Im and Busch (1961b) clearly delineates this system as involving electron transfer with complete retention of configuration for both exchanging species. The electron exchange in the PDTA system can be measured at pH 7, whereas, with the EDTA system, at pH greater than 4 the base-catalyzed racemization of $[Co(EDTA)]^-$ makes electron transfer experiments impracticable (Im and Busch, 1961b). Otherwise the PDTA and EDTA systems of electron transfer reveal a close correspondence.

Irving and Gillard (1961) reacted $(-)_{5461}[Co(+)(PDTA)]^-$ with ethylenediamine and isolated $(+)[Co(en)_3]I_3{\cdot}3H_2O$ in 90% yield with a rotation indicating about 93% retention of configuration. Busch and his associates (1961) maintain that the reaction appears to be surprisingly simple, giving the product directly, without the mediation of detectable concentrations of intermediates. Irving and Gillard (1960) also reacted $(-)_{5461}[Co(+)$-$(PDTA)]^-$ with racemic propylenediamine. They isolated the complex $(+)[Co(+)(pn)_3]I_3$ with $[\alpha]_D = +24.5°$ which agrees excellently with values reported in the literature for the pure compound (see Dwyer *et al.*, 1959). However, the propylenediamine recovered from this complex was reported to have $[\alpha]_D = +4.2°$ for the dihydrochloride in water. To be consistent with the work above this rotation should have been $-4.2°$ since the dihydrochloride of $(+)$propylenediamine has a negative rotation (e.g., see O'Brien and Toole, 1954; Dwyer *et al.*, 1959; Busch and Cooke, 1961).

The quinquedentate complexes $[Co(PDTA)X]^{--}$ have been made for X = Cl, Br, NO_2, H_2O, OH (Dwyer and Garvan, 1961; Swaminathan and

Busch, 1961). Four geometric isomers are possible depending on the position of the foreign substituent with respect to the two nitrogen atoms and the methyl group of PDTA, (XIX) to (XXII). The other four theoretical isomers which maintain the same complex configuration but contain the mirror image of the chelating agent are excluded because of the stereospecificity forced on the complexes, except perhaps in (XXII). Dwyer and Garvan (1961) resolved the chloro and nitro complexes with active $[Co(en)_2(NO_2)_2]^+$. Both complexes were transformed with complete retention of configuration to the pure sexadentate complex. Similarly, the blue chloro complex could be prepared in almost 100% yield from the sexadentate complex with complete retention of configuration.

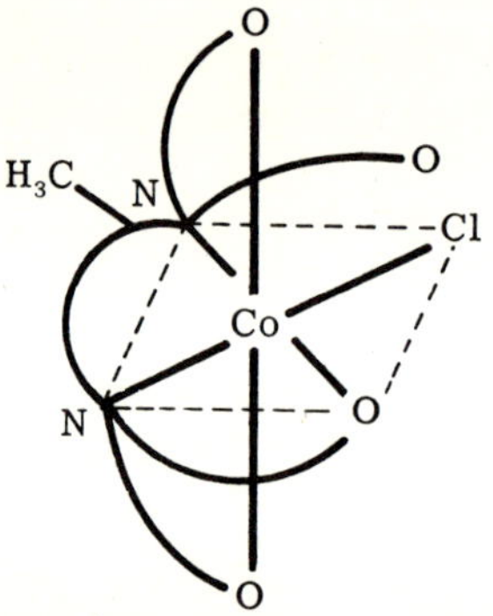

(XIX)

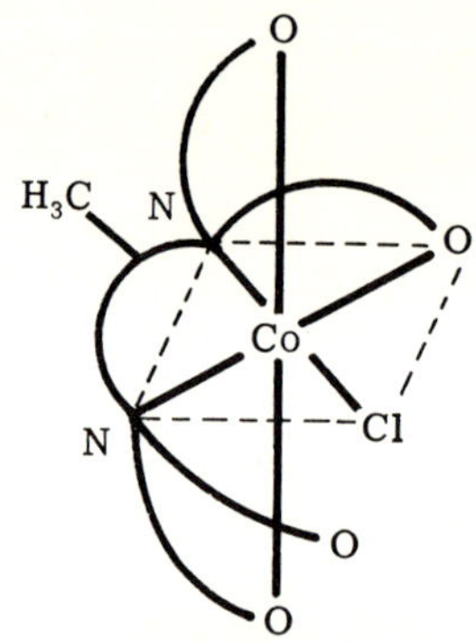

(XX)

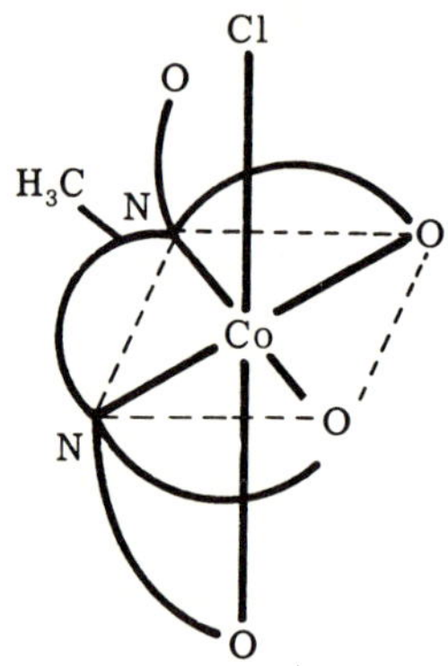

(XXI)

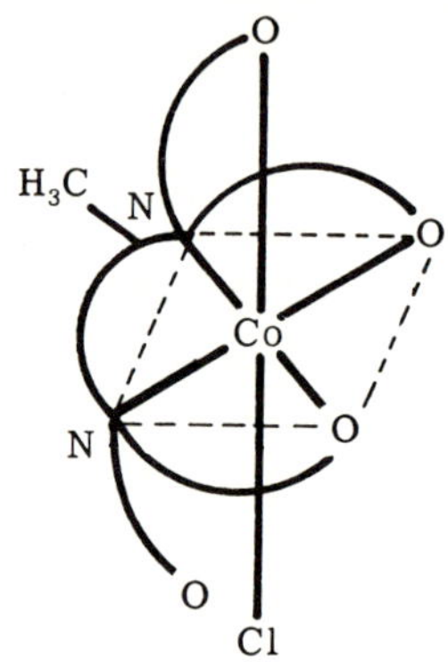

(XXII)

2. *Rhodium and Chromium*

The infrared spectra of $[RhH_2O(H{\cdot}PDTA)]{\cdot}H_2O$ and $[CrH_2O(H{\cdot}PDTA)]{\cdot}H_2O$ show these complexes to be similar to their EDTA analogs.

However, the chromium chelate could be resolved into its optical isomers which could not be racemized even on boiling (Dwyer and Garvan, 1961). This is in marked contrast to the EDTA complex which could not be resolved under a variety of conditions and with a range of resolving agents (Dwyer and Garvan, 1960b). The optical stability of the Cr(III)-PDTA complex arises from the stereospecificity forced on it by the optically active chelating agent.

The Rh(III)-PDTA complex has proved to be one of the most interesting complexes isolated with polyaminocarboxylic acids. The active forms of the complex show no change in rotation when their solutions are heated to 96°C. for several hours. This is to be expected because of the stereospecificity of the chelating agent. However, when a solution of $(-)_{5461}[RhH_2O\text{-}(-)(H\cdot PDTA)]\cdot H_2O$ ($[\alpha]_{5461} = -158°$) is exposed to an ultraviolet lamp for 2 hours it mutarotates until $[\alpha]_{5461} = -50°$, and further exposure does not change the rotation. When this solution is put in the dark it regains its initial rotation of $[\alpha]_{5461} = -158°$ in about 3 days. Likewise the barium salt of the rhodium chelate mutarotates from $[\alpha]_{5461} = -168°$ to $-55°$ on exposure to light, and regains its initial rotation in the dark. This cycle of rotational changes seems to be capable of many repetitions. As with the Rh(III)-EDTA complex, the absorption band at 348 mμ appears to be responsible for the rotational change (Dwyer and Garvan, 1961).

The possible mechanism of the photoracemization of the Rh(III)-EDTA complex has been discussed in Section II. The mutarotation of the PDTA complex may arise in a similar manner from photo-induced breaking of a carboxylate-rhodium bond, followed by aquation. The diaquo isomer(s) would not be inactive, even if the asymmetry about the metal were lost, because of the optical activity of the chelating agent. The absorption spectra of the exposed and unexposed solutions were almost identical, with absorption maxima at 295 and 348 mμ. This requires the not unreasonable postulate that the monoaquo and diaquo species have almost the same absorption spectra. Dilute potassium permanganate is not reduced by the exposed solutions, nor does it affect the rotation of the complex. The rotational changes are therefore not due to photoreduction, followed by slow oxidation in the dark. The solution of this phenomenon should prove to be most intriguing.

3. *Copper*

As a practical means of resolving H_4PDTA the method described by Eqs. (14)–(16) is too long. A more direct method is to resolve the Cu(II) chelate and remove the copper as Cu(II)S thereby eliminating Eqs. (14) and (15). This process was carried out by Dwyer and Garvan (1961) and gave the optical forms of the acid in good yield. The fact that the Cu(II)-

PDTA complex can be resolved again contrasts this complex with the EDTA complex which has so far defied all methods of resolution.

4. *Ligand Exchange Studies*

If equivalent amounts of the metal chelate containing one optical form of PDTA and the free optical antipode are mixed, the rate of loss of optical activity measures directly the rate of ligand exchange:

$$(-)[M(+)(PDTA)]^{n-} + (-)(PDTA) \rightleftharpoons (+)[M(-)(PDTA)]^{n-} + (+)(PDTA)$$

The final value of the rotation will be zero corresponding to a random distribution. Table V shows the times to reach zero rotation (Bosnich *et al.*, 1960). The rate of exchange is clearly a function of pH, being faster in acid solutions. This means that adding protons to $(PDTA)^{4-}$ to reduce its net charge makes it a weaker chelating agent. The preliminary studies of Bosnich *et al.* (1960) indicate that at pH 6 the reaction is first order with respect to both metal chelate and free ligand. Pearson (1961) has discussed these results and offered several interpretations.

TABLE V

EXCHANGE OF METAL CHELATES OF (+)PDTA IN WATER AT 20°C.

Metal chelate[a]	pH	Time for zero rotation
$[Ca(H_2PDTA)]$	7.30	<1 minute
$[Mn(H_2PDTA)]$	3.00	<1 minute
$[Mn(H_2PDTA)]$	6.30	5 minutes
$[Co(H_2PDTA)]$	2.90	<1 minute
$[Co(H_2PDTA)]$	6.30	7 days
$[Ni(H_2PDTA)]$	0.70	30 minutes
$[Ni(H_2PDTA)]$	2.90; 6.40	No exchange in 3 days
$[Cu(H_2PDTA)]$	2.90	<1 minute
$[Cu(H_2PDTA)]$	6.40	16 hours
$[Zn(H_2PDTA)]$	2.90	<1 minute
$[Zn(H_2PDTA)]$	6.32	3 hours
$[Cd(H_2PDTA)]$	2.90	<1 minute
$[Cd(H_2PDTA)]$	6.30	1 hour
$[Al(H\cdot PDTA)]$	2.90	20 hours
$[Y(H\cdot PDTA)]$	2.90	<1 minute
$[Fe(H\cdot PDTA)]$	0.7	7 hours
$[Fe(H\cdot PDTA)]$	6.5	No exchange in 3 days

[a] The stereochemistry of these metal chelates cannot be inferred from the formulas listed.

Bosnich (1962) has studied in detail the exchange of the Cd(II) chelate with PDTA by polarimetric means. Within the pH range 4.75–7.00 the exchange of $[CdH_2O(+)(H \cdot PDTA)]^-$ with $(-)(H_2PDTA)^{--}$ proceeds by three different reaction paths. The molecularity of each path was determined by the kinetic order of the reactions, the pH dependencies, and the use of $(H_2EDTA)^{--}$ as a substituting ligand. At pH 7.00 the reaction is bimolecular and proceeds by the collision of $[CdH_2O(+)(PDTA)]^{--}$ with $(-)(H \cdot PDTA)^{3-}$. Below pH 7.00 the bimolecular reaction component results from the collision of $[CdH_2O(+)(PDTA)]^{--}$ with $(-)(H_2PDTA)^{--}$. The rate-determining step of the unimolecular component, which occurs simultaneously, is the slow dissociation of the Cd^{++} ion from the monoprotonated quinquedentate complex $[CdH_2O(+)(H \cdot PDTA)]^-$.

IV. *Cis*- and *Trans*-1,2-cyclohexanediaminetetraacetic Acids

Trans-1,2-cyclohexanediaminetetraacetic acid was first synthesized by Schwarzenbach and Ackermann (1949), and is now available commercially under the trade name of "Chel 600." The *cis* acid has only recently been prepared by Kroll and Gordon (1960). The pK values are as follows: *trans*-H_4CDTA, $pK_1 = 2.43$, $pK_2 = 3.52$, $pK_3 = 6.12$, $pK_4 = 11.70$; *cis*-H_4CDTA, $pK_1 = 2.44$, $pK_2 = 3.50$, $pK_3 = 5.21$, $pK_4 = 10.26$.

Stability constants of metal chelates are generally in the order *trans*-CDTA > EDTA > *cis*-CDTA (Bjerrum *et al.*, 1957; Holleck and Liebold, 1957; Bond and Jones, 1959; Kroll and Gordon, 1961). The higher stability of the metal chelates of *trans*-CDTA over those of EDTA is explained by the fact that, during chelation, the carbon chain between the nitrogen atoms in EDTA has to be rotated to bring the nitrogen atoms into a favorable position for chelation. In *trans*-CDTA there is a restriction on the rotation of the carbon chain, but, since the ligand nitrogen atoms are placed very close to each other, very little orientation is necessary for chelation to occur (Bond and Jones, 1959). A model of an octahedral sexadentate chelate of *trans*-CDTA shows that the plane of the cyclohexane ring is roughly in the plane of the two nitrogen atoms and the equatorial oxygen atoms from the acetate groups, where it is well away from the rest of the molecule. However, a model of the same chelate with *cis*-CDTA shows that the plane of the cyclohexane ring is oblique to the plane containing the nitrogen and oxygen atoms, where it interferes with the acetate groups coordinated to the metal ion (Kroll and Gordon, 1961). Further information on *cis*-CDTA is not yet available. However, it should be observed that, in contrast to the *trans* acid (XXIII), the *cis* acid (XXIV) cannot be resolved. The remainder of the discussion will be confined to the *trans* acid for which the symbol H_4CDTA will be used.

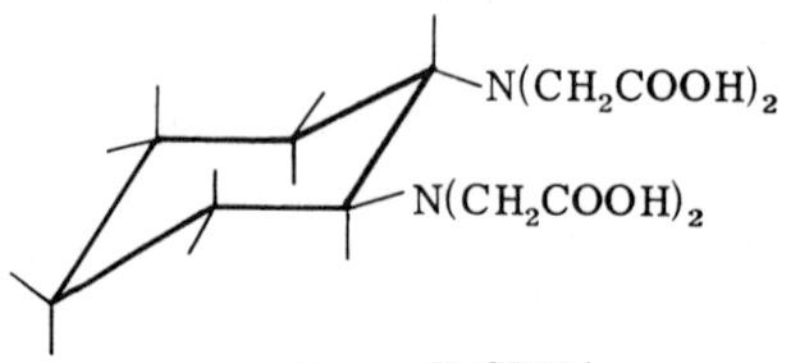

trans-H_4CDTA

(XXIII)

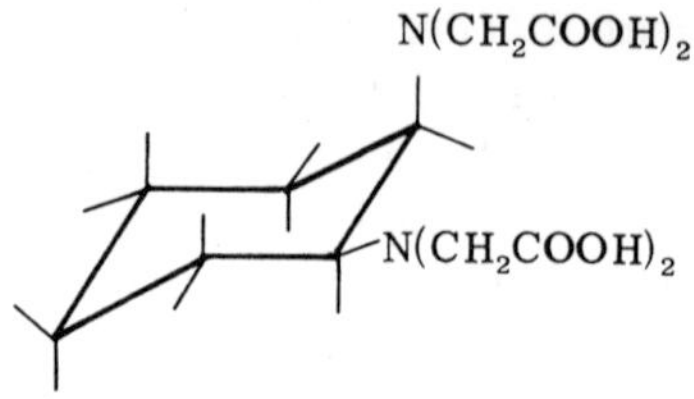

cis-H_4CDTA

(XXIV)

Very few metal chelates of *trans*-CDTA have been isolated but it has been used in analysis for metals (Sir and Pribil, 1955; Pribil, 1955; Goetz and Debrecht, 1955) and in polarographic investigations (Pribil *et al.*, 1953; Eckardt and Holleck, 1955; Matyska *et al.*, 1955). Some lanthanon chelates have been studied (Holleck and Eckardt, 1954; Holleck and Hartinger, 1954; Holleck and Liebold, 1957). Their stereochemistry is thought to be similar to that of the EDTA chelates (Moeller and Horwitz, 1959). The Th(IV) chelate is also believed to be similar to its EDTA analog (Bogucki and Martell, 1958).

Ferrone (1957) attempted the resolution of H_4CDTA without success using optically active phenylethylamine, camphorsulfonic acid, and quinine. He also tried to prepare the optically active acid by synthesis from active *trans*-1,2-cyclohexanediamine but the acid would not separate on acidification. Dwyer and Garvan (1961) also failed to resolve the acid directly with cinchonine, but they were able to effect the resolution by preparing the Co(III) chelate, resolving it with active $[Co(en)_2(NO_2)_2]^+$ and recovering the active acid from the complex in the same way as for the optical forms of H_4PDTA, Eqs. (14)–(16). Unlike the racemic form of the acid, the active isomers are quite soluble in water, thereby explaining Ferrone's failure to precipitate the acid on acidification of his reaction mixture. To obtain the acid a procedure similar to that used by Dwyer and Garvan (1959) for H_4PDTA would have to be employed. The anhydrous active acids have $[\alpha]_D = \pm 53°$.

Dwyer and Garvan (1961) showed that there is complete stereospecificity in the Co(III) chelate. The isomers isolated were $(+)_{5461}[Co(-)(CDTA)]^-$ and $(-)_{5461}[Co(+)(CDTA)]^-$. Models show that it is impossible to obtain the isomers $(+)_{5461}[Co(+)(CDTA)]^-$ and $(-)_{5461}[Co(-)(CDTA)]^-$ because of the fixed stereochemistry of the cyclohexane ring. Failure to racemize solutions of the optically active complexes at 100°C. and on activated charcoal confirmed the inferences made from the models (cf. Chapter 5, Section IV,C). When active samples of the sexadentate Co(III) chelate were heated with hydrochloric acid a blue solution resulted and the rotation changed from $[\alpha]_{5461} = -1100°$ to $+600°$ showing that the chloro complex was formed, but it could not be isolated as the halogen immediately dissociated in the absence of strong acid. Evidently the rigid stereochemistry of the CDTA does not favor quinquedentate chelation.

References

Achard, J. C. (1955). *Compt. rend. acad. sci.* **241,** 800.

Adamson, A. W., and Vorres, K. S. (1956). *J. Inorg. & Nuclear Chem.* **3,** 206.

Ahrens, L. H. (1952). *Geochim. et Cosmochim. Acta* **2,** 155.

Aiken, J. K. (1956). *Chem. & Ind. (London)* p. 1334.

Anderegg, G., Flaschka, H., Sallman, R., and Schwarzenbach, G. (1954). *Helv. Chim. Acta* **37,** 113.

Astakhov, K. V., and Fomenko, M. G. (1957). *Zhur. Fiz. Khim.* **31,** 2110.

Astakhov, K. V., and Verenikin, V. B. (1955). *Chem. Abstr.* **49,** 2929.

Bailar, J. C. (1958). *J. Inorg. & Nuclear Chem.* **8,** 165.

Basolo, F. (1953). *Chem. Revs.* **52,** 459.

Basolo, F., and Pearson, R. G. (1958). "Mechanisms in Inorganic Reactions," pp. 37–41, 157. Wiley, New York.

Beck, M. T., and Gorog, S. (1959). *Magyar Tudományos Akad. Kém. Tudományok Osztályának Közleményei* **12,** 265.

Belcher, R., Gibbons, D., and West, T. S. (1955). *Anal. Chim. Acta* **12,** 107.

Bennett, M. C., and Schmidt, N. O. (1955). *Trans. Faraday Soc.* **51,** 1412.

Betts, R. H., and Dahlinger, O. F. (1959). *Can. J. Chem.* **37,** 91.

Bjerrum, J., Schwarzenbach, G., and Sillen, L. G. (1957). "Stability Constants," Part I. Spec. Publ. No. 6, Chemical Society, London.

Blaedel, W. J., and Knight, H. T. (1954). *Anal. Chem.* **26,** 741.

Blumer, M., and Kolthoff, I. M. (1952). *Experientia* **8,** 138.

Bogucki, R. F., and Martell, A. E. (1958). *J. Am. Chem. Soc.* **80,** 4170.

Bond, J., and Jones, T. I. (1959). *Trans. Faraday Soc.* **55,** 1310.

Bosnich, B. (1962). Ph.D. Thesis, Australian National University, Canberra, A.C.T., Australia.

Bosnich, B., Dwyer, F. P., and Sargeson, A. M. (1960). *Nature* **186,** 966.

Brintzinger, H., and Hesse, G. (1942). *Z. anorg. u. allgem. Chem.* **249,** 113.

Brintzinger, H., and Munkelt, S. (1948). *Z. anorg. Chem.* **256,** 65.

Brintzinger, H., Thiele, H., and Muller, U. (1943). *Z. anorg. u. allgem. Chem.* **251,** 285.

British Patent (1954). 719,901.

British Patent (1955a). 723,317.

British Patent (1955b). 727,465.

British Patent (1955c). 727,482-3.

Brunisholz, G. (1955). *Helv. Chim. Acta* **38,** 455, 1654.

Brunisholz, G. (1957a). *Chimia (Switz.)* **11,** 97.

Brunisholz, G. (1957b). *Chimia (Switz.)* **11,** 363.

Busch, D. H., and Bailar, J. C. (1953). *J. Am. Chem. Soc.* **75,** 4574.

Busch, D. H., and Bailar, J. C. (1956). *J. Am. Chem. Soc.* **78,** 716.

Busch, D. H., and Cooke, D. W. (1961). *J. Inorg. & Nuclear Chem.* **23,** 145.

Busch, D. H., and Im, Y. A. (1958). *Nature* **182,** 1368.

Busch, D. H., Cooke, D. W., Swaminathan, K., and Im, Y. A. (1961). *In* "Advances in the Chemistry of the Coordination Compounds" (S. Kirschner, ed.), p. 139. Macmillan, New York.

Bushra, E., and Johnson, C. H. (1939). *J. Chem. Soc.* p. 1937.

Calkins, R. C., and Hall, N. F. (1958). *J. Am. Chem. Soc.* **80,** 5028.
Calvin, M., and Bailes, R. (1946). *J. Am. Chem. Soc.* **68,** 949.
Care, R. A., and Staveley, L. A. K. (1956). *J. Chem. Soc.* p. 4571.
Carini, F. F., and Martell, A. E. (1954). *J. Am. Chem. Soc.* **76,** 2153.
Cellini, R. F., and Valiente, E. A. (1955). *Chem. Abstr.* **49,** 8030.
Chaberek, S., and Martell, A. E. (1952). *J. Am. Chem. Soc.* **74,** 6228.
Chapman, D. (1955). *J. Chem. Soc.* p. 1766.
Charles, R. G. (1954). *J. Am. Chem. Soc.* **76,** 5854.
Charles, R. G. (1956). *J. Am. Chem. Soc.* **78,** 3946.
Cook, C. M., and Long, F. A. (1951). *J. Am. Chem. Soc.* **73,** 4119.
Cook, C. M., and Long, F. A. (1958). *J. Am. Chem. Soc.* **80,** 33.
Douglas, B. E., and Erdman, W. C. (1957). *J. Am. Chem. Soc.* **79,** 3012.
Dumas, J., and Stas, J. S. (1840). *Ann.* **35,** 129.
Dwyer, F. P., and Garvan, F. L. (1958). *J. Am. Chem. Soc.* **80,** 4480.
Dwyer, F. P., and Garvan, F. L. (1959). *J. Am. Chem. Soc.* **81,** 2955.
Dwyer, F. P., and Garvan, F. L. (1960a). *Inorg. Syntheses* **6,** 192.
Dwyer, F. P., and Garvan, F. L. (1960b). *J. Am. Chem. Soc.* **82,** 4823.
Dwyer, F. P., and Garvan, F. L. (1961). *J. Am. Chem. Soc.* **83,** 2610.
Dwyer, F. P., Gyarfas, E. C., and Mellor, D. P. (1955). *J. Phys. Chem.* **59,** 296.
Dwyer, F. P., Garvan, F. L., and Shulman, A. (1959). *J. Am. Chem. Soc.* **81,** 290.
Dwyer, F. P., Reid, I. K., and Garvan, F. L. (1961). *J. Am. Chem. Soc.* **83,** 1285.
Dyke, R., and Higginson, W. C. E. (1960). *J. Chem. Soc.* p. 1998.
Eckardt, D., and Holleck, L. (1955). *Z. Elektrochem.* **59,** 202.
Ferrone, B. A. (1957). Ph.D. Thesis, University of Illinois, Urbana, Illinois.
Foreman, J. K., and Smith, T. D. (1957). *J. Chem. Soc.* pp. 1752, 1758.
Fuger, J. (1957). *Bull. soc. chim. Belges* **66,** 151.
Furlani, C., Morpurgo, G., and Sartori, G. (1960). *Z. anorg. u. allgem. Chem.* **303,** 1.
Garvan, F. L. (1959). Ph.D. Thesis, University of Sydney, Sydney, Australia.
Geigy Co. (1955). "Sequestrol," p. 8. Technical Bulletin, Manchester, England.
German Patent (1952). 832,889.
Goetz, C. A., and Debbrecht, F. H. (1955). *Anal. Chem.* **27,** 1972.
Goffart, J., Michel, G., and Duychaerts, G. (1953). *Anal. Chim. Acta* **9,** 184.
Hamm, R. E. (1953). *J. Am. Chem. Soc.* **75,** 5670.
Hamm, R. E., and Davis, R. E. (1953). *J. Am. Chem. Soc.* **75,** 3085.
Hill-Cottingham, D. G. (1955). *Nature* **175,** 347.
Hoard, J. L., Smith, G. S., and Lind, M. (1961a). *In* "Advances in the Chemistry of the Coordination Compounds" (S. Kirschner, ed.), p. 296. Macmillan, New York.
Hoard, J. L., Lind, M., and Silverton, J. V. (1961b). *J. Am. Chem. Soc.* **83,** 2770.
Hoard, J. L., Pedersen, B., Richards, S., and Silverton, J. V. (1961c). *J. Am. Chem. Soc.* **83,** 3533.
Holleck, L., and Eckardt, D. (1954). *Z. Naturforsch.* **9a,** 347; **9b,** 274.
Holleck, L., and Hartinger, L. (1954). *Angew. Chem.* **66,** 586.
Holleck, L., and Liebold, G. (1957). *Naturwissenschaften* **22,** 582.
Hynes, W. A., Yanowski, L. K., and Ransford, J. E. (1950). *Mikrochemie ver. Mikrochim. Acta* **35,** 160.
Ichimonji, A. (1955). *Osaka Daigaku Igaku Zassi* **7,** 389.
Im, Y. A., and Busch, D. H. (1961a). *J. Am. Chem. Soc.* **83,** 3357.
Im, Y. A., and Busch, D. H. (1961b). *J. Am. Chem. Soc.* **83,** 3362.
Intorre, B. I., and Martell, A. E. (1960). *J. Am. Chem. Soc.* **82,** 358.
Irving, H., and Gillard, R. D. (1960). *J. Chem. Soc.* p. 5266.

Irving, H., and Gillard, R. D. (1961). *J. Chem. Soc.* p. 2249.
Irving, H., Shelton, R., and Evans, R. (1958). *J. Chem. Soc.* p. 3541.
Iya, V. K., and Loriers, V. (1953). *Compt. rend. acad sci.* **237,** 1413.
Jardin, C. (1959). *Méd trop.* **19,** 703.
Johnson, R. D., and Callis, C. F. (1956). *In* "Chemistry of the Coordination Compounds" (J. C. Bailar, ed.), p. 779. Reinhold, New York.
Jones, M. M. (1955). *Science* **121,** 371.
Jones, S. S., and Long, F. A. (1952). *J. Phys. Chem.* **56,** 25.
Keller, R. N., and Parry, R. W. (1956). *In* "Chemistry of the Coordination Compounds" (J. C. Bailar, ed.), pp. 176–177. Reinhold, New York.
Kirby, R. E., and Freiser, H. (1961a). *In* "Advances in the Chemistry of the Coordination Compounds" (S. Kirschner, ed.), p. 444. Macmillan, New York.
Kirby, R. E., and Freiser, H. (1961b). *J. Phys. Chem.* **65,** 191.
Kirschner, S. (1956). *J. Am. Chem. Soc.* **78,** 2372.
Kirschner, S., Wei, Y. K., and Bailar, J. C. (1957). *J. Am. Chem. Soc.* **79,** 5877.
Klemm, W. (1944). *Z. anorg. Chem.* **252,** 225.
Klemm, W., and Raddatz, K. H. (1942). *Z. anorg. u. allgem. Chem.* **250,** 204.
Klixbull-Jorgensen, C. (1955). *Acta Chem. Scand.* **9,** 1362.
Klygin, A. E., Smirnova, I. D., and Nikol'skaya, N. A. (1959). *Zhur. Neorg. Khim.* **4,** 2766.
Kroll, H., and Gordon, M. (1960). *Ann. N. Y. Acad. Sci.* **88,** 341.
Kroll, H., and Gordon, M. (1961). *Federation Proc.* **20,** 51.
Long, F. A. (1941). *J. Am. Chem. Soc.* **63,** 1353.
Long, F. A., Jones, S. S., and Burke, M. (1948). *Brookhaven Conf. Rept.* No. BNL-C-8.
MacNevin, W. M., and Kriege, O. H. (1954). *Anal. Chem.* **26,** 1768.
MacNevin, W. M., and Kriege, O. H. (1955a). *J. Am. Chem. Soc.* **77,** 6151.
MacNevin, W. M., and Kriege, O. H. (1955b). *Anal. Chem.* **27,** 535.
MacNevin, W. M., and Kriege, O. H. (1956). *Anal. Chem.* **28,** 16.
MacNevin, W. M., McBride, H. D., and Hakkila, E. A. (1958). *Chem. & Ind. (London)* p. 101.
Mann, F. G. (1927). *J. Chem. Soc.* p. 1224.
Mann, F. G. (1928). *J. Chem. Soc.* p. 890.
Margerum, D. W. (1959). *J. Phys. Chem.* **63,** 336.
Marsh, J. K. (1955). *J. Chem. Soc.* p. 451.
Martell, A. E. (1952). *J. Chem. Educ.* **29,** 270.
Martell, A. E. (1956). *Rec. trav. chim.* **75,** 781.
Martell, A. E. (1961). *Federation Proc.* **20,** 35.
Martell, A. E., and Bersworth, F. C. (1950). *J. Org. Chem.* **15,** 46.
Martell, A. E., and Calvin, M. (1952). "Chemistry of the Metal Chelate Compounds," pp. 146–148. Prentice-Hall, Englewood Cliffs, New Jersey.
Martell, A. E., Plumb, R. C., and Bersworth, F. C. (1949). "The Properties and Uses of Ethylenediaminetetraacetic Acid." Versenes, Inc., Massachusetts.
Matyska, B., Dolezal, J., and Roubalova, D. (1955). *Chem. listy* **49,** 1012.
Meinhold, T. F., and Kremers, H. E. (1957). *Chem. Processing* **20,** No. 3, 12–13, 18–19.
Moeller, T., and Brantley, J. C. (1950). *J. Am. Chem. Soc.* **72,** 5447.
Moeller, T., and Graham, E. H. (1957). *J. Inorg. & Nuclear Chem.* **5,** 53.
Moeller, T., and Horwitz, E. P. (1959). *J. Inorg. & Nuclear Chem.* **12,** 49.
Moeller, T., Moss, F. A. J., and Marshall, R. H. (1955). *J. Am. Chem. Soc.* **77,** 3182.
Morgan, L. O., and Justus, N. L. (1956). *J. Am. Chem. Soc.* **78,** 38.

Mori, M., Shibata, M., Kyuno, E., and Nakajima, H. (1956). *Bull. Chem. Soc. Japan* **29,** 887.

Morris, M. L., and Busch, D. H. (1956). *J. Am. Chem. Soc.* **78,** 5178.

Morris, M. L., and Busch, D. H. (1959). *J. Phys. Chem.* **63,** 340.

Moskvin, A. I. (1959). *Radiokhimiya* **1,** 430.

O'Brien, T. D., and Toole, R. D. (1954). *J. Am. Chem. Soc.* **76,** 6009.

Pearson, R. G. (1961). *J. Chem. Educ.* **38,** 164.

Pecsok, R. L. (1952). *J. Chem. Educ.* **29,** 597.

Pecsok, R. L., and Maverick, E. F. (1954). *J. Am. Chem. Soc.* **76,** 358.

Pfeiffer, P., and Offermann, W. (1942). *Ber.* **75B,** 1.

Pfeiffer, P., and Simons, H. (1943). *Ber.* **76B,** 847.

Pribil, R. (1950). *Chimia (Switz.)* **4,** 160.

Pribil, R. (1955). *Collection Czechoslov. Chem. Communs.* **20,** 162.

Pribil, R., and Hornychova, E. (1949). *Collection Czechoslov. Chem. Communs.* **14,** 320, 626.

Pribil, R., and Hornychova, E. (1950). *Collection Czechoslov. Chem. Communs.* **15,** 456.

Pribil, R., Roubal, Z., and Svatek, E. (1953). *Collection Czechoslov. Chem. Communs.* **18,** 43.

Ray, P., and Dutt, N. K. (1941). *J. Indian Chem. Soc.* **18,** 289.

Ray, P., and Dutt, N. K. (1943). *J. Indian Chem. Soc.* **20,** 1943.

Reynolds, W. L., Liu, N., and Mickus, J. (1961). *J. Am. Chem. Soc.* **83,** 1078.

Ringbom, A., and Linko, E. (1953). *Anal. Chim. Acta* **9,** 80.

Ringbom, A., Sutonen, S., and Skrifvars, B. (1957). *Acta Chem. Scand.* **11,** 551.

Rossotti, F. J. C. (1958). *Discussions Faraday Soc.* **No. 26,** 190.

Rossotti, F. J. C. (1960). *In* "Modern Coordination Chemistry" (J. Lewis and R. G. Wilkins, eds.), pp. 1–67. Wiley (Interscience), New York.

Saito, K., and Terrey, H. (1956). *J. Chem. Soc.* p. 4701.

Saito, Y., Nakatsu, K., Shiro, M., and Kuroya, H. (1955). *Acta. Cryst.* **8,** 729.

Saito, Y., Nakatsu, K., Shiro, M., and Kuroya, H. (1957). *Bull. Chem. Soc. Japan* **30,** 795.

Sajo, I. (1958). *Acta Chim. Acad. Sci. Hung.* **16,** 115.

Sawyer, D. T., and Paulsen, P. J. (1958). *J. Am. Chem. Soc.* **80,** 1597.

Schmid, R. W., and Reilley, C. N. (1958). *J. Am. Chem. Soc.* **80,** 2101.

Schwarzenbach, G. (1949). *Helv. Chim. Acta* **32,** 839.

Schwarzenbach, G. (1952). *Helv. Chim. Acta.* **35,** 2344.

Schwarzenbach, G., and Ackermann, H. (1947). *Helv. Chim. Acta* **30,** 1798.

Schwarzenbach, G., and Ackermann, H. (1948). *Helv. Chim. Acta* **31,** 1029.

Schwarzenbach, G., and Ackermann, H. (1949). *Helv. Chim. Acta* **32,** 1682.

Schwarzenbach, G., and Anderegg, G. (1955). *Z. anorg. u. allgem. Chem.* **282,** 286.

Schwarzenbach, G., and Biedermann, W. (1948). *Helv. Chim. Acta* **31,** 459.

Schwarzenbach, G., and Gut, R. (1956). *Helv. Chim. Acta* **39,** 1589.

Schwarzenbach, G., and Heller, H. (1951). *Helv. Chim. Acta* **34,** 576.

Schwarzenbach, G., and Sandera, J. (1953). *Helv. Chim. Acta* **36,** 1089.

Schwarzenbach, G., Kampitsch, E., and Steiner, R. (1945). *Helv. Chim. Acta* **28,** 828.

Schwarzenbach, G., Anderegg, G., and Sallman, R. (1952). *Helv. Chim. Acta* **35,** 1785.

Sen, D. N., Mizushima, S., Curran, C., and Quagliano, J. V. (1955). *J. Am. Chem. Soc.* **77,** 211.

Shimi, I. A. W., and Higginson, W. C. E. (1958). *J. Chem. Soc.* p. 260.

Shimura, Y., and Tsuchida, R. (1956). *Bull. Chem. Soc. Japan* **29,** 643.

Sidgwick, N. V. (1950). "The Chemical Elements and Their Compounds," pp. 259, 260, 285, 286, 799. Oxford Univ. Press (Clarendon), London and New York.
Sir, Z., and Pribil, R. (1955). *Collection Czechoslov. Chem. Communs.* **20,** 871.
Smith, G. S., and Hoard, J. L. (1959). *J. Am. Chem. Soc.* **81,** 556.
Smith, R., Bullock, J. L., Bersworth, F. C., and Martell, A. E. (1949). *J. Org. Chem.* **14,** 355.
Smith, T. D. (1961). *J. Chem. Soc.* p. 2554.
Staveley, L. A. K., and Randall, T. (1958). *Discussions Faraday Soc.* **No. 26,** 157.
Swaminathan, K., and Busch, D. H. (1961). *J. Inorg. & Nuclear Chem.* **20,** 159.
Thomas, W. (1922). *J. Chem. Soc.* p. 196.
Tillotson, M. J. L., and Staveley, L. A. K. (1958). *J. Chem. Soc.* p. 3613.
Topp, N. E. (1956). *Chem. & Ind.* (*London*) p. 1320.
U. S. Patent (1938). 2,130,505.
U. S. Patent (1945a). 2,384,816–7.
U. S. Patent (1945b). 2,387,976.
U. S. Patent (1946). 2,411,019.
U. S. Patent (1947). 2,428,353.
Vickery, R. C. (1952). *J. Chem. Soc.* p. 4357.
Vickery, R. C. (1954). *J. Chem. Soc.* p. 1181.
Vorisek, J. (1959), *Collection Czechoslov. Chem. Communs.* **24,** 3921.
Wald, M. (1961). *Monatsh. Chem.* **92,** (3), 605.
Watters, J. I., Mason, J. G., and Schupp, O. E. (1956). *J. Am. Chem. Soc.* **78,** 5782.
Weakliem, H. A., and Hoard, J. L. (1959). *J. Am. Chem. Soc.* **81,** 549.
Werner, A. (1912). *Ber.* **45,** 3281.
Werner, A., and Bosshart, J. (1914). *Ber.* **47,** 2171.
Wheelwright, E. J., Spedding, F. H., and Schwarzenbach, G. (1953). *J. Am. Chem. Soc.* **75,** 4196.
Yalman, R. G. (1961). *J. Phys. Chem.* **65,** 556.
Yoshino, Y., Yoshihaza, T., and Ouchi, A. (1961). *Bull. Chem. Soc. Japan* **34,** 1194.
Ziemlak, L. W., Bullock, J. L., Bersworth, F. C., and Martell, A. E. (1950). *J. Org. Chem.* **15,** 255.
Zolotov, Y. A., and Novikov, Y. P. (1959). *Zhur. Neorg. Khim.* **4,** 1693.

CHAPTER 8

Enzyme–Metal Ion Activation and Catalytic Phenomena with Metal Complexes

F. P. DWYER

Biological Inorganic Chemistry Section, Australian National University, Canberra, A.C.T., Australia

I. Metal Enzyme Catalysis

A. General Considerations

The presence of metal atoms as essential constituents of some enzymes, and the metal ion requirement of others for maximum activity, provide an obvious link between enzymatic reactions and coordination chemistry. In the first group of enzymes, sometimes called metalloenzymes (Vallee, 1955), specialized bonds which hold the metal atom firmly are provided, and dissociation does not occur at all, or is sluggish under normal physiological conditions. Familiar examples are the heme group of iron-containing enzymes in which the metal is bound covalently to four nitrogen atoms in a macrocyclic ring system, and the zinc enzymes such as alcohol dehydrogenase. The metal-protein molar ratio in the purified products is a small integer. By contrast, the binding of the metal atom in the metal ion–activated enzymes is weak, there is no obvious stoichiometry in the metal ion–protein ratio, and the system metal ion–enzyme–substrate–products is labile.

These categories have been erected for convenience and can be delineated only roughly, since the role of metals in enzymatic processes seems to range from the exertion of nonspecific ionic strength effects to the formation of weak dissociable and some strong nondissociable complexes. The effect of the metal ion in these systems can be likened on the one hand to the effect of the metal on the reactivity of simple organic molecules coordinated in stable metal complexes and on the other to the wider, more uncertain, field of coordination catalysis in which the bonding of the metal atom to one or both of the reactants can only be presumed.

The possible functions of metal ions in enzyme-substrate systems are easily visualized:

(1) the metal may form a complex with donor atoms of either the enzyme or substrate and thereby enhance their tendency towards reaction;
(2) it may serve merely as a bridge through common coordination to bring the enzyme and substrate into proximity;
(3) while serving function (2) it may provide as well a chemical activating influence; and
(4) while coordinated to either the enzyme or the substrate it may appropriately orientate groups undergoing reaction.

All of these have been proposed in reviews and textbooks, and their relative importance weighed. The most obvious property of a metal ion is its positive charge, which makes it effectively an acid in the Lewis sense, and, in common with the proton, it will have a tendency to withdraw electrons from atoms and groups to which it is attached. It has been suggested that the active participation of a metal ion in an enzymatic reaction is most likely to occur through such a mechanism of generalized acid catalysis (Orgel, 1958). Its catalytic efficiency will depend on the effective charge. The latter includes not only the numerical value of the charge but also the size and the shielding effects of the electron cloud. For bivalent ions, which are enzymatically the most significant, the order of polarizing power, for ligands generally, will be approximately the familiar order of complex stabilities.

An important concept, first enunciated by Pauling (1948), is the Principle of Electroneutrality, which states that complex formation itself, whatever the bonds involved, is essentially a process to relieve the electron deficiency of metal ions, and that the positive charge is dissipated over the whole complex entity in accordance with the electronegativities of the component atoms. Further reference will be made to this principle when discussing the reactivity of aniline in trichlorotris(aniline)chromium(III) (Section II,D) and the neuromuscular blocking action of large complex cations. In this way a carbon atom in a metal complex can be rendered more positive and hence more susceptible to nucleophilic reagents. However, the coordination of substrate or enzyme may occur with the detachment of protons whose charge is equal to the ionic charge of the metal atom, and hence the charge distribution over the atoms of the coordinated molecule may not be very different from that of the original free species.

B. Metal Ion Activation of Proteolytic Enzymes

The effect of the charge of the coordinated metal ion can be illustrated by considering the hydrolysis of peptides. This occurs enzymatically in the presence of metal ions and is also acid-catalyzed. The kinetics of hydrolysis of glycylglycine and a number of its *C*-methyl derivatives have been studied in glacial acetic acid solution in the presence of perchloric acid by Martin (1957). The mechanism probably consists of the fast addition of a proton to the peptide nitrogen of an already monoprotonated species. The amide cation then reacts slowly with a water molecule at the carbonyl carbon atom to yield the intermediate (I) which undergoes slow fission at the C—N bond. Glycylglycine can act as a tridentate ligand by utilizing the $-NH_2$, —COOH, and either (but not both) the >NH or —C=O groups. Both mono and bis chelates are known, but the bonding in the four-covalent bis chelates is uncertain (Rabin, 1958; Koltun and Gurd, 1959; Koltun *et al.*, 1960; Martin *et al.*, 1960). Above pH 4, the Cu(II) mono species detaches the NH proton, a reaction which also takes place with Ni(II) above pH 9, but not with Co(II), Mn(II), Zn, or Mg(II,III).

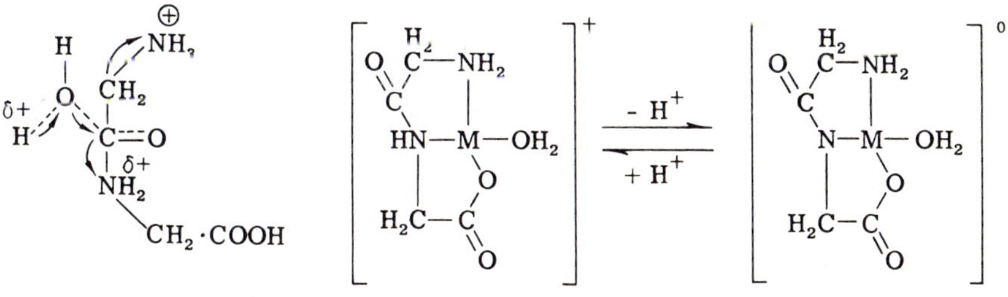

(I) (II) (III)

It will be evident that the mono chelates (II) have an over-all unit positive charge, while the bis chelates and the mono species (III) will be electrically neutral. Insofar as the bond-making step (water or hydroxyl addition) is concerned, a species such as (II) cannot be equated to the amide cation, and in the neutral bis chelates the situation will not be significantly different from that of the free ligand. Bond breaking is inhibited by the stability imposed by the chelate rings, and this factor more than offsets any gain in polarization at the carbonyl carbon atom when strongly interacting metals, such as copper, are coordinated. Furthermore, as Klotz and Loh Ming (1954) have pointed out, the products of peptide hydrolysis are usually stronger chelating agents than the parent peptides. In any event, it has not been possible to demonstrate metal ion–catalyzed peptide hydrolysis in the absence of enzyme. However, the rate of hydrolysis of glycylglycine is accelerated (factor of 1.7) by Co(II) ions in acid solution (Lawrence and Moore, 1951). Presumably, both the —NH_2 and —NH groups are protonated and the Co^{++} ion, attached in a fugitive intermediate of low stability to the oxygen of the carbonyl group, acts as a supplementary electron-withdrawing agent.

The Lewis acid role of metal ions, originally proposed by Smith (1949), and subsequently elaborated by Martell and Calvin (1952) for peptidase action, has been extended to the carboxylases and phosphatases. In addition to its electronic effect, the metal atom may serve to bind the protein through coordination positions not used by the substrate. With the peptidases the function of the enzyme-protein bond is to reinforce the electronic effect of the metal atom by the proximity of polar groups, or, perhaps, to facilitate the dissociation of the products (amino acids) from the metal atom. In a recent very pertinent and critical review, Malmström and Rosenberg (1959) have pointed out that, however plausible chemically, these theories are not tenable biochemically since, at best, they assign a minor enhancement role to the protein which is inconsistent with experimental fact. For example, the metal ions which are most effective enzymatically are often without effect in the absence of enzyme; e.g., in the decarboxylation of oxalacetic acid (Section III,G), Cu^{++} is one of the best catalysts of the nonenzymatic reaction and Mg^{++} is without effect, but the position is reversed enzymatically. Possibly, the only well authenticated instance of a mere enhancement role for the enzyme (and the metal) is found in transaminations involving pyridoxal (Section III,D).

C. Metal Ion Specificity

Insofar as the efficiency of metal ion catalysts is concerned, a kind of quantitative specificity exists; e.g., Mg^{++} is usually the best activating ion for enzymes involving phosphate groups whereas Mn^{++} and Zn^{++} are somewhat less active, and with the peptidases Co^{++} or Mn^{++} are more active than Zn^{++} or Mn^{++} ions. Metal ions high in the order of complex stabilities (Pd^{++}, Cu^{++}) are poor activators and may act as inhibitors. Attempts to relate activating potency to complex stability have not been successful, nor does the inverse order of binding strength give any better correlation. Green *et al.* (1941) have shown that the order of decreasing efficiency of metal ions in the enzymatic decarboxylation of pyruvic acid is Mg > Mn > Co > Cd > Zn > Cu > Fe > Al > Fe(III), in which, for instance, the relative positions of Cu(II) and Fe(II) are quite surprising. It is well known that some metal ions—Cu^{++}, Ni^{++}, Co^{++}—prefer nitrogen donor atoms while others form stronger complexes with oxygen—Mg^{++}, Ca^{++}. Williams (1953) has proposed that metal ion specificity might reside in such preferences which lead to the appropriate kind of complex rather than merely complex stability. Such a concept assumes, as with peptides, for example, that both kinds of donor atom are available, and formation of the wrong kind of complex might be expected to lead to competitive inhibition. The order of complex stabilities emerges from data compiled with O and N donors, generally in bidentate ligands. Sulfur donors (from protein) form strong bonds with the higher members of the series, such as Cu^{++} and Pd^{++}, but would not influence the relative order of the biologically active metals, which have low affinities for sulfur.

The metal ion–catalyzed decarboxylation of oxalacetic acid in the absence of enzyme is roughly in agreement with the order of complex stabilities, but in the enzymatic reaction this order is not followed (Speck, 1949). From a consideration of the catalytic activity versus concentration of the metal ions curves, Eichorn (1961) has suggested a possible rationalization. The stronger binding metal ions, such as Cu^{++}, which are activators at low concentrations become inhibitors at higher concentrations in the presence of enzyme; but if the various ions are compared with respect to the concentrations producing maximum activity, the catalytic order then approximates that of the complex stabilities. It has often been proposed that metal ion specificity resides in part, at least, in the characteristic coordination number and stereochemistry of the metal which constrains the attached molecule to a specific rigid geometry. Most of the active metal ions, in fact, form octahedral complexes, though Zn^{++}, Cd^{++}, Mg^{++}, and Ca^{++} can adopt tetrahedral stereochemistry. Tetrahedral Mn(II), Fe(II), Co(II), and Ni(II) complexes, $[MCl_4]^{--}$, have been prepared (Nyholm, 1961; Gill and Nyholm, 1959). Octahedral Cu(II) complexes, e.g., $[Cu(en)_3]^{++}$, are quite as common as the planar complexes which often have two groups in rather long octahedral positions. Metal atoms show quite an ability to conform to the stereochemistry of the ligand, which almost consistently determines the stereochemistry, and not vice versa (Dwyer, 1961). The mutual stereochemical requirement is, of course, more critical with multidentate ligands (Chapter 6), but quite a number, such as ethylenediaminetetraacetic acid, form complexes with practically all of the metals.

Since the order of complex stabilities for metal ions is subject to qualification in that no steric hindrance for the ligands should occur and no radical change in bond type must take place, its application to enzymatic systems seems dubious. It is also doubtful, except for the metalloenzymes, whether chelate ring formation occurs in enzymatic processes (cf. Klotz and Loh Ming, 1954; Malmström and Rosenberg, 1959).

D. Enzyme-Metal-Substrate Complexes

The bridge mechanism of metal ion activation depends upon the simultaneous binding of the substrate and enzyme because of the multibonding capacity of metal atoms, a possibility not realizable in proton catalysis. Several nonenzymatic reactions are discussed later (Section III) where it is evident that the essential role of the metal ion catalyst is an orientating one. Metal complexes with two or more different ligands, e.g., $[Co(en)_2(NH_3)_2]^{3+}$, are usually not difficult to prepare or isolate provided that the complex is relatively inert to substitution—an important reason why so many cobalt and heavy transition metal complexes are known. The enzymatically important metals necessarily form labile complexes that are in rapid equilibrium with the aquo metal ion and the ligand. Where two different ligands compete for coordination to the aquo ion present in small concentration, the equilibrium concentrations of all possible species are determined by their relative free energies, that is the binding powers of the free ligands as a first approximation. The addition of a mixture of ethylenediamine and ammonia to a solution containing a small amount of Ni^{++} ion thus yields predominantly the $[Ni(en)_3]^{++}$ and $[Ni(en)_2]^{++}$ ions with a very small proportion of species containing coordinated ammonia. If the same experiment is repeated with triethylenetetramine—a quadridentate molecule—instead of ethylenediamine, a significant amount of the mixed species $[Ni(trien)(NH_3)_2]^{++}$ will exist in the equilibrium since two ammonia molecules can neatly complete the octahedral complex.

That many substrates and enzyme proteins form complexes with metals is well authenticated (Schubert, 1954; Gurd and Wilcox, 1956). In the molecules of the latter, the common donor atoms O, N, and S occur in various chemical combinations, and one protein molecule may bind many metal atoms. It is, however, very unlikely that the stereochemistry of the functional groups permits coordination at more than three positions about a metal atom. Provided that the substrate does not bind the metal very much more strongly than the protein, quite a reasonable concentration of a ternary complex, formed from substrate, metal ion, and protein, could exist in equilibrium mixtures of complexes. Kinetically, of course, not more than quite a small concentration of the ternary complex is necessary provided that it implicates the active center of the enzyme. It is the last condition that is difficult to establish experimentally.

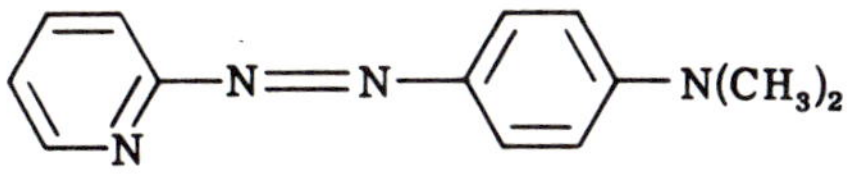

(IV)

The dye pyridine-2-azo-*p*-dimethylaniline (IV) can act as a bidentate ligand through the heterocyclic and one of the azo nitrogen atoms, yielding bis or tris chlates which are cations. The dye generally does not combine with proteins, but, in the presence of a metal ion, binding is mediated and a ternary complex, dye–metal ion–protein (DMP), results (Klotz and Loh Ming, 1954). The analogous pyridine-4-azo-*p*-dimethylaniline, which cannot chelate, is not bound to protein in the presence of metal ions. Though the ternary complex would appear to be a "mixed" complex there are some difficulties associated with this concept. The binding constant K_{DMP} has a substantial temperature coefficient, whereas, in conformity with the usual experience, the temperature coefficient for the metal complex itself and for the binding of metal ions by proteins is small. Although K_{DM} for Cu^{++} is >500 times greater than for Zn^{++}, the values of K_{DMP} are almost equal (6×10^3 and 4×10^3 respectively). The mediated binding increases with increasing pH values, and this would be consistent with the formation of a complex. However, at higher pH values there would also be a larger number of anionic sites to which large cationic complexes could associate by van der Waals forces. The large stable cations discussed in Chapter 9 adsorb strongly on protein (cf. Jensen *et al.*, 1958). Studies by other workers have shown that metal ions can act as bridges in various "model" systems such as the binding of various chelating molecules to serum albumin in the presence of Be^{++} ion (Lindenbaum and Schubert, 1956). The kinetic approach to bridging mechanism has been discussed critically by Malmstrom and Rosenberg (1959) who conclude that direct evidence for the enzymatic participation of ternary complexes is still lacking.

The transfer of the phosphoryl group from phosphorylcreatine to adenosinediphosphate (ADP) by the enzyme creatinephosphoryltransferase in the presence of magnesium ion has been studied recently by Morrison *et al.* (1961). It has been supposed (Griffiths *et al.*, 1957) that the reaction occurs via an active enzyme-metal-substrate complex arising from an initial enzyme-metal complex. A kinetic approach has been used to determine the dissociation constants for all possible intermediate steps leading to the enzyme-metal-substrate active intermediate. It is concluded that the Mg^{++} and both substrates are capable of being bound independently to the enzyme and that the metal ion does not seriously affect the binding of the substrates. The attack of the nucleophile ADP^{3-} on the phosphorus atom of phosphorylcreatine might be facilitated by the proximity of the positively charged metal atom, which, for instance, could partially neutralize the negative charges in the phosphorylcreatine. Alternatively, the magnesium ion might induce a particular configuration at the active center of the enzyme.

E. Summary

The available experimental evidence does not permit the clear assignment of a chelating role to metal ions in enzymatic reactions. The fact that those that form the weakest complexes are usually catalytically the most active seems inconsistent with such a role. Since a metal ion is primarily a small, positively charged entity, the most consistent interpretation of its role, at our state of knowledge, is that it exercises a supplementary polarizing influence. It may induce a favorable conformation of the enzyme; it may, by acting in conjunction with polar centers on the enzyme, provide an additional electronic distortion of the substrate; it may assist the bond-making or the bond-breaking reaction steps. These functions may well be served without the formation of metal chelates or even coordination compounds as we commonly know them.

II. The Reactivity of Coordinated Molecules

A. Introduction

There is an increasing number of reactions in which metal ions participate, presumably as catalysts, and in which it is supposed that the catalytic action takes place through metal complexes involving one or all of the reactants. Alternatively, preformed metal complexes with free or labile coordination positions may act as catalysts. One may enquire whether or not these are true catalysts and are recoverable unchanged at the end of the reaction, but it is more profitable to consider whether the mechanisms of polarization or orientation are involved. The interpretation of metal ion catalysis, as we have seen in enzymatic systems, is complicated by the lack of precise knowledge of the participating species, an inherent difficulty where labile complexes are concerned. Reactions of simple chelated molecules in stable, relatively inert complexes of known composition will, therefore, be considered first, and then the more speculative metal ion–catalyzed reactions.

B. Acetylacetone Chelates

β-Diketones, of which acetylacetone is the most familiar example, react with practically all of the metals to form neutral complexes of high chemical and thermal stability. Anionic complexes of lower stability, e.g., Na[Co-(aca)$_3$], are known and acetylacetone may be implicated in cationic complexes such as [Co(en)$_2$aca]Cl$_2$. It will be evident that various equivalent structures differing only in the positions of the double bonds can be written for the neutral chelates (V). The high chelate stability is usually correlated

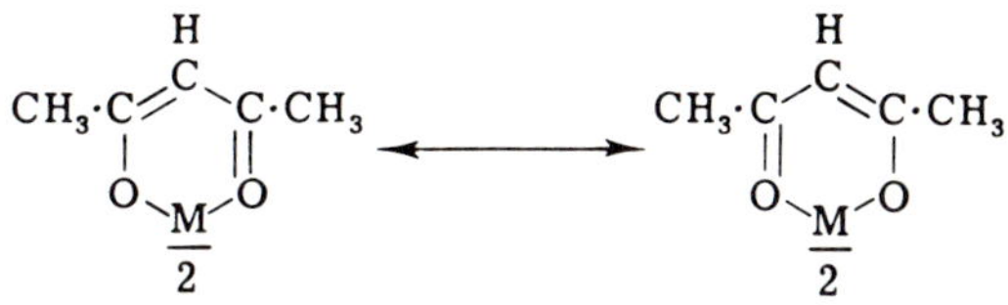

(V)

with the strong resonance effect implicit in such structures (Martell and Calvin, 1952). Comparison of the stability constants of the Cu(II) chelates with a wide variety of substituents, e.g., acetylacetone, benzoylacetone, trifluoroacetone, and the chelates with salicylaldehyde and 2-hydroxy-1-naphthaldehyde, led Calvin and Wilson (1945) to the conclusion that enolate resonance is more significant in the bonding of copper than in the bonding of hydrogen. The higher stability of the Cu(II) bonding was explained in terms of double-bonding between the donor oxygen atoms and the metal, leading to the proposition that completely conjugated six-membered benzenoid rings were important contributors to the resonance in bis(acetylacetonato)copper(II). The refined crystal structure analysis of tris(acetylacetonato)iron(III) has shown that the chelate rings are planar. The C—C bond distances, 1.39 Å, are the same as in benzene, and the C—O bond distances, 1.28 Å, are also intermediate between double and single bonds (Roof, 1956). Refined data are not available for the Cu(II) chelate. Holm and Cotton (1958) have pointed out that the above structural data must imply some degree of resonance in the chelate rings but do not necessarily require benzenoid resonance (VI). The $d\pi$–$p\pi$ component of

(VI)

the double bond must of necessity transfer further negative charge from the oxygen atoms to the metal since the acetylacetonate anion lacks vacant π-orbitals, and the resulting large charge separation makes benzenoid resonance unlikely. Nuclear magnetic resonance and ultraviolet studies have not substantiated aromaticity in any of the chelates (Holm and Cotton, 1958; see, however, Orgel *et al.*, 1959). Similarly, electron spin resonance measurements with the copper chelate showed that the in-plane σ and π bonding was appreciably covalent, but the out-of-plane π bonding was ionic. This was interpreted to mean that there is little interaction with the π system of the ligands (Maki and McGarvey, 1958). The implications of the concept of aromaticity are beyond the scope of this book. Two aspects of the concept, "typical reactivity," e.g., towards electrophilic substitution, and thermochemical resonance energy, have been discussed by Ingold (1954) and Craig (1959).

Early work on the action of bromine, sulfur monochloride, and acetylchloride on acetylacetone chelates suggested that the products were the results of ring cleavage (Auwers and Auffenberg, 1917; Valliant, 1894; Michael and Carlson, 1936). However, Reihlen *et al.* (1925) showed that tris(acetylacetone)chromium(III) could be brominated to yield the tribromo chelate. Djordjevic *et al.* (1959) attempting to prepare five-covalent complexes, treated bis(acetylacetonato)nickel(II) and palladium-(II) with nitrite ion in acid solution, and found that, instead, the chelate rings were attacked yielding substances of the formula $M(C_5H_7N_2O_2)_2$. Nitrogen tetroxide reacted with bis(acetylacetonato)copper(II) in benzene solution with the production of $Cu(C_5H_6O_2NO_2)_2$—probably bis(3-nitro-2,4-pentanediono)copper(II).

A large number of electrophilic substitutions at the —CH= group have been carried out by Collman *et al.* (1960, 1961a,b), including halogenation, acetylation, and nitration. Trinitro derivatives, $[M(C_5H_6O_2NO_2)_3]$, were prepared by the use of a mixture of acetic anhydride and copper(II) nitrate as the nitrating agent. The usual nitrating agents tend to destroy acid-labile complexes—a difficulty also encountered in halogenation because of the formation of halogen acid (VII). Some of these substitutions

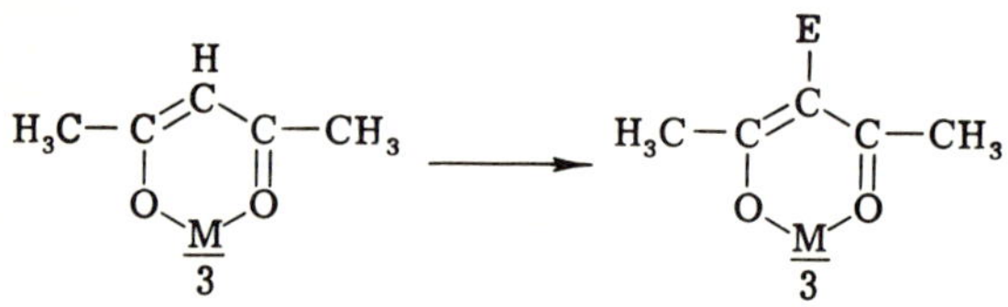

(M = Co,Rh,Cr: E = CH_3CO—, CHO—, NO_2,Br,Cl,SCN)

(VII)

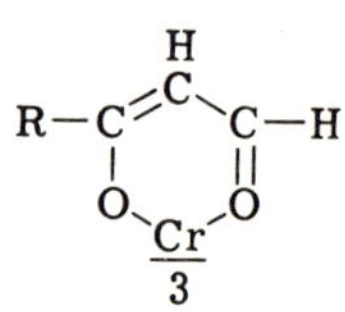

(a) R = H
(b) R = CH_3

(VIII)

have been carried out with the partially resolved Co(III) and Cr(III) chelates, and the substituted products found to be still optically active (Collman *et al.*, 1961b). This suggests, though it does not prove, that substitution occurs while the acetylacetone is attached to the metal. The chemically analogous (*a*) tris(malonaldehyde) and (*b*) tris(formylacetone)-Cr(III) chelates (VIII) have also been converted to the trinitro and tribromo derivatives (Collman and Kittleman, 1961). The reaction of bromine with acetylacetone and allied substances is base-catalyzed and independent of the bromine concentration. The rate-determining process is the production of the mesomeric anion which undergoes electrophilic attack at the carbanion (IX). The mechanism of substitution in the metal

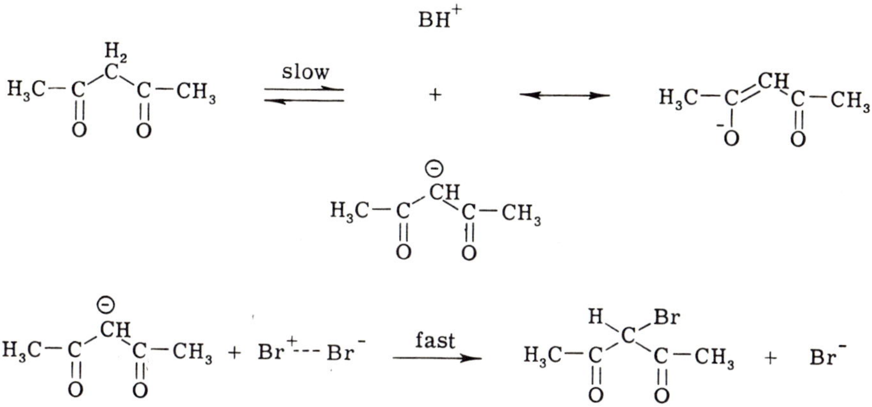

(IX)

complexes is not known. There would be an electronic displacement from the —CH= group, through the conjugated system to the metal, weak in the neutral chelates but stronger in cations such as $[Co(en)_2aca]^{++}$ or $[Ru(II)(phen)_2(aca)]^+$. As with the free ligand, bromination may be base-catalyzed

$$(\text{—CH=} \underset{}{\overset{\text{slow}}{\rightleftharpoons}} \text{—}\overset{(-)}{\text{C}}\text{=} + BH^+)$$

and independent of bromine concentration. The mechanism of substitution proposed by Collman and Blair (1961) assigns a pseudoaromatic character to the metal chelate which undergoes an S_N2 attack by the electrophile, the addition of which is slow and rate-determining (X).

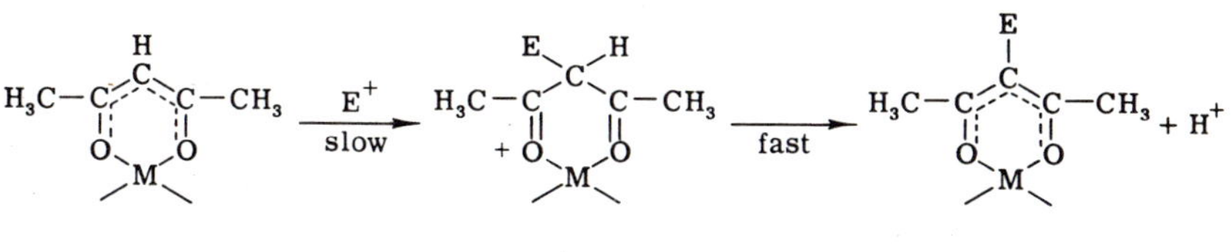

(X)

The relevance of these reactions, whatever their mechanism, to metal ion–catalyzed enzymatic processes involves the demonstration that the reactivity of the ligand is enhanced in the coordinated environment. Kluiber (1960), by the use of C^{14}-labeled acetylacetone, showed that ligand exchange did not occur with the Co(III) complex, and with only about 1% with the Cr(III) complex, during bromination with *N*-bromosuccinimide. When an equivalent mixture of $[Cr(aca)_3]$ and free acetylacetone was treated in chloroform solution with a *deficiency* of *N*-bromosuccinimide, both substances were brominated in approximately equivalent proportions. In other words, chelation in a neutral complex did not appreciably affect the reactivity of acetylacetone.

C. Amino Acid Chelates

The Knoevenagel reaction occurs between aldehydes and compounds with a reactive methylene group in the presence, generally, of an organic base, and involves a nucleophilic attack at a carbanion. The synthesis of threonine from bis(glycinato)copper(II) and acetaldehyde in the presence of sodium carbonate *at 50°* has been carried out by Sato *et al.* (1957), who reasoned that the amino group would be protected by coordination, and sufficient activation of the methylene group would result from the polarizing effect of the metal atom (XI).

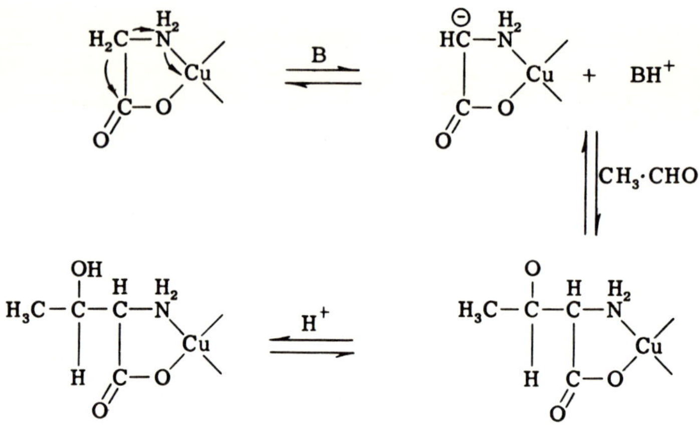

(XI)

A mixture of threonine and allothreonine (64% yield) was obtained after removal of the copper. More significant is the comparison of the reactions using tris(glycinato)cobalt(III) and *l*-glycinatobis(ethylenediamine)cobalt(III) iodide. In the former reaction the yield of the mixed amino acids was lower (34%) and some kinetic discrimination of the attacking electrophile towards the α and β geometrical isomers was evidenced by different ratios in the mixture (Ikutani *et al.*, 1959). The levorotatory cationic complex $[Co(en)_2gly]^{++}$ reacted *at room temperature* to give a slightly dextrorotatory mixture of the amino acids (total yield 80%, asymmetric yield 8%) (Murakami and Takahasi, 1959). These syntheses leave no doubt that the methylene group is activated in the complexes, but do not clearly establish that the metal atom is the activating agent, or, as far as the neutral complexes are concerned, is even a notable contributor to activation. Comparison with free uncoordinated glycine is difficult since, in the alkaline solution in which the condensation occurs, it would be present as the glycinate ion. In the neutral complexes which have arisen by the displacement of two or three protons, the carboxylate is bound to the metal in much the same way as an ester group and for this reason carbanion formation is facilitated. As far as additional activation of the methylene group is concerned, it should be appreciated that both the oxygen and nitrogen donors are bound to an essentially neutral metal atom. Electronic displacement towards the metal atom would not be significant except with the most electronegative metals and, in any event, the hydrogen atoms of the amino group would carry much of the resulting fractional positive charge. Protection of the amino group might be an important function of coordination to a metal atom. In the cation $[Co(en)_2gly]^{++}$ movement of negative charge from the ligands to the metal atom is to be expected, and, though much of the resulting positive charge would reside on the hydrogen atoms of the five amino groups, it is possible that both the carbonyl group and the metal atom contribute to the activation of the methylene group.

The formation of the Cu(II) complex can be used to mask amino groups that would otherwise be attacked in syntheses. Kurtz (1938), for example, prepared *d,l*-citrulline (XII) from the copper complex of ornithine by condensation with urea in hot aqueous solution, followed by elimination of the metal as the sulfide. The protected α-amino group did not react.

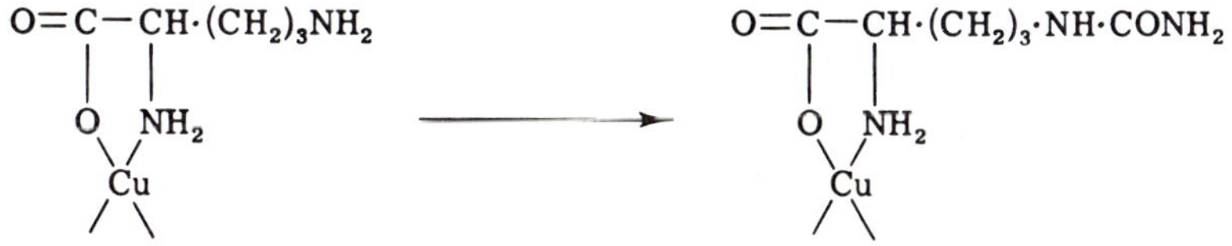

(XII)

Ornithine, lysine, and α,γ-diaminobutyric acid have been benzoylated and phenylsulfonated as their copper complexes without any indication that the metal atom behaves other than as a masking agent (Kurtz, 1949). It would be of interest to substitute malonato complexes for diethylmalonate in the alkylation reaction:

$$CH_2(COOC_2H_5)_2 + Na(OC_2H_5) + RX \rightarrow R{\cdot}CH{\cdot}(COOC_2H_5)_2 + NaX$$

In this series of complex ions $[CoCH_2(COO)_2en_2]^+$, $[Co(CH_2(COO)_2)_2en]^-$, $[Co(CH_2(COO)_2)_3]^{3-}$, the activation of the methylene group should decrease progressively, despite the electron-withdrawing tendencies of the adjacent carbonyl groups, as a result of the increasing over-all negative charge. Notwithstanding the positive charge on the cationic complex, the ethylenediamine ligands are deactivating. Enhanced activation would be anticipated in complexes such as $[Co(phen)_2CH_2(COO)_2]^+$ in which the extraneous ligands have smaller electron availabilities and lack hydrogen atoms attached to the donor nitrogen atoms.

D. Complexes with Coordinated Aniline and Pyridine

Metal complexes containing pyridine and aniline are well known. Complexes containing the former ligand are not only more abundant but are also more stable. One reason for this greater stability is the greater basic strength of pyridine; an additional, though as yet unproved, reason is that pyridine may be attached to the metal by double bonds. In addition to the well known effect of protonation on the substitution reactions of aniline, there is the enhanced reactivity of pyridine as the *N*-oxide for electrophilic substitution in the *para* position that might be simulated in their metal complexes. As a result of activation by the amino group, aniline is highly sensitive to electrophilic attack at the *ortho* and *para* positions, bromination yielding 2,4,6-tribromaniline. The ring is strongly deactivated in the anilinium ion present in strongly acid solution, so that nitration with potassium nitrate in concentrated sulfuric acid yields *m*-nitraniline. The effect of the proton is to attact electrons with considerable deactivation of the *para* position, less of the *meta* positions, and still less of the *ortho* positions. The location of the substitution may well be the result of strong steric repulsion of the electrophile by the $-[NH_3]^+$ group away from the favored *ortho* to the less favored *meta* positions (Ross, 1961). Any general concept of the equivalence of the anilinium ion and coordinated aniline is an oversimplification. It is difficult to envisage polarization and electron drift of the magnitude induced by a proton even in a multicharged cation with a single coordinated aniline molecule. However, ideally, sufficient deactivation of the *para* position might be brought about to cause appreciable attack at the *meta* position.

The resistance of pyridine to direct attack by electrophilic reagents is well known and has been compared to that of nitrobenzene (Tucker, 1959).

The electron cloud is concentrated about the nitrogen atom leaving charge densities about all the carbon atoms less than unity (Coulson, 1952). Halogenation, nitration, and sulfonation thus require extreme conditions and substitution occurs at the 3,5 positions. Protonation decreases the susceptibility to attack as discussed with the anilinium ion above. On the other hand, pyridine-*N*-oxide is readily nitrated at 95° (Den Hertog and Overhoff, 1950; Oichiai, 1953), attack occurring primarily in the 4 position and less in the 2 positions. The enhanced electron densities in the 4 and 2 positions have been ascribed by the above authors to the participation of double-bonded structures in the resonance, of which (XIIIB) and (XIIIC) would be most important. Oichiai has proposed that contributions due to

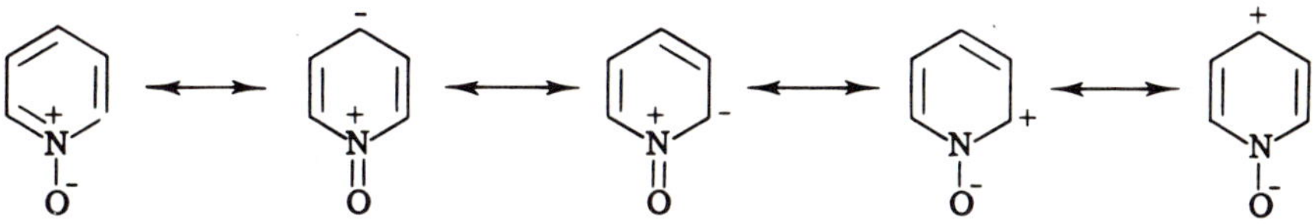

(A) (B) (C) (D) (E)

(XIII)

structures (B) and (C) could account for the low dipole moment compared with that of aliphatic amine oxides and its marked resistance to reduction. Pyridine-*N*-oxide is a weak base but would be present as the cation $[C_5H_5N{\cdot}OH]^+$ in the strongly acidic media in which the electrophilic substitutions are performed. Protonation, as well as reducing the number of structures contributing to the resonance, would tend to draw electrons away from the 4 position but obviously does not do this to the same extent as with aniline. The charge distribution has been evaluated by Jaffé (1954) and is compared with pyridine and aniline (XIV) (Coulson, 1952). Evalua-

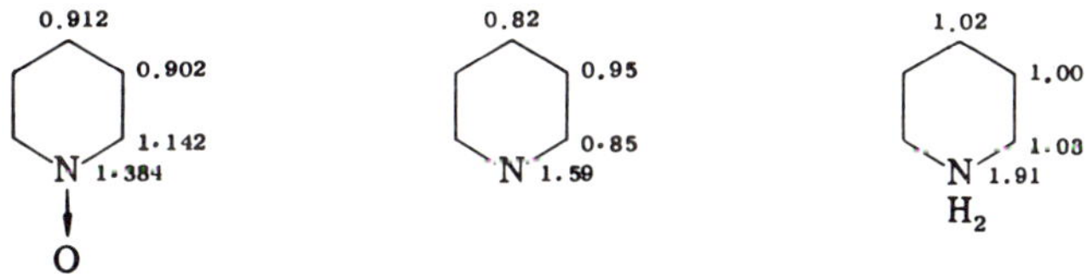

(XIV)

tion of the localization energy showed that the 4 position had the lowest activation energy for either electrophilic or nucleophilic substitution in agreement with experimental fact (Jaffé, 1954).

The coordination of pyridine to a metal atom as far as the σ bond is concerned may be crudely likened to protonation subject to the same considerations as with aniline. Since protonation does not alter the relative order or reactivity of substituent positions or decrease the activation energies (for example, pyridine resists bromination under alkaline, neutral, or acidic conditions), coordination to a metal atom is not likely to change the order either.

Though pyridine itself is attacked by a variety of nucleophilic agents—NH_2^-, OH^-, etc.—the reaction conditions are often so drastic that metal complexes would not survive. Substituted pyridines, notably the 2- and 4-halogeno derivatives, readily undergo nucleophilic substitution and enhanced reactivity is manifested, as expected, with alkyl and aryl pyridinium salts. Some of these react with bases as weak as water (XV). The analogy

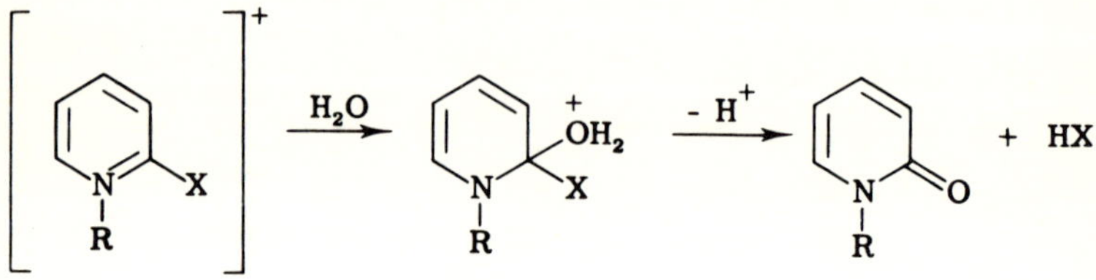

(XV)

between quaternization and coordination to a metal is closer than protonation though certainly not equivalent. As before, in complex cations only would there be any notable enhancement of nucleophilic attack at the substituted pyridine.

Double-bonded structures can be written for metal complexes containing pyridine, and experimental observations of the stronger bonding of this ligand than would be expected from its base strength have been rationalized according to such structures (Murmann and Basolo, 1955).

The very high stability of the iron triad chelates of 2,2′-bipyridine and 1,10-phenanthroline can also be correlated with analogous double-bonded structures which, *inter alia*, probably contribute to the higher stability of the +2 oxidation states over the +3 states (Brandt *et al.*, 1954). Little is known of the reactivity of the ligands in these complexes, or whether, for instance, the phenanthroline ring is sensitized or stabilized to disruptive oxidation. Bipyridine has a strong resistance to electrophilic substitution like pyridine, but the *N*-oxide is rather easily nitrated (Haginiwa, 1955). The contributory double-bonded structures (XVI) become of increasing importance as the oxidation state of the metal and the over-all positive

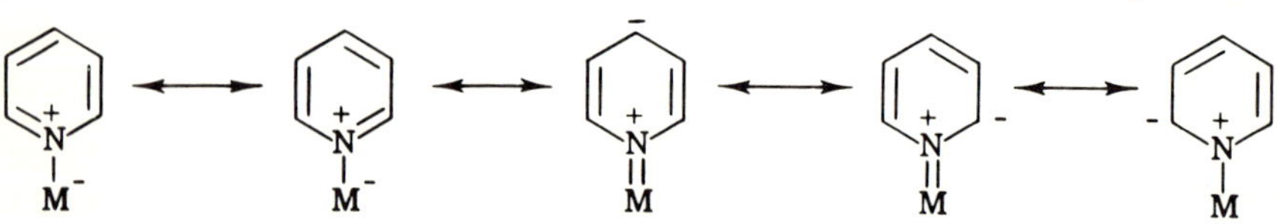

(XVI)

charge of the complex is reduced, and operate increasingly against nucleophilic agents. However, it is very doubtful if the double-bonded structures can be equated to similar structures with pyridine-and 2,2′-bipyridine-*N*-oxides.

A familiar reaction with 2- and 4-methylpyridines is the condensation with formaldehyde which is facilitated by quaternization (XVII). It would

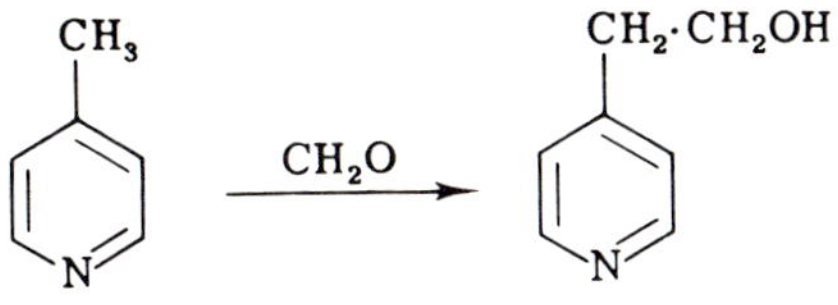

(XVII)

be of interest to compare this condensation with that of the cationic tris chelates of 4,4′-dimethyl-2,2′-bipyridine and 4,7-dimethyl-1,10-phenanthroline and the free molecules.

Taft and Jones (1960) compared the reactivity of coordinated and uncoordinated aniline and pyridine towards electrophilic substitution. Trichlorotris(pyridine)chromium(III), an inert neutral complex, could neither be nitrated by 100% nitric acid in the presence of borontrifluoride at 93° nor chlorinated at 150° in nitrobenzene solution. Bromination of trichlorotris(aniline)chromium(III) in acteic acid gave the usual 2,4,6-tribromaniline in comparable yield and at about the same rate as free aniline, though it was shown conclusively that dissociation of the complex did not occur. The results were interpreted in accordance with the principle of electroneutrality (Pauling, 1948) as showing that little movement of electrons took place between metal and ligand. The corollary proposed: "the reactivity of simple, electrically neutral monodentate aromatic ligands towards electrophilic substitution is relatively slightly affected by coordination to a metal atom," is probably valid for neutral complexes of the type studied. Comparison of the ease of *N*-methylation of free tertiary nitrogen atoms in neutral and cationic metal complexes helps to put these experiments in the correct perspective. The neutral Cu(II) and Ni(II) derivatives of 5-salicylideneaminoquinoline (XVIII) are readily methylated with methyl iodide (XIX) whereas the cationic species derived from 5α-pyridylmethyleneaminoquinoline (XX) are quite resistant (Lions and MacDermott, 1958). In (XVIII) but not in (XX) the nitrogen atoms are still sufficiently basic to react with the positively charged methyl radical.

E. Bis(salicylaldimine) Chelates

As well as by direct reaction of the Schiff base with the metal acetate, these may be made by condensation of the amine with the bis(salicylalde-

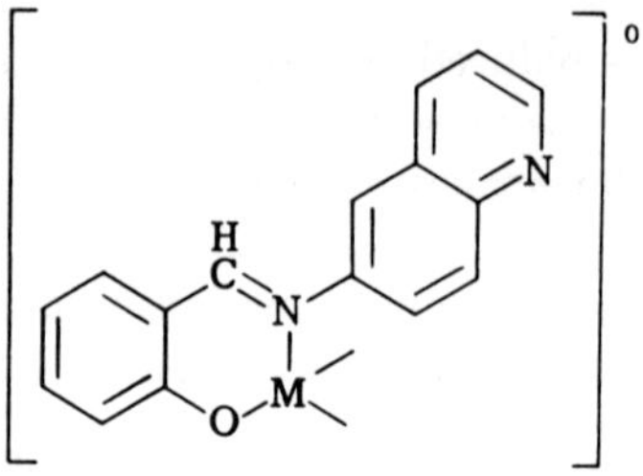

(XVIII)

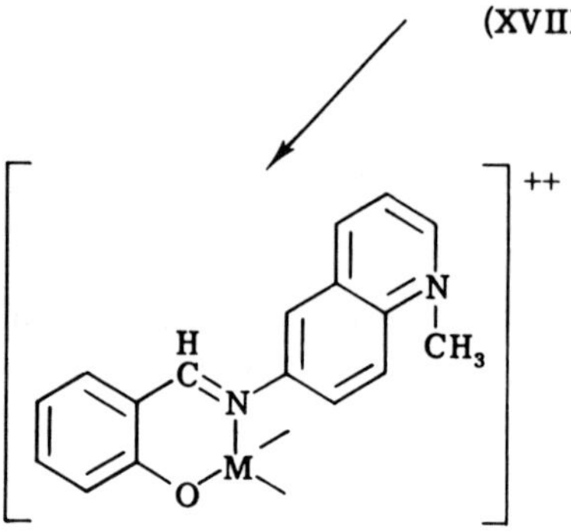

(XIX)

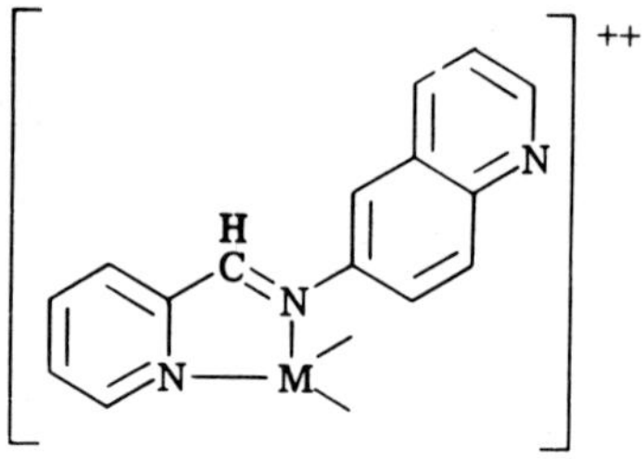

(XX)

hyde) chelate (Pfeiffer *et al.*, 1931). There is no evidence of preliminary dissociation of the aldehyde moiety nor is there any reason to suppose that the presence of the metal atom alters the usual reaction mechanism (XXI) → (XXII) → (XXIII). The presence of the metal atom may well

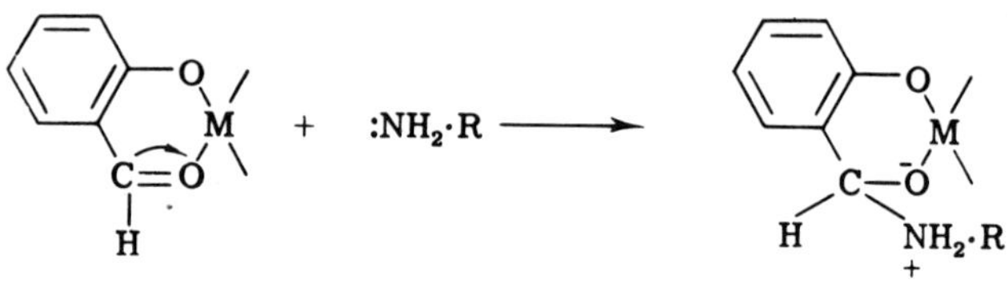

(XXI) (XXII)

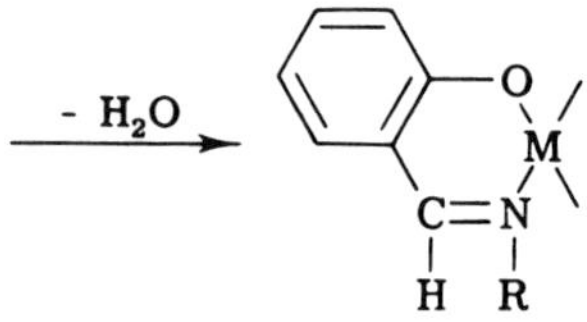

(XXIII)

enhance the activation of the carbon atom, but comparative studies of the rates of Schiff base formation in the presence and absence of metal atoms are lacking. Amines with *ortho* substituents, e.g., *o*-carbomethoxyaniline, do not combine directly with (XXI), although the Schiff bases can be made directly and then coordinated to the metal (Verter and Frost, 1960). Steric factors rather than reduced activity of the carbon atom towards weak bases in the presence of metal atoms may be important (West, 1954).

The acid-catalyzed hydrolysis of Schiff bases might be expected to be simulated by metal ions in some circumstances. The situation, however, differs from that of most of the systems discussed already in that the metal atom is not relatively remote from the reactive site, which, instead, is usually part of a chelate ring. Schiff bases are well known chelating agents which may be bidentate but are often multidentate ligands. Since hydrolysis usually disrupts the coordinated structures, any factor which operates to increase the stability of the coordinated structure, notably the number of chelate rings involved, lessens the tendency to hydrolysis. For this reason chelates derived from metals high in the stability series—Pd(II) and Cu(II)—are likely to be more resistant to hydrolysis than those that form weak chelates such as Zn(II) and Mn(II).

Many Schiff bases, even in the absence of acid, are hydrolyzed in warm aqueous solution—a reaction which clearly takes place by attack of a water molecule at the azomethine carbon atom and is due to the electronic shift associated with the —CH=N— grouping (XXIV). Further displacement of charge consequent upon the formation of even a relatively weak metal—nitrogen bond will, in the absence of chelate stabilization, obviously facilitate hydrolysis in some measure.

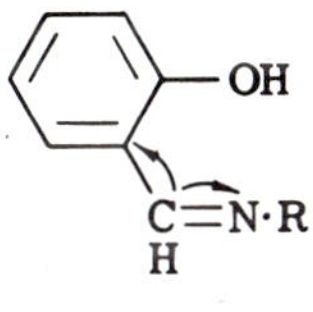

(XXIV)

In addition to the stabilization imposed by chelation, metal ligand double-bonded structures (feasible for most transitional metals of zero or low charge) will increase the stability and reduce the positive charge at the azomethine center. Electron spin resonance studies of the neutral complex bis(salicylideneimine)copper(II) [(XXV) and (XXVI)] indicate some de-

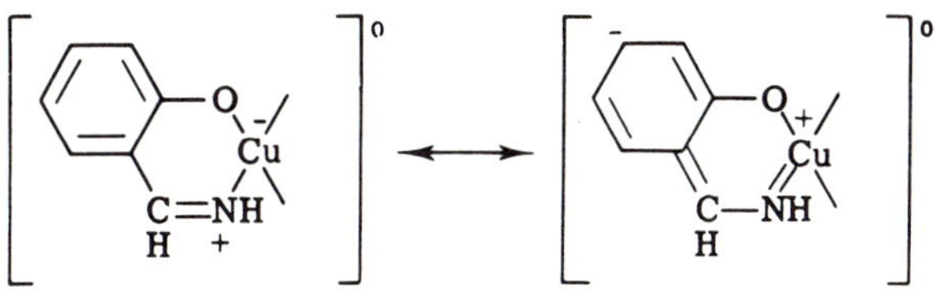

(XXV) (XXVI)

gree of covalent character in the in-plane π bonds as a result of metal-ligand interaction (Maki and McGarvey, 1958).

Eichorn and Bailar (1953) have shown that the potential quadridentate molecule ethylenediaminebis(2-thiophenal) coordinates to Cu(II) through the nitrogen atoms only (XXVII). Rapid hydrolysis ensues with the forma-

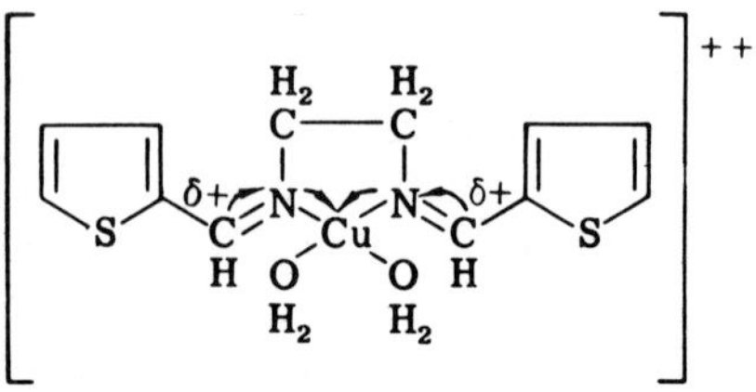

(XXVII)

tion of $[Cu(en)_2]^{++}$ and $[Cu(en)(H_2O)_2]^{++}$ species. The situation here is especially favorable for hydrolysis because of the double positive charge, as well as the single point attachment of each thiophenal moiety, and closely simulates proton catalysis. The activation energies for the hydrolysis of the Cu(II) and Ni(II) complexes, 11.3 and 12.5 kcal./mole respectively, are in accord with the known greater polarizing power of the copper ion (Eichorn and Trachtenberg, 1954). Another example is the rapid hydrolysis in cold water of the cation salicylaldiminebis(1,10-phenanthroline)nickel(II) to salicylaldehydebis(1,10-phenanthroline)nickel(II) ion and ammonia (Harris *et al.*, 1961), in which one five-membered ring is substituted for another [(XXVIII) and (XXIX)]. The tridentate ligand salicylideneglycine

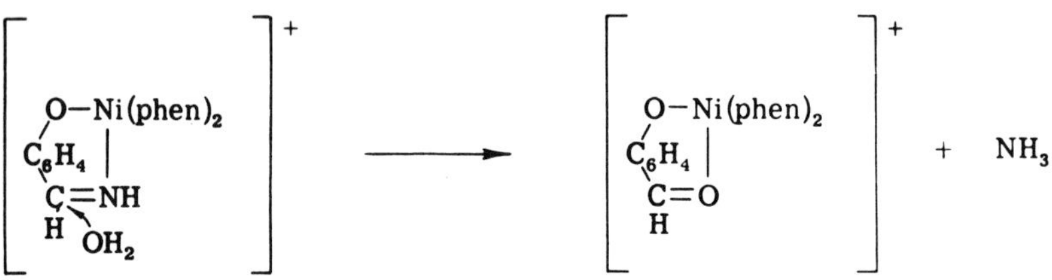

(XXVIII) (XXIX)

yields a neutral Cu(II) complex which by contrast with the parent Schiff base is stable in acid solution (pH 3) (Eichorn and Marchand, 1956; Nakahara, 1959). The enhanced stability to hydrolytic cleavage may be attributed to the bicyclic chelation as well as to the zero over-all charge (XXX). Hydrosalicylamide, $HO{\cdot}C_6H_4{\cdot}CH{=}N{\cdot}CH(C_6H_4OH){\cdot}N{=}CH{\cdot}C_6H_4OH$, the condensation product of three molecules of salicylaldehyde and two molecules of ammonia, can act as a quinquedentate chelating agent. In the presence of Cu(II) and Ni(II) ions, a molecule of salicylaldehyde is removed and the ordinary planar salicylaldimine chelates result (Pfeiffer

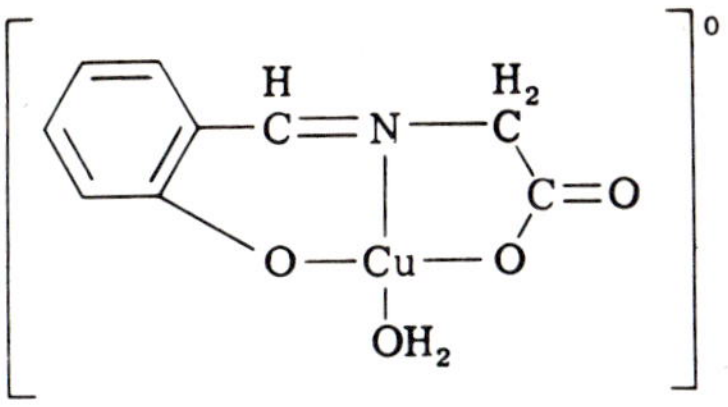

(XXX)

et al., 1931). The molecule, however, is stabilized in the quinquedentate iron(III) chelate which has an extra chelate ring and utilizes the third hydroxyl group (Tsumaki *et al.*, 1960).

Bromination at the carbon atom of the azomethine group has been carried out by the action of *N*-bromosuccinimide on the neutral complexes bis(*N*-*n*-butylsalicylaldimine)nickel(II) and ethylenediiminebis(acetyl-acetone)copper(II), and the latter chelate was acetylated by the use of acetic anhydride and sodium bicarbonate (Kluiber, 1960). No data are available on the behavior of the free ligands, and speculation about the effect of the coordinated metal atom is therefore fruitless.

In the course of studies of the Cotton Effect in metal complexes, Pfeiffer *et al.* (1942) observed a number of reactions of salicylaldimines which were either catalyzed in the metal chelate or did not occur at all with the free ligands. The copper chelates, derived from the Schiff bases prepared from salicylaldehyde and optically active amino acid ethyl esters, were found to have no optical activity in a series of wavelengths; i.e., the optical activity at the carbon atom had been lost. This could be confirmed by hydrolysis and recovery of the inactive amino acid ester. It was proposed that racemization had occurred by a tautomeric change induced by the coordinated metal atom [(XXXI) and (XXXII)]. Salicylal-1-phenylalanine

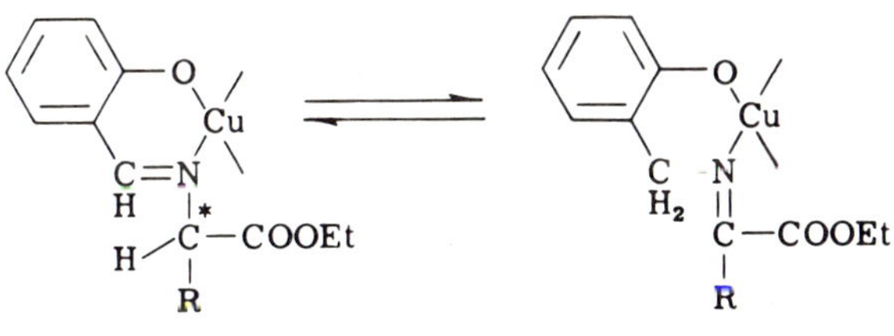

(XXXI) (XXXII)

ester prepared in the same manner as the Cu(II) chelate in hot alcohol in the presence of acetic acid, was optically active but racemized in alcohol at 22° in 7 days. The free ligand salicyl-1-phenylalanine ester also racemizes and the question arises as to whether racemization occurs by dissociation of a proton and subsequent rearrangement of the carbanion formed or whether the activity is lost by tautomerization to a symmetrical compound as in the system studied by Ingold *et al.* (1935) [(XXXIII) and (XXXIV)]. Ingold found that the rates of racemization and tautomerization for the change (XXXIII) → (XXXIV) were the same and hence the activity

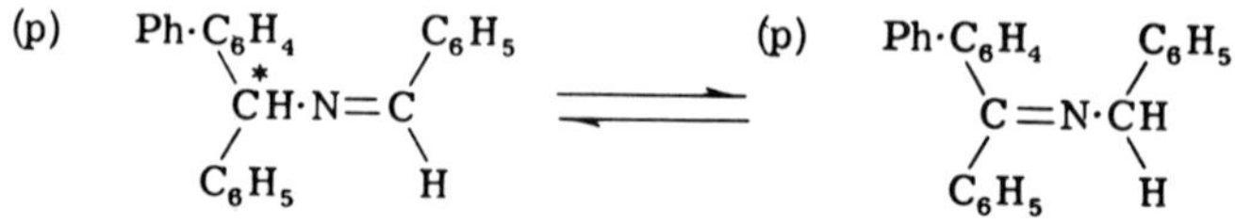

(XXXIII) (XXXIV)

was not lost in the intermediate carbanion. However, unlike (XXXIV) there is no evidence for the existence of (XXXII) and, in the absence of a detailed kinetic study, racemization of the carbanion intermediate is to be preferred to prototropic change for the more acidic metal complexes containing ester groups (cf. Metzler *et al.*, 1954a,b).

Pfeiffer also observed that the copper and nickel chelates of salicylal-leucine and salicylal-phenylalanine ethyl esters absorbed oxygen in alcoholic solution with elimination of the keto acid esters. The reaction mechanism may be inferred from the many transamination reactions carried out with pyridoxal and amino acids (Metzler *et al.*, 1954a,b; see also Section III,D). Hydrolysis occurs at the C—N bond of the amino acid leaving the chelate ring intact, to yield the copper complex of *O*-hydroxybenzylamine and the keto acid, followed by oxidation to the bis(salicylideneimine) chelate (XXV) (Chaberek and Martell, 1959). It was found also that the rate of

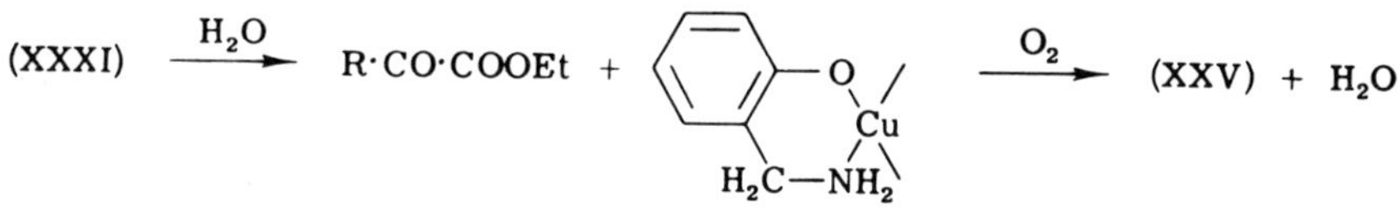

(XXXV)

ester exchange was notably catalyzed in the metal chelates. The Cu(II) and Ni(II) complexes similar to (XXXI) could be converted rapidly to methyl, propyl, or isoamyl esters simply by refluxing with the appropriate alcohol. Recently, Verter and Frost (1960) have confirmed that transesterification does occur, and have drawn attention to the important stereochemistry of the reaction. Planar complexes similar to (XXXI) exist usually in the stable *trans* form and the higher energy *cis* forms are present in the equilibrium in very small amounts. However, transesterification does not occur with (1-carbomethoxy)pentamethylenebis(salicylaldimine-1,5)Cu(II) which is constrained by the pentamethylene ring to exist in the *cis* configuration only (XXXVI). The alternative mechanism proposed

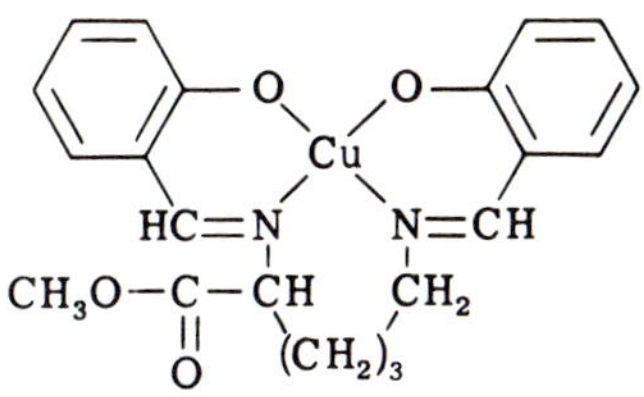

(XXXVI)

assigns to the metal, primarily, the role of orientating reactive centers in proximity, and any electronic effect is supplementary. The ester carbonyl linkage is attacked by the oxygen of the adjacent salicylaldehyde portion of the molecule to form an intermediate lactone-type structure with elimination of the ester group. This is a reversible process and attack by the alcoholic solvent (present in excess) opens the lactone to yield the new ester (XXXVII).

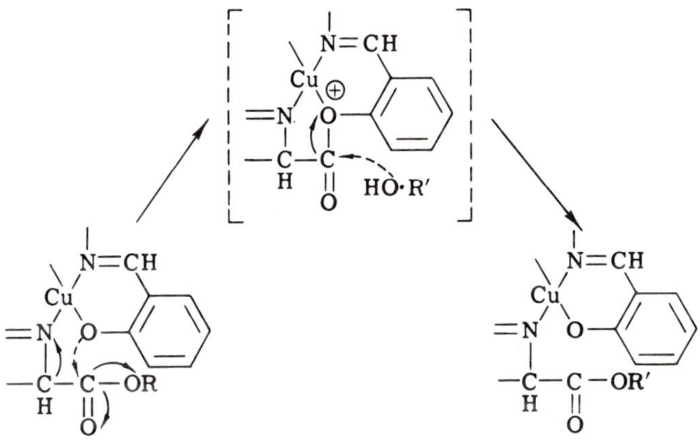

(XXXVII)

When the Cu(II) complexes similar to (XXXI) are heated with primary amines, amidation with displacement of the ester group occurs (Verter and Frost, 1960). The mechanism is probably similar to the transesterification process above. A competitive reaction is the displacement of the amine group implicated in the Schiff base, and this proceeds much faster than amidation when relatively weak bases have been incorporated in the parent aldimine. The mechanism proposed by Verter and Frost involves attack at the polarized carbon atom of the azomethine group by the lone electron pair of the amine (XXXVIII). The possibility that the entering amine attaches to the copper atom to form a five-covalent pyramidal intermediate, which then rearranges intramolecularly with the elimination of the weaker

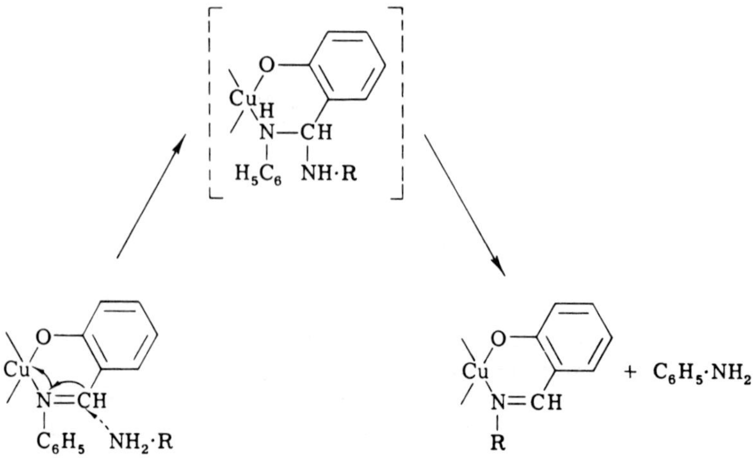

(XXXVIII)

amine, cannot be overlooked. Such intermediates are important in the ligand exchange reaction of planar metal complexes.

F. Platinum Chelates

A few observations have been made of the reactivity of organic molecules in planar platinum(II) complexes, which, in general, are comparatively inert. The organic molecule remains attached and it is feasible that the reactant coordinates in the 1 or 6 position prior to the reaction but the weak bonding characteristics of these positions render the proposed mechanism uncertain.

The uncoordinated amide groups of bis(asparagine)platinum(II) are readily hydrolyzed by hot dilute alkali at a rate which is almost double that of free asparagine (XXXIX) (Volshtein and Anokhova, 1959). If it

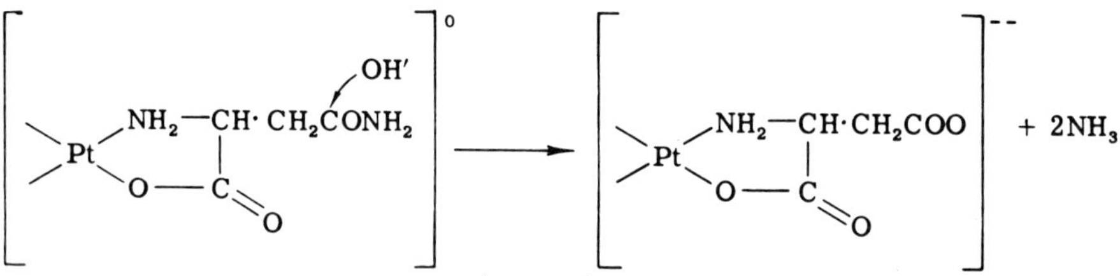

(XXXIX)

is assumed that hydrolysis occurs by the usual nucleophilic carbonyl attack mechanism, this would be facilitated in some degree by the coordination of the electronegative, though remote, platinum atom. A more obvious role of coordination is the destruction of the acidity of the asparagine, so that, in effect, we are comparing the rates of hydrolysis of an amide group in an almost neutral molecule and in an anion. The carbonyl group is more positive in the former than in the latter situation, and the slight polarization induced by the platinum atom is a subsidiary effect. The abstraction of a proton from the adjacent —CH_2— group in alkaline solution cannot be entirely discounted, but the complex is certainly a weaker acid than free asparagine.

When dichlorobis(acetonitrile)platinum(II) is treated with silver sulfate solution, silver chloride precipitates, and the yellow solution gradually becomes deep blue because of the formation of the neutral acetamide derivative $[Pt(NHCOCH_3)_2]^0$ (Hoffman and Bugge, 1908). The four-membered chelate ring structure has been established by Cernjaev and Nazarova (1951), who prepared the substance directly from chloroplatinate(II) ion and acetamide. The removal of the chlorine atoms yields the diaquobis(acetonitrile) complex, which undergoes rearrangement (hydrolysis of the nitrile). Sodium acetate and other basic substances may be substituted for silver salts in replacing the Cl groups (XL). Hydrolysis takes

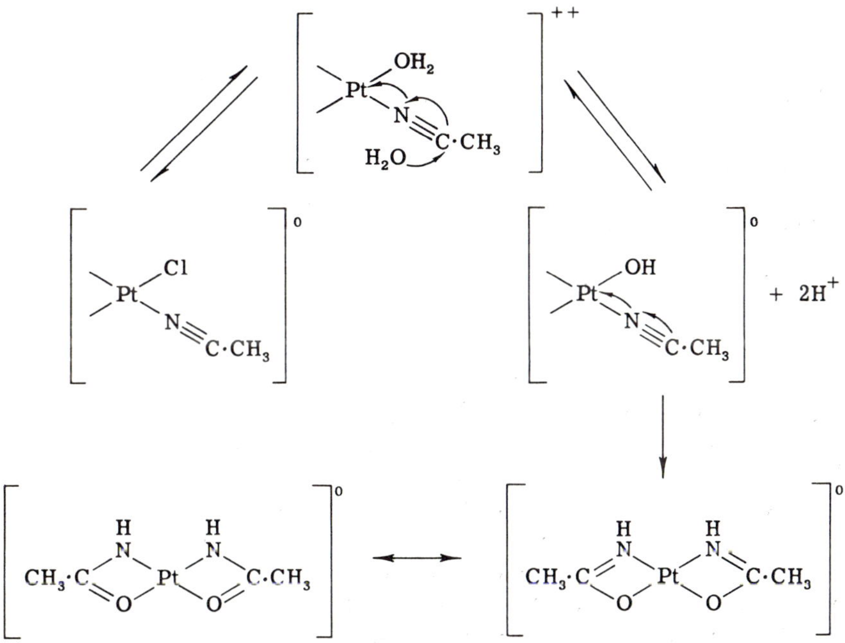

(XL)

place by a nucleophilic attack by a water molecule at the positively charged carbon atom, and elimination of the hydroxyl or aquo group attached to the platinum. The linear stereochemistry of the Pt—N≡C·CH_3 grouping makes an intramolecular reaction unlikely. Unlike the previous example, hydrolysis does not go beyond the amide stage. Hydrolysis to acetic acid would break the strong Pt—N bond and the chelate ring, and is opposed for just these reasons.

With hot concentrated aqueous ammonia (or, better, liquid ammonia), $[PtCl_2 \cdot (CH_3CN)_2]$ gives a substance to which the formula $[Pt(CH_3CN)_2(NH_3)_4]Cl_2$ has been assigned, and which has been quoted as an authentic example of six-covalent Pt(II) (Sidgwick, 1950). The recent crystal structure analysis has shown that in reality it is diamminobis(methylamidine)-platinum(II) chloride (XLI) (Stephenson, 1961). Amidines are usually pre-

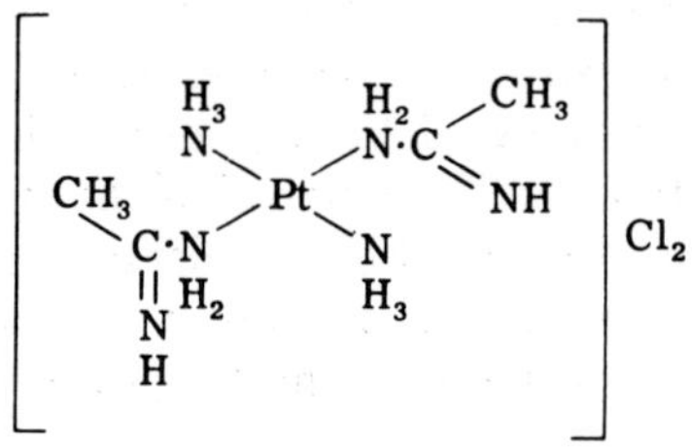

(XLI)

pared by heating a nitrile with solid ammonium chloride, but are not formed when aqueous ammonia is used. The first product of the reaction with ammonia is probably $[Pt(CH_3CN)_2(NH_3)_2]Cl_2$ analogous to the diaquo cation in(XL), and the amidine complex results by attack at the positive carbon atom by ammonia or amide ion. With very dilute aqueous ammonia, as expected, the blue acetamide chelate is formed to some extent.

III. Metal Ion Catalysis

A. Introduction

In this section are discussed a number of reactions that take place more rapidly in the presence of metal ions, in which one or all of the reactants contain donor atoms suitably disposed for attachment to the metal ion. The identity of the most stable chelate formed can usually be inferred from coordination chemical concepts of ring size, donor atom preference, stereochemistry, etc., or from studies of absorption spectra. However, it cannot be inferred that the most stable species is also the kinetically active species. This may instead be a minor component, of low stability, whose existence would not be seriously considered on classical concepts. The mechanisms ascribed to the catalyses are, to this extent, speculative.

B. Phosphate Group Reactions

The inorganic polyphosphates are analogous to a large group of biologically important substances containing —P—O—P— links, and information about the hydrolysis and metal complexing tendencies of the former can be extended in some measure to the latter. The inorganic polymeric structures are based upon the tetrahedral arrangement of four oxygen atoms about each phosphorus atom—the triphosphate structure, for instance, consisting of three such tetrahedra linked together. The linking may be such that the tetrahedra form a cyclic structure, as has been demonstrated in the crystalline state with cyclotrimetaphosphate anion $[P_3O_9]^{3-}$. The higher polymers with the general formula $[(PO_3)_nPO_4]^{(n+4)-}$ exist as glasses with a more or less random distribution of the cyclic or long chain molecular units (Audrieth and Hill, 1948; Quimby, 1947; Topley, 1949). The metal complexing power increases with increasing polymerization and is especially evident in the glassy polymers. Besides calcium and magnesium ions, sodium and potassium ions may also be bound to the polymeric anion, though much more weakly (Van Wazer and Campanella, 1950). The polyphosphates are thermodynamically unstable with respect to orthophosphate ion in aqueous solution and undergo hydrolysis at elevated temperatures and low pH values (Van Wazer *et al.*, 1955). Protonation of the oxygen atoms increases electron movement away from the phosphorus atoms and facilitates nucleophilic attack by water molecules (XLII). Since there are

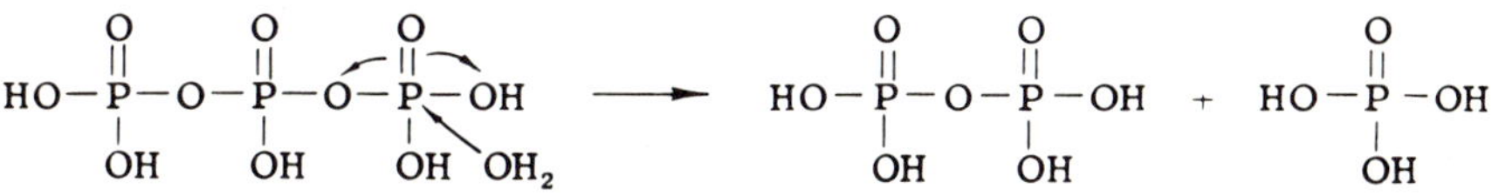

(XLII)

a large number of available donor oxygen atoms disposed in such a way that metal chelate rings may be formed it is usually supposed that the metals are bound in this way. It can be seen from (XLII) that attachment to two oxygen atoms of the same phosphorus atom can yield a relatively unstable four-membered ring, whereas, if oxygen atoms of adjacent phosphorus atoms are used, a six-membered ring would result. It has been proposed that with suitable orientation the polyphosphate chain may coordinate as a tridentate (Chaberek and Martell, 1959). The glassy polyphosphates show ion exchange behavior, and as well as possessing additional sites for bonding (statistical effect) have an additional negative charge per molecular unit. These facts may be more important in metal binding than in the possibility of multidentate function, especially with calcium and magnesium ions where the bonds are largely electrostatic.

The hydrolysis of polyphosphates is catalyzed by calcium and magnesium ions especially in alkaline solution (Green, 1950; Topley, 1949). The latter ion (5×10^{-4} M) increases the rate of hydrolysis of polymetaphosphate (8×10^{-2} M) at 100° by a factor of ten. Coordination (or association) with the metal atom increases the positive charge at the phosphorus atom facilitating attack by water or hydroxyl ion. Where the metal atom is chelated, the P—O—P link within the chelate ring should be stabilized and hydrolysis probably occurs at proximate phosphorus atoms (XLIII).

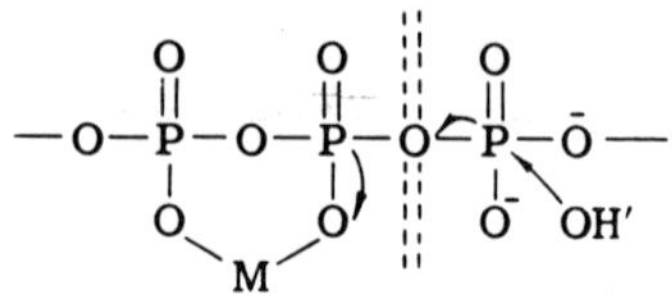

(XLIII)

The biologically important transfer of a phosphoryl group from one molecule to another is an enzymatic process with what appears to be an absolute metal ion requirement: Mg^{++}, Mn^{++}, Ca^{++}. A nonenzymatic transphosphorylation which occurs between adenosine triphosphate (ATP) and orthophosphate ion has recently been described. It is catalyzed by bivalent metal ions, notably Mn^{++}, Ca^{++}, and Cd^{++} (Lowenstein, 1958). Instead of water or hydroxyl ion, the nucleophilic attacking agent at the activated phosphorus atom is the orthophosphate (or orthophosphate monoester) ion. The proposed mechanism visualizes the binding of the metal ion to the ATP, which acts as a bidentate or tridentate ligand, and also to the entering phosphate ion. The metal ion has the dual function of orientating the nucleophile and at the same time rendering the phosphorus atom of the ATP more susceptible to attack. The enzymatic process is considered to be a more specific, or more highly orientated, form of the nonenzymatic reaction, rendered so by the binding of the metal phosphate chelate to the enzyme surface through vacant coordination positions. It is probably not necessary to bind the metal to the ATP at more than one oxygen atom and the orientating function of the enzyme would be more important thereby (XLIV).

The hydrolytic detoxification of alkylated fluorophosphates and related substances, which are powerful anticholinesterase drugs and are used as insecticides, is brought about *in vivo* by an enzyme widely distributed in tissues and plasma. Diisopropylfluorophosphate (DFP) is excreted as diisopropylphosphate ion as a result of the replacement of the fluorine atom byhydroxyl. Hydrolysis *in vitro* is strongly catalyzed by a large number of coordinately unsaturated metal complexes. The first group, discovered by Wagner-Jauregg *et al.* (1955) comprises the Cu(II) 1:1 chelates of 2,2′-bipyridine, 1,10-phenanthroline, histidine, ethylenediamine, and simple amino acids. The rate of hydrolysis increases considerably with increasing pH, and, since the hydrated Cu(II) ion has some catalytic effect, it is probable that an important function of the chelating agent is the retention of

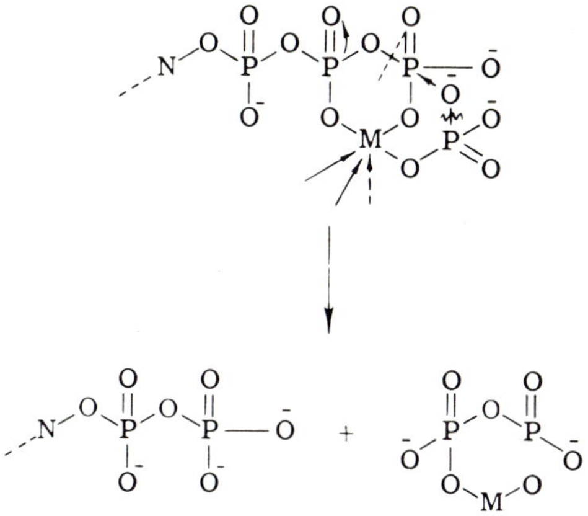

(XLIV)

the metal in solution at high pH values. The substrate may be bound directly to the metal chelate or through hydrogen bonds, but, in any event, polarization of the P=O and P=F bonds facilitates nucleophilic attack by hydroxyl ion (XLV, XLVI). Like the hydrated Cu(II) ion, the diaquo

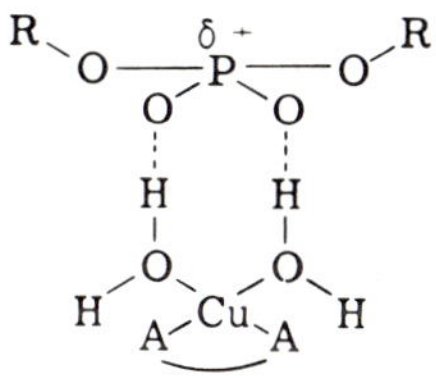

(XLV)

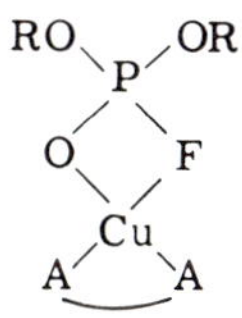

(XLVI)

chelates (1:1 complexes with bidentate ligands) are acidic and at the reaction pH (values 6–8) exist partly, at least, in the hydroxoaquo form

$$[Cu(bipy)(H_2O)_2]^{++} \rightleftharpoons [Cu(bipy)OHH_2O]^{+} + H^{+}$$

(Martell *et al.*, 1957). An alternative intermediate is, thus, the combination of substrate with the hydroxoaquo chelate (XLVII), the hydroxyl group

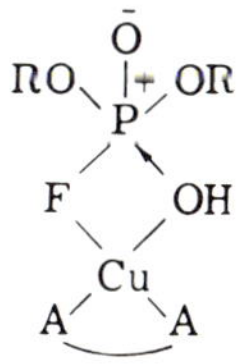

(XLVII)

of which would constitute the nucleophile. A detailed study of the structure of the complex in relation to catalytic activity has been made by Courtney *et al.* (1957) using Cu(II) ion in association with a very large group of bi-, tri-, and quadridentate ligands. In the latter categories it is probable that some proportion of the chelate exists in the hydroxo or aquo tridentate form. The highest activity was found with bidentate ligands, notably N,N,N′,N′-tetramethylethylenediamine. Chelating agents which reduce the positive charge of the copper chelate (amino acids) or form negatively charged complexes (EDTA) have lowered or no catalytic activity. The complexes derived from oxo metal cations UO_2^{++}, ZrO^{++}, ThO^{++}, and MoO_2^{++} also have considerable activity. Since these may well exist in solution partly in the dihydroxo form, their presence provides some support for the active intermediate (XLVII). Dimethylaminoethoxycyanophosphate is hydrolyzed, by replacement of the cyano group, in the presence of a number of aquo cations and chloro complexes which have the relative orders of activity Cu(II) > *P*d(II) > Au(III) > Ag(I) > Ni(II) > Co(II) > Zn (Augustinsson and Heimburger, 1955).

The catalyzed hydrolysis of α-glycerylphosphate in the presence of metal hydroxide gels, notably lanthanum hydroxide, was observed by Bamann and Meisenheimer (1938). More recently, Butcher and Westheimer (1955) studied the lanthanum hydroxide–promoted catalysis of a group of simple phosphate esters, and showed, for example, a catalytic acceleration of more than a thousandfold with hydroxyethylphosphate at pH 8.5. By the use of optically active methoxypropyl-2-phosphate and O^{18} tracer studies it was shown that both the acid-catalyzed reaction, which has a maximum rate at pH 4, and the lanthanum hydroxide reaction (pH 8.5) proceed with cleavage of the P—O and not of the C—O bonds. On the assumption that the lanthanum hydroxide reacts in the form $[LaOH]^{++}$ and because a Lewis acid can be equated to a proton, similar intermediates were proposed (XLVIII). The considerably higher rate of hydrolysis of

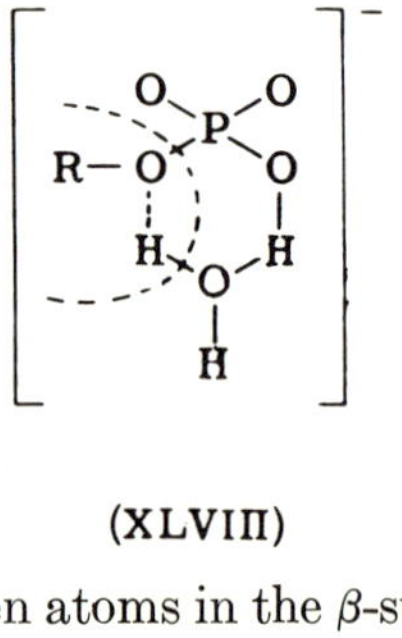

(XLVIII)

esters with oxygen or nitrogen atoms in the β-substituent has been ascribed to coordination of these atoms to the metal (XLIX), i.e. stronger binding of the substrate. Alternatively, it may be proposed that the hydroxo-

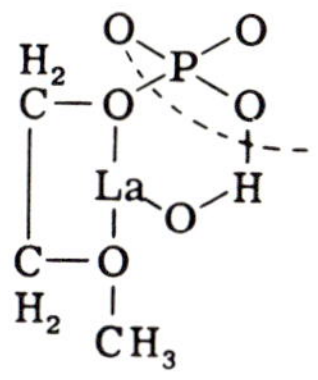

(XLIX)

lanthanum ion, while rendering the phosphorus atom more susceptible to attack, also provides the hydroxyl group as the nucleophile. Donor atoms in the β-substituent raise the rate of hydrolysis by orientation of the nucleophile and the phosphorus atom (L). Since the reaction occurs hetero-

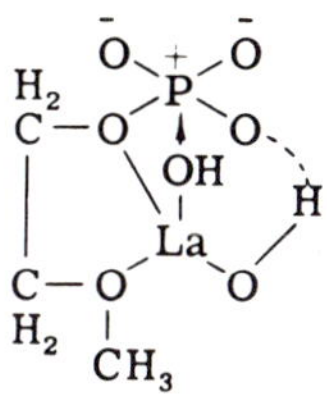

(L)

geneously, proposals about the nature of the intermediate are, of course, speculative. Enhanced rates of ester hydrolysis have been observed with trialkylphosphites coordinated to Pt(II) (Troitskaya, 1953; Troitskaya and Itskovich, 1954). The complexes were of uncertain composition, possibly $[Pt(P(OMe)_3)_4]Cl_2$ and $[Pt(P(OEt)_3)_3 \cdot C_6H_5NH_2]Cl_2$. The direct Pt—P bond and the over-all positive charge, which also operates against charge reduction at the phosphorus atom by π bonding, promote hydroxyl ion or water addition.

C. Amino Acid Esters

Reference has already been made to the hydrolysis of the amino group from asparagine in its platinum complex (Section II,F). The hydrolysis of amino acid esters and amides catalyzed by bivalent metal ions has been studied by Kroll (1952) and Meriwether and Westheimer (1956). The proposed mechanism is in accord with the principle enunciated by Schwarzenbach (1949) that metal complexes are stronger Lewis acids than the free ligand molecules and the increase in acid strength is related to the complexing tendency (and the charge) of the central metal atom. Kroll concluded that the amino acid esters formed the 1:1 metal complexes in the pH range 7.5–8.5 and the rates of hydrolysis followed the stability sequence Cu > Co(II) > Mn ≫ Ca, Mg. The latter two ions had little or no effect on the rate. The proposed mechanism shown in (LI), in essence, follows the accepted scheme for the acid-catalyzed bimolecular hydrolysis of esters. Unlike the latter, the reaction is irreversible because of the high stability of the amino acid chelates, and this may be an important consideration.

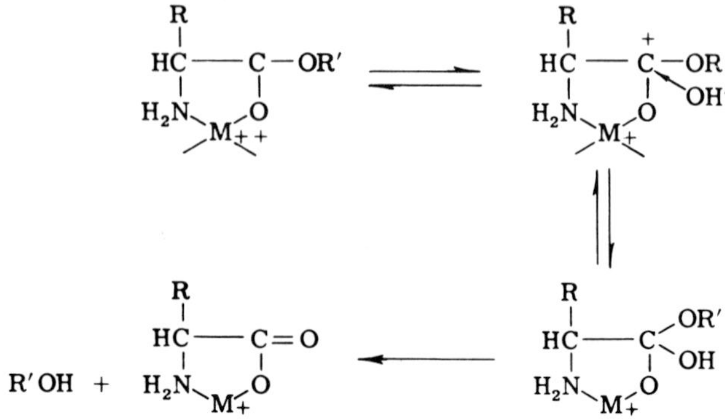

(LI)

A similar mechanism may be applied to the amide hydrolysis (—OR′=NH_2) which occurs in acid solution (pH 5), but the structure of the metal substrate intermediate is in some doubt. Rising and Yang (1933) have shown that the Cu(II) bis complex of glycine amide prepared under alkaline conditions is a nonelectrolyte in aqueous solution, and hence two protons must have been detached from the amide group. Recent work has confirmed the acidic function of amide groups in peptides, and the detachment of protons from metal-peptide complexes in alkaline solution (Koltun *et al.*, 1960; Koltun and Gurd, 1959; Martin *et al.*, 1960). At pH 5, however, it seems likely that a substantial proportion of the protonated species is present. Meriwether and Westheimer (1956) have pointed out, however, very pertinently that the effect of metal ions on the hydrolysis of glycine amide, though large, compares very unfavorably with the enormous catalytic effects obtained with metal ion–catalyzed peptidases.

D. Amino Acid–Pyridoxal Reactions

A large group of essential amino acid reactions are catalyzed by pyridoxal-phosphate enzyme, and nonenzymatically, though much less efficiently, by pyridoxal in the presence of metal ions. The reactions include transamination, racemization, and decarboxylation and are usually related to Pfeiffer's observations on stable azomethine metal chelates of definite composition (Section II,E), which are often considered to mirror the enzymatic role of metal ions. Careful appraisal of the data summarized by Snell (1958) leaves no doubt that the catalytic properties of pyridoxal-containing enzymes are but a potentiated form of their prosthetic group:

pyridoxal. It may be supposed that the metal ion in the nonenzymatic reactions performs, in an inferior fashion, the same function as the protein in the enzyme. The well authenticated enzymatic reaction, glutamate ion + pyruvate ion ⇌ α-ketoglutarate + alanine, is believed to occur in two steps:

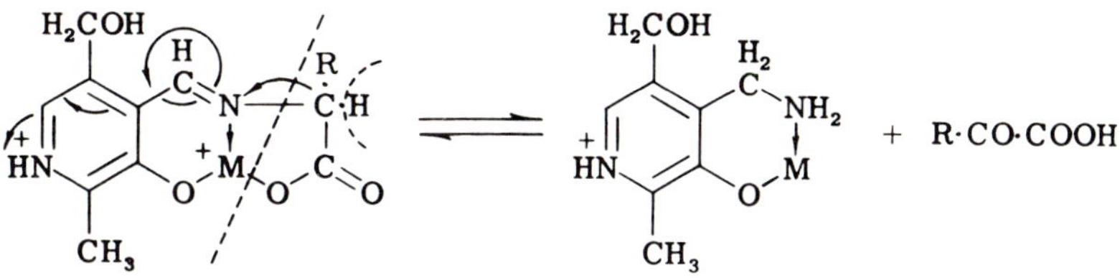

(LII) (LIII)

(1) glutamate ion + pyridoxal ⇌ pyridoxamine + α-ketoglutarate ion ($RCO{\cdot}COO^-$) [(LII) and (LIII)]
(2) pyruvate ion ($R'CO{\cdot}COO^-$) + pyridoxamine ⇌ alanine + pyridoxal

Detailed nonenzymatic investigation of step (1) has shown that the reaction proceeds slowly to the 47% equilibrium concentration of pyridoxamine in the absence of metal ions, and even in the presence of ethylenediaminetetraacetic acid, added to suppress trace metal ion impurities. The reaction is catalyzed by metal ions in the decreasing order $Ga^{3+} > Cu^{++} > Al^{3+} > Fe^{++} > Fe^{3+} > In^{3+} > Ni^{++} > Co^{++}$. With Al^{3+} ion, which has been generally used, a tenfold enhancement occurs, and Cu^{++} ion is about twice as effective as Al^{3+} ion. The essential structural and electronic features of pyridoxal, notably the strong electronic withdrawal by the nitrogen atom, can be duplicated largely by other aldehydes: 2-formyl-3-hydroxypyridine, ω-methylpyridoxal, 4-nitrosalicylaldehyde, etc., and these molecules can be transformed to the amino acid via the Schiff base through a series of conjugated double bonds (Metzler *et al.*, 1954a). It will be seen that in step (1) the NH_2 group is transferred to the pyridoxal moiety. In step (2) the Schiff base chelate is reformed from $R'CO{\cdot}COO^-$, and $R'CH(NH_2)COO^-$ results on dissociation. The formation of Schiff base is evident from the yellow color that develops immediately pyridoxal and the amino acid are mixed, and various metal complexes (1:1 or 1:2) have been isolated from alcoholic solutions (Banks *et al.*, 1961) or their existence inferred from spectrophotometric studies (Eichorn and Dawes, 1954). Molecular models show that the Schiff bases may behave as bi- or tridentate ligands [(LIV) and (LV)]. Two molecules, acting as bidentates, cannot be coordinated in a planar configuration (Cu^{++}) in the usual stable *trans* arrangement because of group clashing, and in the *cis* arrangement severe

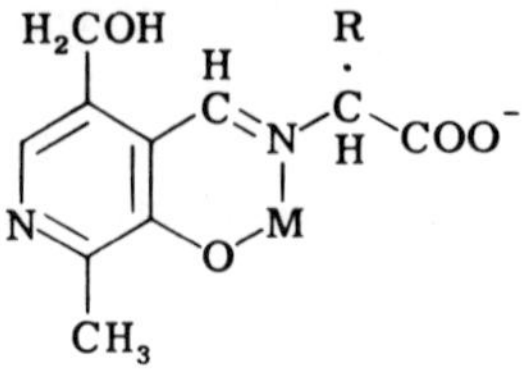

(LIV)

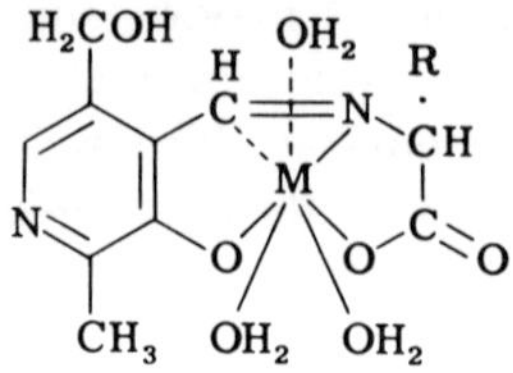

(LV)

interference occurs between the α-carbon atoms of the amino acid and the coordinated nitrogen atoms. Strainless octahedral coordination with one or two of the ligand molecules in the angular configuration (LV) is possible. With one tridentate ligand a molecule of coordinated water is sufficiently close to the α-carbon atom to be possibly kinetically significant.

It has been suggested that the catalytic metal atom (1) promotes Schiff base formation, (2) maintains the coplanarity necessary for maximum conjugation, and (3) acts synergistically with the heterocyclic nitrogen atom in withdrawing electrons. The first proposal has been inferred from the higher concentration of the Schiff base obtained in alcoholic solution wherein hydrolysis is limited and transamination occurs rapidly without a metal ion requirement. Similarly, the presence of the metal ion does not seem necessary when the reactants are absorbed on paper. In aqueous solution the Schiff bases are probably soon in equilibrium with their generators, and, because of the electronegative heterocyclic nitrogen atom, are relatively weak complexing agents. The unusual trend in catalytic potency, $Ga^{3+} > Cu^{++} > Al^{3+}$, suggests that weak complexes are the active intermediates, and this is consistent with the principle that trace metal catalysis must necessarily involve labile complexes with both the reactants and the products. As Snell (1958) has pointed out, the demonstration that Schiff base complexes can be formed does not establish them as the reactive species. Mixed complexes containing pyridoxal and/or amino acid, as well as species with other than five- or six-membered ring structures may be present. A small optical specificity has been observed in the catalyzed transamination between *d*-alanine and α-ketoglutamate ion (Longenecker and Snell, 1956). This implies that the transition metal complex itself is dissymmetric and one optical form predominates. However, whatever the mode of attachment, the true role of the metal atom remains obscure, though the catalytic efficiency of small, highly polarizing cations like Al^{3+} supports at least the idea of a supplementary electron-withdrawing function.

E. Bromination of Ethylacetoacetate

The metal ion–catalyzed bromination of ethylacetoacetate and the cyclic ketoester 2-carbethoxycyclopentanone illustrates the effect of metal ions, in suitable circumstances, in facilitating the detachment of protons from a more remote part of the molecule. In the absence of metal ions, both brominations are base-catalyzed and follow the mechanism discussed previously (Section II,B). In acidic solution in the presence of metal ions in excess, the cationic mono chelate is rapidly formed, and this undergoes slow conversion to the mesomeric dipolar species, which then brominates rapidly (LVI). The relative order of catalytic power is approximately the

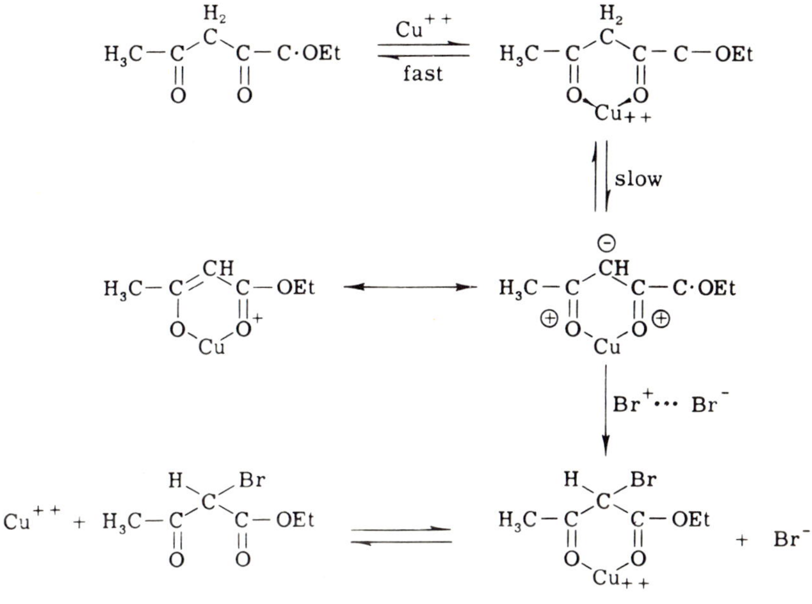

(LVI)

order of stability of the metal complexes: Cu > Ni > La > Zn (Pedersen, 1934, 1948).

F. Beckmann Rearrangement of Aldoximes

Generally, under the conditions that cause the rearrangement of ketoximes, aldoximes are either converted to nitriles or regenerated rapidly. The rearrangement of a number of aldoximes to the amides has been carried out with Raney nickel in various solvents at 100°. α-(*anti*)-Furfuraldoxime [(LVII), H·Ox] thus gives furamide (LVIII) in 80% yield, as well as a reddish substance which is probably a nickel complex (Paul, 1937). Whereas (LVII) may form chelates involving the oxime and furan oxygen atoms, the metal chelates are somewhat uncertain in composition and may be polymeric, and coordination to nitrogen alone in the usual way seems

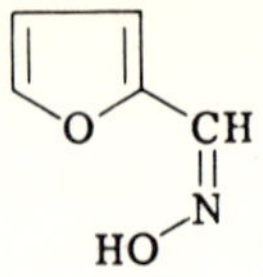

(LVII) = H·Ox

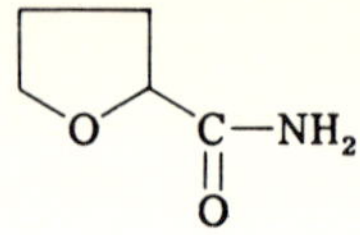

(LVIII)

probable (Bryson and Dwyer, 1940a). The neutral tetrakis complex [$NiOx_2$·$(H{\cdot}Ox)_2$] eliminates one or two molecules of the ligand in benzene or acetone solution as furamide (Bryson and Dwyer, 1940b). In benzene at room temperature in the presence of excess oxime, the rearrangement is catalyzed, presumably by elimination of the amide and reformation of the tetrakis complex. Molecular models of the planar nickel complex show that the oxygen atom (or hydroxyl group) of one oxime ligand is very close to the —CH=N— group of its coordinated neighbor, and suggest the feasibility of attack by the oxygen atom at the carbon atom. The mechanism of rearrangement (LIX) is very similar to that proposed for the Beckmann

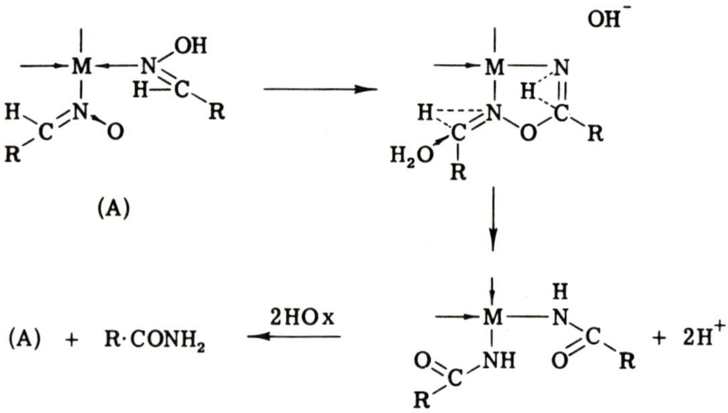

(LIX)

conversion of ketoximes to alkyl acid amides, but is cyclic and cooperative, two oxime molecules rearranging simultaneously. We may suppose that the oxygen-carbon attack occurs with synchronous migration of the H atom to the N atom and dissociation of an hydroxyl group. At the same time a water molecule (or hydroxyl ion) attacks the other carbon atom (lower left-hand side of diagram) from which the H atom is migrating to the N atom. Two amide ions dissociate from the metal and take up protons, and the vacant coordination positions are filled with fresh oxime molecules. This type of metal-catalyzed conversion has been shown recently to take place with a number of other *anti*-aldoximes (Field *et al.*, 1961).

G. Decarboxylation

The decarboxylation of oxalacetic acid—a typical β-ketodicarboxylic acid—occurs spontaneously in aqueous solution (pH 4). It is catalyzed by enzymes (β-carboxylases) and many metal ions (Krebs, 1942; Speck, 1949). It is significant that metal-catalyzed decarboxylation does not occur with ketomonocarboxylic acids, e.g., acetoacetic acid, or with the monoesters of ketodicarboxylic acids and a comparison with nitroacetic acid may be made with profit. Both the undissociated acid and the metal chelates are stable but the anion decomposes into carbon dioxide and the anion of *aci*-nitromethane:

$$O_2N\cdot CH_2COO \longrightarrow CH_2NO_2^- + CO_2$$

In the metal chelate the carboxylate group forms part of the ring and is thereby stabilized against cleavage; for another reason, the metal atom can be expected to hinder the displacement of electrons towards the nitro group (Pedersen, 1949, 1952). A study of the decarboxylation of dimethyloxalacetic acid and its monoethyl ester in the presence and absence of metal ions has been made by Steinberger and Westheimer (1951). The decarboxylation of the ester is not catalyzed by metal ion and the acid, from pH studies, appears to form the intermediate metal complex in the dianion form. Although sufficient of the keto acid is present to form coordinately saturated metal chelates, the active form appears to be the monochelated form. This is shown, for instance, by the competitive effects of anions such as citrate which can displace water molecules attached at the vacant coordination positions, and thereby reduce the positive charge at the metal. A not unreasonable role is ascribed to the metal in the mechanism shown in (LXII). The metal chelate of the enolic form of α-ketoisovalerate (LXI) must of necessity be capable of rapid conversion (probably via dissociation) to (LX). The replacement of the coordinated water molecules by neutral ligands of lower electron availability, preferably with some π bonding, should increase the positive charge at the metal atom, and hence promote decarboxylation. This effect, indeed, is shown well by pyridine, and would probably be shown better by 2,2′-bipyridine. Addition of enzyme to the metal ion catalyzed reaction does not notably enhance the rate; thus the enzyme could conceivably function by binding of the metal atom to sites of low electron availability.

H. Oxidation-Reduction Reactions

Though spontaneous reaction is thermodynamically possible, the oxidation (or reduction) of many substrates (chiefly organic molecules) is often

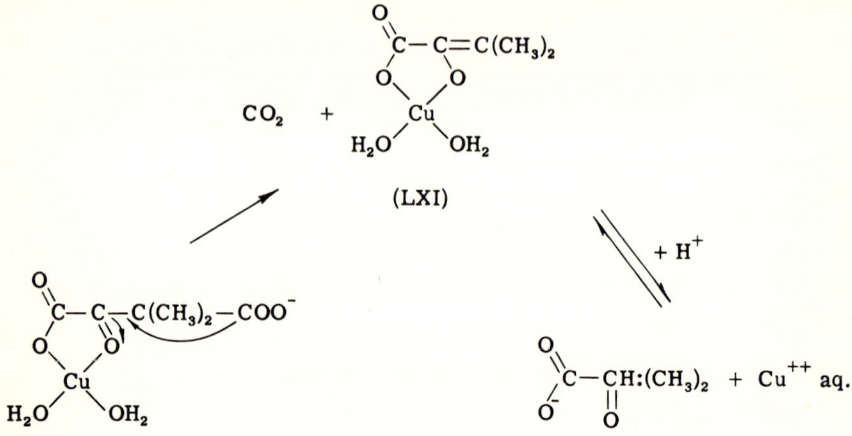

(LXI)

(LX) (LXII)

very slow. Metal ions or metal complexes can sometimes catalyze these reactions by providing energetically easier paths or by reason of special steric factors. In generalized form we may write:

$$S + A^{+} \overset{\text{slow}}{\rightleftarrows} S^{+} + A$$

in which S and A^+ represent the reducing substrate and the oxidizing agent respectively. The intervention of a metal complex [Mc], which may be an aquo transition metal ion but usually has groups other than water molecules attached in order to have a suitable redox potential (cf. Chapter 6), may occur in two main ways:

$$(1) \qquad \begin{cases} [\mathrm{Mc}] + A^{+} \rightleftharpoons [\mathrm{Mc}]^{+} + A \\ [\mathrm{Mc}]^{+} + S \rightleftharpoons [\mathrm{Mc}] + S^{+} \end{cases}$$

$$(2) \qquad S + [\mathrm{Mc}] + A^{+} \rightleftharpoons S\text{—}[\mathrm{Mc}]\text{—}A^{+} \rightleftharpoons S^{+} + [\mathrm{Mc}] + A$$

Numerous variations upon the essential types of catalytic action shown in (1) and (2) are possible. In Fenton's reagent, which is one of the most powerful inorganic oxidizing reagents, the actual oxidizing species, hydroxyl radical, is obtained from hydrogen peroxide by the use of ferrous ion:

$$HO\text{—}OH + Fe^{++} \rightarrow [FeOH]^{++} + OH$$

A weaker oxidizing system arises when hydrogen peroxide, instead, adds an electron to ferric ion:

$$H_2O_2 + Fe^{3+} \rightleftarrows Fe^{++} + H^{+} + HO_2$$

The catalytic agent may be Cr^{++}, V^{++}, or Cu^{+} ions, or complexes with suitable redox potentials which can even provide sites for taking up hydroxyl ion. In scheme (2) an intermediate which binds both the electron acceptor and donor in suitable proximity is visualized. In general the metal atom itself will mediate the electron transfer through a higher or lower oxidation state which may be even in the nature of a strongly polarized intermediate between the oxidation states.

Many direct oxidations with molecular oxygen are slow. The first reaction product with a metal ion or complex is usually an addition compound which may be formed reversibly (cf. Chapter 10) or go on by the substraction of an electron to the higher oxidation state of the metal ion or complex and the production of hydrogen peroxide or a reactive species HO_2 or a peroxo metal radical:

$$Co^{++} + O_2 \rightleftharpoons CoO_2$$

Salicylic acid and benzoic acids are oxidized by molecular oxygen (or hydrogen peroxide) to mixtures of 2,3- and 2,5-dihydroxybenzoic acids and *o*-, *m*- and *p*-hydroxybenzoic acids, respectively, in the presence of the iron(II) ethylenediaminetetraacetato chelate. The latter is oxidized first to the Fe(III) chelate with the production of OH and HO_2 radicals. A coordination position for the attachment of oxygen already exists if the ligand is quinquedentate, or can be provided by dissociation of an acetic acid residue (Grinstead, 1960a). In the presence of a suitable reducing agent (ascorbic acid), the Fe(II) state can be regenerated, and the system behaves as a model peroxidase system (Udenfriend *et al.*, 1954; Grinstead, 1960b). The potential of the system $[Fe(EDTA)]^{--}$–$[Fe(EDTA)]^{-}$ ($E^0 = -0.12$ volt) is much less negative than the aquo Fe^{++}–Fe^{3+} couple ($E^0 = -0.771$ volt).

The autoxidation of ascorbic acid (H_2A) is strongly catalyzed by copper(II) ion (Weissberger *et al.*, 1943; Weissberger and Lu Valle, 1944). The first intermediate is the Cu(II) chelate (LXIII) with ascorbate ion (HA^- or A^{--}), which undergoes internal oxidation-reduction with the production of a Cu(I)-semiquinone chelate (LXIV). The latter dissociates to the Cu^+ ion and the semiquinone, both of which are oxidized by molecular oxygen yielding HO_2 radical and/or hydrogen peroxide.

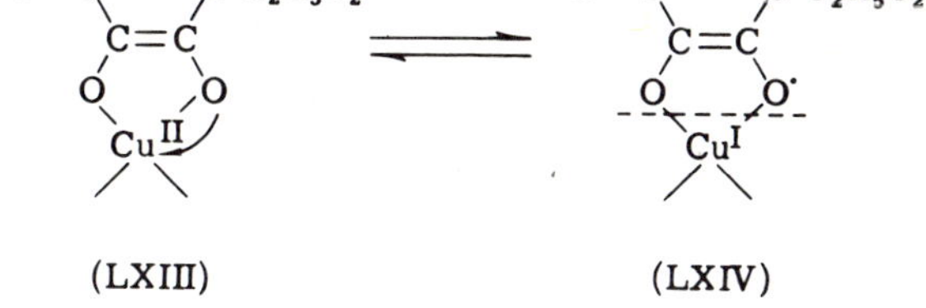

(LXIII) (LXIV)

When air is passed through a solution of copper(I) chloride in pyridine, rapid oxidation to a Cu(II) complex ensues, and, if the oxidation is performed in the presence of aromatic amines, good yields of azo compounds result (Terent'ev and Mogilyanski, 1955). In the same manner, benzoin can be converted to benzil and thence to benzoic acid (Kinoshita, 1954). The $[Cu(py)_2]^+$ ion appears to be rapidly converted to an hydroxy(pyridine)Cu(II) ion which may be (LXV) or the dimer (LXVI).

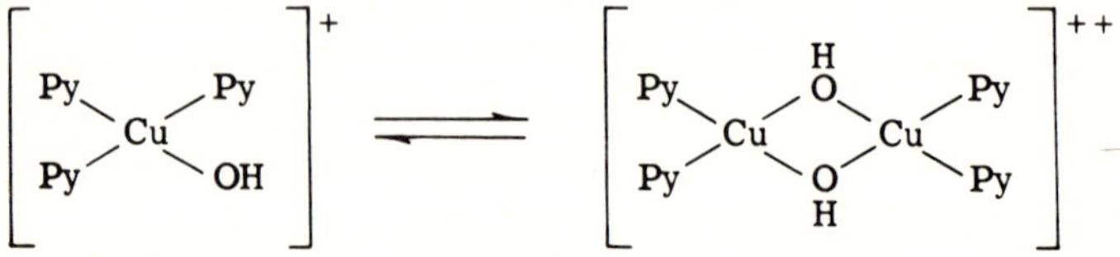

(LXV) (LXVI)

In the presence of excess pyridine the redox potential of the couple Cu^+–Cu^{++} would be about −0.2 to −0.3 volt, and reaction with oxygen might be expected to yield OH and HO_2 radicals, the actual oxidants. However, Kinoshita (1960) has shown that aniline is oxidized in the presence of nitrogen gas by a mixture of pyridine and copper(I) chloride that has been previously fully aerated. The presence of oxidizing substances (H_2O_2) in this solution cannot be excluded. Since the passage of air through a mixture of copper(II) chloride, potassium hydroxide, and pyridine also caused the oxidation reaction, it was concluded that the hydroxy(pyridine)Cu(II) complex was the actual oxidant, which was simply regenerated by the oxygen. It was visualized that the aromatic amine coordinated to the Cu(II) atom lost a proton, which was taken up by the OH group, and was then converted to the $\cdot NH\cdot C_6H_5$ radical. Dimerization yielded hydrazobenzene and the process was repeated. Since the potential of the hydrazobenzene-azobenzene system is about −0.4 volt (Latimer, 1952), this mechanism is not feasible, and a radical oxidation mediated through the Cu(III) state or a pyridine–Cu(II)–oxygen adduct is more probable.

The oxidation of sulfhydryl compounds with molecular oxygen:

$$4RSH + O_2 \rightarrow 2R\cdot S\cdot S\cdot R + 2H_2O$$

is catalyzed by transition metal ions, notably Fe^{++} and Cu^{++} (Michaelis and Schubert, 1930; Schubert, 1931, 1932, 1933). Two reactions will be considered: the spontaneous oxidation-reduction decomposition of the Fe(III) chelate of thioglycolic acid, and then the iron(II)-catalyzed autoxidation. The almost colorless anion $[Fe^{II}(S\cdot CH_2COO)_2]^{--}$ formed under anaerobic conditions develops a blue or red color on the admission of air (acid or alkaline conditions, respectively). The color gradually fades—it may be regenerated with more air—and ultimately either a basic iron(III) salt precipitates or the pale yellow iron(III) complex of dithioglycolic acid ($HOOC\cdot CH_2\cdot S\cdot S\cdot CH_2\cdot COOH$) remains (Cannan and Richardson, 1929).

The red color is due to the monomeric anion $[Fe^{III}(S{\cdot}CH_2COO)_2OH]^{--}$ with a molecule of water occupying the sixth coordination position in *cis* or *trans* relationship to the hydroxyl group (Leussing and Kolthoff, 1953; Leussing and Newman, 1956). Spectrophotometric kinetic studies of the bleaching of the red substance show that the reaction is second order with respect to the total iron(III) content in the pH range 8.4–10.1, and can be explained by the formation of dimeric species containing $[Fe(S{\cdot}CH_2COO)_2$-$OH]^{--}$ with another similar species or $[Fe(S{\cdot}CH_2COO)(OH)_2]^{-}$. Electron transfer from each iron(III) atom to a coordinated sulfur atom occurs within the dimer (LXVII). The resulting two thyl radicals combine and

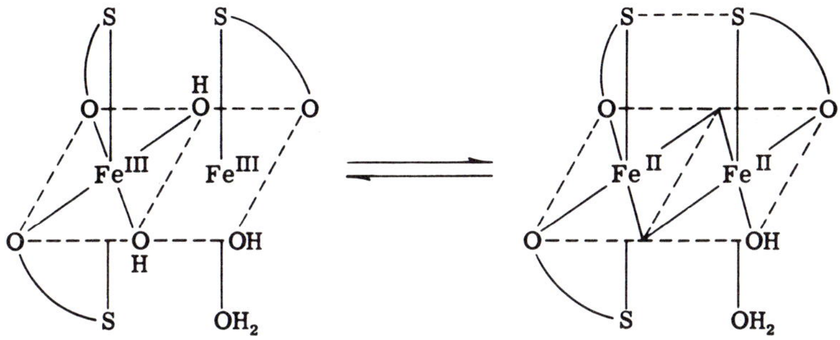

(LXVII) (LXVIII)

split off as the disulfide acid, leaving an iron(II) complex (Leussing and Newman, 1956). It will be evident that the thyl radical can be formed from the monomeric species, and after dissociation might then dimerize. Orgel (1954) has pointed out that the characteristic absorption band of the red complex is due to charge transfer, i.e., the formation of an iron(II) complex and a thyl radical. The first order reaction is evidently not observed because of the rapid back reaction. Molecular models show that with *cis*-hydroxyl bridging in the dimer two sulfur atoms can approach very closely, and if there is any rocking about the diol bridge they can come into contact. Both sulfur atoms can transfer an electron simultaneously to each iron atom, and then link (LXVIII) in competition with the back reaction.

The blue complex, formed in acid solution, has been shown to be the dimeric species $[Fe_2(SCH_2{\cdot}COO)_3(OH_2)^{--}$. Decomposition in acid solution is first order in accordance with the scheme proposed above (Lamfrom and Nielsen, 1957). These authors also concluded from kinetic studies that the Fe(II)-catalyzed oxidation by molecular oxygen does not take place by the same mechanism but probably the free radicals $^{\cdot}O_2^{-}$ and $^{\cdot}OH$ are involved:

$$[Fe(S{\cdot}CH_2COO)_2]^{--} + O_2 + H_2O \rightarrow [Fe(S{\cdot}CH_2COO)_2OH]^{--} + {}^{\cdot}O_2^{-}(OH)$$

$${}^{\cdot}O_2^{-}(OH) + HS{\cdot}CH_2COO^{-} \rightarrow HO_2^{-}(H_2O) + {}^{\cdot}S{\cdot}CH_2COO^{-}$$

$${}^{\cdot}S{\cdot}CH_2COO^{-} + [Fe^{III}(S{\cdot}CH_2COO)_2OH]^{--} \rightarrow (S{\cdot}CH_2COO)_2^{--} + Fe^{II}\text{ complex}$$

In the presence of acetic anhydride, tertiary amine oxides containing at least one *N*-methyl group are converted into secondary amines and formaldehyde

$$R_2CH_3N\cdot O \rightarrow R_2NH + H\cdot CHO$$

(Polonovski and Polonovski, 1927). In aqueous solution these oxidative demethylations are catalyzed at pH 2–7 by iron(III) salts in the presence of chelating agents such as oxalate or tartrate ions (Fish *et al.*, 1955, 1956; Sweeley and Horning, 1957). Trimethylamine-*N*-oxide thus rearranges to formaldehyde and dimethylamine, but a secondary reaction produces, as well, some trimethylamine and formic acid:

$$H\cdot CHO + (CH_3)_3N\cdot O \rightarrow H\cdot COOH + (CH_3)_3N$$

Since the rearrangement is essentially an internal two-electron oxidation of the methyl group at the expense of the nitrogen atom, an obvious role for the metal ion is to assist the transfer of electrons from the methyl group to the $—N^+—O^-$ portion of the molecule. Trimethylamine-*N*-oxide is an oxidizing agent, but its action is slow and irreversible. In hot acid solution (pH 1) it oxidizes a variety of reducing agents, e.g., $[Fe(H_2O)_6]^{++}$, Br^-, $[Fe(bipy)_3]^{++}$, and from oxidation experiments of this kind it can be inferred that the potential of the reaction

$$(CH_3)_3NO + 2H^+ \rightarrow (CH_3)_3N + H_2O + 2e$$

is of the order of -1.1 volts. Rearrangement is catalyzed by a variety of hydrated cations, e.g., Fe^{++}, VO^{++}, as well as a number of transition metal complexes. Catalytic action depends upon the ability of the catalyst to exist, even momentarily, in a higher oxidation state, and, as well, to provide a coordination position for attachment of the amine oxide adjacent to a coordinated water molecule or hydroxyl group (Craig *et al.*, 1961). The optimum redox potential appears to be about -1.0 volt.

Ferric ion does not act as a catalyst except in the presence of chelating agents, e.g., oxalate, tartrate, citrate, or aspartate ions. The nature of the active chelate species is unknown, but it must contain aquo or hydroxo groups, since, for example, excess oxalate ions suppress the reaction. It is now recognized that the higher oxidation states of iron are capable of existence (Kleinberg, 1950), and the ferryl ion FeO^{++} is considered to be involved in the catalase and peroxidase reactions (King and Winfield, 1959). Iron(III) complexes of oxalic or hydroxy acids have not yet been oxidized to a higher oxidation state but, on general principles, oxidation of diaquobis (chelate) species should be possible since not only is the detachment of electrons facilitated by zero or a negative complex charge but the dissociation of a proton from an aquo group would tend to stabilize the

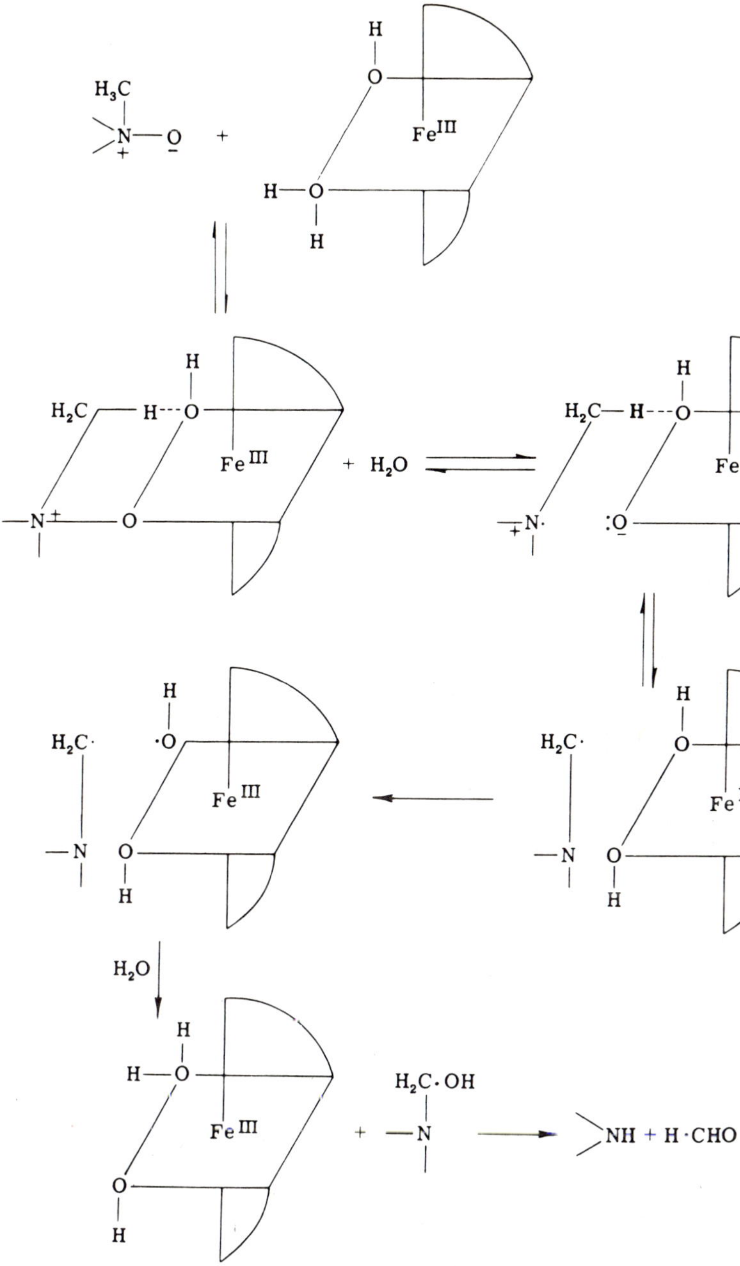
H₃C
N
O
+
H
O
Fe III
H—O
H
H₂C
H---O
Fe III
+ H₂O
N
O
H₂C— H---O
Fe IV
N·
:O
H₂C·
·O
Fe III
N
O
H
H₂C·
O
Fe IV
N
O
H
H₂O
H—O
Fe III
O
H
H₂C·OH
N
NH + H·CHO

(LXIX)

oxidized form. However, the catalytic mode of action envisaged requires only a momentary existence of the oxidized form, which is more in the nature of a highly polarized state of the Fe(III), or other metal atom, than a discrete oxidation state. It will be realized that two competing reactions are always involved: one leading merely to oxidation of the catalyst and the other of the methyl group.

Coordination of the *N*-oxide must occur through the oxygen atom as with phosphine and arsine oxides (Cotton and Bannister, 1960; Nyholm, 1961). With molecular models it can be seen that when coordinated in this way one H atom of a methyl group is quite close to the oxygen atom of an adjacent coordinated hydroxyl group or water molecule. When the N—O—M bond angle is about 120°, the atoms are close enough to form a hydrogen bond, which is sufficiently strong to hold the methyl group in the correct orientation by inhibiting free rotation of the trimethylamine group about the N—O bond. The mechanism proposed by Craig *et al.* (1961) is shown in (LXIX), and for simplicity in the form of discrete steps rather than in the more probable form of an electron flow. The metal atom, e.g., Fe(III), loses an electron to the oxygen atom of the *N*-oxide and becomes oxidized to Fe(IV). An aquo group attached in the Fe(III) complex would tend to shed a proton at this stage. The N—O bond ruptures, the oxygen atom taking with it one electron from the nitrogen atom. In this way the oxygen atom now attached as an oxo group to the iron has been reduced in two one-electron steps. A proton, possibly detached from the methyl group, transforms the oxo to an hydroxo group, and the odd electron on the N atom becomes paired with an electron which has migrated from the carbon atom. The metal atom accepts an electron from the coordinated hydroxyl group, and reverts to the Fe(III) state. The hydroxyl group has been oxidized to an hydroxyl radical which attacks the methylene carbon atom, which has an unpaired electron, forming the methanolamine. The latter rearranges to form the products, and addition of an aquo group to the metal regenerates the original complex. In the mechanism, which may well be of wider applicability, the metal atom performs two functions simultaneously: orientating the molecule and mediating the electron movement.

References

Audrieth, L. F., and Hill, O. F. (1948). *J. Chem. Educ.* **25,** 80.
Augustinsson, K. B., and Heimburger, G. (1955). *Acta Chem. Scand.* **9,** 383.
Auwers, K. V., and Auffenberg, E. (1917). *Ber.* **50,** 929.
Bamann, E., and Meisenheimer, M. (1938). *Ber.* **71,** 1711.
Banks, B. E., Diamantis, A. A., and Vernon, C. A. (1961). *J. Chem. Soc.* p. 4235.
Brandt, W. W., Dwyer, F. P., and Gyarfas, E. C. (1954). *Chem. Revs.* **10,** 960.
Bryson, A., and Dwyer, F. P. (1940a). *J. Proc. Roy. Soc. N.S.Wales* **74,** 455.

Bryson, A., and Dwyer, F. P. (1940b). *J. Proc. Roy. Soc. N.S.Wales* **74,** 471.
Butcher, W. W., and Westheimer, F. H. (1955). *J. Am. Chem. Soc.* **77,** 2420.
Calvin, M., and Wilson, K. (1945). *J. Am. Chem. Soc.* **67,** 2003.
Cannan, R. K., and Richardson, G. M. (1929). *Biochem. J.* **23,** 1242.
Cernjaev, I. I., and Nazarova, L. A. (1951). *Izvest. Sektora Platiny i Drug. Blagarod. Metal. Inst. Obshchei i Neorg. Khim. Akad. Nauk S.S.S.R.* **26,** 101.
Chaberek, S., and Martell, A. E. (1959). "Organic Sequestering Agents," p. 433. Wiley, New York.
Collman, J. P., and Blair, R. P. (1961). Abstr. of papers, *Am. Chem. Soc.* 139th Meeting, St. Louis, Missouri.
Collman, J. P., and Kittleman, E. T. (1961). *J. Am. Chem. Soc.* **83,** 3529.
Collman, J. P., Moss, R. A., Goldby, S. D., and Trahanovsky, W. S. (1960). *Chem. & Ind. (London)* p. 1213.
Collman, J. P., Moss, R. A., Maltz, H., and Heindel, C. C. (1961a). *J. Am. Chem. Soc.* **83,** 531.
Collman, J. P., Moss, R. A., Goldby, S. D., Marshall, R. L., and Young, W. L. (1961b). Abstr. of Papers, *Am. Chem. Soc.* 139th Meeting, St. Louis, Missouri, 4-M.
Cotton, F. A., and Bannister, E. (1960). *J. Chem. Soc.* p. 1873.
Coulson, C. A. (1952). "Valence," pp. 240–249. Oxford Univ. Press (Clarendon), London and New York.
Courtney, R. C., Gustafson, R. L., Westerback, S. T., Hyytiainen, H., Chaberek, S., and Martell, A. E. (1957). *J. Am. Chem. Soc.* **79,** 3030.
Craig, D. P. (1959). "Theoretical Organic Chemistry; Kekulé Symposium," p. 20. Butterworths, London.
Craig, J. C., Dwyer, F. P., Glazer, A. N., and Horning, E. C. (1961). *J. Am. Chem. Soc.* **83,** 1871.
Den Hertog, H. J., and Overhoff, J. (1950). *Rec. trav. chim.* **69,** 468.
Djordjevic, C., Lewis, J., and Nyholm, R. S. (1959). *Chem. & Ind. (London)* p. 122.
Dwyer, F. P. (1961). *Australian J. Sci.* **24,** 97.
Eichorn, G. L. (1961). *Federation Proc.* **20,** 40.
Eichorn, G. L., and Bailar, J. C. (1953). *J. Am. Chem. Soc.* **75,** 2905.
Eichorn, G. L., and Dawes, J. W. (1954). *J. Am. Chem. Soc.* **76,** 5663.
Eichorn, G. L., and Marchand, N. D. (1956). *J. Am. Chem. Soc.* **78,** 2688.
Eichorn, G. L., and Trachtenberg, I. M. (1954). *J. Am. Chem. Soc.* **76,** 5183.
Field, L., Hughmark, P. B., Shumaker, S. H., and Marshall, W. S. (1961). *J. Am. Chem. Soc.* **83,** 1983.
Fish, M. S., Johnson, N. M., Lawrence, E. P., and Horning, E. C. (1955). *Biochim. et Biophys. Acta* **18,** 564.
Fish, M. S., Johnson, N. M., and Horning, E. C. (1956). *J. Am. Chem. Soc.* **78,** 3668.
Gill, N. S., and Nyholm, R. S. (1959). *J. Chem. Soc.* p. 3997.
Green, D. E., Herbert, D., and Subrahmanyan, V. (1941). *J. Biol. Chem.* **138,** 327.
Green, J. (1950). *Ind. Eng. Chem.* **42,** 1542.
Griffiths, D. E., Morrison, J. F., and Ennor, A. H. (1957). *Biochem. J.* **65,** 153.
Grinstead, R. R. (1960a). *J. Am. Chem. Soc.* **82,** 3472.
Grinstead, R. R. (1960b). *J. Am. Chem. Soc.* **82,** 3464.
Gurd, F. R. N., and Wilcox, P. E. (1956). *Advances in Protein Chem.* **11,** 311.
Haginiwa, J. (1955). *J. Pharm. Soc. Japan* **75,** 731, 733.
Harris, C. M., Lenzer, S. L., and Martin, R. L. (1961). *Australian J. Chem.* **14,** 420.
Hoffman, K. A., and Bugge, G. (1908). *Ber.* **41,** 312.
Holm, R., and Cotton, F. (1958). *J. Am. Chem. Soc.* **80,** 5658.

Ikutani, Y., Okuda, T., Sato, M., and Akabori, S. (1959). *Bull. Chem. Soc. Japan* **32,** 203.
Ingold, C. K. (1954). "Chemistry of Carbon Compounds," (E. H. Rodd, ed.), Vol. III, Part A, pp. 3–41. Elsevier, Amsterdam.
Ingold, C. K., Hsu, S. K., and Wilson, C. L. (1935). *J. Chem. Soc.* p. 1778.
Jaffe, H. H. (1954). *J. Am. Chem. Soc.* **76,** 3527.
Jensen, A., Basolo, F., and Neumann, H. M. (1958). *J. Am. Chem. Soc.* **80,** 2354.
King, N. K., and Winfield, M. E. (1959). *Australian J. Chem.* **12,** 47.
Kinoshita, K. (1954). *J. Chem. Soc. Japan* **75,** 48.
Kinoshita, K. (1960). *Bull. Chem. Soc. Japan* **32,** 777, 780, 783.
Kleinberg, J. (1950). "Unfamiliar Oxidation States." Univ. of Kansas Press, Lawrence, Kansas.
Klotz, I. M., and Loh Ming, W. C. (1954). *J. Am. Chem. Soc.* **76,** 805.
Kluiber, R. W. (1960). *J. Am. Chem. Soc.* **82,** 4839.
Koltun, W. L., and Gurd, F. R. N. (1959). *J. Am. Chem. Soc.* **81,** 301.
Koltun, W. L., Fried, M., and Gurd, F. R. N. (1960). *J. Am. Chem. Soc.* **82,** 233.
Krebs, H. A. (1942). *Biochem. J.* **36,** 303.
Kroll, H. (1952). *J. Am. Chem. Soc.* **74,** 2036.
Kurtz, A. C. (1938). *J. Biol. Chem.* **122,** 477.
Kurtz, A. C. (1949). *J. Biol. Chem.* **180,** 1253.
Lamfrom, H., and Nielson, S. O. (1957). *J. Am. Chem. Soc.* **79,** 1966.
Latimer, W. M. (1952). "The Oxidation States of the Elements and Their Potentials in Aqueous Solution," p. 136. Prentice-Hall, Englewood Cliffs, New Jersey.
Lawrence, L., and Moore, W. J. (1951). *J. Am. Chem. Soc.* **73,** 3973.
Leussing, D. L., and Kolthoff, I. M. (1953). *J. Am. Chem. Soc.* **75,** 3904.
Leussing, D. L., and Newman, L. (1956). *J. Am. Chem. Soc.* **78,** 552.
Lindenbaum, A., and Schubert, J. (1956). *J. Phys. Chem.* **60,** 1663.
Lions, F., and McDermott, T. E. (1958). B.Sc. Thesis, University of Sydney, Australia.
Longenecker, J. B., and Snell, E. E. (1956). *Proc. Natl. Acad. Sci. U. S.* **42,** 221.
Lowenstein, J. M. (1958). *Biochem. J.* **70,** 222.
Maki, A., and McGarvey, B. (1958). *J. Chem. Phys.* **29,** 31.
Malmström, B. G., and Rosenberg, A. (1959). *Advances in Enzymol.* **21,** 131.
Martell, A. E., and Calvin, M. (1952). "Chemistry of the Metal Chelate Compounds," pp. 160–167, 400. Prentice-Hall, Englewood Cliffs, New Jersey.
Martell, A. E., Chaberek, S., Courtney, R. C., Westerback, S., and Hyytiainen, H. (1957). *J. Am. Chem. Soc.* **79,** 3036.
Martin, R. J. L. (1957). *Australian J. Chem.* **10,** 256.
Martin, R. B., Chamberlin, M., and Edsall, J. T. (1960). *J. Am. Chem. Soc.* **82,** 495.
Meriwether, L., and Westheimer, F. H. (1956). *J. Am. Chem. Soc.* **78,** 5119.
Metzler, D. E., Longenecker, J. B., and Snell, E. E. (1954a). *J. Am. Chem. Soc.* **76,** 639.
Metzler, D. E., Ikawa, M., and Snell, E. E. (1954b). *J. Am. Chem. Soc.* **76,** 648.
Michael, A., and Carlson, G. H. (1936). *J. Am. Chem. Soc.* **58,** 353.
Michaelis, L., and Schubert, M. P. (1930). *J. Am. Chem. Soc.* **52,** 4418.
Morrison, J. F., O'Sullivan, W. J., and Ogston, A. G. (1961). *Biochim. et Biophys. Acta* **52,** 82.
Murakami, M., and Takahashi, K. (1959). *Bull. Chem. Soc. Japan* **32,** 308.
Murmann, R. K., and Basolo, F. (1955). *J. Am. Chem. Soc.* **77,** 3484.
Nakahara, A. (1959). *Bull. Chem. Soc. Japan* **32,** 1195.
Nyholm, R. S. (1961). *Proc. Chem. Soc.* p. 273.
Oichiai, E. (1953). *J. Org. Chem.* **18,** 534.

Orgel, L. E. (1954). *Quart. Revs. (London)* **8,** 422.
Orgel, L. E. (1958) "Metals and Enzyme Activity" (E. M. Crook, ed.), p. 8. Cambridge Univ. Press, London and New York.
Orgel, L. E., Forman, A., and Murrell, J. N. (1959). *J. Chem. Phys.* **31,** 1129.
Paul, R. (1937). *Compt. Rend. acad. sci.* **204,** 363.
Pauling, L. (1948). *J. Chem. Soc.* p. 1461.
Pedersen, K. J. (1934). *J. Phys. Chem.* **38,** 581, 601.
Pedersen, K. J. (1948). *Acta. Chem. Scand.* **2,** 252, 385.
Pedersen, K. J. (1949). *Acta. Chem. Scand.* **3,** 676.
Pedersen, K. J. (1952). *Acta. Chem. Scand.* **6,** 285.
Pfeiffer, P., Bucholz, E., and Bauer, O. (1931). *J. prakt. Chem.* **129,** 163.
Pfeiffer, P., Offermann, N., and Werner, H. (1942). *J. prakt. Chem.* **160,** 313.
Polonovski, M., and Polonovski, M. (1927). *Bull. soc. chim. France* **41,** 1190.
Quimby, O. T. (1947). *Chem. Revs.* **40,** 141.
Rabin, B. R. (1958). "Metals and Enzyme Activity" (E. M. Crook, ed.), p. 21. Cambridge Univ. Press, London and New York.
Reihlen, H., Illig, R., and Wittig, R. (1925). *Ber.* **58,** 12.
Rising, M. M., and Yang, P. S. (1933). *J. Biol. Chem.* **99,** 755.
Roof, R. (1956). *Acta Cryst.* **9,** 781.
Ross, I. G. (1961). Private communication.
Sato, M., Okawa, K., and Akabori, S. (1957). *Bull. Soc. Chem. Japan* **30,** 937.
Schubert, J. (1954). *In* "Chemical Specificity in Biological Interactions" (F. R. N. Gurd, ed.), p. 116. Academic Press, New York.
Schubert, M. P. (1931). *J. Am. Chem. Soc.* **53,** 3851.
Schubert, M. P. (1932). *J. Am. Chem. Soc.* **54,** 4077.
Schubert, M. P. (1933). *J. Am. Chem. Soc.* **55,** 4563.
Schwarzenbach, G. (1949). *Chimia.* **3,** 1.
Sidgwick, N. V. (1950). "The Chemical Elements and Their Compounds," p. 1583. Oxford Univ. Press (Clarendon), London and New York.
Smith, E. L. (1949). *Federation Proc.* **8,** 581.
Snell, E. E. (1958). *Vitamins and Hormones* **16,** 77.
Speck, J. (1949). *J. Biol. Chem.* **178,** 315.
Steinberger, R., and Westheimer, F. H. (1951). *J. Am. Chem. Soc.* **73,** 429.
Stephenson, N. (1961). *Acta Cryst.* (in press).
Sweeley, C. C., and Horning, E. C. (1957). *J. Am. Chem. Soc.* **76,** 2620.
Taft, J. C., and Jones, M. M. (1960). *J. Am. Chem. Soc.* **82,** 4196.
Terent'ev, A. P., and Mogilyanski, Y. D. (1955). *Doklady Akad. Nauk U.S.S.R.* **103,** 91; *Chem. Abstr.* **50,** 4807.
Topley, B. (1949). *Quart. Revs.* **3,** 345.
Troitskaya, A. D. (1953). *Zhur. Priklad. Khim.* **26,** 781.
Troitskaya, A. D., and Itskovich, T. V. (1954). *Trudy Kazansk. Khim. Teckhnol.* **79,** 19
Tsumaki, T., Antoku, S., and Shito, M. (1960). *Bull. Chem. Soc. Japan* **33,** 1096.
Tucker, S. H. (1959). "Electronic Outline of Organic Chemistry," pp. 449–452. Univ. of London Press, London.
Udenfriend, S., Clark, C. T., Axelrod, J., and Brodie, B. B. (1954). *J. Biol. Chem.* **208,** 731.
Vallee, B. L. (1955). *Advances in Protein Chem.* **10,** 318.
Valliant, V. (1894). *Compt. rend. acad. sci.* **119,** 648.
Van Wazer, J. R., and Campanella, D. A. (1950). *J. Am. Chem. Soc.* **72,** 655.
Van Wazer, J. R., Griffith, E. J., and McCullogh, J. F. (1955). *J. Am. Chem. Soc.* **77,** 287.

Verter, H. S., and Frost, A. E. (1960). *J. Am. Chem. Soc.* **82,** 85.
Volshtein, L. M., and Anokhova, L. S. (1939). *Zhur. Neorg. Khim.* **4,** 325, 1734.
Wagner-Jauregg, T., Hackley, B. E., Lies, T. A., Owens, O. O., and Proper, R. (1955). *J. Am. Chem. Soc.* **77,** 922.
Weissberger, A., and Lu Valle, J. E. (1944). *J. Am. Chem. Soc.* **66,** 700.
Weissberger, A., Lu Valle, J. E., and Thomas, D. S. (1943). *J. Am. Chem. Soc.* **65,** 1934.
West, B. O. (1954). *Nature* **173,** 1187.
Williams, R. J. P. (1953). *Biol. Revs.* **28,** 381.

CHAPTER 9

Metal Chelates in Biological Systems

A. SHULMAN AND F. P. DWYER

Department of Physiology, University of Melbourne, Melbourne, Victoria, Australia, and Department of Biological Inorganic Chemistry, Australian National University, Canberra, A.C.T., Australia

I. Metal Ion Toxicity

A. INTRODUCTION

Approximately 75% of the chemical elements are metals whose distribution in animals can be placed in three categories: (1) the bulk metals—Na, K, Ca, Mg—present as the hydrated ions in relatively large amounts that account for about 99.5% of the total metal content, and which are associated chiefly with the skeleton of vertebrates and the osmotic equilibria; (2) iron and the "essential" trace metals—Zn, Cu, Mn, Mo—that function in metalloenzymes or as enzymatic activators. Most of the remaining metals can be placed in the third category if we include those that may be accumulated because of specific environmental factors. They are ubiquitous in tissues and organs, occur in very small but variable amounts, and no biological role has been assigned to them. Examples are lead and aluminium which occur to the extent of 0.03 mg. and 0.06 mg./100 ml. in the whole blood of civilized man.

The ingestion of many soluble metal salts as well as metal-containing anions such as $(CrO_4)^{--}$ can produce local irritation and tissue damage, or even systemic poisoning when the metal ion is absorbed in sufficient amount into the circulation. Zinc and copper salts cause emesis and considerable gastrointestinal irritation, but the absorption normally is low and the excretory mechanism efficient, and hence systemic poisoning is rare. In Wilson's disease the copper level control mechanisms are deranged with the result that absorption is increased, excretion cannot dispose of the excess and accumulation occurs, especially in the liver.

With the exception of Na^+, K^+, and Ca^{++}, the absorption of metal ions from the gastrointestinal tract is generally poor. Slow absorption is also characteristic of many organic cations such as cetylpyridinium and cetyltrimethylammonium ions. As we shall see later, complex cations are not absorbed or poorly absorbed. Mercuric ion quickly forms the neutral molecule $HgCl_2$ and this is probably the reason for its more rapid absorption. Though the increased uptake of copper in Wilson's disease would suggest an active carrier mechanism, the existence of such carriers operating by coordination remains to be established. With iron, whose absorption has been extensively studied, the ferrous ion appears to be absorbed directly from the duodenum and the jejunum and is captured in the mucosal cells by the protein apoferritin. In the form of ferritin, which is an hydroxoiron(III) phosphate or more probably a polymeric diol bridged complex, the iron is in equilibrium with the ferrous ion of the mucosal cells and the plasma iron (an iron(III) hydroxy-protein adduct or complex with serum protein) of the blood stream. A lowered plasma iron level causes migration of iron from the mucosal cells, depletion of the ferritin-stored iron, and, as a result, increased uptake from the gastrointestinal tract (Granick, 1946). The nature of transferrin, the iron-transporting β-globulin, and the other storage protein complex hemosiderin has been discussed by Eichhorn (1961).

All metal ions which have gained access to the circulation in sufficient amount and are not rapidly excreted are toxic, to an extent depending on the metal itself, by deranging the electrolyte balance, as irritants that damage specific organs, notably the kidney, by affecting the central nervous system, or by interfering with enzymatic processes. Although there is considerable variation in the toxic concentration levels, there is no suggestion that the organism is tolerant to high concentrations of metal ions that serve a useful biological purpose. The normal extracellular potassium ion level is 16–20 mg.% but if doubled by parenteral administration marked effects on nerve and muscle activity occur that can cause death by cardiac depression (Goodman and Gilman, 1955). Lead ion is absorbed poorly from the gastrointestinal tract, though much better from the lungs, but the renal excretion is very limited. Toxic symptoms are evident when the blood level reaches about 0.08 mg.% compared with the normal value of 0.03 mg.%. Aluminium ion, which occurs in the blood of animals in the range 0.05–0.1 mg.% is especially low in toxicity. The single parenteral lethal dose for a dog is 5 gm./kg. body weight of hydrated aluminium sulfate or an increase of more than 5000 in the normal Al^{3+} level (Underhill and Peterman, 1929).

B. Mode of Toxic Action

The toxicity of metal ions of low complexing ability, e.g., Ba^{++}, K^{+}, and Li^{+}, results primarily from electrolyte disturbance or resides in other phenomena associated with the positive charge: competition for or adsorption on negative sites, or the displacement of protons or other cations with consequent alteration of the properties of the protein. Important aspects of the binding of small cations to protein have been discussed by Scatchard *et al.* (1954). The specificities of the interactions are related to the magnitude of the ionic charges and the hydrated ion sizes. However, most metal ions, as well, can form complexes with the many available donor atoms of protein; some have distinct donor atom preferences, e.g., Hg^{++} for sulfur; strongly complexing ions may displace the weaker from their normal coordination sites.

The connection between toxic action and metal coordination was proposed by Voegtlin *et al.* (1923), who advanced the view that the toxic action of the metalloid arsenic on living cells is due to its combination with certain essential thiol compounds present in protoplasm. These conclusions have been supported by other workers and elaborated notably by Peters and Stocken during their studies of the toxicity of warfare arsenicals (Peters, 1936; Peters and Wakelin, 1946; Peters *et al.*, 1946; Stocken and Thompson, 1946). The pyruvate oxidase system is especially sensitive both to arsenical vesicants and arsenite ion, and is inhibited, with the result that the pyruvate blood level is increased. Skin respiration is also reduced, and vesication, the primary cell damage when Lewisite comes into contact with tissues, results from pyruvate enzyme inhibition (Thompson, 1946). The "biochemical lesion" characteristic of toxic arsenic compounds irrespective of whether they are inorganic or organic and of the oxidation state is the result of the formation of an As(III) chelate with suitably disposed —SH groups of essential enzymes[1] (I). This is the prototype of similar reactions with mercury, cadmium, gold, zinc, possibly lead, bismuth, and the metalloid antimony. It should be remembered that arsenic (and probably antimony) have a specific preference for —S^- donor atoms and do not complex with nitrogen, while mercury, for instance, forms strong bonds to nitrogen and

(I)

could bond to the —NH— groups of peptides. The bidentate nature of the arsenic enzyme chelate has been inferred from detoxification studies with mono- and dithiols. Whereas the inhibition can be reversed with the latter, the monothiols are relatively poor competitors for the metalloid (Stocken and Thompson, 1946). Complexes with kerateine, a typical sulfhydryl protein, and 2,3-dimercaptopropanol (BAL), prepared from arsenite ion, chlorovinyldichlorarsine (Lewisite), and phenyldichlorarsine, show that the arsenic is present in the proportion 1As:2SH (Stocken and Thompson, 1946). It has been pointed out, however, that this ratio is not conclusive evidence for the existence of a dithiol and hence of a chelate ring, since the arsenic atom could be bridging two molecules of kerateine (Eagle and Doak, 1951).

Brain pyruvic oxidase is inhibited by the antimonyltartrate and tetrachloraurate(III) ions and by mercuric chloride and the inhibition is reversed by dithiols (Thompson and Whittaker, 1947). Inhibition of another sulfhydryl enzyme succinoxidase *in vitro* has been shown with mercuric chloride, bismuthyltartrate, cadmium, lead, and vanadium(II) ions with effective reversal with dithiols (Barron and Kalnitsky, 1947). The symptoms of mercury, cadmium, gold, and antimony poisoning are consistent with the proposition that the inhibition of sulfhydryl enzymes by chelate formation is the basic mechanism of toxicity; but with lead and most other metals unknown sites in addition to the oxidase enzymes appear to be implicated. Though cobalt(II) ion inhibits —SH enzymes and its toxicity can be abolished with cysteine, the most characteristic reaction from small parenteral or large oral doses is a marked polycythemia or increase in the number of red blood cells.

Industrial poisoning, characterized by very severe pulmonary damage,

[1] The substance lipoic acid, 1,2-ditholane-3-valeric acid is recognized as a coenzyme in many oxidases that are arsenite ion-sensitive. In the reduced form (dithiol) it could chelate As(III), and is known to be inhibited by arsenite ion with BAL reversal (Gunsalus, 1954).

occurs when beryllium-containing dusts are inhaled, and a skin lesion also results when beryllium compounds come into contact with broken tissue. The lethal parenteral dose for small animals is very low (<1 mg./kg.) (White *et al.*, 1951). Since the bonds formed between beryllium and oxygen are strong whereas those to sulfur and especially nitrogen are weak, this metal forms its most stable chelate with oxygen. *In vitro* the beryllium ion is known to inhibit the alkaline phosphatases—a group of enzymes specifically associated with the hydrolysis of phosphate esters at around pH 9. However, it has been concluded that the inhibitory action of beryllium ion on magnesium ion–activated phosphatases does not appear to be related directly to the acute toxic effects (Schubert, 1954).

Uranyl and uranium(IV) ions may gain access to the body by a variety of routes—even absorption through the skin—and are bound temporarily in bone and various organs, but they gradually pass into the kidney, it is thought, as a bicarbonate complex. The complex is excreted in the glomerular filtrate, with subsequent resorption of bicarbonate ion. Uranium, either as the uranium ion (U^{4+}) or as the uranyl ion (UO_2^{++}), is thereby deposited in high concentration on the cells at the site of resorption with destruction of the resorbing cells. It has been suggested that both ions of uranium are generally toxic to the enzymes located in the kidney cells (Barnett and Metcalf, 1949; Dounce and Tien Ho Lan, 1949).

If metal ion toxicity is associated with a chelation mechanism, an obvious, if naive, correlation should exist between toxic potency and the order of metal chelate stabilities. From studies with aquatic organisms—fish, tadpoles, etc.—it has been proposed that cation toxicity is directly related to the stability of transition metal complexes in the order Mn $<$ Fe(II) $<$ Co(II) $<$ Ni(II) $<$ Cu(II) $>$ Zn, and this order also applies for a group of enzymes such as urease and diastase (Shaw, 1961).

C. Metal Ion Detoxification

The use of chelating agents in medicine depends upon the formation of soluble easily excretable metal chelates by sequestering metal ions in the circulation or competing with chelating biological sites for bound metal ion. The chelating agent must be of low toxicity, not readily metabolized, and ideally should be capable of penetrating to metal storage sites. An obvious condition is that the metal chelate should be less toxic than the free metal ion, and, since the function of the chelating agent is the elimination of all binding positions about the metal so that chemical bonds to essential enzymes cannot be formed, low toxicity of the chelate is usually taken for granted. As an overriding principle, as we shall see later, this is erroneous.

Many of the familiar chelating agents, e.g., ethylenediaminetetraacetic acid and its analogs, exist in the form of hydrophilic anions and as a result penetrate cells poorly. While their inherent toxicity is thereby limited, they are effective chiefly against circulating metal ions, and the removal of stored metal ions whose equilibrium concentration in the blood may be quite low is often necessarily a protracted process needing many courses of treatment. Because of the large number and variety of available donor groups, biological material is, in effect, a rather concentrated chelating agent, and, what is more important, possesses specialized bonds for binding at least the essential trace metals and iron. The proposal that the antibacterial action of 8-hydroxyquinoline, a familiar analytical chelating agent, is due to the removal of essential trace metal ions seems, for the latter reason, especially dubious (Zentmyer, 1944). Successful competition for toxic metal ions requires primarily a high stability of the metal complex with the therapeutic chelating agent, which, therefore, should preferably be multidentate in function, in accordance with the principle that chelate stability is enhanced in multiple ring structures, and of appropriate stereochemistry. Multidentate chelating agents, by occupying a greater number of the potential coordinating positions about the metal, decrease the opportunity of ternary complex (Chapter 8, p. 340) formation with donor groups of enzymes. Their therapeutic use thus minimizes the risk of metal toxicity.

The chelating agent 2,3-dimercaptopropanol (BAL) (II) was developed on a rational basis to protect —SH enzymes from war gas arsenicals (Peters *et al.*, 1945) and is used for the treatment of poisoning by compounds of mercury, arsenic, antimony, gold, and bismuth. The metals and metalloids that inhibit —SH enzymes combine preferentially with BAL through complex formation. It is ineffective against selenium poisoning—which is due to oxidation of —SH groups, not complex formation—and against metal ions such as Be^{++} and UO_2^{++} that form relatively weak bonds to sulfur.

Mono- and bis(2,3-dimercaptopropanol)mercury(II) complexes, (III) and (IV), are known—the former a nonelectrolyte and the latter an acid which at physiological pH exists as a very soluble anion. Similar water-

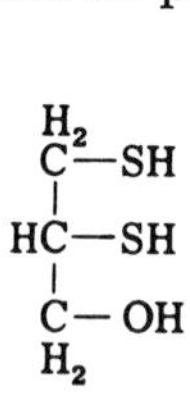

(II)

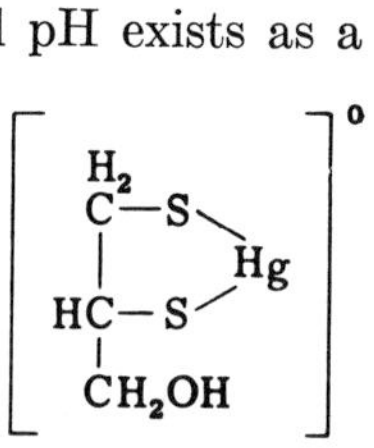

(III)

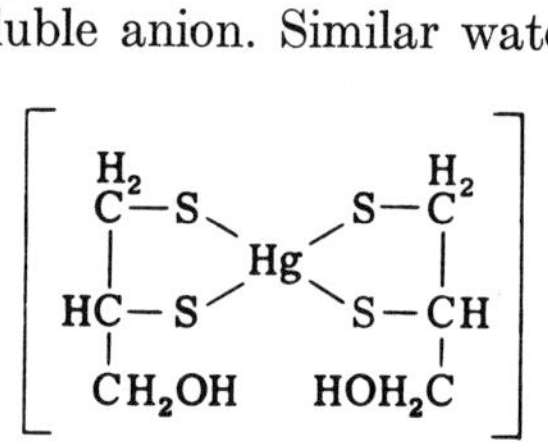

(IV)

soluble alkali and alkaloid bis(chelates) of 1-methyl- and 1-chlorobenzene-3,4-dithiol were prepared by Mills and Clark (1936) in attempts to resolve a tetrahedral Hg(II) chelate. The arsenic(III) and antimony(III) chelates, similar to the monomercury(II) chelate, are derived from arsenious or antimonious acid by elimination of two hydroxyl groups in favor of the metalloid —S bonds following reduction from the pentavalent state. These are acidic by virtue of the hydroxyl group(s).

The organic mercurials used in medicine as diuretics, such as chlormerodrin Cl—Hg·CH_2·CH(OCH_3)·CH_2·NH·$CONH_2$, which are mostly mercurated organic acids, act, partly at least, in the form $(R{\cdot}Hg)^+$. The mechanism of action appears to be the inhibition of —SH enzymes of the renal tubular cells. The diuretic action and extraneous toxic reactions are abolished by BAL through the formation of a mono(BAL) chelate. There is a general tendency of mercury(II) to form four-covalent tetrahedral complexes. This tendency is suppressed in the linear dialkyl and diaryl derivatives, which apparently do not form coordination compounds (Coates, 1956), but is merely decreased in the monoalkyl or aryl derivatives. The capacity of the mercury atom in the $(R{\cdot}Hg)^+$ ion to accept electrons from a donor (or to provide electrons for π bonding) is determined by the properties of the attached alkyl or aryl groups and their substituents. Biological specificity or lowered toxicity obviously resides in these electronic factors, in addition to enhanced or diminished permeability of cells and accessibility to the —SH site by reason of the shape and size of the R group.

Pertinent aspects of the vast literature on the therapeutic organic arsenicals (and antimonials) have been reviewed by Eagle and Doak (1951), Albert (1960), and Goodman and Gilman (1955). The antispirochetal and antiprotozoan activity of all these drugs is mediated by the form R·$As(OH)_2$, following oxidation and/or degradation and —SH enzyme inactivation. It will be evident that in the reactive form these differ from arsenious acid only in the replacement of a hydroxyl group by a substituted aromatic ring. Like the organic mercurials discussed above, site and host-parasite specificity resides in the nature of the organic substituent with an additional influence due to the penetrability and reactivity (stability) of the original drug. Parasites are protected, by the use of BAL and suitable —SH-containing compounds, from the action of these drugs both *in vivo* and *in vitro*. The risk of clinical toxicity in patients receiving mercurial treatment is also decreased by the use of BAL and related substances.

Ethylenediaminetetraacetic acid and various analogous aminopolycarboxylic acids are usually administered in the form of the sodium salt of the calcium chelate Na_2 (CaEDTA) by injection, since the rate of absorption from the gastrointestinal tract is very low. Rapid parenteral administration of the sodium salt immobilizes extracellular calcium ion with the production of the symptoms of hypocalcemia (calcium tetany), but when administered slowly mobilization of calcium ion from the skeleton keeps pace with the chelation and excretion (Foreman *et al.*, 1953; Spencer, 1960; Rubin *et al.*, 1960). The calcium chelate is the drug of choice, rather than BAL, for the treatment of acute and chronic lead poisoning, the calcium being replaced by the more strongly bound lead atom which is then rapidly excreted (Foreman *et al.*, 1953; Rieders *et al.*, 1955). It is also valuable for acute iron poisoning (Rubin and Princiotto, 1960) and for the mobilization and excretion of radioactive elements (Foreman, 1960). Penicillamine (V) and 2,2-dimethylthiazolidine-4-carboxylic acid (VI), both of which are bidentate chelating agents of low toxicity, have been used to mobilize the stored copper which is associated with Wilson's disease (Walshe, 1956; Uzman, 1960).

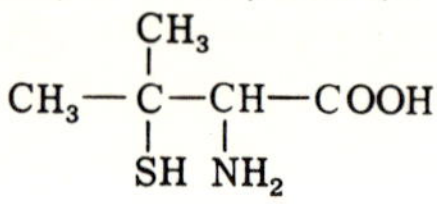

(V)

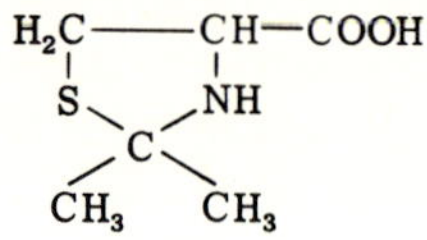

(VI)

The dye aurintricarboxylic acid (VII) contains three oxygen atoms *ortho* to carboxylic acid groupings in the aromatic rings, acts as a bidentate chelating agent, and is effective against poisoning by beryllium ion (Schubert and Lindenbaum, 1960).

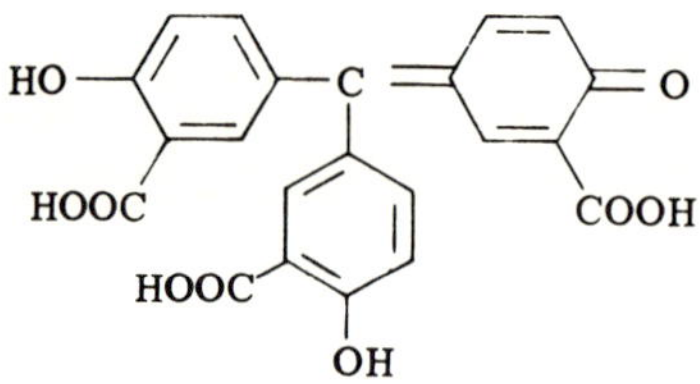

(VII)

Although chelating agents cause the mobilization and excretion of metal ions, many of the chelates so formed are quite toxic. Renal tubular damage due to zinc and cadmium complexed with BAL in the presence of excess BAL is often more serious than with the simple ions. Rapid excretion of beryllium ion can be achieved with citrate ion which forms a complex, but the untreated animals have a higher survival rate than the treated (Schubert, 1954). Although BAL protects against poisoning by mercury(II) compounds the preformed bis(BAL) chelate when injected into animals shows the same high toxicity as the free metal ion. It has been proposed that dissociation and oxidation of the thiol ligands frees the metal ion in high concentration, especially at the kidney (Goodman and Gilman, 1955). Attempts to use heavy metal chelates, notably the $[PbEDTA]^{--}$ ion, as safe X-ray contrast agents have been frustrated as yet by toxic kidney and liver reactions. Shapiro and Papa (1959) have suggested that whereas *in vitro* there is little free Pb^{++} ion in the equilibrium, dissociation occurs *in vivo* with metabolism and/or excretion of the chelating agent and release of Pb^{++} ions which are competed for by the biological chelating sites.

In addition to the hypothesis of complete dissociation with liberation of the free metal ion, two other explanations can be offered for the toxicity of these complexes: (1) partial dissociation by the loss of one chelating group, or the freeing of some coordination positions about the metal in multidentate chelates, which may form a "mixed" or ternary complex with a chelating enzyme; (2) the whole chelate itself is the toxic agent which binds to essential enzymes or other important charged sites by electrostatic or van der Waals forces without any covalent bond formation. This latter aspect will be discussed in detail in Section II, but is probably of greatest importance only with cationic complexes. The first dissociation product of the bis(BAL)mercurate(II) ion is the nonelectrolytic mono complex which can revert to the favored four-covalent state by recombination with BAL, especially if the latter is present in excess, or by combination with a biological chelating site with S and/or N donor atoms. The free energies and stabilities of "mixed" complexes have been discussed in general terms earlier (Chapter 8, Section I,D). It should be emphasized that the mono complex is essentially similar to many un-ionized simple Hg(II) compounds such as mercuric iodide, which forms the anion $[HgI_4]^{--}$ very readily without prior ionization. The mono complex can be expected to have a higher lipid solubility than the anionic bis(BAL) mercury chelate and to penetrate to sites with about the same facility as undissociated mercuric chloride; its attachment to —SH enzymes should be relatively nonspecific. Gold(I) complexes such as the bis(thiomalato) and bis(thiosulfato)aurate(I) ions are used for the treatment of arthritis. These tetrahedral four-covalent anions have no residual binding capacity and probably act as the mono complexes which could bind to S and/or N donor atoms of receptor sites. The mono complexes disproportionate with the formation of gold(I) chloride which is very sparingly soluble. The severe general toxic reactions that often occur are similar to —SH enzyme inactivation and are alleviated by BAL. Most of the ethylenediaminetetraacetic acid chelates encountered in biology are labile and exchange rapidly with the free ligand (Bosnich *et al.*, 1960). It has been shown, for instance, that the anions 1-[Cd-*d*-PDTA]$^{--}$ and [CdEDTA]$^{--}$ exchange with 1-propylenediaminetetraacetic acid, (H_4PDTA), by an S_N2 mechanism above pH 6.5, the transition state involving a "mixed" chelate (Bosnich, 1962). Although many of the EDTA chelates have high stabilities with respect to dissociation to the *free* metal ion, one or more of the chelate rings are often not attached, leaving coordination positions occupied by water, hydroxyl, or halide ions (cf. Chapter 7). It is not unreasonable that in the kidney, for example, where the EDTA chelate of lead is concentrated, sufficient "mixed" chelate may be formed to account for the toxic action.

D. Chelating Agents *in Vivo*

We have already seen that the aminopolycarboxylic acids rapidly chelate free calcium ion and mobilize calcium from the skeleton. The calcium chelates, when injected, cause the rapid excretion of zinc which has displaced calcium from the complex (Perry and Perry, 1959), but, under normal circumstances where the iron-binding capacity of the plasma is only partially saturated, iron is not taken up by synthetic chelating agents. Experiments with various iron chelates have shown that the binding capacities of some aminopolycarboxylic acids and the plasma chelating agents are of the same order (Rubin and Princiotto, 1960). Diphenylthiocarbazone, which is commonly used to remove trace metals from biological culture media, is itself toxic when administered to animals in high concentrations; primary toxicity is probably mediated in the central nervous system. It does, however, gradually complex in the body with zinc which is removed from the pancreas and prostate glands, and may inhibit the zinc-containing enzymes such as carbonic anhydrase. An interesting effect of diphenylthiocarbazone is the partial destruction of the highly reflecting surface in the eyes of carnivores, a surface consisting mainly of a crystalline zinc cysteine complex (Philips, 1961). Zinc is also removed from enzymatic sites and storage organs by 8-hydroxyquinoline (Kadota and Kawachi, 1959). The sequestering action of chelating agents is generally slow except in relation to metal ions weakly bound to the plasma proteins or free in the circulation. The essential metal ion store of the organism is protected by poor accessibility and the provision of specific strong bonding sites.

The detailed studies of Albert and co-workers have shown that 8-hydroxyquinoline (VIII) exerts its antifungal and antibacterial action through

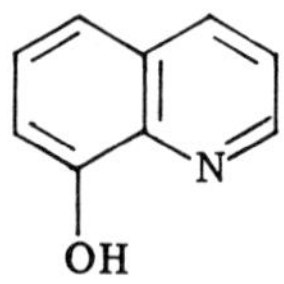

(VIII)

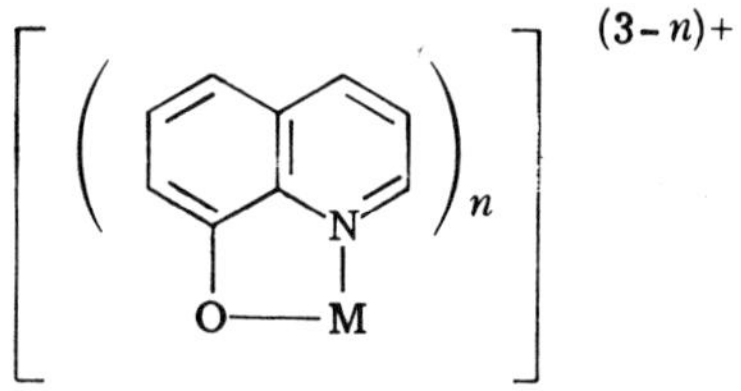

(IX)

a metal chelate (Albert *et al.*, 1947; Albert, 1958, 1960). At many times the antibacterial concentration both 8-hydroxyquinoline and Fe^{3+} ion are innocuous, but together, especially in the molar ratio 1:1 they are strongly antibacterial. Only the 1,8 isomer is effective, and *O*- or *N*-methylation destroys the antibacterial action, leaving no doubt that the chelate is the actual toxic agent. Mono, bis, and tris(8-hydroxyquinoline)iron(III) chelates (IX) are possible and would be bi- and univalent cations and a nonelectrolyte, respectively. It is possible, however, that the unsaturated mono or bis chelates are hydroxyl-bridged and hence the positive charge is reduced; it is also possible that the tris complex gives rise to a polymeric compound. On the assumption that reduced charge increases lipid solubility and hence penetration of the cell, Albert concluded that the 1:1 and 1:2 complexes are toxic but cannot penetrate, and the 1:3 complex can penetrate but is nontoxic. These conclusions are consistent with the fact that the extent and rate of toxic reaction are reduced by either excess 8-hydroxyquinoline or Fe^{3+} ion and that the chelates of the hydrophilic 8-hydroxyquinoline-5-sulfonic acid have no antibacterial properties. Because of the lability of the iron chelates and uncertainty about their exact chemical composition, a clear interpretation of the site of action or the mechanism is not possible. Apparently, the tris chelate itself is not the active species since the protective action of small amounts of Co^{++} ion must be due to the formation of the analogous but inert tris(8-hydroxyquinoline)cobalt(III). The iron(III)–8-hydroxyquinoline system may serve simply to transport the components across the cell membrane into the cell where dissociation occurs, and either free 8-hydroxyquinoline or even Fe^{3+} ion, as proposed by Albert, or some intracellular derivative of these components, is the toxic agent (cf. Section II,E). The active chelate may interfere with redox reactions within the cell, or if in the form of the mono or bis chelates may attach to enzymatic chelating sites through the free coordination positions about the iron atom. Metal ion activation or potentiation is also shown by a number of other antibacterial and antifungal substances such as 2-mercaptopyridine-*N*-oxide and dimethyldithiocarbamic acid. The antifungal action of the latter is potentiated especially by Cu^{++} ion, which itself is antifungal, the mono chelate being the more toxic form and the bis chelate the probable mode of entry into the fungal cell (Albert, 1958).

Conclusions about the mechanism of action of these substances are necessarily speculative since the chelates are labile. A more definitive approach, assuming that cell penetration can be achieved, would be to use suitable inert chelates of known composition derived preferably from the heavier transitional elements, for example, the substitution of Ru(II) or Ru(III) for Fe(III).

An obvious extension of these observations is that many drugs exert their action in a similar manner by utilizing trace metal ions *in vivo*. Isonicotinic acid hydrazide (X), which is used in the treatment of tuberculosis, is approximately as good a chelating agent as aminoacetic acid, and

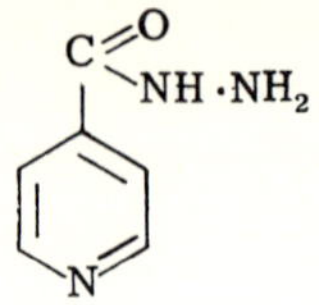

(X)

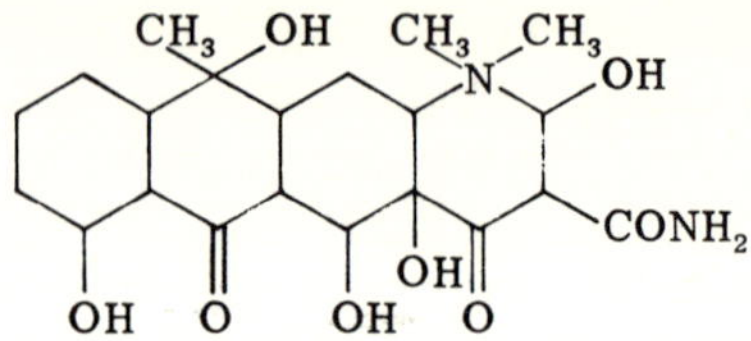

(XI)

its antitubercular action is potentiated by Cu^{++} ion (Albert, 1953; Cymerman-Craig *et al.*, 1955). Tetracycline (XI) and its analogs have a number of oxygen donor atoms suitably disposed and do, in fact, form chelates of fairly high stability (Albert and Rees, 1956). It has been suggested that the antipyretic action of salicylates, aurintricarboxylic acid, and aminopyrine is associated with the chelated transport of plasma copper ion, which is released from intracellular sites in the process of pyrexia, back into the cell (Schubert, 1960). A review of the association of potential chelating action with antimicrobial action has been written by Weinberg (1960) who discusses the possible mode of action of the chelates presumed to be formed. Metal binding by neomycin and penicillin is very dubious and does not occur at all with chloramphenicol and erythromycin. On present experimental evidence it is difficult to decide if chelating ability is merely an incidental chemical property of many drugs and the favorable effects of added metal ions, where observed, due to contemporaneous action at another site. Very few of the antimicrobial drugs are sufficiently strong chelating agents to be able to compete for free metal ions, though they may attach to free coordination positions about bonded metal atoms (XII).

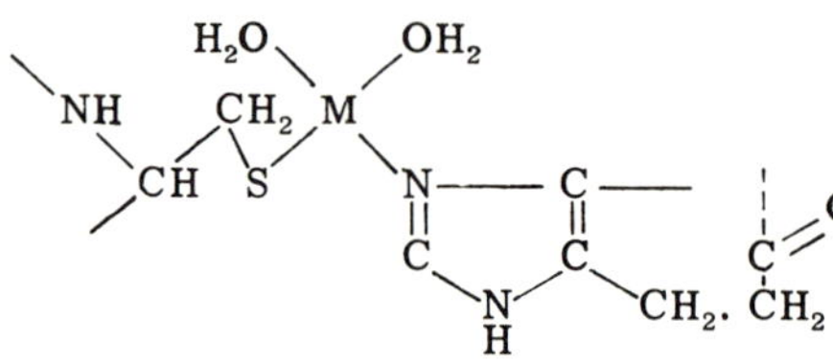

(XII)

As pointed out by Chenoweth (1961) there is an enormous number of drugs that *can* bind metals *in vitro* and, although there is no conclusive evidence one way or the other, it is likely that such binding also occurs *in vivo*.

II. Biological Effects of Synthetic Chelates

A. Introduction

Inasmuch as stable, coordinately saturated metal chelates possess no residual bonding capacity, it might be expected that they would exert little effect per se in biological systems. Complex anions and cations in high concentration might interfere temporarily with the electrolyte balance, and the latter might show neuromuscular action similar to quaternary ammonium cations (cf. Sections I,B and II,C).

The large literature that already exists on the use of anionic chelates containing aminopolycarboxylic acids for the treatment of iron, manganese, and zinc deficiency in plants has been summarized by Chaberek and Martell (1959). The iron(III) chelate is absorbed through many plant roots along with some free metal ion which may have been extracted during absorption (Wallace, 1960). Labilization of the iron(III) chelate—and hence availability of the metal—is promoted by sunlight which causes reduction to the iron(II) chelate; the freed chelating agent may be toxic to plants by immobilizing other trace metal ions. Thus, as discussed above for the lead EDTA chelate (Section I,C), the biological effect may not be a function of the intact chelates. The mode of nephrotoxic action of the anionic calcium chelate of EDTA has not been elucidated (Seven, 1960). Little appears to be known about the effects of stable nonelectrolyte chelates. Koch (1955) has reported that tris(glycinato)cobalt(III) causes hyperglycemia when injected into rats.

Inert, fully complexed cations simulate metal ions, though they are larger, but generally possess no capacity to bind to protein by chemical means and can be regarded simply as vehicles of positive charge that must act purely by physical means. The possible exceptions are the square-planar Cu(II), Ni(II), Pt(II), Pd(II), and Au(III) chelates which possess weak residual coordinate bonding ability at right angles to the plane. This may be of considerable importance as a binding force to supplement electrostatic and van der Waals bonding. Even neutral complexes of these metals may be held to protein surfaces by the van der Waals forces to be anticipated from their sizes as well as by the residual chemical bond. Tetrahedral and six-covalent chelates should behave as rather large, positively charged pseudo spheres that may similarly be held to surfaces by van der Waals forces and electrostatically to negative sites. As a result of the positive charge they may alter the isoelectric point of proteins and may depolarize charged surfaces. Changes in the zeta potential may lead to an alteration in the dispersity of colloids. By occupying negatives sites they may block active spots on enzymes (Brandt *et al.*, 1954; Dwyer, 1959). As will be evident from later sections, the proposition that the inactivation of enzymes by chelating agents is due simply to removal of the metal may not be entirely valid [cf. Vallee (1955)].

The coordination of ligands to a metal ion effectively reduces the positive charge at the metal atom by withdrawing electrons from the ligand. Familiar consequences are the acidity and hydrolysis of multivalent hydrated cations and the formation of mercury(II) amides (XIII). With

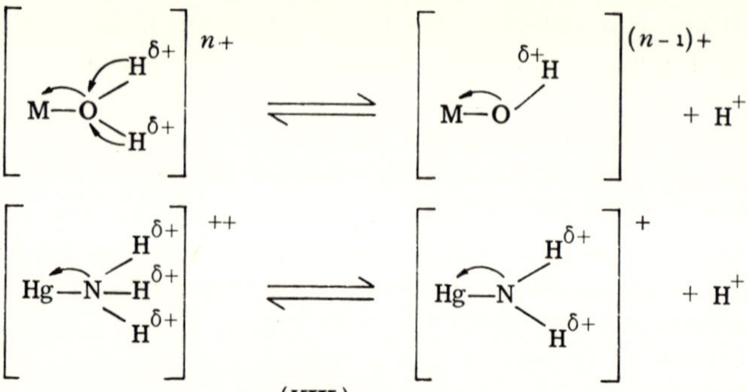

(XIII)

anionic ligands greater charge reduction naturally takes place and consequently the ligand polarization is smaller. According to the concept of electroneutrality (Pauling, 1948) the charge is distributed over the whole of the complex unit in accordance with the electronegativities of the constituent atoms (and groups) including the metal atom. Since the charge of complex cations is not concentrated at the metal atom, an effect of chelation is to increase the charge volume. In the coordinated pyridine molecules (XIV) which have no displaceable hydrogen atoms the positive

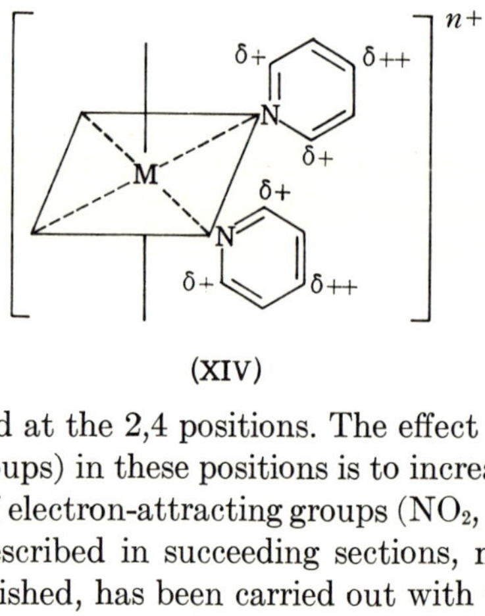

(XIV)

charge is concentrated at the 2,4 positions. The effect of electron-donating substituents (CH_3 groups) in these positions is to increase the local positive charge density, and of electron-attracting groups (NO_2, Cl) to do the reverse. Most of the work described in succeeding sections, much of which until now has been unpublished, has been carried out with the cationic chelates derived from 2,2′-bipyridine [bipy, (XV)], 1,10-phenanthroline [phen, (XVI)], and their substituted derivatives. These substances form stable

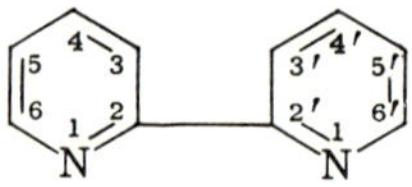

(XV)

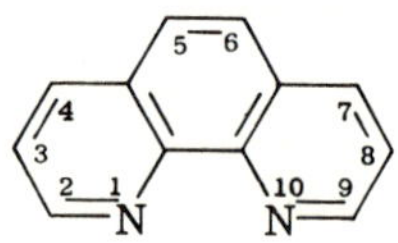

(XVI)

chelates with practically all of the transitional elements, the tris chelates MB_3^{++} (M = Fe, Ru, Os, Ni) being noteworthy for their resolution into optically active forms. The fundamental chemistry has been discussed in a review (Brandt *et al.*, 1954). The rate of racemization and dissociation of the Fe(II) chelate is rather high, the Ni(II) chelate low, but the Ru(II) and Os(II) chelates are stable in boiling concentrated acids and alkalis. The possibility that the biological effects of the less expensive and more readily available Fe(II) and Ni(II) chelates, which have been generally used, may be ascribed to dissociation, and hence to free metal ion or chelating agent, can be readily checked by reference to the highly inert Ru(II) and Os(II) chelates. A specific metal effect would not be anticipated for stable chelates of the same over-all charge because the metal atom is deeply buried and inaccessible and the large cations are approximately of the same size. A quantitative difference in effect may result from the variation in the electronegativity of the metals (Os is the most electronegative) or increased π bonding with the ligand which will reduce the positive peripheral charge. The redox potentials of the Ru, Os, and Fe chelates (-0.9 to -1.35 volts) lie outside the biological range of oxidizing power (cf. Chapter 6 for the effect of ligand substitution).

With the exception of the 6,6′ and 2,9 positions of bipyridine and phenanthroline, respectively, where steric effects intervene, the substitution of alkyl groups in the aromatic rings increases the stability of the resulting metal chelates. Alkylation in all positions increases the solubility of the chelates in organic solvents. Thus the tris chelates of tetramethyl-substituted phenanthrolines can be completely extracted from aqueous solution by chloroform. The same effect is achieved by charge reduction, and salts of the cation acetylacetonebis(3,5,6,8-tetramethyl-1,10-phenanthroline)ruthenium(II) are readily soluble in olive oil. The substance 4,7-diphenyl-1,10-phenanthroline has been used analytically for some years to complex ferrous ion. The chelate so formed is extractable into an organic liquid (Smith *et al.*, 1952). With increasing alkyl substitution, the chelates show pronounced surface-active properties and the aqueous solutions foam readily. Substitution not only creates zones of positive charge concentration about the periphery of the chelate but substantially increases its surface area. It is probable that such chelates more readily penetrate biological membranes and bind more strongly to biological surfaces since stronger van der Waals binding or adsorption now supplements electrostatic binding. It will be evident that these chelates are electrically not unlike the alkaloid drugs, which are either quaternary bases or act in the cationic form. *N*-alkylation of the substituted phenanthrolines and bipyridines creates a cation the positive charge of which is not entirely concentrated at the N atom. The electronic distribution of such a cation thus simulates that of a metal chelate in which the charge is not localized at the metal but distributed over the whole surface. The flat cations would be expected similarly to attach to negative sites of biological surfaces and the attachment to be reinforced by van der Waals forces.

B. Action in the Intact Animal

1. *Absorption*

Metal complexes containing bivalent metals (Fe, Ni, Co, Zn, Ru, Os) coordinated to a variety of ligands (ammonia, ethylenediamine, 2,2′-bipyridines, 1,10-phenanthrolines, 2,2′,2″-terpyridine) are rapidly absorbed following intraperitoneal, intramuscular, or subcutaneous injection in several animal species (mice, rats, and guinea pigs). However, only minimal absorption from the gastrointestinal tract seems to occur since ten times the intraperitoneal LD_{50}[2] dosage of stable complexes [e.g., tris(3,5,6,8-tetramethyl-1,10-phenanthroline)ruthenium(II) chloride] has been administered orally to mice for several days without signs of toxicity. The feces of the animals showed the characteristic orange color of the chelate which appeared to pass through unchanged while the urine appeared to be virtually free of the complex. Absorption of metal chelates of relatively low stability may occur under conditions which favor their dissociation. Thus, a considerable quantity of tris(5-nitro-1,10-phenanthroline)iron(II) sulfate, $[Fe(5\text{-}NO_2phen)_3]SO_4$, is found in the urine of mice following oral administration of this chelate together with ferrous sulfate. The tris metal chelate dissociates to form the un-ionized molecule dichloro(5-nitro-1,10-phenanthroline)iron(II), $[Fe(5\text{-}NO_2phen)Cl_2]^0$, and free base, 5-nitro-1,10-phenanthroline (Reaction 1), both of which can be isolated from saline solution.

$$[Fe(5\text{-}NO_2phen)_3]SO_4 \overset{NaCl}{\rightleftarrows} [Fe(5\text{-}NO_2phen)Cl_2]^0 + 2(5\text{-}NO_2phen) + Na_2SO_4 \quad (1)$$

The un-ionized molecule is unstable and rapidly disproportionates to reform the tris chelate (Reaction 2).

$$3[Fe(5\text{-}NO_2phen)Cl_2]^0 \rightleftarrows [Fe(5\text{-}NO_2phen)_3]Cl_2 + 2FeCl_2 \quad (2)$$

Although the equilibrium lies strongly in the direction of the cationic tris chelate, it can be forced in the direction of the un-ionized molecule by the simultaneous administration of excess ferrous ion. Since the mono chelate is uncharged it should readily penetrate the barrier membranes of the intestinal mucosa and thus act to transport the tris chelate from the lumen of the intestinal tract to the blood stream and hence to the urine. The simultaneous administration of tris(5-nitro-1,10-phenanthroline)-

[2] LD_{50} is defined as the dosage which is lethal to 50% of a group of animals.

iron(II) sulfate and ferrous sulfate may thus also provide a means of increasing iron transport from the gastrointestinal tract.

The poor absorption from the gastrointestinal tract (Dwyer *et al.*, 1961a) of multicharged complex cations containing metal atoms of high atomic number should prove useful in permitting radiography of this tract. Similarly, inert and multicharged metal complexes with high antibacterial activity should prove useful in the treatment of infective disorders located here. These topics are still under investigation (Dwyer *et al.*, 1961a).

The mechanism of absorption from the peritoneal cavity appears to be associated with a stereoselective biological surface. For example, when the (+) and (−) forms of the tris(1,10-phenanthroline)ruthenium-106 ion $[Ru^{106}(phen)_3]^{++}$, are administered intraperitoneally in equivalent dosages to rats and mice, the (+) form reaches the blood in approximately twice the concentration of the (−) form during the same time interval (Koch *et al.*, 1957). As would be expected, the (+) form is also approximately twice as toxic as the (−) form.

2. *Distribution, Metabolism, and Excretion*

Following intraperitoneal absorption, $[Ru^{106}(phen)_3]^{++}$ ion has been found in high concentration in liver, kidney, and frequently diaphragm, pancreas, and spleen. Varying small amounts were found in other tissues (lung, intestine, suprarenals, testis, heart, skeletal muscles, eye, and skin) whereas no radioactivity could be detected in any part of the central nervous system (Koch *et al.*, 1957). A similar pattern of distribution has been reported for tris(2,2′-bipyridine)iron(II) sulfate (Beccari, 1941). Koch and her co-workers also reported that after 24 hours 97–99% of the dose of $[Ru^{106}(phen)_3]^{++}$ ion administered to mice was recovered in the urine apparently as unchanged $[Ru^{106}(phen)_3]^{++}$ ion.

Renal excretion of metal complexes is rapid: colored iron chelates of substituted 1,10-phenanthrolines and 2,2′-bipyridines appear in the urine 5–10 minutes after intraperitoneal administration to mice. This is also demonstrated by the rapid renal clearance of the radio-opaque salt tris(ethylenediamine)platinum(IV) chloride when administered intravenously to rabbits (Dwyer *et al.*, 1961a). There is the same differential rate for excretion of the (+) and (−) forms of $[Ru^{106}(phen)_3]^{++}$ ion as for absorption from the peritoneal cavity, the (+) form being excreted more rapidly than the (−) form (Koch *et al.*, 1957).

3. *Toxicity*

Beccari (1941) and Dwyer *et al.* (1952) have demonstrated that the signs of acute toxicity of tris(2,2′-bipyridine)iron(II) sulfate and tris(1,10-phenanthroline)ruthenium(II) iodide resemble those of curariform drugs.

Within a few minutes of receiving a lethal intraperitoneal dose of such chelates, i.e., when an adequate blood concentration is reached, mice and rats showed a syndrome of labored respiration and ataxia which became progressively more severe and terminated within 5 to 15 minutes in complete respiratory paralysis and asphyxial convulsions; lethal doses of such substances do not cause cardiac arrest or produce signs of sympathetic or parasympathetic activity (Dwyer *et al.*, 1961c).

A more gradual and as yet unidentified type of death of a noncurariform type is produced following intraperitoneal administration of subcurarimimetic doses of complex cations such as acetylacetonebis(1,10-phenanthroline)ruthenium(II) chloride. In this case, the animals became anorexic, weak, prostrate, and died within 12 to 24 hours; in some cases Gram-negative microorganisms could be cultured from the blood prior to death (Dwyer *et al.*, 1961a).

Intraperitoneal LD_{50} values for some typical complex cations have been indicated by Dwyer *et al.* (1952) and by Koch *et al.* (1956). The dose/toxicity curves in mice are generally very steep (Dwyer *et al.*, 1961c).

High concentrations of stable metal chelates derived from substituted 1,10-phenanthrolines have been applied, for long periods, to skin surfaces, mucous surfaces, and mucous cavities of rabbits (Laycock *et al.*, 1961) and clinically (Section IV) with no signs of acute or chronic peripheral irritation or systemic toxicity. The lack of irritation by such substances is probably due to their stability, to the neutrality of their solutions, and to their slow rate of penetration of biological surfaces.

C. Curariform Action

"Curare" is a generic name for various South American arrow poisons which bring about paralysis of skeletal muscle. The main component, *d*-tubocurarine,[3] has been identified and synthesized; it is a large molecule containing two positively charged centers (Goodman and Gilman, 1955). Skeletal muscle is composed of many motor units which consist of a muscle fiber joined by its stimulating motor nerve fiber at a specialized region of the muscle called the neuromuscular junction or motor end-plate. The transmitter substance, acetylcholine, is liberated at the motor end-plate with each nerve impulse and sets up a series of reactions which results in muscle contraction. Following the passage of the impulse, acetylcholine is rapidly hydrolyzed by an enzyme acetylcholinesterase, and the receptor site is ready to respond to the next impulse.

d-Tubocurarine acts by decreasing the sensitivity of the motor end-plate to acetylcholine. Many other drugs are known which are capable of

[3] This is the accepted biological terminology for (+)tubocurarine.

producing muscular paralysis by other means such as direct action on nerve or muscle or on the release of acetylcholine. The earliest evidence relating the charge of a drug with its ability to paralyze voluntary muscle was produced by Brown and Fraser (1869) who converted inactive tertiary bases such as brucine and morphine to active quaternary methyl salts. Similar activity has been shown by onium compounds (Ing and Wright, 1931, 1933), by alkyltrimethylammonium ions (Alles and Knoefel, 1939) and by complex metal cations (Höber, 1917; Beccari, 1941, 1949). Some basic principles concerning such action have been discussed by Taylor (1951) and Riker (1953).

A detailed analysis of the factors responsible for the curariform activity of complex cations and of the mechanism whereby such substances produce muscular paralysis has been carried out (Dwyer *et al.*, 1957; Dwyer *et al.*, 1961a). In the latter study the action of metal complexes derived from ammonia, ethylenediamine, 2,2′-bipyridines, 1,10-phenanthrolines, and 2,2′,2″-terpyridine has been investigated on nerve-muscle preparations, both isolated and in the intact animal. Of these complexes only the cations showed curariform activity.

After transmission at the neuromuscular junction of the isolated rat diaphragm muscle–phrenic nerve preparation had been completely blocked by a complex cation, conduction in the phrenic nerve and contraction of the diaphragm muscle when directly stimulated were still present. The action potentials of the nerve and muscle as well as the resting potential of the muscle were normal. Such junctional block was readily reversed by the anticholinesterase drugs, eserine and prostigmine, and by potassium ion, but it was potentiated by atropine. The type of block produced in these *in vitro* systems thus resembled that produced by *d*-tubocurarine. A similar type of block and reversal by the same anticholinesterase drugs has been demonstrated in the spinal cat preparation in which the flexor tibialis anterior muscle is stimulated through its nerve. It is highly probable that a *d*-tubocurariform type of paralysis of neuromuscular transmission is the prime cause of death in mice given lethal doses of complex cations, since death can be prevented by the administration of the anticholinesterase drugs, eserine and prostigmine.

The pattern of neuromuscular block and recovery from block produced by different bivalent metal complexes has been compared with that of well known blocking agents such as *d*-tubocurarine, succinylcholine, decamethonium, and gallamine (Table I). Generally the larger complex cations, derived from 1,10-phenanthroline and 2,2′,2″-terpyridine, resemble *d*-tubocurarine, both in potency as blocking agents and in the time required for full recovery on washing out of the complex, whereas the smaller complex cations, derived from ethylenediamine and ammonia, resemble

TABLE I
EFFECT OF DIVALENT NEUROMUSCULAR BLOCKING AGENTS ON THE INNERVATED RAT DIAPHRAGM MUSCLE

Substance	Average dose (mg./25 ml.) producing complete block in 3 minutes	Molarity	Recovery time (minutes)
d-Tubocurarine (chloride)	0.2	$1.2 \times 10^{-5} M$	20–25
Succinylcholine chloride	0.25	$2.5 \times 10^{-5} M$	4–6
$[Ru(trpy)_2](ClO_4)_2$	2.5	$1.3 \times 10^{-4} M$	16–24
d,l-$[Ru(phen)_3]I_2$	8	$3.4 \times 10^{-4} M$	14–22
Decamethonium iodide	6	$4.1 \times 10^{-4} M$	4–8
Gallamine triethiodide	8	$3.6 \times 10^{-4} M$	5–9
d,l-$[Ru(bipy)_3]I_2$	12	$5.6 \times 10^{-4} M$	8–12
$[Co(NH_3)_5(NO_2)]Cl_2$	16	$2.5 \times 10^{-3} M$	4–7

decamethonium, gallamine, and succinylcholine in these characteristics. It seems likely that the rate of dissociation of the larger cations from the receptors at the neuromuscular junction is slower than that of the small cations and this may well explain their greater potency as blocking agents. The presence of electronegative groups (NO_2, Cl) or electropositive groups (CH_3) in the phenanthroline ring greatly increases the blocking potency of the metal chelate in both the *in vitro* and *in vivo* preparations. Induction of block is quicker and recovery from block is slower with these complexes (Fig. 1). Such observations are consistent with the postulate that there are receptors at the neuromuscular junction which contain polar binding sites (Wilson and Bergmann, 1950).

It has been shown that an increase in the blocking potency of complex cations is associated with an increase in the over-all peripheral charge of the metal complex. Thus, the trivalent cation hexamminecobalt(III) has two to four times the potency of the bivalent cation nitropentammine-cobalt(III). Further, the bivalent cations tris(2,2′-bipyridine)osmium(II) and tris(1,10-phenanthroline)ruthenium(II) show consistently a slightly greater potency than the corresponding zinc and iron chelates; the former cations have a somewhat greater peripheral charge due to the greater electronegativities of the ruthenium and osmium ions. Increased blocking potency may also be associated with an increased capacity of the molecule to form van der Waals attachments with the receptor surface; the sixteen-fold increase in the blocking potency of the bivalent cation tris(1,10-phenanthroline)ruthenium(II) when compared with the smaller bivalent cation nitropentamminecobalt(III) (Table I) supports this contention.

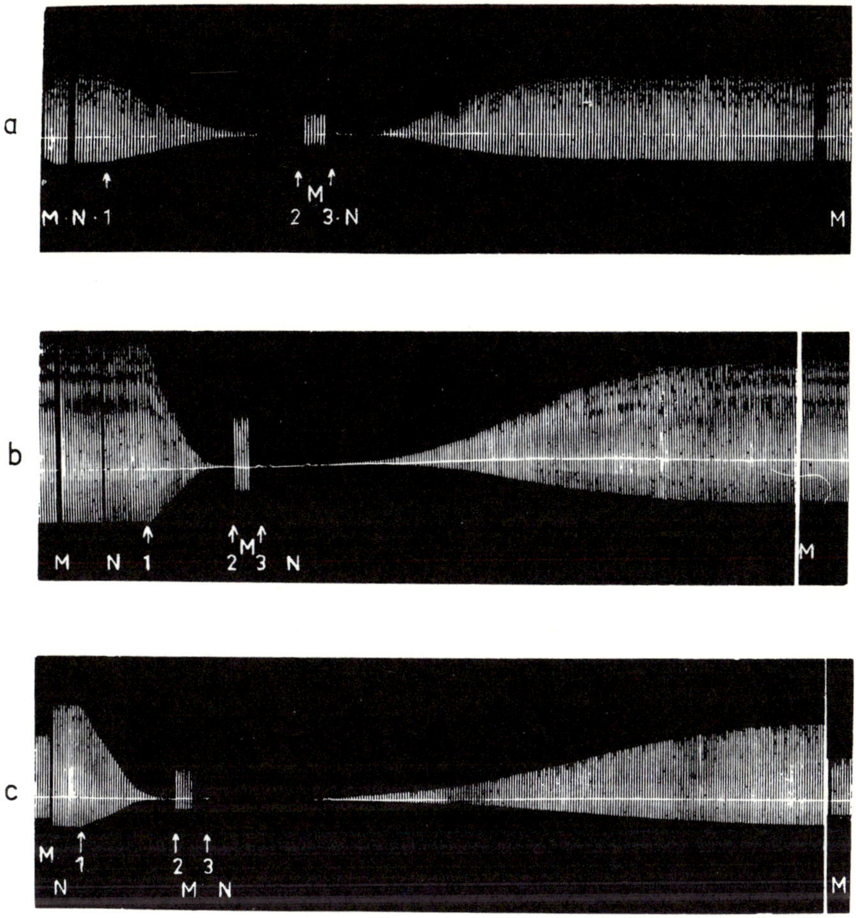

FIG. 1. Action of $3 \times 10^{-4}\,M$ (*a*) $[Ni(1,10\text{-phenanthroline})_3]^{++}$, (b) $[Ni(5\text{-nitro-}1,10\text{-phenanthroline})_3]^{++}$, and (*c*) $[Ni(5\text{-chloro-}1,10\text{-phenanthroline})_3]^{++}$ on the rat diaphragm–phrenic nerve preparation. Time scale: 8 stimuli/minute. M = direct stimulation of muscle; N = maximal stimulation of nerve; 1 = addition of drug; 2 = complete block; 3 = drug washed out.

The same quantitative trend has been shown for the direct competitive antagonism of this series of complex cations against acetylcholine on the toad rectus abdominis muscle preparation suggesting that the metal complex cations may show similar competitive antagonism to acetylcholine liberated at the end-plate of the neuromuscular junction in the rat diaphragm muscle–phrenic nerve preparation.

The complex cations most likely owe their competitive action towards acetylcholine to their peripheral charge, thus suggesting that acetylcholine may also initiate the events leading to muscular activity by its charged onium head (Dwyer *et al.*, 1957).

D. Action on Acetylcholinesterase and on Mitochondrial Enzyme Systems

A similar correlation between increasing peripheral charge of the same series of complex metal cations and their increasing potency (10^{-5} to 10^{-6} M) in inhibiting the ability of the enzyme acetylcholinesterase to hydrolyze its substrate, acetylcholine, has been demonstrated in an *in vitro* system by Koch *et al.* (1956). Further, these workers have shown that the inhibition is reversible and competitive and have suggested that these substances may interfere with hydrolysis of acetylcholine at the anionic site of the enzyme model proposed by Wilson and Bergmann (1950). Since the enzyme acetylcholinesterase occurs in high concentration at the motor end-plate region, the powerful inhibition of this enzyme shown by complex metal cations could be a factor contributing to their neuromuscular blocking action. This has not yet been fully evaluated.

Further support for the thesis that the biological activity of complex metal cations is mediated by their peripheral charge has been presented by Koch and Gallagher (1959, 1960) who showed that *d*-tubocurarine and tris(1,10-phenanthroline)ruthenium(II) iodide, which show similar blocking activity at the neuromuscular junction, also inhibit, in a similar manner, oxidations due to liver mitochondrial enzyme systems (citrate, 1-glutamate, α-oxoglutarate, 1-malate, and octanoate) which require pyridine nucleotide for electron transport. These authors have suggested that these charged agents both act by increasing the permeability of mitochondrial membranes resulting in an accelerated loss of diphosphopyridinenucleotide from the particles associated with such oxidations. Since at each site selective semipermeable membranes, undoubtedly susceptible to the action of charged molecules, are involved, it is considered that the mechanism of their action may be similar in both places and that the biological effects may be initiated by the peripheral charge common to each molecule.

E. Effect on the Oxygen Consumption and Glutamine Synthesis in Guinea Pig Brain Cells *in Vitro*

It has been shown that the synthesis of glutamine takes place in the intact brain cell by the action of the enzyme glutamine synthetase. This enzyme converts glutamate in the presence of glucose and ammonia to glutamine (Krebs, 1935). The process requires the presence of an adequate concentration of adenosinetriphosphate (ATP) and is sensitive to the lack of this substance (Speck, 1947); hence inhibition of ATP synthesis will result in inhibition of glutamine synthesis. The synthesis of glutamine by intact brain cell is thus a convenient system for testing the effect of drugs on an energy-requiring process (Messer, 1958). Thus, 2,4-dinitrophenol and salicylates depress glutamine synthesis without depressing oxygen consumption whereas compounds such as cyanide and barbiturates depress both glutamine synthesis and oxygen consumption. The former compounds exert their effect by uncoupling phosphorylation from oxidation, whereas the latter are respiratory depressants, their depressant effect on glutamine synthesis being secondary to depression of respiration.

The effects on this biological system of a series of metal chelates, representatives of which are shown in Table II, have been studied by Dwyer

TABLE II
COMPARATIVE EFFECTS OF METAL TRIS CHELATES OF 1,10-PHENANTHROLINES AND 2,2′-BIPYRIDINES ON THE OXYGEN CONSUMPTION AND GLUTAMINE SYNTHESIS OF INTACT GUINEA PIG BRAIN CELLS[a]

	Ligand			
	1,10-Phenanthroline (percentage of controls)		2,2′-Bipyridine (percentage of controls)	
Metal	Oxygen consumption	Glutamine synthesis	Oxygen consumption	Glutamine synthesis
Osmium	93	91	99	92
Ruthenium	96	93	99	98
Nickel	95	91	103	101
Iron	108	47	106	58
Cobalt	49	48	84	72
Zinc	56	43	50	25

	Ligand			
	3,5,6,8-Tetramethyl-1,10-phenanthroline (percentage of controls)		4,4′-Dimethyl-2,2′-bipyridine (percentage of controls)	
Metal	Oxygen consumption	Glutamine synthesis	Oxygen consumption	Glutamine synthesis
Iron[b]	84	17	92	43
Ruthenium[c]	14	17	—	—

[a] Concentration of all compounds was 0.1 mM.
[b] Sulfates.
[c] Chloride; all other compounds were iodides.

et al. (1961b). Ruthenium(II), osmium(II), and nickel(II) chelates derived from 1,10-phenanthroline or 2,2′-bipyridine were inactive up to concentrations of 0.1 mM while active substances showed effects similar to those of respiratory depressants or uncoupling agents described above.

Some chelates depress both oxygen uptake and glutamine synthesis, there being a close correlation between the degrees of depression of these two indices. Substances which resemble cyanide and barbiturates in their effect will be called respiratory depressants. These compounds must penetrate the membranes of the brain cell to reach active sites and in this case inhibition of glutamine synthesis can be explained in terms of inhibition of oxygen uptake leading to decreased ATP formation; i.e., the effect on glutamine synthesis is secondary to the effect on respiration. Although direct inhibition of the enzyme glutamine synthetase cannot be excluded, it is made unlikely by the close correlation between depression of oxygen uptake and glutamine synthetase activity. The cobalt(II) and zinc chelates derived from 1,10-phenanthroline and 2,2′-bipyridine as well as the tris-(3,5,6,8-tetramethyl-1,10-phenanthroline)ruthenium(II) chelate fall into the above group.

Other substances including all iron(II) chelates derived from 1,10-phenanthrolines and 2,2′-bipyridines had little or no inhibitory effect on oxygen consumption while producing marked inhibition of glutamine synthesis. The effect on glutamine synthesis cannot be secondary to respiratory depression and must therefore be due either to direct inhibition of the enzyme glutamine synthetase or to the uncoupling of phosphorylation from oxidation. Although the effect of the chelates on glutamine synthetase has not yet been investigated, the uncoupling explanation seems the more plausible since two of these chelates, like 2,4-dinitrophenol and salicylates, produced respiratory stimulation. This second group of chelates must also penetrate the cell.

The effects on the intact brain cell of the cations $[Fe(phen)_3]^{++}$ and $[Ru(phen)_3]^{++}$ were found to be increased by methyl substitution in the ligand although the qualitative nature of the response was different in the two series of chelates. Methyl substitution in such substances may have four principal effects, namely:

(*i*) to increase the lipophilic properties of the resulting chelate and hence facilitate its penetration of barrier membranes;
(*ii*) to alter the redox potential of the chelates;
(*iii*) to change the pattern of charge distribution around the periphery of the chelate resulting in the formation of zones of increased positive charge and also increasing the capacity for binding by van der Waals' forces; and
(*iv*) to alter the steric configuration of the resulting chelate.

We are unable at present to say with any certainty which of these properties is the one of fundamental importance. That methyl substitution produces the same trend in two series of substances producing different biological effects suggests that a physical mechanism such as increasing lipophilia may be the one of prime importance although each of the others may contribute to varying degrees.

The higher olive oil/phosphate buffer (pH 7.4) and chloroform/phosphate buffer (pH 7.4) partition coefficients of 3,5,6,8-tetramethyl-1,10-phenanthroline and its tris ruthenium chelate compared with those of 1,10-phenanthroline and its tris ruthenium chelate, respectively, are shown in Table III (Shulman and Vaughan, 1961). The ability of the methylated

TABLE III
THE PARTITIONING OF 1,10-PHENANTHROLINE, 3,5,6,8-TETRAMETHYL-1,10-PHENANTHROLINE, AND THEIR TRIS RUTHENIUM(II) CHELATES BETWEEN PHOSPHATE BUFFER (pH 7.4) AND CHLOROFORM OR ARACHIS OIL

Compound	Partition coefficient	
	Chloroform	Arachis oil
1,10-Phenanthroline	15	1
3,5,6,8-Tetramethyl-1,10-phenanthroline	26	12
Tris(1,10-phenanthroline)ruthenium(II) iodide	0.6	0.05
Tris(3,5,6,8-tetramethyl-1,10-phenanthroline)-ruthenium(II) chloride	13	1.5

derivatives to partition preferentially into the organic phase is consistent with the above suggestion.

It appears that while the Ru(II), Os(II), and Ni(II) chelates of 1,10-phenanthroline and 2,2′-bipyridine are inactive, the corresponding Fe(II) chelates act like uncoupling agents. The Co(II) and Zn chelates as well as the Ru(II) chelate derived from 3,5,6,8-tetramethyl-1,10-phenanthroline show respiratory depression. Since the responses affected are concerned with respiration and energy production, it would appear that the active chelates must be able to penetrate the mitochondria of the brain cells where these processes are known to occur, although other sites may be involved as well.

Why do chelates which differ only in the metal ion produce apparently different qualitative responses? With respect to the inactive chelates it has been shown that the ruthenium(II) and osmium(II) chelates are extremely stable substances (Brandt *et al.*, 1954). Further, it has been shown that $[Ru^{106}(phen)_3]^{++}$ ion is not broken down within the rat (Koch *et al.*, 1957). It has also been demonstrated that diamminobis(1,10-phenanthroline)-ruthenium(II) ion is stable in the presence of intact brain cells (Dwyer *et al.*, 1961b). Hence, whatever biological activity is produced by the ruthenium and osmium chelates must be a function of the cation as a whole. Provided penetration is adequate, the cation may combine nonspecifically with vital intracellular sites of the correct steric configuration and which contain groups capable of forming electrostatic or van der Waals bonds, with the subsequent inactivation of the biochemical processes associated with these sites.

Is the inactivity of such stable chelates due to their failure to elicit a biological response or to inability to penetrate to susceptible sites? This question cannot be answered with certainty at present but there are two pieces of evidence which suggest that the latter is the case. The ion $[Ru(phen)_3]^{++}$ has been shown to inhibit oxidations carried out by isolated rat liver mitochondria. Thus it might be anticipated that, if it penetrated the intact brain cell and reached the mitochondria, the same effect would be produced, reflected by depression of respiration of the brain cell. On the other hand, tris(3,5,6,8-tetramethyl-1,10-phenanthroline)ruthenium(II) ion, $[Ru(Me_4phen)_3]^{++}$, produces marked respiratory depression, showing that a stable complex cation can penetrate the brain cell and produce an effect. The difference in activity between the $[Ru(phen)_3]^{++}$ and $[Ru(Me_4phen)_3]^{++}$ ions is most likely due to difference in penetration; $[Ru(Me_4phen)_3]^{++}$ is considerably more lipophilic than $[Ru(phen)_3]^{++}$ (Table III) which would account for such differences. The same conclusion is probably true for the other inactive ruthenium and osmium chelates as well as for the corresponding inactive nickel chelates which also have high stability.

A consideration of the relative over-all stability constants of the nickel, cobalt, zinc, and iron chelates derived from 1,10-phenanthroline shows the order to be $Ni > Fe^{++} > Co^{++} > Zn > Fe^{3+}$; the order is similar for chelates of 2,2′-bipyridine (Albert, 1961), and presumably for many substituted bipyridines and phenanthrolines in the octahedral chelates.

The stabilities and rates of dissociation of the iron, cobalt, and zinc tris chelates of 1,10-phenanthroline and 2,2′-bipyridine are such that the biological effect may be due to the free metal ion, to the ligand, to any of the equilibrium forms of the metal chelate, or to metal chelates formed from the liberated metal ion and ligand and physiological constituents present at their site of action (ternary complexes). This rather complex situation will be more readily understood from a consideration of Fig. 2, which depicts the theoretical distribution of the equilibrium forms of tris(1,10-phenanthroline)iron(II) sulfate between the external solution and the cytoplasm and mitochondria of the intact guinea pig brain cell.

When tris(1,10-phenanthroline)iron(II) sulfate is dissolved in physiological saline an equilibrium mixture of several substances is obtained. While the equilibrium is strongly towards the tris iron(II) chelate a small

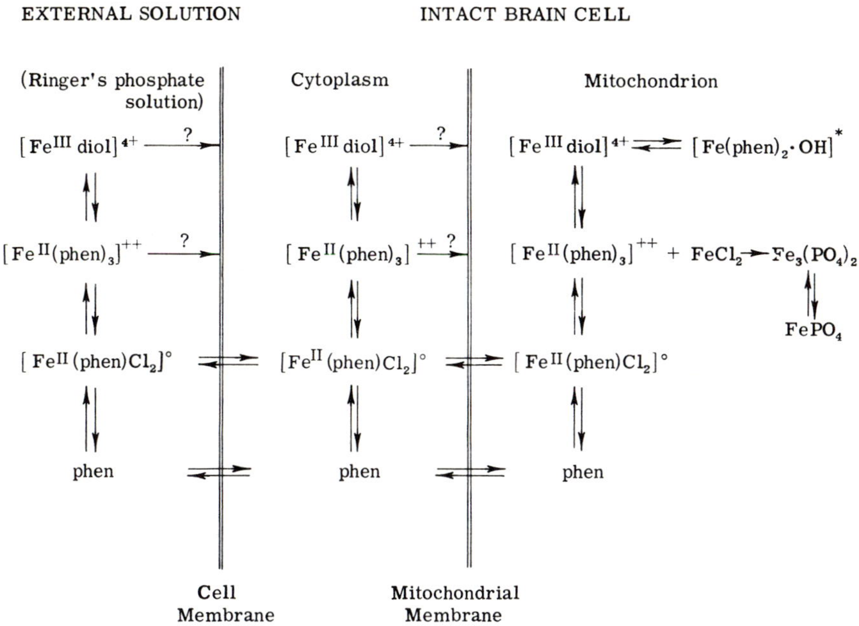

FIG. 2. Theoretical distribution of tris(1,10-phenanthroline)iron(II) ion between the external solution and the intact brain cell. The asterisk indicates an intermediate, possibly phosphorylated, reduced or part of a ternary complex.

degree of progressive dissociation results in the formation of the undissociated molecule $[Fe(phen)Cl_2]^0$ and the base 1,10-phenanthroline. The former molecule, being uncharged, should readily penetrate membranes and so theoretically act as a carrier for the transportation of Fe(II) ion and 1,10-phenanthroline into the cytoplasm and mitochondria of the intact brain cell. However, the molecule $[Fe(phen)Cl_2]^0$ is unstable and rapidly disproportionates according to Eq. (3) to produce the tris chelate and iron(II) chloride.

$$3[Fe(phen)Cl_2]^0 \rightleftarrows [Fe(phen)_3]Cl_2 + 2FeCl_2 \tag{3}$$

Consequently, the neutral chelate can serve to transfer progressively $[Fe(phen)_3]^{++}$ ion from the external solution to the interior of the brain cell and an equilibrium between the two sites will be reached. The base having considerable lipophilia can penetrate the membrane barriers in its own right. Hence, as indicated in Fig. 2, all components of the equilibrium mixture may reach the mitochondrial sites at which most probably the biological response is produced. It is likely that neither the $[Fe(phen)_3]^{++}$ ion nor the dimeric $[(Fe(phen)_2{\cdot}OH)_2]^{4+}$ cation, tetrakis(1,10-phenanthroline)iron(III)diol, penetrates the biological membranes, but theoretically they could be formed within the mitochondria.

We have been unable, so far, to identify unequivocally the active component of this equilibrium mixture. However, the following considerations may be helpful in reaching such a decision:

(1) If the Fe, Co, and Zn chelates of 1,10-phenanthroline (or 2,2′-bipyridine) all dissociate into metal ion and the same base, it should follow that the apparent differences in the biological response produced by the zinc and cobalt chelates on the one hand and by the iron complexes on the other are associated with the metal or its complexes and not with the base.

(2) Notwithstanding the inference in (1), 1,10-phenanthroline and 2,2′-bipyridine, at a concentration comparable to that which theoretically could be formed from the concentration of tris chelates used in these experiments, have been shown to uncouple phosphorylation from oxidation in rat heart mitochondria (Yang *et al.*, 1958).

It is not known at present whether such activity is due to the base itself or to a complex formed with a metal normally in the mitochondrion, or indeed if such a response would be obtained with brain mitochondria.

(3) Panimon *et al.* (1941) have shown that 1,10-phenanthroline and 2,2′-bipyridine and their Fe(II) chelates stimulate the respiration of rat brain cells whereas the Cu, Mn, Ni, and Co chelates of 1,10-phenanthroline do not possess such stimulant activity; the Zn complex produces slight respiratory stimulation but only in high concentrations. These authors have also shown that both Fe^{++} and Fe^{3+} ions stimulate oxygen uptake in rat brain tissue and that cyanide inhibits the stimulant effect of Fe^{3+} but not of Fe^{++} ion; cyanide also inhibits the stimulant effect of the ferrous chelate of 1,10-phenanthroline. The direct combination of Fe^{3+} and 1,10-phenanthroline in aqueous solution (pH 3) yields the dimeric diol Fe(III) chelate $[(Fe(phen)_2{\cdot}OH)_2]^{4+}$ which is hydroxo-bridged. Small amounts of this dimer are sometimes found in specimens of the Fe(II) tris chelate, but it is not known whether they arise by oxidation of the latter ion or from Fe^{3+} impurity when the Fe(II) chelate is prepared. The dimeric Fe(III) chelate is sensitive to reduction and subsequently disproportionates yielding, *inter alia*, the ordinary Fe(II) tris chelate:

$$3[(Fe(phen)_2{\cdot}OH)_2]^{4+} + 6e \leftrightarrows 4[Fe(phen)_3]^{++} + 2Fe^{++} + 6OH' \qquad (4)$$

There is a possibility that the $[(Fe(phen)_2{\cdot}OH)_2]^{4+}$ ion is the component of the Fe^{++} ion–phenanthroline equilibrium mixture which is inactivated by cyanide. Panimon *et al.* have also shown that the addition of splenic extract, red blood cells, or hemoglobin, all of which are sources of iron protoporphyrin, to rat brain tissue blocks the stimulant effect of 1,10-phenanthroline, its ferrous chelate, and Fe^{3+}, but not of Fe^{++} ion.

It would appear then that the iron porphyrin component of the inhibitory additives can prevent the respiratory stimulant effect of the active component of the Fe^{++} ion–phenanthroline equilibrium system. If such inactivation were due to competition for electrons between two metal chelates capable of existing as a redox pair, then it would suggest that the active component of the Fe^{++} ion–phenanthroline system should have a redox potential close to that of the iron porphyrin system, i.e., about −0.3 volt. The Fe(III) diol chelate would most closely approach this theoretical requirement.

(4) Lenta and Riehl (1960) investigated the action of trivalent Fe and bivalent Ni, Co, Cu, Zn, Mn, Ca, and Mg chelates of EDTA (ethylenediamine tetraacetic acid) on the components of the respiratory chain of isolated liver mitochondria and found that the ferric chelate was able to stimulate oxidation of DPNH (reduced diphosphopyridinenucleotide) in the presence of the DPNH-oxidase system. Moreover, they demonstrated competitive antagonism between the ferric chelate and cytochrome c reductase indicating that the ferric chelate could compete with the cytochrome component for electrons. Having demonstrated the ability of the ferric chelate to reverse the inhibitory action of Antimycin A and sodium Amytal on the DPNH-oxidase system, Lenta and Riehl suggested that the most likely site of action of the ferric chelate in the respiratory chain was at the level of flavoproteins; this suggestion was consistent with the existence of the similar oxidation-reduction potentials of the flavoprotein systems and the ferric/ferrous chelate system.

Lenta and Riehl also demonstrated at least two differences in the biological activity of the Fe(III) chelate of EDTA and that of the related chelates containing different metals. It was shown that only the Fe(III) chelate, in the presence of the DPNH-oxidase system, was able to produce a progessive increase in oxidation of DPNH with increasing concentration of the chelate and, further, that only the Fe(III) chelate was able to renew DPNH oxidation by the liver mitochondrial preparations in which the cytochrome *c* oxidase component of the DPNH-oxidase system had been blocked by cyanide. These results suggested a difference in the site and/or mode of action of the ferric chelate and the other closely related chelates. Lenta and Riehl have thus shown in a study using components of the respiratory chain of isolated liver mitochondria that the Fe(III) chelate of EDTA acts near the level of flavoprotein oxidation, that it competitively antagonizes cytochrome *c* reductase in accepting electrons, and that it can replace the cyanide-blocked cytochrome *c* oxidase component of the DPNH-oxidase system in permitting DPNH oxidation to proceed. This specific action of the Fe(III) chelate is not shared by related chelates containing different metals.

It seems likely that the mechanism of action of the active component of the Fe(II)-phenanthroline system and of the Fe(III) EDTA chelate is similar although the site of action within the respiratory chain may be different since this would depend upon the magnitude of the redox potential of the active component. The active component cannot be tris(1,10-phenanthroline)iron(II) ion since its redox potential (−1.06 volts, Brandt and Smith, 1949) is well outside the range found in biological systems. The component of the Fe(II)-phenanthroline equilibrium mixture responsible for respiratory stimulation in rat and guinea pig brain cells may be tetrakis-(1,10-phenanthroline)iron(III) diol ion or a closely related breakdown product.

If it is true that the Fe(III) diol chelate is the substance responsible for the respiratory stimulation of the intact brain cell, then it follows that this substance must penetrate the cell to reach the susceptible respiratory components (or be formed at this site which is far more likely), that its steric configuration must be such as to permit combination with or replacement of this component, and that its oxidation-reduction potential must be closely related to that of the respiratory component(s) whose activity it is replacing.

The strict metal specificity shown by the Fe(II) phenanthroline chelate in the intact rat and guinea pig brain systems brings us to a consideration of this topic. There is every reason to believe that the following considerations may also apply to other biological systems. Strict metal specificity has also been demonstrated in a model metal-phenothiazine system. Borg (1961) has shown that, of the bivalent metals Fe, Ni, Co, Cu, Zn, Mn, Mg, and Ca, and of the trivalent metals Fe, Mn, Al, and Cr, only bivalent Mn and trivalent Fe and Mn were able to interact with phenothiazine. He suggested that the specificity of the metal reaction, which is an electron transfer associated with phenothiazine semiquinone radical formation, derived from the need to match the oxidation potential of the metal-redox couple with that of the phenothiazine oxidation step; i.e., from all the metal cations, phenothiazine selects only those with an appropriate oxidation potential. Free radical formation associated with metalloprotein oxidations has also been demonstrated by Commoner *et al.* (1958), Commoner and Lippincott (1958), and Bray *et al.* (1959).

Mason (1959) has pointed out that, since electronic orbital states of atoms of individual elements are distinct, specificity in such a charge-transfer process would derive from an "energy match" between donor (ligand) and acceptor (metal) levels. If the orbital spacing is narrow, the energy levels in the two components (donor and acceptor) must be closely matched, thus resulting in a very high specificity for the metal. The similarity in the response of Fe and Mn to phenothiazine has been mimicked by their almost identical binding by suspensions of *Staphylococcus pyogenes* (Robinson *et al.*, 1960).

The basis of the differential effects on oxygen uptake and glutamine synthesis of a series of related metal chelates, which differ only in the metal ion, is not yet clear. It seems most likely on the evidence cited that such differences may reside in the involvement of different components of the respiratory chain. Other hypotheses may be concerned with a differential capacity of metal ions to bind inorganic phosphate, ADP, ATP, or high energy intermediates which are also involved in oxidative phosphorylation (Mandl and Neuberg, 1956; Chance, 1959; Slater and Hülsmann, 1959; Jayson, 1961) or with a direct effect on glutamine synthetase.

While it seems unlikely that the respiratory depressant activity of the inert chelate tris(3,5,6,8-tetramethyl-1,10-phenanthroline)ruthenium(II) chloride is mediated primarily in the same manner as that of the Co(II) and Zn chelates of 1,10-phenanthroline and 2,2′-bipyridine discussed above, the cation may well have an effect on the respiratory system. It seems probable that an important factor in this activity could be its ability to increase permeability of the mitochondrial membrane with a subsequent loss of DPN from the mitochondrion (Koch and Gallagher, 1959).

F. Effect on the Blood Glucose Concentration of the Rat

The concentration of blood glucose in the intact animal is carefully regulated, under normal physiological conditions, principally by secretions from several endocrine glands. A fine balance exists between the actions of secretions from the anterior and posterior pituitary gland, the adrenal cortex and medulla, the thyroid gland, and the α cells of the pancreas, all of which tend to increase blood glucose concentration, and those of the secretion arising from the β cells of the pancreas which tend to depress blood glucose levels. Other factors such as diet, exercise, and renal excretion also play an important part in this regulatory mechanism. It is clear that an increase in the blood glucose concentration (hyperglycemia) or a decrease (hypoglycemia) can be brought about in many ways by physiological or drug induced imbalance of the controlling mechanisms (Best and Taylor, 1961).

It has been shown that simple Co(II) salts can elicit a transitory rise in the blood glucose concentration in animals and damage the α cells of the pancreas (van Campenhout and Cornelis, 1951; Goldner *et al.*, 1952). A more powerful and prolonged hyperglycemia, in the absence of damage to the α cells of the pancreas, has been demonstrated by Koch (1955) in rats given the stable cations $[M(en)_3]^{3+}$ [M = Co(III), Rh(III), Os(III)]. Although unable to suggest a mechanism whereby these metal complexes exerted their effect, she suggested that in the case of the tris(ethylenediamine)cobalt(III) ion the action was due to the complex cation as a whole and not to cobalt ion dissociated from it.

The mechanism whereby selected metal complexes produce hyperglycemia has been studied in normal rats and in those following surgical removal of the adrenal glands, the adrenal medullae, or the pituitary gland (hypophysis) (Nelson *et al.*, 1961). Typical metal complexes used in these studies and their effects on the blood glucose level are shown in Table IV.

TABLE IV
THE EFFECT OF SELECTED COMPOUNDS ON BLOOD GLUCOSE CONCENTRATION IN NORMAL (N), ADRENALECTOMIZED (A), ADRENODEMEDULLATED (D), AND HYPOPHYSECTOMIZED (H) RATS

Compound	Effective dose range (mg./kg.)	State of animals	Mean mg.% rise in blood glucose		
			1 hour	2 hours	4 hours
Cobalt(II) chloride	12.5	N	86	36	4
		A	−5	8	8
		D	17	12	0
		H	64	46	3
Hexamminecobalt(III) chloride	20–25	N	61	72	50
		A	4	6	1
		D	−5	1	−5
		H	55	55	18
Tris(ethylenediamine)cobalt(III) chloride	80–120	N	50	78	93
		A	8	14	9
		D	1	19	9
		H	45	66	88
Tris(3,5,6,8-tetramethyl-1,10-phenanthroline)ruthenium(II) chloride	10–15	N	40	62	46
		A	−5	1	−30
		D	−3	2	1
		H	67	103	—

It may be seen that the hyperglycemic response produced by the salts cobalt(II) chloride, hexamminecobalt(III) chloride, tris(ethylenediamine)-cobalt(III) chloride, and tris(3,5,6,8-tetramethyl-1,10-phenanthroline)-ruthenium(II) chloride is abolished by adrenodemedullation or adrenalectomy but is little reduced by hypophysectomy. The hyperglycemic response thus appears to be mediated by way of the adrenal medulla and most likely follows an increase in the concentration of adrenaline circulating in the blood with a secondary effect of this substance on the breakdown of liver glycogen to blood glucose. There is as yet no unequivocal proof that such a process is involved but its operation is suggested by the similar time course of hyperglycemia in the rat which follows the administration of metal complex or adrenaline tartrate and by the observation that the adrenaline antagonist, dihydroergotamine, inhibits hyperglycemia produced by cobalt-(II) chloride (Ellis *et al.*, 1953). The possible operation of a stress mechanism due to the toxic drug doses administered may make a significant contribution to the degree of hyperglycemia but it is not considered to be the factor of prime importance.

It has been shown that the type or the dosage of the metal complex has no detectable effect on the renal threshold for glucose. The degree of glycosuria was approximately commensurate with the degree of hyperglycemia suggesting that there had been negligible effect on the renal mechanism concerned with the regulation of the blood glucose concentration.

III. Metal Chelates in Microbiological Systems

A. Action on Gram-Positive, Gram-Negative and Acid-Fast Bacteria

The bacteriostatic effect of 1,10-phenanthroline either alone or in the presence of metal ions has been demonstrated on rumen bacteria (McNaught and Owen, 1949), on acid-fast bacteria (Turian, 1951), and on a number of Gram-positive microorganisms (Feeney *et al.*, 1957).

The bacteriostatic activities of a series of 1,10-phenanthroline and 2,2′-bipyridine bases, their quaternary salts, and their metal chelates containing identical or mixed ligands have been compared on Gram-positive (*Staphylococcus pyogenes* var. *Oxford*, *Streptococcus pyogenes*, and *Clostridium welchii*), Gram-negative (*Escherichia coli* and *Proteus vulgaris*), and acid-fast (*Mycobacterium tuberculosis* H37Rv) organisms (Dixson *et al.*, 1961). Comparative bacteriostatic activities, on a molar basis, of representatives of these four classes of compounds derived from the parent base, 1,10-phenanthroline, and from its 3,5,6,8-tetramethyl- and 5-nitro- derivatives are shown in Table V.

The activity of these compounds may be summarized in the following generalizations:

(1) For each class of compound, highest activity was shown against *M. tuberculosis* and least activity against the Gram-negative organisms where activity of all substances was low. In the case of the Gram-positive organisms, activity generally was highest against *Staph. pyogenes* and lowest against *Cl. welchii* although differences were not marked.

(2) 1,10-Phenanthroline and its derivatives showed only slight activity

TABLE V

THE BACTERIOSTATIC ACTIVITY OF 1,10-PHENANTHROLINE BASES, QUATERNARY SALTS, AND METAL CHELATES ON SELECTED GRAM-POSITIVE,[a] GRAM-NEGATIVE,[a] AND ACID-FAST[b] BACTERIA

Compound	Bacteriostatic concentration ($-\log_{10} M$)					
	Staphylococcus pyogenes	*Streptococcus pyogenes*	*Cl. welchii*	*E. coli*	*Proteus vulgaris*	*M. tuberculosis*
Bases						
1,10-Phenanthroline	4.2	3.6	3.9	3.6	3.6	4.2
3,5,6,8-Tetramethyl-1,10-phenanthroline	4.8	4.5	4.2	3.9	3.9	6.0
5-Nitro-1,10-phenanthroline	3.6	3.6	3.9	<3.6	<3.6	5.1
Quaternary salts						
N-Methyl-1,10-phenanthrolinium iodide	<3.6	<3.6	<3.6	<3.6	<3.6	3.9
N-Methyl-3,5,6,8-tetramethyl-1,10-phenanthrolinium iodide	4.2	4.5	4.2	<3.6	<3.6	4.8
N-Methyl-5-nitro-1,10-phenanthrolinium iodide	<3.6	<3.6	<3.6	<3.6	<3.6	4.8
Metal chelates (identical ligands)						
Tris(1,10-phenanthroline)iron(II) iodide	3.6	3.6	3.6	<3.6	<3.6	4.5
Tris(1,10-phenanthroline)ruthenium(II) chloride	<3.9	<3.9	<3.9	<3.9	<3.9	<3.9
Tris(3,5,6,8-tetramethyl-1,10-phenanthroline)iron(II) sulfate	5.4	5.1	4.5	<3.6	<3.6	5.7
Tris(3,5,6,8-tetramethyl-1,10-phenanthroline)ruthenium(II) chloride	5.4	5.1	4.8	3.9	3.9	5.4
Tris(5-nitro-1,10-phenanthroline)iron(II) sulfate	3.9	3.9	<3.6	<3.6	<3.6	7.5
Tris(5-nitro-1,10-phenanthroline)ruthenium(II) chloride	<3.9	<3.9	<3.9	<3.9	3.9	5.1
Metal chelates (mixed ligands)						
Acetylacetonebis(1,10-phenanthroline)ruthenium(II) chloride	4.8	4.5	4.2	<3.9	<3.9	4.2
Acetylacetonebis(3,5,6,8-tetramethyl-1,10-phenanthroline)-ruthenium(II) chloride	5.4	4.5	4.2	<3.9	<3.9	4.8
Acetylacetonebis(5-nitro-1,10-phenanthroline)-ruthenium(II) chloride	4.8	4.5	4.2	4.2	4.2	4.8

[a] Incubated for 48 hours at 37°C. in Difco Heart-Infusion Broth + 10% serum + test compound.

[b] Incubated for 14 days at 37°C. in Youman's Medium + 10% serum + test compound.

against all test organisms, the one exception being acetylacetonebis(1,10-phenanthroline)ruthenium(II) chloride which showed moderate activity against Gram-positive organisms. Methyl substitution of 1,10-phenanthroline increased the activity of the base and its derivatives against the Gram-positive and acid-fast organisms whereas nitro substitution was only beneficial against *M. tuberculosis.* Acetylacetonebis(5-nitro-1,10-phenanthroline)ruthenium(II) chloride also showed moderate activity against Gram-positive and some activity against Gram-negative microorganisms.

(3) *M. tuberculosis* was sensitive to both methyl and nitro derivatives but, whereas the methyl-substituted base was more active than its metal chelates and the quaternized base, the 5-nitro base was less active than its iron(II) tris chelate although equally as active as its Ru(II) chelates and quaternary salt against this organism. On the other hand, *Staph. pyogenes* had high sensitivity only to the methyl derivatives the metal chelates of which were all equally active; the base had an activity intermediate between those of the metal chelates and the quaternary salt.

Although accurate information concerning the mode of action of these four classes of substances can only follow identification of the active compound and its site of action, the following suggestions may provide a stimulus for thought and further work.

The activities of the inert metal complexes [Ru(II) chelates] and of the stable *N*-methyltetramethylphenanthrolinium ion undoubtedly depend on the properties of the cation itself (Section II,A) whereas those of the bases, of the less stable iron(II) chelates, and of the quaternary salts (derived from 1,10-phenanthroline or nitrophenanthroline) may be due in part to the base itself or to some metal chelate formed by it with trace metals either in the medium or intracellularly. In the case of the less stable iron(II) chelates, activity may be due to a metal complex formed by the ferrous ion, following its intracellular transportation (Section II,E). With the stable cations, differences in activity may be associated with differences in the penetration by the compound of the bacterial cell wall structure. With *M. tuberculosis* such differences will be minimized by the extremely slow growth rate of this organism which would allow several hours for the attainment of an active concentration of the test compound at a susceptible intracellular site.

The activity of the free bases, the unstable metal chelates, and the quaternary salts is generally greater against *M. tuberculosis* than it is against *Staph. pyogenes*, suggesting that such activity may be due primarily to the base, either added or liberated, since growth of acid-fast organisms is known to be extremely dependent upon the presence of trace metals, especially iron (Turian, 1951).

The most outstanding differences in activity against *M. tuberculosis* and *Staph. pyogenes* are those shown by the tetramethylphenanthroline base and the iron(II) tris chelate of 5-nitrophenanthroline. The higher activity of the former substance against *M. tuberculosis* is most likely due to the greater sensitivity of this organism to a chelating agent which strongly binds iron, whereas, with the latter substance, similar activity cannot be due only to the liberated base but must depend also on other of the equilibrium species of the Fe(II) chelate, which undoubtedly reach vital intracellular sites. The components of this equilibrium mixture and their manner of transport into the tubercle bacillus are probably similar to those discussed for tris(1,10-phenanthroline)iron(II) chelate in the case of the brain cell (Section II,E). Their transport across the mucosa of the gastrointestinal tract has been considered in Section II,B.

One of the problems which faces the synthetic chemist in the design of effective antibacterials (Shulman, 1959) is the development of drug resistance which is pronounced with the microorganisms *Staph. pyogenes* and *M. tuberculosis* (Abraham, 1953). It is thus of considerable interest that neither of these organisms develops significant resistance to highly active metal chelates (*vide infra*).

Preliminary investigations have shown that neither the active substituted 1,10-phenanthroline bases nor their tris metal chelates are effective in controlling the growth of a strain of *Staph. pyogenes* in mice or of *M. tuberculosis* (H37Rv) in mice and guinea pigs (Dixson *et al.*, 1961).

B. Effect on *Staphylococcus pyogenes*

An organism frequently cultured from infections commonly found in clinical practice is *Staph. pyogenes*. The remarkable capacity of this organism to survive a battery of synthetic antibacterials and antibiotics lies chiefly in its capacity to produce variants which may show at the same time a high degree of resistance to one or several of these agents in current clinical use. Infections due to the organism may thus be protracted and very difficult to eradicate.

The action of metal chelates on *Staph. pyogenes* has been studied as a means of permitting more effective control and in the hope of determining some of the fundamental properties of this organism responsible for its clinical effects (Harris *et al.*, 1961). Some important biological effects of two typical metal chelates, tris(3,5,6,8-tetramethyl-1,10-phenanthroline)iron(II) ion $[Fe(Me_4phen)_3]^{++}$ and acetylacetonebis(3,5,6,8-tetramethyl-1,10-phenanthroline)ruthenium(II) ion $[Ru(Me_4phen)_2(aca)]^+$, on two strains of *Staph. pyogenes* will be considered. The two organisms were *Staph. pyogenes* var. *Oxford*, a standard laboratory strain of low virulence and high sensitivity to penicillin (0.05 U./ml.) and other antibiotics, and *Staph. pyogenes* var.

Phillips, a highly virulent strain isolated clinically and showing marked resistance to penicillin (10,000 U./ml.) and other antibiotics. The chelates have been used as the sulfate and chloride respectively.

1. *The Bacteriostatic and Bactericidal Activity of Metal Chelates*

The bacteriostatic activity of a representative series of metal chelates and related substances of different structural types on *Staph. pyogenes* var. *Oxford* has been considered in Section III,A. This has been confirmed with many other strains of *Staph. pyogenes* showing varied degrees of antibiotic resistance; the range of minimal bacteriostatic concentrations of the $[Fe(Me_4phen)_3]^{++}$ and $[Ru(Me_4phen)_2(aca)]^+$ ions is 1.5–12.5 μg./ml. in 10% serum broth. Drug concentrations of 25 and 100 μg./ml. in 10% serum broth of $[Fe(Me_4phen)_3]^{++}$ ion are bactericidal to *Oxford* in approximately 4 hours and 1 hour, respectively, and to *Phillips* in 8 hours and 2 hours, respectively.

2. *Development of Resistance*

Both organisms were subcultured twenty-five times at 48-hour intervals in the presence of either metal chelate or sodium benzylpenicillin. A fifteen-fold increase in resistance to $[Fe(Me_4phen)_3]^{++}$ ion and a twofold increase in resistance to $[Ru(Me_4phen)_2(aca)]^+$ ion was obtained with both organisms at a time when there was a 10,000-fold increase in the penicillin resistance of *Oxford* and a fivefold increase in that of *Phillips*. The development of resistance to penicillin was not associated with increased resistance to the chelates. Similarly increased chelate resistance was without effect on resistance to penicillin. Similar results have been obtained with a number of different metal chelates and a variety of strains of *Staph. pyogenes*, including some highly resistant to various combinations of antibiotics. The lack of ability of *Staph. pyogenes* to develop more than minimal resistance to such metal chelates suggests that this may be a general phenomenon for other organisms; minimal resistance has also been demonstrated for *M. tuberculosis*.

3. *Production and Some Characteristics of Variant Strains*

After a single 48-hour incubation of *Oxford* or *Phillips* with a sub-bacteriostatic concentration of either $[Fe(Me_4phen)_3]^{++}$ or $[Ru(Me_4phen)_2(aca)]^+$ ions, a high proportion of pale miniature and larger pigmented (intermediate-type), as well as apparently normal colonies, was obtained. The percentage of the above variants produced from *Oxford* was much greater than that from *Phillips*.

It should be realized that even under normal conditions similar small-colony variants occur in cultures of *Staph. pyogenes* but the incidence is very low (Hale, 1947; Browning and Adamson, 1950) increasing under adverse conditions of nutrition or temperature or in the presence of antibacterials (Wise and Spink, 1954; Elek, 1959).

The presence of metal chelates may give rise to staphylococcal variants in two ways (see Abraham, 1953): (*a*) by a mechanism of *selection* whereby they create conditions which favor the growth of the miniature and intermediate-type colonies; or (*b*) by a mechanism of *induction* whereby the metal chelates may effect fundamental chemical changes at a genetic level in the parent strain with production of a mixed population of colony types showing subtle or gross changes in many characteristics.

The miniature colonies of *Oxford* were usually very unstable giving rise to further variants, whereas miniature colonies derived from *Phillips* appeared to be more stable, one showing no change after twenty or more subcultures in drug-free medium. Generally, it was found amongst a number of strains of *Staph. pyogenes* that the most stable miniatures were derived from organisms possessing, like *Phillips*, high antibiotic resistance. The intermediate-type colonies of both *Oxford* and *Phillips* were generally unstable and frequently gave rise to a mixture of colonies including miniature, intermediate-type, and apparently normal colonies.

A more rapid reversion of *Oxford* but not *Phillips* from miniature to apparently normal colonies was produced by growing *Bacillus subtilis* or a series of normal strains of *Staph. pyogenes* on the same blood agar plate; this suggests that diffusion of a metabolite or possibly transformation of genetic material may occur. (Attempts to produce transformation with culture filtrates or extracts have so far failed.)

Some miniature variants of *Phillips* were found to differ from the parent strain in the following characteristics: sensitivity to the metal chelate, growth rate, hemolysin activity, pigment production, antibiotic sensitivities, virulence, and phage patterns. As a rule, no changes were observed in the coagulase, catalase, and penicillinase activities nor in the response to Gram-staining.

a. Growth Rate. The differential growth rates of the parent *Phillips* strain and its two variants in a nutrient medium containing 2.5% glucose are shown in Table VI: the growth rate of the miniature strain is much slower than that of the intermediate or parent strains.

In the absence of glucose the onset of the stationary phase of growth of the miniature colonies occurred sooner and consequently the viable count of these organisms was decreased; the absence of glucose had no effect on the growth of the other two organisms.

b. The Cytochrome System. The miniature variant of *Phillips*, unlike the parent strain, grew no further when exposed to aerobic conditions following anaerobic growth. This, together with the slower growth rate and

TABLE VI
COMPARATIVE GROWTH RATES OF THE PARENT AND TWO VARIANT STRAINS OF *Staphylococcus pyogenes* var. *Phillips*

Strain	Lag phase (hours)	Time for onset of stationary phase (hours)	Viable count (organisms/ml.)		Mean generation time (minutes)
			Inoculum	Stationary phase	
Parent	4	11	2×10^5	$>7.5 \times 10^9$	35
Intermediate	6	11	2×10^5	$>5.0 \times 10^9$	35
Miniature	17	26	1.3×10^5	5.0×10^8	120

greater glucose requirement of the miniature variant, suggested a decreased capacity of the miniature to utilize oxygen, possibly resulting from dysfunction of the cytochrome system. Other small-colony variants which were obtained by growing *Staph. pyogenes* strain 209 in the presence of Albomycin have been described (Gauze *et al.*, 1959); the respiratory-deficient character of these variants was demonstrated manometrically by their decreased oxygen uptake.

A comparison has been made of the cytochrome absorption bands of the parent, intermediate-type, and miniature strains of *Phillips*, the latter two strains being obtained by growing the parent strain in the presence of $[Fe(Me_4phen)_3]^{++}$ ion. The parent and intermediate-type strains showed the same two absorption bands in the orange (600 Å.) and green (560 Å.) areas of the visible spectrum, whereas no absorption bands could be detected for the miniature strain. This marked difference demonstrates a major lesion in the cytochrome system of the miniature strain which could well explain many of its altered characteristics, since ATP formation would be greatly impaired with the attendant loss of many metabolic processes.

The miniature strain could arise by the action of the metal chelate at a genetic level, thus inducing a cytochrome-deficient mutant, or by direct action on the cytochrome system, thus selecting cytochrome-deficient mutants previously present as a very small proportion of the parent strain. Genetic changes induced in a strain of *Staph. pyogenes* by chloramphenicol, penicillin, terramycin, nitrogen mustard, and hydrogen peroxide, leading to the production of small-colony mutants similar to *Phillips* miniatures, have been described by Voureka (1952).

c. Virulence. The number of organisms required to kill 50% of a group of mice (LD_{50}) in 7 days following intravenous administration has been determined for the miniature, intermediate-type, and parent strains of *Phillips*. The LD_{50} of the parent strain was three times greater than that of the intermediate-type strain whereas no comparable LD_{50} figure could be determined for the miniature strain since no number of organisms could be administered which produced an LD_{50} in 7 days. A number of miniature organisms, a thousandfold greater than the LD_{50} of the parent strain, produced an LD_{50} in 3 weeks. Thus, the systemic virulence to mice of the miniature staphylococci is extremely low; their ability to form localized abscesses in the mouse is also very poor. Hence treatment with metal chelate induced or selected a variant of extremely low virulence. The miniature strain appears to be very stable in the host since only organisms displaying the characteristics of miniature staphylococci could be recovered from mice to which lethal doses of this organism had been administered.

At present there is no conclusive evidence in favor of either selection or induction as the mechanism whereby the action of these metal chelates produce the staphylococcal variants; it is possible that both mechanisms may operate.

In summary, the lack of effective resistance to the metal chelates shown by highly virulent and antibiotic-resistant strains of *Staph. pyogenes*, the lack of cross-resistance to such metal chelates and antibiotics, and the ability of the metal chelate to attenuate markedly the virulence of such staphylococcal strains by favoring the growth of an almost avirulent miniature variant have great clinical importance in both the prophylaxis and therapy of staphylococcal infection. It would appear that the miniature variant of *Staph. pyogenes* might be a form adapted to withstand the rigors of adverse conditions, reversion occurring, where possible, when such adverse conditions are removed. It lacks the cytochrome system which makes for bountiful existence but greater susceptibility to drug action and appears to retain principally those mechanisms most necessary for survival. Since its virulence is so low and its needs probably small, the miniature staphylococcus may be approaching the perfect parasite (Butler and Shulman, 1961).

C. Action on the Yeast *Saccharomyces cerevisiae*

Normal yeast cells during their growth constantly give rise to miniature colonies called "petites" with an incidence of about 1% (Ephrussi *et al.*, 1949). These "petites" are stable mutants and are respiratory-deficient because of lack of several respiratory enzymes and components bound to particles present in the cytoplasm of the yeast cell (Slonimski and Ephrussi, 1949; Ephrussi and Slonimski, 1955; Yotsuyanagi, 1955).

Such mutation in yeast cells may arise from the loss or functional inactivation of a cytoplasmic factor which may be related to or identical with normal yeast mitochondria in which some fraction carrying components of the respiratory chain is lost (Ephrussi and Slonimski, 1955; Yotsuyanagi,

1955; Nagai *et al.*, 1961). The dependence of this cytoplasmic factor on nuclear gene control in the formation of respiratory enzymes has been discussed by Nagai *et al.* (1961).

It has been shown that the incidence of "petite" mutants can be increased greatly by exposure of normal yeast cells to a variety of organic substances—e.g., acriflavine (Ephrussi *et al.*, 1949; Mudd *et al.*, 1951), euflavine (Ephrussi and Hottinguer, 1950; Marcovich, 1951), triphenyltetrazolium chloride (Laskowski, 1954), *p*-nitrophenol (Yanagishima, 1956), propamidine isethionate (Lindegren, 1958), caffeine (Nagai and Nagai, 1958), and a series of triphenylmethane and xanthene dyes (Nagai, 1959)—or to a series of metal salts such as copper(II) sulfate, cobalt(II) chloride, nickel(II) chloride, and manganese(II) chloride (Lindegren *et al.*, 1958). In all cases tested so far "petite" mutation following exposure of yeast cells to drugs is an inductive rather than a selective phenomenon and the possibility that a primary site of such action may be nucleic acids has been discussed by Nagai *et al.* (1961). Further, it has been demonstrated that chelating agents such as oxalate, diethyldithiocarbamate, and ethylenediaminetetraacetate ions produce miniature respiratory-deficient variants by releasing mitochondria from the cytoplasm of the yeast cell *Candida albicans*, and it has been suggested that these substances may act by removing metal-linked barriers which restrain the mitochondria within the yeast cell (Merkel and Nickerson, 1953).

Free metal ions and chelating agents both produce in yeast "petite" mutants which lack components of the cytochrome system. In both cases the active agent may be a metal complex formed by combination of the metal ion or chelating agent with a physiological substituent. Since the presence of stable metal chelates results in the production of miniature variants which are cytochrome-deficient in *Staph. pyogenes* (Section III,B) it was considered that such substances might have a similar effect on yeast cells. The activities of a representative series of 1,10-phenanthroline bases and of metal chelate cations and quaternary salts derived from them have been compared with those of selected metal ions and euflavine (2,8-diamino-10-methylacridinium chloride) on the yeast *Saccharomyces cerevisiae* (Harris *et al.*, 1961a). Table VII shows the maximal subinhibitory concentration of each drug, the percentage growth, and the percentage of "petites" obtained after the yeast had been exposed to this drug concentration.

Generally, with the exception of some metal chelates derived from tetramethylphenanthroline, the inhibitory concentrations of the bases are considerably lower than those of their corresponding quaternary salts and metal chelates. There is an eightfold difference in activity between the various metal chelates of tetramethylphenanthroline. For all four

TABLE VII

THE INHIBITORY AND MUTAGENIC ACTIVITY OF 1,10-PHENANTHROLINE BASES, QUATERNARY SALTS, METAL CHELATES, AND MISCELLANEOUS SUBSTANCES ON THE YEAST *Saccharomyces cerevisiae*[a]

Compound	Maximal subinhibitory concentration[b] ($-\log_{10} M$)	"Petite" colonies (%)	Growth (% of controls)
Bases			
1,10-Phenanthroline	4.8	2	35
3,5,6,8-Tetramethyl-1,10-phenanthroline	5.1	49	38
5-Nitro-1,10-phenanthroline	4.8	1	81
Quaternary salts			
N-Methyl-3,5,6,8-tetramethyl-1,10-phenanthrolinium iodide	3.6	99	105
N-Methyl-5-nitro-1,10-phenanthrolinium iodide	3.6	13	56
Metal chelates (identical ligands)			
Tris(1,10-phenanthroline)iron(II) iodide	4.5	2	35
Tris(1,10-phenanthroline)nickel(II) iodide	3.9	1	47
Tris(1,10-phenanthroline)cobalt(II) iodide	3.9	1	28
Tris(1,10-phenanthroline)ruthenium(II) iodide	3.9	2	50
Tris(3,5,6,8-tetramethyl-1,10-phenanthroline)-iron(II) sulfate	4.5	99	45
Tris(3,5,6,8-tetramethyl-1,10-phenanthroline)-nickel(II) sulfate	4.5	95	65
Tris(3,5,6,8-tetramethyl-1,10-phenanthroline)-cobalt(II) sulfate	5.4	94	112
Tris(3,5,6,8-tetramethyl-1,10-phenanthroline)-ruthenium(II) chloride	5.1	73	60
Bis(3,5,6,8-tetramethyl-1,10-phenanthroline)-copper(II) sulfate	5.1	49	49
Bis(3,5,6,8-tetramethyl-1,10-phenanthroline)-manganese(II) sulfate	5.4	99	130
Tris(5-nitro-1,10-phenanthroline)iron(II) sulfate	3.9	8	5
Tris(5-nitro-1,10-phenanthroline)nickel(II) sulfate	3.9	4	27
Metal chelates (mixed ligands)			
Acetylacetonebis(1,10-phenanthroline)-ruthenium(II) chloride	4.2	51	11
Acetylacetonebis(3,5,6,8-tetramethyl-1,10-phenanthroline)ruthenium(II) chloride	5.7	98	16
Acetylacetonebis(5-nitro-1,10-phenanthroline)-ruthenium(II) chloride	5.1	3	42

(*Continued*)

TABLE VII (*Continued*)

Compound	Maximal subinhibitory concentration[b] ($-\log_{10} M$)	"Petite" colonies (%)	Growth (% of controls)
Miscellaneous substances			
$NiCl_2 \cdot 6H_2O$	3.9	30	11
$CoCl_2 \cdot 6H_2O$	2.7	88	5
$CuSO_4 \cdot 5H_2O$	3.3	47	26
$MnSO_4 \cdot 4H_2O$	1.8	77	8
Euflavine (2,8-diamino-10-methylacridinium chloride)	3.6	100	84

[a] Incubated in synthetic medium + test compound for 72 hours at 30°C.
[b] This was half the inhibitory concentration.

classes of compounds, the maximal subinhibitory concentration is generally lowest and the capacity to induce "petites" greatest among the derivatives of tetramethylphenanthroline. The derivatives of phenanthroline and 5-nitrophenanthroline generally show comparable activities on both of these parameters.

The greatest capacity to produce "petites" is shown by the charged substances, especially tetramethylphenanthrolinium methiodide, the various metal chelates containing the tetramethylphenanthroline ligand and a group of miscellaneous substances. In most cases, these charged tetramethylphenanthroline derivatives, in comparable or much higher dilutions, show the same low toxicity and high "petite"-inducing capacity as euflavine. Whereas simple metal salts are some two to fifty times less effective than euflavine as mutagenic agents, the corresponding metal chelates derived from tetramethylphenanthroline are some eight to one hundred and twenty times more effective. The compound of highest mutagenic activity is acetylacetonebis(3,5,6,8,-tetramethyl-1,10-phenanthroline)ruthenium(II) chloride. The enhanced activity of this metal chelate may be a function of its redox potential, increased lipophilia, charge localization, and an increased binding capacity to susceptible biological sites.

The action of metal chelates and the quaternary salts may take place both at the surface of the yeast cell and intracellularly. The yeast cell possesses surface receptors and a specific transport mechanism for univalent and bivalent cations (Rothstein, 1959) which may be possible sites of interference by charged compounds. Support for an intracellular site of action of metal chelates is provided by the demonstration that stable "petite" colonies, produced by growing the parent yeast cells in the presence of tris(3,5,6,8-tetramethyl-1,10-phenanthroline)iron(II) sulfate, have lost the absorption bands due to cytochromes *a* and *b* but retain that of cytochrome *c* which is indeed more intense; this finding is characteristic of "petite" mutants (Slonimski and Ephrussi, 1949; Laskowski, 1954). The loss of absorption bands due to cytochromes *a* and *b* has also been demonstrated in miniature variants obtained by growing *Staph. pyogenes* in the presence of the same metal chelate (Section III,B). This suggests that this metal chelate may be producing common effects in both species of microorganisms.

It would seem possible then that chelating agents, quaternary salts, metal ions, metal chelates, and other drugs (e.g., basic dyes, amidines, guanidines) capable of existing as cations which produce, probably by induction, "petite" yeast mutants lacking similar components of the cytochrome system may all owe their effects to some action on the yeast cell mitochrondria or on a cytoplasmic factor controlling the synthesis of respiratory enzymes. The active agent may in each case be a charged molecule but this has not yet been established with any degree of certainty. The primary site of action may indeed be the nucleic acids, such action being mediated by way of the cytoplasmic factor on the mitochondria (Nagai *et al.*, 1961). It is possible that several of these mechanisms could be involved simultaneously and that similar effects are produced by such substances in other cell systems, e.g., in *Staph. pyogenes* (Section III,B) and other microorganisms (Nagai *et al.*, 1961).

D. Action on Pathogenic Fungi

The relatively low biological activity of most antifungal drugs when compared with that of antibacterials (Miller and McCallan, 1956) may depend largely upon the greater resistance to drug penetration of the various membranes of the fungal cell and the large reserve of alternate biochemical pathways available to support the survival of the fungus in the presence of the drug. It seems likely that many active drugs produce their effects in a nonspecific manner (Ferguson, 1939; Horsfall, 1956) such as inactivation of many different enzyme systems (Owens, 1953; McCallan, 1957) or by alteration in membrane permeability of the cell wall and inclusions of the fungus (Miller and McCallan, 1957) with ensuing biochemical dysfunction and death of the organism. The fungistatic activity of 1,10-phenanthroline bases has been reported previously (Blank, 1951). The activity of a series of substances derived from 1,10-phenanthrolines on pathogenic fungi is shown in Table VIII (Maslen and Shulman, 1961).

With the exception of tris(5-nitro-1,10-phenanthroline)nickel(II) sulfate, the activities of the metal complexes containing substituted 1,10-phenanthroline ligands were greater than those of tris(1,10-phenanthroline)-nickel(II) sulfate; generally the corresponding iron and ruthenium complexes showed similar quantitative activities. Variation of the substituent

TABLE VIII
FUNGISTATIC ACTIVITY OF SELECTED 1,10-PHENANTHROLINE COMPOUNDS[a]

Compound	Inhibitory concentration ($-\log_{10} M$)					
	A	B	C	D	E	F
Tris(1,10-phenanthroline)nickel(II) chloride	<2.3	<2.3	<2.3	2.3	2.6	<2.3
Tris(5-chloro-1,10-phenanthroline)-nickel(II) sulfate	3.0	3.0	3.0	3.0	>3.3	3.0
Tris(5-nitro-1,10-phenanthroline)-nickel(II) sulfate	<2.3	<2.3	<2.3	<2.3	<2.3	<2.3
Tris(5-phenyl-1,10-phenanthroline)-nickel(II) sulfate	2.7	2.7	>3.0	>3.0	>3.0	2.7
Tris(3,5,6,8-tetramethyl-1,10-phenanthroline)nickel(II) sulfate	3.3	3.3	3.3	3.3	3.6	3.0
3,5,6,8-Tetramethyl-1,10-phenanthroline hydrochloride	>3.9	>3.9	>3.9	>3.9	>3.9	>3.9
N-Methyl-3,5,6,8-tetramethyl-1,10-phenanthrolinium iodide	>3.0	2.7	>3.0	2.7	>3.0	<2.7

[a] Fungi incubated for 28 days at 30°C. on malt agar containing test compound:

A = *Microsporum canis*
B = *Trichophyton mentagrophytes*
C = *Trichophyton rubrum*
D = *Trichophyton sulphureum*
E = *Epidermophyton floccosum*
F = *Candida albicans*

group in the 1,10-phenanthroline molecule amongst the more active nickel chelates did not greatly alter the fungistatic end point which was generally in the range 0.5–1.0 mg./ml. The most active substance, tris(3,5,6,8-tetramethyl-1,10-phenanthroline)nickel(II) sulfate, is about eighty times less active against the pathogenic fungi than against *Staph. pyogenes* (Table V). These results suggest the operation of a physical mechanism of action, a contention supported by the marked stability especially of the ruthenium chelates which appear to act merely as large charged pseudo spheres (cf. Sections II,A and II,E).

A comparison, on a molar basis, of the activities of the base hydrochloride, the quaternary salt, and the tris metal chelate derived from 3,5,6,8-tetramethyl-1,10-phenanthroline shows that the metal chelate is approximately three times as active as the quaternary salt and the base hydrochloride is about five times as active as the metal chelate. The higher activity of the base is probably due to its greater ability to penetrate to the site of action compared with that of the metal chelate or the charged quaternary salt. However, its capacity to form chelates with essential trace metals at intracellular sites may also be an important factor regulating its activity (Albert, 1958). This contention is not supported, however, by the observation of Blank (1951) who demonstrated high fungistatic activity of 2,9-dimethyl-1,10-phenanthroline which forms metal chelates of low stability. Chelation cannot explain the activity of the quaternary salt since it is unable to form metal chelates unless *N*-dealkylation occurs in the presence of the fungus (Section III,A).

The most resistant fungus to the action of the substances tested was *Candida albicans* although it markedly accumulated colored metal chelates which was evident by the color of the fungal growth.

E. Action on Viruses and Cells in Culture

A virus particle may be envisaged as a core of nucleic acid enclosed in a protective coating of protein and occasionally lipid and polysaccharide. In the extracellular state it is metabolically inert and completely ncapable of self-propagation. Viral reproduction takes place only inside living cells and is completely dependent on the materials and energy sources of those cells. It is generally true that once a virus has entered a cell its identity is lost and it behaves very much as a component of the cell; indeed, it directs the metabolism of the cell towards a single end, the synthesis of more virus.

The process of virus infection may be traced through a number of discrete and sequential steps: *Adsorption* of virus to the cell is initially a process of electrostatic attraction and may subsequently involve a specific union of viral enzyme and complementary cell receptors. This is followed by cell *penetration* by the operation of a process akin to phagocytosis, and subsequently, immediately the cell wall has been traversed, by *breakdown of the virus* with liberation of its nucleic acid. Then follows the *eclipse phase* (or dark interval) during which it must be presumed that the viral nucleic acid directs the replication not only of itself but of all other viral constituents as well. No evidence of the infectious virus particle nor of any of its parts or properties can be detected by any known physical or chemical means during this interval. The *appearance of new viral precursors* some hours later brings the eclipse phase to an end and is the first outward sign of activity. With the well studied influenza virus this takes the form of complement-fixing antigen in the nucleus and is followed, an hour later, by the appearance of hemagglutinating material in the cytoplasm. *Assembly of viral parts* into the final infectious form possibly takes place very close to the cell wall. A large number of particles appear within a very short time and are then *released* almost immediately following assembly.

Each cell whether infected initially by a single virus particle or by a number releases upwards of a thousand progeny one at a time by a process of local self-healing lysis of the cell wall. Each liberated particle goes on to infect new cells while the host cell, with few exceptions, is irrevocably damaged and eventually dies.

Clearly an antiviral drug may succeed either by direct action on the virus in its extracellular state or by inhibiting the reproductive process at any one of the stages described above. Since the events of virus replication are so closely integrated with those of normal cell metabolism, the chance of obtaining a drug showing high selective toxicity to the intracellular virus particle is remote and the greatest hope of success lies at present in the exploitation of quantitative differences in the requirements of host and virus for particular biochemical pathways. Tamm (1958) has indicated that all chemical compounds shown so far to inhibit virus multiplication also affect, at virus-inhibitory concentrations, nucleic acid or protein synthesis or energy-yielding processes of the host cell.

Since it is known that a virus particle may contain components (nucleic acid, protein, lipid, and polysaccharide) capable of forming strong bonds with cations and since cations derived from 1,10-phenanthroline bases have been shown to exert a marked inhibitory effect on important biochemical mechanisms concerned with respiration and energy production in a variety of biological systems (Sections II,E, III,B, and III,C), it was anticipated that such substances would inhibit the multiplication of viruses. A study has been made of the activity of a range of metal chelates and quaternary salts derived from 1,10-phenanthroline bases on a selection of viruses ("Melbourne" strain of influenza A (MEL), herpes simplex, vaccinia, and poliomyelitis) and susceptible normal and malignant cells (chick embryo allantois, monkey kidney epithelium, human amnion, chick embryo fibroblasts, and HeLa cells (Harris *et al.*, 1961b). Some important aspects of this study are now briefly considered.

It has been shown that metal chelates and quaternary salts prevent virus multiplication in two quite distinct ways, namely, by direct "inactivation" of the virus or by primary action on the host cell.

Direct "inactivation" seems to follow adsorption of the chelate to the virus to render it noninfective. Influenza and herpes simplex viruses are much more readily "inactivated" in this way than poliomyelitis with vaccinia occupying an intermediate position. The most active substance tested $[Fe(Me_4phen)_3]^{++}$ ion, shows some activity at concentrations as low as 10^{-5} M. The mechanism of this action is currently under investigation.

The effect of selected phenanthroline chelates and quaternary salts on the multiplication of influenza virus (MEL) in tissue suspensions of allantois-on-shell (Fazekas de St. Groth and White, 1958) is shown in Table IX.

It may be seen that all the quaternary salts and metal chelates inhibit to varying degrees the multiplication of influenza virus. Generally, the greatest effect is shown by derivatives of tetramethylphenanthroline. Similar substances derived from phenanthroline and

5-nitrophenanthroline have, in most cases, about equal activity. Each quaternary salt derived from phenanthroline and 5-nitrophenanthroline is more active than its corresponding inert ruthenium(II) chelate containing identical ligands whereas the reverse is the case with the quaternary salt and the ruthenium(II) chelate derived from tetramethylphenanthroline.

TABLE IX

THE INHIBITORY EFFECT OF 1,10-PHENANTHROLINIUM QUATERNARY SALTS AND METAL CHELATES ON THE MULTIPLICATION OF INFLUENZA VIRUS (MEL) IN CHICK CHORIOALLANTOIC MEMBRANE[a]

Compound	Minimal inhibitory concentration ($-\log_{10} M$)
Quarternary salts	
N-Methyl-1,10-phenanthrolinium iodide	4.9
N-Methyl-3,5,6,8-tetramethyl-1,10-phenanthrolinium iodide	5.0
N-Methyl-5-nitro-1,10-phenanthrolinium iodide	4.9
Metal chelates (identical ligands)	
Tris(1,10-phenanthroline)iron(II) iodide	4.5
Tris(1,10-phenanthroline)ruthenium(II) iodide	4.0
Tris(3,5,6,8-tetramethyl-1,10-phenanthroline(iron(II) sulfate	5.5
Tris(3,5,6,8-tetramethyl-1,10-phenanthroline)ruthenium(II) chloride	4.4
Tris(5-nitro-1,10-phenanthroline)ruthenium(II) chloride	3.9
Tris(5-nitro-1,10-phenanthroline)iron(II) sulfate	4.5
Metal chelates (mixed ligands)	
Acetylacetonebis(1,10-phenanthroline)ruthenium(II) chloride	5.3
Acetylacetonebis(3,5,6,8-tetramethyl-1,10-phenanthroline)-ruthenium(II) chloride	6.2
Acetylacetonebis(5-nitro-1,10-phenanthroline)ruthenium(II) chloride	4.5

[a] 100 ID_{50} MEL virus + tissue + test compound incubated at 36°C. for 72 hours.

The activity of the ruthenium(II) chelate with mixed ligands is, in each case, greater than that of the corresponding chelate with identical ligands. This observation may well be associated with the decreased redox potential and increased penetrability of the monovalent mixed-ligand chelate permitting the accumulation of a higher intracellular concentration. It should be noted that in a series of different bivalent metal chelates containing identical tetramethylphenanthroline ligands the tris chelates of Co(II) and Fe(II) were more effective than those of Ni and Ru(II), whereas there was no significant difference between the activities of the bis chelates of Cu(II), Cd, Mn(II), and Zn which were the most active.

A detailed study has been made of the effect of phenanthroline chelates on each of the known stages of multiplication of influenza virus in tissue suspensions of allantois-on-shell. The most potent inhibitor, acetylacetonebis(3,5,6,8-tetramethyl-1,10-phenanthroline)ruthenium(II) chloride, is active at an extremely low concentration, 6×10^{-7} *M*. It penetrates allantoic cells and acts within 20 minutes; its antiviral effect is progressive and cannot be reversed by repeated rinsing. Multiplication of very large doses of virus is completely suppressed by minimal doses of this chelate, and the dose/effect curve is very steep (Fig. 3). It may be seen that whereas a concentration of 4×10^{-7} *M* is completely ineffective, a fourfold increase in dosage inhibits the multiplication of no less than 10^5 infectious doses.

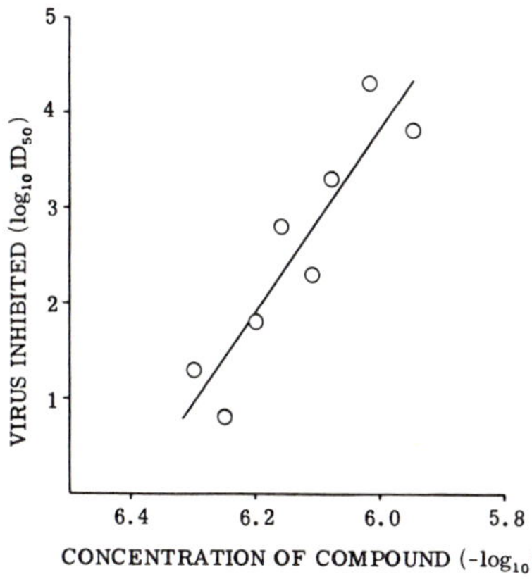

FIG. 3. Relationship between concentration of acetylacetonebis(3,5,6,8-tetramethyl-1,10-phenanthroline)ruthenium(II) chloride, $[Ru(Me_4phen)_2(aca)]^+$, and dose of influenza virus (MEL) prevented from multiplying in chick embryo allantoic cells.

When $[Ru(Me_4phen)_2(aca)]^+$ ion is added simultaneously with influenza virus to tissue suspensions of allantois-on-shell, the chelate fails to inhibit the stages of adsorption, penetration, and breakdown of virus but does block production of hemagglutinin and complete virus. Since multiplication of very large doses of virus is completely suppressed by minimal doses of chelate, added as late as 3 hours after infection, the chelate must act on some process or processes occurring during the eclipse phase. As all concentrations of chelate which inhibit viral reproduction simultaneously depress oxygen uptake of the host tissue, the chelate almost certainly produces its antiviral effects by direct or indirect action on mechanisms concerned with cell respiration. Some of these have been considered in Sections II,E, III,B, and III,C.

The toxicity of the chelates for cells in culture has been assessed in terms of the minimum concentration affecting the microscopic appearance of cells or the multiplication of single cells to form clones. The two methods give identical results. Most chelates tested are lethal for cells at concentrations of 10^{-5} to 10^{-7} M with no apparent differences between normal and malignant, adult and embryonic, or epithelial and fibroblastic cells. The dose/toxicity curve by either method is extremely steep resembling in this respect the dose/effect curve of chelate inhibition of influenza virus multiplication (Fig. 3) and the dose/toxicity curve in mice (Section II,B,3). Some chelates like $[Ru(Me_4phen)_2(aca)]^+$ ion induce marked stimulation of acid production; others like the $[Fe(Me_4phen)_3]^{++}$ ion do not.

In order to determine whether any concentration of any chelate showed selective toxicity, dilutions of several chelates were inoculated into cultures of HeLa cells, monkey kidney epithelium, human amnion, or chick embryo fibroblasts together with 100 ID_{50} of influenza, vaccinia, or poliomyelitis virus. It could be anticipated that unprotected cultures would show signs of cellular damage resulting from virus multiplication, while cultures receiving a toxic dose of chelate would present a similar appearance. If there were selective toxicity there should exist a concentration of chelate at which neither virus- nor chelate-induced cell damage occurred. However no such concentration was found.

Clearly, therefore, the chelates can inhibit virus multiplication by initiating changes which ultimately lead to the death of the cell. Consequently, it is not surprising that a chelate such as $[Ru(Me_4phen)_2(aca)]^+$ ion was unable to protect against intranasal infection of mice with influenza virus or intradermal infection of rabbits, intracerebral infection of mice, or chorioallantoic infection of embryonated eggs with vaccinia virus.

It would seem then that inert metal chelates can "inactivate" virus by occupying sites on the surface of the virus apparently associated with initiating the process of host cell infection. It is thus of interest that influenza virus can be partially protected from chelate-induced "inactivation" by the presence of other charged molecules such as Ca^{++}, Mg^{++}, and gelatin. High concentrations of $[Ru(Me_4phen)_2(aca)]^+$ and $[Fe(Me_4phen)_3]^{++}$ ions agglutinate fowl red blood cells, the former substance subsequently lysing these cells, presumably following its penetration.

The chelate cations may also prevent virus reproduction by penetrating the host cell to produce progressive and irreversible biochemical dysfunctions which deprive the virus of the metabolites and energy sources required for its multiplication.

IV. Clinical Uses of Metal Chelates

It has been demonstrated that inert, highly stable metal chelates have considerable activity in the presence of tissue, body fluid, and infective exudates against a wide range of microorganisms including some which show high resistance to various antibiotics. Resistance to these chelates is not developed by microorganisms such as *Staph. pyogenes*, the virulence of which to mice is, indeed, attenuated by contact with the chelates. It has also been shown that almost all the chelates are nonirritant to skin surfaces, mucous surfaces (eye, ear, nose), mucous cavities (vagina, gut), and subcutaneous tissues (muscle), following their prolonged administration at concentrations well above therapeutic levels. Hence it might be anticipated that such chelates should be effective in the prophylaxis and control of topical infections found in clinical and veterinary practice. Systemic administration is prevented by the curariform activity shown by these substances when administered in doses necessary for a therapeutic effect.

Clinical trials, at present under way in Melbourne (Victoria), have shown that such substances may be used to greatest advantage in the control of infection due to Gram-positive microorganisms especially *Staph. pyogenes*, but control of infections due to Gram-negative microorganisms, trichomonads and pathogenic fungi, has also shown considerable promise.

The chelates appear to have prophylactic value in:

(*a*) neonatal infections of the skin and
(*b*) wound infections at body surfaces, the chelate being applied preoperatively and postoperatively.

They have therapeutic value topically applied in the following conditions:

(*a*) wound infections due particularly to *Staphylococcus pyogenes;* control is rapid with removal of slough and with the promotion of healthy granulation tissue;
(*b*) dermatological infections (boils, folliculitis, acne vulgaris, pustular dermatitis, infective eczema, dermatomycoses);
(*c*) vaginal infection due to *Trichomonas vaginalis* and *Candida albicans;* in these cases symptomatic relief may be rapid and sometimes lasting but remission is frequent;
(*d*) ear and nose infection (infected mastoid cavities, otitis externa and interna, nasal furunculosis, and mucopurulent discharge);
(*e*) eye infections (acute and chronic conjunctivitis);

V. Conclusions

This work has covered certain aspects of the biological activity of metals, chelating agents, and metal chelates. Considerable attention has been paid to fully coordinated and inert metal chelates (derived principally from ruthenium and osmium) which owe their biological activity solely to the physicochemical properties of the cation as a whole. Such substances have shown marked activity in a large variety of biological systems such as intact and isolated cell systems, intracellular, intraparticulate and isolated enzyme systems, the neuromuscular junction, the adrenal medulla, a variety of microbiological systems ranging from Gram-positive, Gram-negative and acid-fast bacteria to pathogenic fungi, yeast and viruses, and neoplastic tissue.

Qualitative and quantitative differences in biological activity have been observed among metal chelates which differ only in the metal ion or in the ligand, and the basis of such differences has been discussed.

Greatest biological activity is generally associated with complex cations whose considerable lipid solubility *in vitro* would favor penetration of biological membranes and which at the same time would have a high capacity to bind to diverse biological surfaces of suitable steric complementariness by Coulombic and van der Waals forces.

In all systems so far investigated a common effect of active metal chelates has been depression of cellular respiration which in some cases was progressive and irreversible, a finding consistent with the strong binding expected of large charged pseudo spheres to intracellular macromolecules. Action of such substances on cytochrome components of the respiratory chain has been demonstrated but it is likely that they act as well on more fundamental cellular components, e.g., nucleic acids.

It may be anticipated that a stable metal chelate will produce its effect either by structural or functional activation or inactivation of a susceptible biological site, and the type of response, either stimulant or depressant, may depend on the rate of association of the metal chelate with the susceptible site and on its rate of dissociation from this site in accordance with the theory of drug action proposed by Paton (1961). A stable metal chelate could exert a powerful inhibitory effect on an intracellular biological process by concentrating adequately at a susceptible site from which it dissociates slowly. If the rate of dissociation of the chelate from the biological site is sufficiently slow, then inhibition induced by metal chelate will be irreversible. If this site is essential to the cell for transport, respiration, energy production, protein synthesis, etc., then the resulting permanent dysfunction of the biochemical mechanisms associated with the site may lead to death of the cell or to the development of alternate biochemical mechanisms which compensate for those irreversibly impaired by metal chelate.

It is expected that stable metal chelates will produce a variety of biological effects in different cell systems as well as multiple actions within the one cell.

Since stable complex cations can only produce their effects by physical means, a correlation of their more important physical properties with their biological activity should provide a great deal of information concerning the physicochemical characteristics of the biological systems with which they interact. Since the size, charge distribution, stereochemistry, redox potential, and other physical properties of metal chelates can be varied readily during chemical synthesis, these substances would seem to be ideal pharmacological tools with which to investigate many functional systems of the living cell.

ACKNOWLEDGMENTS

We wish to thank the National Health and Medical Research Council of Australia for their support of this work and Professor A. Haddow, F.R.S., and Professor F. Bergel, F.R.S., Chester Beatty Research Institute, London, in whose laboratories one of us (A.S.) was working while this chapter was written, for their kind interest.

We especially wish to thank our colleagues in the Departments of Bacteriology, Biochemistry, and Physiology of the University of Melbourne, in the Victorian Pharmacy College, and in the Royal Melbourne Hospital, Royal Women's Hospital, and the Eye and Ear Hospital, Melbourne, who have made available to us the results of their unpublished work, and assisted in the preparation of the chapter.

We are grateful to Professor R. D. Wright for his advice and guidance during the conduct of this work.

References

Abraham, E. P. (1953). *In* "Adaptation in Micro-organisms." *Symposium Soc. Gen. Microbiol.* **3,** 201.

Albert, A. (1953). *Experientia* **9,** 370.

Albert, A. (1958). *In* "The Strategy of Chemotherapy." *Symposium Soc. Gen. Microbiol.* **8,** 112.

Albert, A. (1960). "Selective Toxicity." Methuen, London.

Albert, A. (1961). *Federation Proc.* **20,** Suppl. 10, p. 137.

Albert, A., and Rees, C. W. (1956). *Nature* **177,** 433.

Albert, A., Rubbo, S. D., Goldacre, R. J., and Balfour, B. G. (1947). *Brit. J. Exptl. Pathol.* **28,** 69.

Alles, G. A., and Knoefel, P. K. (1939). *Univ. Calif. (Berkeley) Publ. Pharmacol.* **1,** 187.

Barnett, T. B., and Metcalf, R. G. (1949). *In* "Pharmacology and Toxicology of Uranium Compounds" (C. Voegtlin and H. C. Hodge, eds.), p. 207. McGraw-Hill, New York.

Barron, E. S. G., and Kalnitsky, G. (1947). *Biochem. J.* **41,** 346.

Beccari, E. (1941). *Boll. soc. ital. biol. sper.* **16,** 216.

Beccari, E. (1949). *Arch. sci. physiol.* **3,** 611.

Best, C. H., and Taylor, N. B. (1961). "The Physiological Basis of Medical Practice," 7th ed., p. 811. Williams & Wilkins, Baltimore, Maryland.

Blank, F. (1951). *Nature* **168,** 516.

Borg, D. C. (1961). *Federation Proc.* **20,** Suppl. 10, p. 104.

Bosnich, B. (1962). Thesis, Australian National University, Canberra.

Bosnich, B., Dwyer, F. P., and Sargeson, A. M. (1960). *Nature* **186,** 966.

Brandt, W. W., and Smith, G. F. (1949). *Anal. Chem.* **21,** 1313.

Brandt, W. W., Dwyer, F. P., and Gyarfas, E. C. (1954). *Chem. Revs.* **54,** 959.

Bray, R. C., Malmström, B. G., and Vänngård, T. (1959). *Biochem. J.* **73,** 193.

Brown, A. C., and Fraser, T. R. (1869). *Proc. Roy. Soc. Edinburgh* **6,** 556.

Browning, C. H., and Adamson, H. S. (1950). *J. Pathol. Bacteriol.* **62,** 499.

Butler, H. M., and Shulman, A. (1961). Unpublished observations.

Chaberek, S., and Martell, A. E. (1959). "Organic Sequestering Agents," p. 455. Wiley, New York.

Chance, B. (1959). *In* "CIBA Foundation Symposium on Regulation of Cell Metabolism (G. E. W. Wolstenholme, ed.), p. 91. Churchill, London.
Chenoweth, M. B. (1961). *Federation Proc.* **20,** Suppl. 10, p. 125.
Coates, G. E. (1956). "Organo-Metallic Compounds," p. 48. Methuen, London.
Commoner, B., and Lippincott, B. B. (1958). *Proc. Natl. Acad. Sci. U. S.* **44,** 1110.
Commoner, B., Lippincott, B. B., and Passonneau, J. V. (1958). *Proc. Natl. Acad. Sci. U. S.* **44,** 1099.
Cymerman-Craig, J., Rubbo, S. D., Willis, D., and Edgar, J. (1955). *Nature* **176,** 34.
Dixson, S., Dwyer, F. P., Laycock, G. M., Reid, I. K., Rubbo, S. D., Sassa, W., and Shulman, A. (1961). Unpublished observations.
Dounce, A. L., and Tien Ho Lan (1949). *In* "Pharmacology and Toxicology of Uranium Compounds," (C. Voegtlin and H. C. Hodge, eds.), p. 759. McGraw-Hill, New York.
Dwyer, F. P. (1959). *Australian J. Sci.* **22,** 240.
Dwyer, F. P., Gyarfas, E. C., Rogers, W. P., and Koch, J. H. (1952). *Nature* **170,** 190.
Dwyer, F. P., Gyarfas, E. C., Wright, R. D., and Shulman, A. (1957). *Nature* **179,** 425.
Dwyer, F. P., Laycock, G. M., Ray, L. J., Shulman, A., and Wright, R. D. (1961a). Unpublished observations.
Dwyer, F. P., Messer, M., Shulman, A., and Wright, R. D. (1961b). Unpublished observations.
Dwyer, F. P., Laycock, G. M., Shulman, A., Traill, A., and Wright, R. D. (1961c). Unpublished observations.
Eagle, H., and Doak, G. O. (1951). *Pharmacol. Revs.* **3,** 107.
Eichhorn, G. L. (1961). *Federation Proc.* **20,** Suppl. 10, p. 40.
Elek, S. D. (1959). "Staphylococcus Pyogenes and Its Relation to Disease," p. 21. Livingstone, Edinburgh and London.
Ellis, S., Anderson, H. L., Jr., and Collins, M. C. (1953). *Proc. Soc. Exptl. Biol. Med.* **84,** 383.
Ephrussi, B., and Hottinguer, H. (1950). *Nature* **166,** 956.
Ephrussi, B., and Slonimski, P. P. (1955). *Nature* **176,** 1207.
Ephrussi, B., Hottinguer, H., and Chimenes, A. M. (1949). *Ann. inst. Pasteur* **76,** 351.
Fazekas de St. Groth, S., and White, D. O. (1958). *J. Hyg.* **56,** 151.
Feeney, R. E., Petersen, I. M., and Sahinkaya, H. (1957). *J. Bacteriol.* **73,** 284.
Ferguson, J. (1939). *Proc. Roy. Soc.* **B127,** 387.
Foreman, H. (1960). *In* "Metal Binding in Medicine" (M. J. Seven, ed.), p. 160. Lippincott, Philadelphia, Pennsylvania.
Foreman, H., Hardy, H. L., Shipman, T. L., and Belknap, E. L. (1953). *A. M. A. Arch. Ind. Health* **7,** 148.
Gauze, G. F., Kochetkova, G. V., and Sorbaeva, N. A. (1959). *Nature* **184,** 1821.
Goldner, M. G., Volk, B. W., and Lazarus, S. S. (1952). *Metabolism* **1,** 544.
Goodman, L. S., and Gilman, A. (1955). "The Pharmacological Basis of Therapeutics," pp. 596, 799, 942, 948. Macmillan, New York.
Granick, S. (1946). *Chem. Revs.* **38,** 379.
Gunsalus, I. C. (1954). *Federation Proc.* **13,** 715.
Hale, J. H. (1947). *Brit. J. Exptl. Pathol.* **28,** 202.
Harris, A. W., Laycock, G. M., Shulman, A., and Wright, R. E.* (1961). Unpublished observations.
Harris, A. W., Laycock, G. M., Shulman, A., and Wright, R. E.* (1961a). Unpublished observations.

* Deceased, 1960.

Harris, A. W., Shulman, A., and White, D. O. (1961b). Unpublished observations.
Höber, R. (1917). *Arch. ges. Physiol. Pfüger's* **166,** 531.
Horsfall, J. G. (1956). "Principles of Fungicidal Action." Chronica Botanica, Waltham, Massachusetts.
Ing, H. R., and Wright, W. M. (1931). *Proc. Roy. Soc.* **B109,** 337.
Ing, H. R., and Wright, W. M. (1933). *Proc. Roy. Soc.* **B114,** 48.
Jayson, G. G. (1961). *Nature* **190,** 144.
Kadota, I., and Kawachi, Y. (1959). *Proc. Soc. Exptl. Biol. Med.* **101,** 365.
Koch, J. H. (1955). *Nature* **175,** 856.
Koch, J. H., and Gallagher, C. H. (1959). *Nature* **184,** 1039.
Koch, J. H., and Gallagher, C. H. (1960). *Biochem. Pharmacol.* **3,** 231.
Koch, J. H., Gyarfas, E. C., and Dwyer, F. P. (1956). *Australian J. Biol. Sci.* **9,** 371.
Koch, J. H., Rogers, W. P., Dwyer, F. P., and Gyarfas, E. C. (1957). *Australian J. Biol. Sci.* **10,** 342.
Krebs, H. A. (1935). *Biochem. J.* **29,** 1951.
Laskowski, W. (1954). *Heredity* **8,** 79.
Laycock, G. M., Shulman, A., and Wright, R. D. (1961). Unpublished observations.
Lenta, M. P., and Riehl, M. A. (1960). *J. Biol. Chem.* **235,** 859.
Lindegren, C. C. (1958). *Bacteriol. News* **24,** 8.
Lindegren, C. C., Nagai, S., and Nagai, H. (1958). *Nature* **182,** 446.
McCallan, S. E. A. (1957). *In* "Proceedings of the Plant Protection Conference 1956," p. 77. Butterworths, London.
McNaught, M. L., and Owen, E. C. (1949). *Intern. Congr. Biochem., 1st Congr., Cambridge, Engl., 1949, Abstr. Communs.* p. 340.
Mandl, I., and Neuberg, C. (1956). *Advances in Enzymol.* **17,** 135.
Marcovich, H. (1951). *Ann. inst. Pasteur* **81,** 452.
Maslen, M., and Shulman, A. (1961). Unpublished observations.
Mason, R. (1959). *Discussions Faraday Soc.* **No. 27,** 129.
Merkel, J. R., and Nickerson, W. J. (1953). *Proc. Natl. Acad. Sci. U. S.* **39,** 1008.
Messer, M. (1958). *Australian J. Exptl. Biol.* **36,** 65.
Miller, L. P., and McCallan, S. E. A. (1956). *Proc. Intern. Conf. on Peaceful Uses Atomic Energy, Geneva, 1955* **12,** 170.
Miller, L. P., and McCallan, S. E. A. (1957). *J. Agr. Food. Chem.* **5,** 116.
Mills, W. H., and Clark, R. E. D. (1936). *J. Chem. Soc.* p. 175.
Mudd, S., Brodie, A. F., Winterscheid, L. C., Hartman, P. E., Beutner, E. H., and McLean, R. A. (1951). *J. Bacteriol.* **62,** 729.
Nagai, S. (1959). *Science* **130,** 1188.
Nagai, S., and Nagai, H. (1958). *Naturwissenschaften* **45,** 577.
Nagai, S., Yanagishima, N., and Nagai, H. (1961). *Bacteriol. Revs.* **25,** 404.
Nelson, J. F., Shulman, A., and Wright, R. D. (1961). Unpublished observations.
Owens, R. G. (1953). *Contribs. Boyce Thompson Inst.* **17,** 221.
Panimon, F., Horwitt, M. K., and Gerard, R. W. (1941). *J. Cellular Comp. Physiol.* **17,** 17.
Paton, W. D. M. (1961). *Proc. Roy. Soc.* **B154,** 21.
Pauling, L. (1948). *J. Chem. Soc.* p. 1461.
Perry, H. M., and Perry, E. F. (1959). *J. Clin. Invest.* **38,** 1452.
Peters, R. A. (1936). *Nature* **138,** 327.
Peters, R. A., and Wakelin, R. W. (1946). *Biochem. J.* **40,** 513.
Peters, R. A., Stocken, L. A., and Thompson, R. H. S. (1945). *Nature* **156,** 616.
Peters, R. A., Sinclair, H. M., and Thompson, R. H. S. (1946). *Biochem. J.* **40,** 516.

Philips, F. S. (1961). *Federation Proc.* **20,** Suppl. 10, p. 129.
Rieders, F., Dunnington, W. G., and Brieger, H. (1955). *Ind. Med.* **24,** 195.
Riker, W. F. (1953). *Pharmacol. Revs.* **5,** 1.
Robinson, A. E., Beckett, A. H., and Dar, R. N. (1960). *J. Pharm. and Pharmacol.* **12,** 385.
Rothstein, A. (1959). *Bacteriol. Revs.* **23,** 175.
Rubin, M., and Princiotto, J. V. (1960). *Ann. N. Y. Acad. Sci.* **88,** 450.
Rubin, M., Alexander, R., and Lindenblad, G. (1960). *Ann. N. Y. Acad. Sci.* **88,** 474.
Scatchard, G., Hughes, W. L., Gurd, F. R. N., and Wilcox, P. E. (1954). *In* "Chemical Specificity in Biological Interactions" (F. R. N. Gurd, ed.), p. 193. Academic Press, New York.
Schubert, J. (1954). *In* "Chemical Specificity in Biological Reactions" (F. R. N. Gurd, ed.), p. 145. Academic Press, New York.
Schubert, J. (1960). *In* "Metal Binding in Medicine" (M. J. Seven, ed.), p. 325. Lippincott, Philadelphia, Pennsylvania.
Schubert, J., and Lindenbaum, A. (1960). *In* "Metal Binding in Medicine" (M. J. Seven, ed.), p. 68. Lippincott, Philadelphia, Pennsylvania.
Seven, M. J., ed. (1960). "Metal Binding in Medicine," p. 95. Lippincott, Philadelphia, Pennsylvania.
Shapiro, R., and Papa, D. (1959). *Ann. N. Y. Acad. Sci.* **78,** 756.
Shaw, W. H. R. (1961). *Nature* **192,** 754.
Shulman, A. (1959). *Proc. Roy. Australian Chem. Inst.* **26,** 489.
Shulman, A., and Vaughan, G. (1961). Unpublished observations.
Slater, E. C., and Hülsmann, W. C. (1959). *In* "CIBA Foundation Symposium on Regulation of Cell Metabolism" (G. E. W. Wolstenholme, ed.), p. 58. Churchill, London.
Slonimski, P. P., and Ephrussi, B. (1949). *Ann. inst. Pasteur* **77,** 47.
Smith, G. F., McCurdy, W. H., and Diehl, H. (1952). *Analyst.* **77,** 418.
Speck, J. F. (1947). *J. Biol. Chem.* **168,** 403.
Spencer, H. (1960). *Ann. N. Y. Acad. Sci.* **88,** 435.
Stocken, L. A., and Thompson, R. H. S. (1946). *Biochem. J.* **40,** 529, 535.
Tamm, I. (1958). *In* "The Strategy of Chemotherapy." *Symposium Soc. Gen. Microbiol.* **8,** 178.
Taylor, D. B. (1951). *Pharmacol. Revs.* **3,** 412.
Thompson, R. H. S. (1946). *Biochem. J.* **40,** 525.
Thompson, R. H. S., and Whittaker, V. P. (1947). *Biochem. J.* **41,** 342.
Turian, G. (1951). *Helv. Chim. Acta* **34,** 917.
Underhill, F. P., and Peterman, F. I. (1929). *Am. J. Physiol.* **90,** 40, 62.
Uzman, L. L. (1960). *In* "Metal Binding in Medicine" (M. J. Seven, ed.), p. 269. Lippincott, Philadelphia, Pennsylvania.
Vallee, B. L. (1955). *Advances in Protein Chem.* **10,** 318.
van Campenhout, E., and Cornelis, G. (1951). *Compt. rend. soc. biol.* **145,** 933.
Voegtlin, C., Dyer, A., and Leonard, C. S. (1923). *U. S. Public Health Repts.* **38,** 1882.
Voureka, A. (1952). *J. Gen. Microbiol.* **6,** 352.
Wallace, A. (1960). *Ann. N. Y. Acad. Sci.* **88,** 361.
Walshe, J. M. (1956). *Lancet* **I,** 25.
Weinberg, E. D. (1960). *In* "Metal Binding in Medicine" (M. J. Seven, ed.), p. 329. Lippincott, Philadelphia, Pennsylvania.
White, M. R., Finkel, A. J., and Schubert, J. (1951). *J. Pharmacol. Exptl. Therap.* **102,** 88.

Wilson, I. B., and Bergmann, F. (1950). *J. Biol. Chem.* **186,** 683.
Wise, R. I., and Spink, W. W. (1954). *J. Clin. Invest.* **33,** 1611.
Yanagishima, N. (1956). *J. Inst. Polytech. Osaka City Univ. Ser .D* **7,** 131.
Yang, W. C., Yanasugondha, D., and Webb, J. L. (1958). *J. Biol. Chem.* **232,** 659.
Yotsuyanagi, Y. (1955). *Nature* **176,** 1208.
Zentmyer, G. A. (1944). *Science* **100,** 294.

CHAPTER 10

Physical and Coordination Chemistry of the Tetrapyrrole Pigments

J. E. FALK AND J. N. PHILLIPS

Division of Plant Industry, C.S.I.R.O., Canberra, A.C.T., Australia

I. Porphyrins as Chelating Agents

A. Introduction

The tetrapyrrole nucleus (I) has unique chelating properties; through the effects of substituents on the nucleus and through the effects of the extra ligands which may be added perpendicularly to the porphyrin plane, the physicochemical properties of metalloporphyrins are subject to almost infinite modification. In the natural compounds, modifications of these kinds lead to the wide range of specific biological activities of the complexes of iron (the hemoproteins), of magnesium (the chlorophylls), and of cobalt (vitamin B_{12} and the related "cobamide coenzymes").

Metalloporphyrins are essential to the life of bacteria, fungi, plants, and animals, and extensive studies have been made of their organic chemistry (cf. Fischer and Orth, 1937; Fischer and Stern, 1940; Willstätter and Stoll, 1913), their biochemistry (cf. Lemberg and Legge, 1949; Falk *et al.*, 1961; Falk, 1963), and their biosynthesis (cf. Wolstenholme and Millar, 1955; Shemin, 1956; Rimington, 1958; Bogorad, 1960; Margoliash, 1961; Granick and Mauzerall, 1961).

We are concerned here mainly with the physical and coordination chemistry of the porphyrins and metalloporphyrins, exclusive of their protein complexes, but in the final section some aspects of the physicochemical properties of the natural hemoproteins are discussed.

Porphin (I) is the parent nucleus of the porphyrins, which carry substituents in some of the positions 1 to 8 or α to δ. Chlorin (II) and phorbin (III) are important related nuclei.

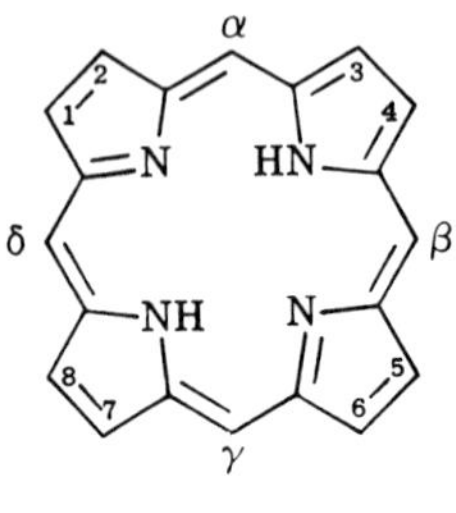

Porphin

(I)

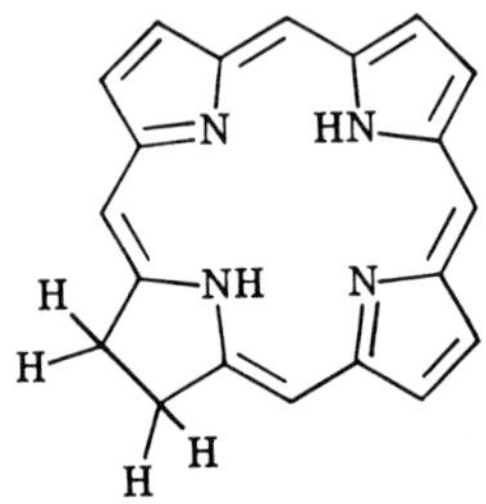

Chlorin

(II)

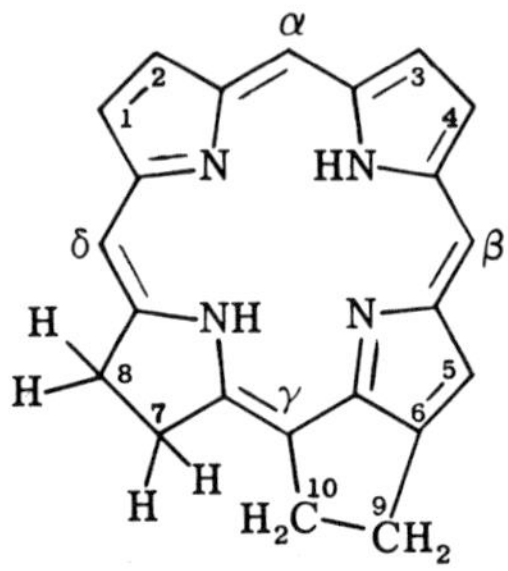

Phorbin

(III)

The names of various porphyrins and related compounds mentioned in this article are defined in Table I.

That the porphin nucleus is likely to be planar had long been evident from the X-ray crystallographic analysis of the related tetraazaporphyrin, phthalocyanine (Robertson, 1936); the X-ray analysis of Ni etioporphyrin (Crute, 1959) indicates that the true porphyrin nucleus also is planar, and that the chelated Ni atom lies in the same plane (IV).*

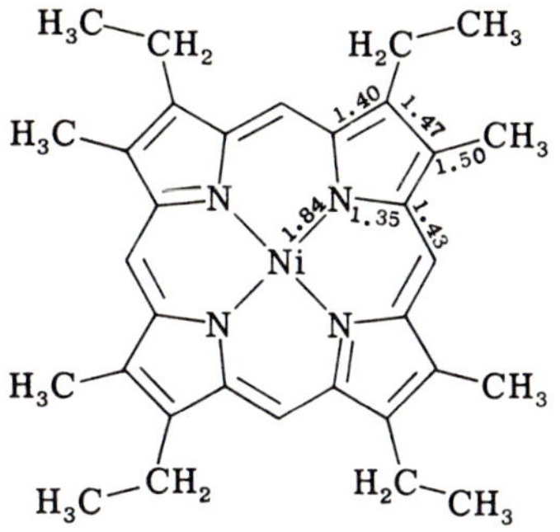

(IV)

Ni(II) etioporphyrin (isomer II) drawn to scale from the Fourier diagram of Crute, 1959. Bond lengths are shown in Å.

The chelation of a metal ion by a porphyrin involves the incorporation of the metal ion into the center of the tetrapyrrole nucleus with the simultaneous displacement of two protons from the secondary (pyrrole-type) nitrogen atoms. Porphyrins form tetradentate chelates of the inner complex type; thus with a divalent metal ion the positive charges on the metal ion are exactly compensated by the negative charges on the porphyrin nucleus, so that in the absence of ionized substituents on the porphyrin the resulting chelate has no net charge. Most of the naturally occurring porphyrins, however, contain carboxylic acid side chains which contribute negative charges to their chelates in the neutral and alkaline pH region.

The planar nature of the porphyrin nucleus tends to force a square-planar configuration on its metal chelates (V); with certain metals, a further one or two ligands may coordinate perpendicularly to the plane of the porphyrin ring and so form square-pyramidal (VI) or octahedral (VII)

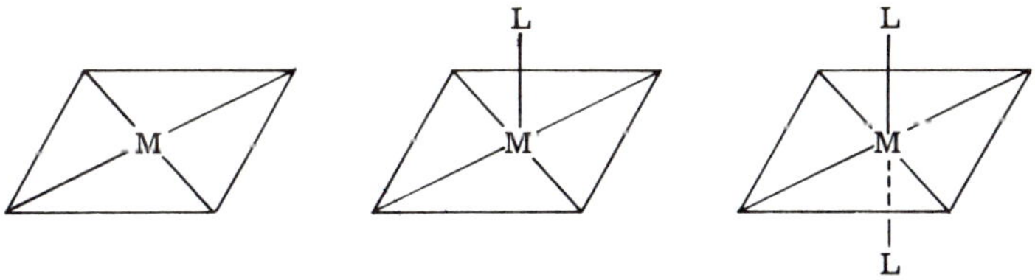

(V) (VI) (VII)

complexes (see Section III). The unusually low thermodynamic stability of porphyrin chelates with ions such as Hg(II), Pb(II), and Ba(II) suggests that in these cases only a distorted planar configuration can be attained

* Significant deviations from planarity have recently been reported (Fleischer, 1963a, 1963b; Kendrew, 1963).

TABLE I
COMPOUNDS MENTIONED IN THE TEXT

Type	Compound[a]	Substitution position 1	2	3	4	5	6	7	8	Other substituents
Porphin	Etioporphyrin I	M	E	M	E	M	E	M	E	—
	Etioporphyrin II	M	E	E	M	M	E	E	M	—
	$\alpha,\beta,\gamma,\delta$-Tetraphenylporphin[b]	H	H	H	H	H	H	H	H	$\alpha = \beta = \gamma = \delta = \phi$
	Pyrroporphyrin XV	M	E	M	E	M	H	P	M	—
	γ-Phylloporphyrin XV	M	E	M	E	M	H	P	M	γ = M
	Desoxophylloerythrin	M	E	M	E	M	—	P	M	6-(—CH_2—CH_2—)-γ
	Phylloerythrin	M	E	M	E	M	—	P	M	6-(—CO—CH_2—)-γ
	Mesoporphyrin IX	M	E	M	E	M	P	P	M	—
	Deuteroporphyrin IX dimethyl ester disulfonic acid	M	S	M	S	M	P_m	P_m	M	—
	Diacetyldeuteroporphyrin IX	M	Ac	M	Ac	M	P	P	M	—
	Hematoporphyrin IX	M	B	M	B	M	P	P	M	—
	Protoporphyrin IX	M	V	M	V	M	P	P	M	—
	Rhodoporphyrin XV	M	E	M	E	M	C	P	M	—
	Coproporphyrin I	M	P	M	P	M	P	M	P	—
	Coproporphyrin III	M	P	M	P	M	P	P	M	—
	Uroporphyrin I	A	P	A	P	A	P	A	P	—
Chlorin	Etiochlorin II	M	E	E	M	M	E	E	M	—
	$\alpha,\beta,\gamma,\delta$-Tetraphenylchlorin[b]	H	H	H	H	H	H	H	H	$\alpha = \beta = \gamma = \delta = \phi$
	Pyrrochlorin	M	V	M	E	M	H	P	M	—
	γ-Phyllochlorin	M	V	M	E	M	H	P	M	γ = M
	Purpurin 18	M	V	M	E	M	—	P	M	6-(—CO—O—CO—)-γ
	Pheopurpurin 18	M	V	F	E	M	—	P	M	6-(—CO—O—CO—)-γ

	Rhodochlorin	M	V	M	E	M	C	P	M		—
	b-Rhodochlorin	M	V	F	E	M	C	P	M		—
	Chlorin p_6	M	V	M	E	M	C	P	M	$\gamma = C$	
	Chlorin e_6	M	V	M	E	M	C	P	M	$\gamma = A$	
	Rhodin g_7	M	V	F	E	M	C	P	M	$\gamma = A$	
Phorbin	Methylpheophorbide *a*	M	V	M	E	M	—	P	M	9(—C = O) 10(—$COOCH_3$)	
	Methylpheophorbide *b*	M	V	F	E	M	—	P	M	9(—C = O) 10(—$COOCH_3$)	
	Pyropheophorbide *a*	M	V	M	E	M	—	P	M	9(—C = O)	
	Pheophytin *a*	M	V	M	E	M	—	P_p	M	9(—C = O) 10(—$COOCH_3$)	
	Pheophytin *b*	M	V	F	E	M	—	P_p	M	9(—C = O) 10(—$COOCH_3$)	
	Chlorophyll *a*	Magnesium pheophytin *a*									—
	Chlorophyll *b*	Magnesium pheophytin *b*									—

KEY:

H = Hydrogen
M = CH_3
E = CH_2CH_3
V = $CH{=}CH_2$
B = $CH(OH)CH_3$
F = CHO
ϕ = C_6H_5
C = COOH
A = CH_2COOH
P = CH_2CH_2COOH
P_m = $CH_2CH_2COOCH_3$
P_p = $CH_2CH_2COOC_{20}H_{39}$ (phytyl ester)
S = SO_3H
Ac = $COCH_3$

[a] Where the literature reference does not specify the particular porphyrin isomer the naturally occurring isomer has been assumed.

[b] Referred to throughout the text as tetraphenyl porphin and tetraphenyl chlorin.

because of the difficulty of fitting such large ions into the center of the nucleus. On the other hand the X-ray structural data for Ni(II) etioporphyrin and Ni(II) phthalocyanine, and also the very great thermodynamic stability of the Ni(II), Co(II), and Cu(II) chelates, indicate that ions of the first transition series can be accommodated coplanar with the ring.

An unusual feature of porphyrin chelates is that the metal ion is bonded to four nitrogen atoms which are themselves linked together in a conjugated system. It is this high degree of mesomerism, whereby a substituent anywhere on the porphyrin nucleus can relay its electron-attracting or donating tendency to all four coordinating atoms, that makes metalloporphyrins so sensitive to the influence of substituents. Moreover, such effects can be transmitted to and from the perpendicularly coordinated ligands. No doubt this is one reason why in nature metalloporphyrins are so favored as prosthetic groups, since only subtle modifications in the biosynthetic pathway are required to yield molecules differing little in structure but greatly in biological behavior, particularly when coupled with (that is coordinated to) specific proteins.

B. Kinetics of Metalloporphyrin Formation

In view of the extremely high thermodynamic stability of the chelates formed (see Section I,D) it is somewhat surprising that the reaction between a porphyrin and a metal ion in aqueous solution proceeds very slowly at normal temperatures. The high activation energy which has to be overcome to form metalloporphyrins in an aqueous environment may be an important biological factor, enabling the cell to control through enzymatic specificity the particular metal to be incorporated.

The rate of incorporation of a metal ion (M^{++}) into a porphyrin (PH_2) has been studied in aqueous detergent solutions and shown to follow classical kinetics for the bimolecular reaction expressed in Eq. (1) (Dempsey *et al.*, 1961, p. 30; Lowe and Phillips, 1961).

$$M^{++} + PH_2 \rightleftharpoons MP + 2H^+ \qquad (1)$$

The rate of this reaction is particularly dependent on the nature of the porphyrin and on the nature and coordination state of the metal ion. With respect to the porphyrin, increasing the electron-withdrawing character of groups substituted in positions 1 to 8 or α to δ of the porphin nucleus tends to decrease the rate of metal incorporation (Table II). With respect to the metal ion, it would appear that tetracoordinated ions are more rapidly incorporated than those which are hexacoordinated (cf. Dempsey *et al.*, 1961, pp. 34, 214). Thus cobalt(II) is rapidly incorporated from blue solutions where the cobaltous ion has a tetrahedral configuration but only slowly from pink solutions where it has an octahedral configuration.

TABLE II
THE INFLUENCE OF ELECTRONEGATIVE SUBSTITUENTS ON RATE OF FORMATION OF METALLOPORPHYRINS

Porphyrin[a]	pK_3[b]	Relative rate of formation of Cu(II) complex[b]
Mesoporphyrin	5.94	7
Protoporphyrin	4.89	2.5
Diacetyldeuteroporphyrin	3.50	1

[a] For side chains see Table I.
[b] At 20°, in 2.5% aqueous sodium dodecylsulfate.
Lowe and Phillips (unpublished); cf. Dempsey *et al.* (1961).

Recently, two rather unusual ways of catalyzing the metal incorporation reaction have been reported (Lowe and Phillips, 1961). The first involves carrying out the reaction in an aqueous ionic detergent solution in which the charge of the detergent ion is opposite in sign to that of the metal ion. Thus the incorporation of Cu^{++} into protoporphyrin dimethyl ester takes place 20,000 times faster in an anionic than in a cationic or nonionic detergent solution. It has been suggested that in the detergent solution the reactive porphyrin species is a solubilized complex for which a doughnut-like structure has been postulated (Fig. 1). The pyrrole rings are embedded in the hydrocarbon interior of the doughnut, the outer surface of which is covered with the detergent head groups; the porphyrin nitrogens are situated in the inner ring of the doughnut in contact with the aqueous phase. Such a structure would tend to attract or repel ions electrostatically from the vicinity of the central nitrogen atoms according to the sign of the charge on the detergent and on the metal ion, and the reaction rate would be influenced accordingly.

It has also been observed that the rate of incorporation of metals into porphyrins can be greatly influenced by the presence of certain chelating agents. Figure 2 illustrates the effect of increasing quantities of the chelator 8-hydroxyquinoline-5-sulfonic acid (oxine-5-sulfonic acid) on the incorporation of copper into protoporphyrin dimethyl ester in 5% aqueous cetyltrimethylammonium bromide (CTAB) solution. Figure 3 illustrates the effect of varying the concentration of oxine itself on the same reaction carried out in 2.5% sodium dodecylsulfate (SDS) solution; it is apparent that the 1:1 copper-oxine chelate is the catalytically active species and that the 1:2 chelate is inhibitory to the reaction. Table III summarizes the catalytic effect of a number of chelating agents in this reaction.

The mechanism of the metal incorporation reaction is believed to be of

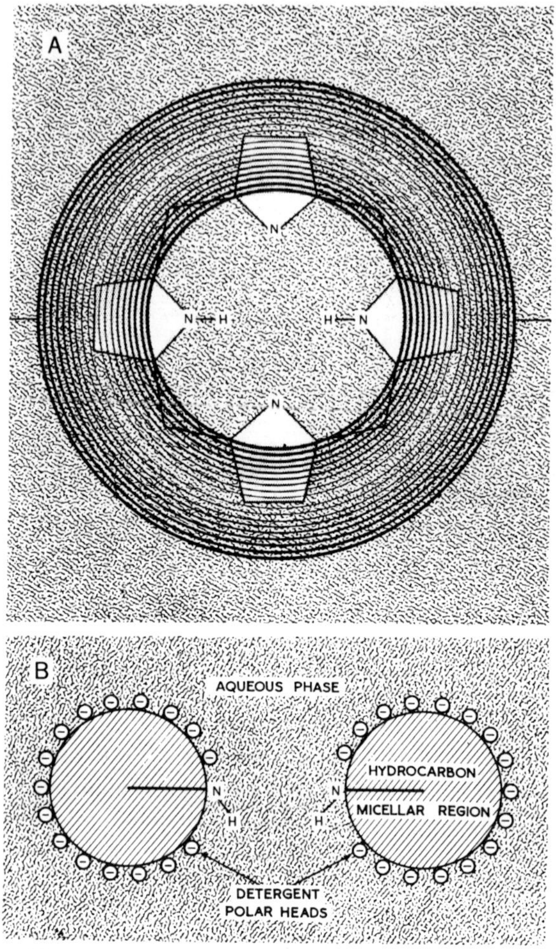

FIG. 1. A possible structure for the porphyrin micelle complex, with the porphyrin mainly immersed in the hydrocarbon region of the micelle and the nitrogen atoms in contact with the aqueous phase: (A) plan; (B) section. From Lowe and Phillips, 1961.

the displacement (S_N2) type (Dempsey *et al.*, 1961), and it has been suggested (Wang, 1961) that the intermediate complex involves the coordination of the metal ion to the lone pairs on the pyrrole type of nitrogen atoms. The catalytic effect of certain chelating agents has been attributed to their ability to reduce the charge on the metal ion (Fig. 4), thus weakening the attachment of the water molecules in the remaining coordination positions and so facilitating their replacement by the porphyrin nitrogen atoms.

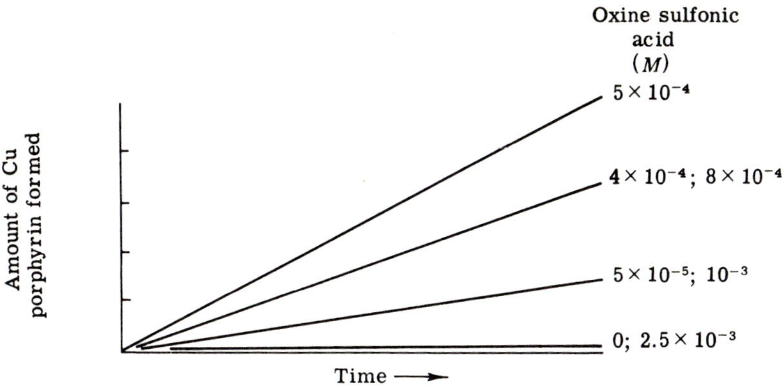

FIG. 2. Effect of oxinesulfonic acid upon initial rate of formation of Cu protoporphyrin (dimethyl ester). Measured at 60°C., in 2.5% CTAB solution; Cu(II) concentration, 5×10^{-4} *M*. Lowe and Phillips, unpublished.

TABLE III
EFFECT OF CHELATING AGENTS ON RATE OF COPPER INCORPORATION INTO A PORPHYRIN

Chelating agent	$k \times 10^{-1}$ [a] (liters/mole—minutes)
Sodium diethyldithiocarbamate	3300
Potassium ethylxanthate	1100
2-Hydroxypyridine-*N*-oxide	1000
8-Hydroxyquinoline	750
Kojic acid	550
Salicylaldehyde	110
Control (no chelating agent)	20
Salicylic acid	10
Glycine	10
Terramycin	10
Nitrilotriacetic acid	10
o-Phenanthroline	5
2,2′-Dipyridyl	5
Histidine	5
Ethylenediaminetetraacetic acid	2

[a] Second order rate constant at 20°C. for formation of mesoporphyrin copper complex. Equimolar amounts of chelating agent and copper ion, in the presence of 2.5% sodium dodecylsulfate (Lowe and Phillips, 1962).

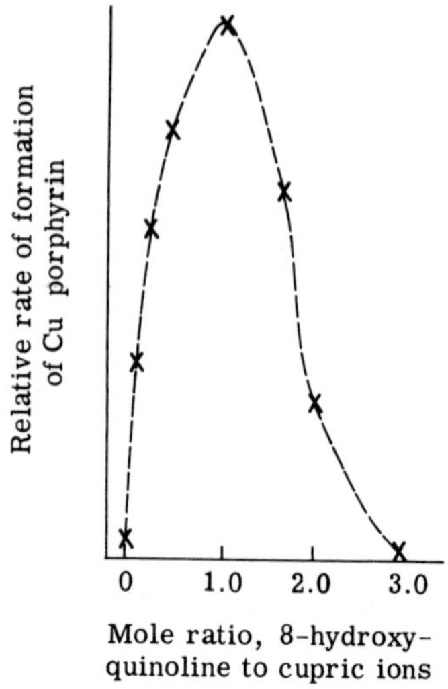

FIG. 3. Effect of varying the oxine:Cu ratio upon initial rate of formation of Cu(II) protoporphyrin (dimethyl ester). Measured at 40°C. in 2.5% SDS; pH ~ 7. From Lowe and Phillips, 1961.

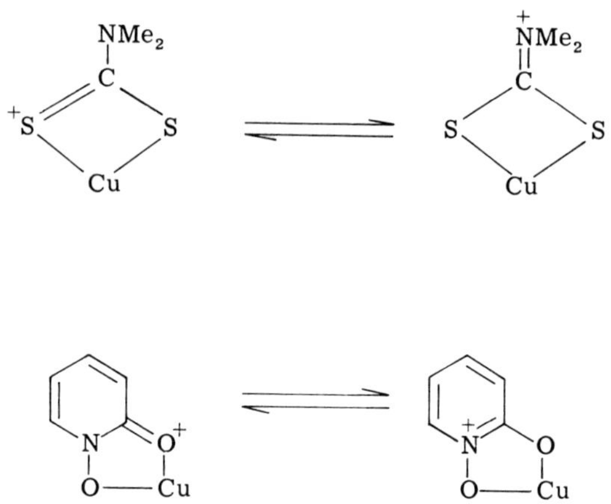

FIG. 4. Charge-delocalization in Cu dimethyldithiocarbamate and Cu-2-hydroxy-pyridine-*N*-oxide.

C. Enzymatic Incorporation of Iron

An enzyme system, "ferrochelatase" (cf. Rimington, 1958), catalyzing the incorporation of iron into porphyrins, has been obtained in soluble form from liver mitochondria (Minakami, 1958; Minakami *et al.*, 1958; Nishida and Labbe, 1959; Lochhead and Goldberg, 1961) and from avian red cell particulates (Goldberg *et al.*, 1956; Krueger *et al.*, 1956; Nevé, 1961).

It is not yet clear whether a single enzyme or an enzyme complex is involved in the reaction, nor has the mechanism of the enzymatic reaction been elucidated. Some interesting facts have, however, emerged.

The erythrocyte enzyme catalyzes the incorporation into protoporphyrin of Co(II) and Zn(II), but only some 2 and 10% as effectively as Fe(II) (Oyama *et al.*, 1961), while Co(II) is incorporated by the liver enzyme at least as efficiently as is Fe(II) (Labbe and Hubbard, 1961). The latter workers found that a number of other metals (Mg, Ca, Ni, Cd, Pb, Cu, Mn, Zn, Hg) were not incorporated and could also inhibit Fe incorporation. Some other enzyme or perhaps another biosynthetic mechanism (cf. Granick, 1961; Tait and Gibson, 1961) may be involved in the incorporation of Mg into the chlorophylls.

Labbe and Hubbard (1961) have shown that the liver enzyme catalyzes the insertion of iron into deutero-, proto-, hemato-, meso-, and 2,4-dibromodeuteroporphyrins (for structures see Table I), demonstrating that vinyl groups are not essential for the enzymatic reaction, though this had been suggested by Granick and Gilder (1946) as a result of growth factor studies with *Haemophilus influenzae*. Labbe and Hubbard (1961) have shown also that the liver enzyme cannot utilize copro- or uroporphyrins as substrates. This no doubt explains the absence of the iron complexes of these porphyrins in nature in spite of the high concentrations of the corresponding porphyrins found in certain pathological conditions.

Using a soluble enzyme preparation from liver mitochondria, Porra and Jones (1963) have made a quantitative study of substrate specificity. The following initial rates were found for the iron incorporation reaction (mμmoles Fe complex formed per milliliter enzyme per 20 minutes): deuteroporphyrin (183); mesoporphyrin (59); hematoporphyrin (29); 2-(-β-hydroxyethyl)-4-vinyldeuteroporphyrin (6); protoporphyrin (4); 2-formyl-4-vinyldeuteroporphyrin (+); 2,4-diformyldeuteroporphyrin (0). These results are consistent with the suggestion (cf. Section I,B) that incorporation of iron becomes more difficult as porphyrin basicity decreases; rates found with monoformyldeuteroporphyrin (12) and monohydroxymethyldeuteroporphyrin (9) do not, however, fit this hypothesis. The models for cytochrome prosthetic group precursors, porphyrin *a* (Clezy and Barrott, 1961) and porphyrin *c* (Neilands and Tuppy, 1960), were not converted to hemes by the liver enzyme.

The inefficient utilization of the main biological substrate, protoporphyrin, in comparison with a number of synthetic substrates, is intriguing.

In view of the importance of porphyrinogens (hexahydroporphyrins) in heme biosynthesis it has been a matter of speculation (cf. Bogorad, 1960) whether iron is incorporated enzymatically into protoporphyrin or into protoporphyrinogen. Mechanisms to permit the incorporation of metals into reduced porphyrins have been proposed by Orlando (1958), Lemberg (1961, p. 216) and Dempsey *et al.* (1961, p. 214). It has been shown (Porra and Falk, 1961, 1963), that protoporphyrinogen is formed during the conversion of coproporphyrinogen to protoporphyrin. Recent evidence (Porra and Jones, 1963), however, shows conclusively that mesoporphyrin and protoporphyrin are used by the iron-incorporating enzyme, but indicates that the corresponding porphyrinogens are not used.

D. The Stability of Metal Porphyrin Chelates

A considerable amount of qualitative evidence indicates that some metalloporphyrins have a very high degree of thermodynamic stability. Nothing less than concentrated sulfuric acid, to which the porphin nucleus is stable, will remove Ni(II), Cu(II), Co(II), Mn(II), Ag(II), or Fe(III) from their chelates with the naturally occurring porphyrins or their simple analogs. The powerful stabilizing influence of the porphyrin ring upon these metal chelates is no doubt due in part to the fact that the ligand is tetradentate and in part to factors associated with its aromaticity, since, once coordinated, the metal becomes an integral part of a highly resonant aromatic system (cf. Section I). On the other hand there are a number of metalloporphyrins, such as the Hg, Pb, Na, K, and Mg chelates, which are so unstable that they dissociate in dilute mineral acid.

Due partly to the extremes in stability encountered among the metalloporphyrins, and partly to their low solubility in suitable ionizing solvents such as water, there is little quantitative information available on the equilibrium of the metal ion–porphyrin reaction. Several criteria, however, have been used to assess qualitatively the relative stabilities of various metal porphyrin chelates. Such criteria (cf. Phillips, 1960) include:

(1) The degree of dissociation of metalloporphyrins as a function of the acidity of the environment;
(2) The extent to which one metal can be replaced by another by exchange from solution;
(3) The relative position of the λ_{max} of the Soret band—it being assumed that for a given porphyrin series the more the Soret band (cf. Section tion IV,C) is displaced to shorter wavelengths the more stable the chelate; and
(4) The relative intensities of the α and β visible bands, the assumption being that the greater the α/β intensity ratio the more stable the complex.

It should be noted that none of these criteria has a sound thermodynamic basis—(1) and (2) both involve kinetic as well as thermodynamic considerations and (3) and (4) are purely empirical correlations. Neverthe-

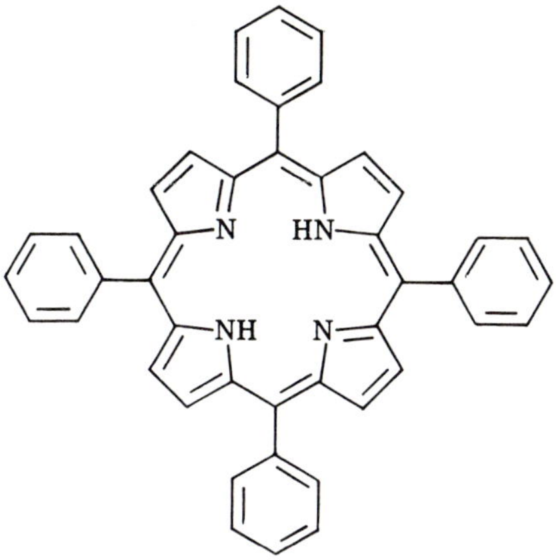

Mesotetra-
phenylporphin

(VIII)

less such criteria lead to a reasonable and self-consistent order of relative stabilities for a series of square-planar metal porphyrin chelates in which the porphyrin nucleus is kept constant and the metal varied. Thus for metal chelates of *meso*-tetraphenylporphin (VIII) one finds Pt(II) > Pd(II) > Ni(II) > Co(II) > Ag(II) > Cu(II) > Zn(II) > Mg(II) > Cd(II) > Sn(II) > Li_2 > Na_2 > Ba(II) > K_2 > $Ag(I)_2$; for references, see Phillips (1960).

The first quantitative study (Fig. 5) of the equilibrium of a porphyrin–metal ion reaction has recently been reported for the reaction between mesoporphyrin dimethyl ester and zinc sulfate in 0.25% (wt. vol.) aqueous

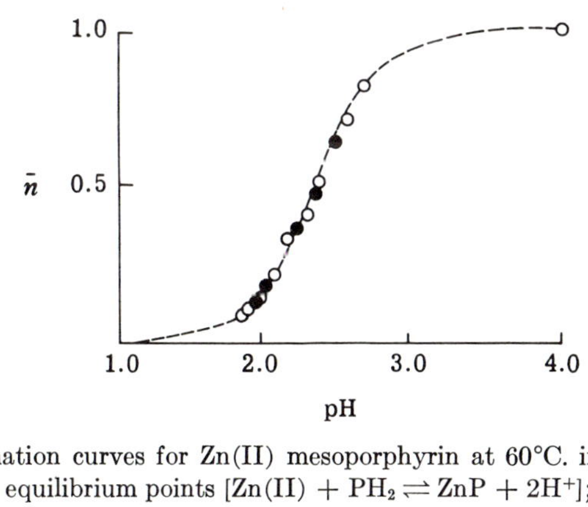

FIG. 5. Formation curves for Zn(II) mesoporphyrin at 60°C. in 0.25% CTAB. ○, forward reaction equilibrium points [Zn(II) + $PH_2 \rightleftharpoons$ ZnP + $2H^+$]; ●, backward reaction equilibrium points [$\bar{n}$ = ZnP/T_{PH_2}, where T_{PH_2} is the total porphyrin concentration].

cetyltrimethylammoniumbromide solution (Dempsey *et al.*, 1961). By making assumptions as to the acidic ionization values of the porphyrin nucleus one calculates a value for the stability constant of zinc mesoporphyrin dimethyl ester of the order of 10^{29} at room temperature. This value is enormous when compared with the corresponding β values for zinc complexes with other chelators, such as 8-hydroxyquinoline-5-sulfonic acid (10^{16}), ethylenediamine (10^{11}), and glycine ($10^{9.5}$) under comparable conditions.

II. Perpendicular Coordination by Metal Porphyrin Chelates

The biologically important metalloporphyrins, that is the hemoproteins and probably also the chlorophylls, depend in large part for their remarkable biological reactivities upon the specific proteins with which they are linked. The primary bond is from a protein ligand atom to the metal ion already coordinated in the porphyrin, and the bond must be more or less perpendicular to the plane of the latter. The sixth coordination position, on the other side of the plane, may be occupied by another ligand atom from the same protein, as is apparently the case in cytochromes *b* and *c* (cf. Section V,D), or it may be filled by a water molecule which is replaceable by other small ligands such as O_2, CO, etc., in hemoglobin or myoglobin.

Vitamin B_{12} and its analogs and the related "cobamide" coenzymes (cf. Barker *et al.*, 1960) are cobalt chelates of a tetrapyrrole macrocycle closely analogous to a porphyrin; little is known about the detailed role of the cobalt (cf. Hodgkin *et al.*, 1957; Eggerer *et al.*, 1960; Plant, 1961).

Evidence has been advanced that a Mn porphyrin chelate may occur in blood (Borg and Cotzias, 1958). This is the only suggestion of the occurrence in living tissue of a Mn chelate of this type.

Extensive studies of the reactions of metalloporphyrins with additional ligands have been stimulated by their obvious relevance to our understanding of the natural compounds. Although there are still many gaps in our knowledge of both natural and model compounds, a degree of understanding is now possible. In addition, among the model compounds excellent examples are emerging of the way in which the nature of extra ligands can influence the bond-type and the physical properties of the chelated metal.

The further coordination of metalloporphyrins is primarily determined by the nature of the metal involved, and for purposes of discussion it is convenient to classify the perpendicular complexes of porphyrin chelates according to the number of outer *d*-electrons associated with the metal ion. The following treatment is mainly descriptive, the physicochemical properties, bond-type, etc., of these complexes being discussed in Sections III and IV.

A. d^{10} Systems

The closed shell metal ion–porphyrin chelates, for example Mg(II), Zn(II), and Cd(II), readily add ligands such as water, alcohols, amines, etc., to give square-pyramidal–type complexes in which the metal ion is pentacoordinated. This coordination is associated with marked spectroscopic changes, the complex becomes fluorescent, and there is a 5–10 mμ shift to longer wavelength in the UV and visible absorption spectra. Table IV summarizes values for the stability constant for the complexes of chlorophyll *a* with various ligands.

TABLE IV
Stability Constant Values for Chlorophyll *a* Complexes in Benzene

Ligand	Log K_{S^1} [a]
Dimethylaniline	+1.02
Phenol	+1.19
Aniline	+1.66
Benzyl alcohol	+3.46
Octyl alcohol	+3.66
Quinoline	+4.12
Benzylamine	+4.43
Water	+4.47
n-Heptylamine	+5.19

[a] Calculated on the assumption that one ligand molecule combines with one chlorophyll molecule (data from Livingstone *et al.*, 1949; Livingstone and Weil, 1952). K_{S^1} = [MP.L]/[MP][L], where MP represents the metalloporphyrin (chlorophyll) and L the ligand.

An interesting example of perpendicular coordination of a special type has been observed with the Zn chelate of mesotetra(4-pyridyl)porphin (IX); it appears that a pyridine nitrogen of a second molecule of the same compound is able to coordinate to the chelated Zn ion, as shown by the change in spectrum as solutions in chloroform are made more concentrated.

B. d^9 Systems

As might be expected, metalloporphyrins such as the Cu(II) and Ag(II) chelates show an extremely low affinity for extra ligands. Miller and Dorough (1952) concluded from spectroscopic evidence that even in pure pyridine there was little interaction between Cu(II) tetraphenylporphin and pyridine.

C. d^8 Systems

It had generally been considered that d^8 metalloporphyrins behaved similarly to their d^9 analogs in resisting perpendicular coordination, no

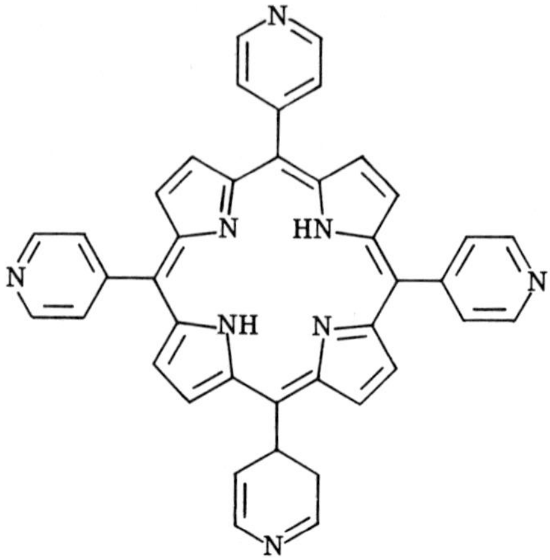

Meso(4-tetra-
pyridyl)porphin

(IX)

evidence of this having been found with, for example, Ni(II) meso- or protoporphyrins. Recently however, Caughey *et al.* (1962) have shown that nickel chelates of weakly basic porphyrins of the 2,4-disubstituted deuteroporphyrin ester series do interact with donor ligands. The coordination, which presumably involves the addition of two ligands to give a distorted octahedral complex, is associated with a large shift (approximately 30 mμ) to longer wavelength in the visible and ultraviolet absorption bands and with a change from diamagnetic to paramagnetic. It would appear that the more weakly basic the porphyrin the greater the tendency for perpendicular coordination among such chelates.

D. d^7 Systems

It is known that Co(II) porphyrins add two extra ligands to form either pure or mixed complexes, as in the bispyridine or cyanopyridine Co(II) mesoporphyrin complexes (McConnell *et al.*, 1953). Many derivatives of vitamin B_{12} are also mixed perpendicular complexes of cobalt already chelated in a square porphyrin-like nucleus. Little quantitative information is, however, available.

E. d^6 Systems

Metalloporphyrins of this type which have been studied include the Co(III) and Fe(II) chelates—the latter quite extensively because of their direct biological relevance. These d^6 metalloporphyrins show a high affinity for extra ligands, particularly π-acceptor–type ligands such as pyridine.

The Co(III) and Fe(II) porphyrins show a much greater affinity for pyridine than the corresponding Co(II) or Fe(III) chelates. The Co(III) chelates differ from the Fe(II) chelates by having a residual positive charge on the metal ion and thus a moderately strong affinity for anionic ligands such as halide and hydroxide ions.

Both Co(III) and Fe(II) porphyrins, but particularly the former, show a strong affinity for the cyanide ion which is both a π-acceptor and an anionic ligand (cf. Fig. 7).

The effect of both porphyrin and ligand structure on the stability of the Fe(II) porphyrin–ligand complexes has recently been investigated in aqueous detergent solutions at 20°C. The equilibrium constants for the reaction between Fe(II) meso- and protoporphyrin dimethyl esters and some substituted pyridines are summarized in Table V. It will be noted

TABLE V

AFFINITY OF SUBSTITUTED PYRIDINES FOR FE(II) PORPHYRINS; EFFECT OF π-BONDING AND OF PORPHYRIN SIDE CHAINS

Ligand	pK_a (H_2O) of ligand ring N	pK_S[a]	
		Fe(II) meso	Fe(II) proto
4-Aminopyridine	9.35	4.0	5.4
Pyridine	5.17	4.6	5.4
4-Cyanopyridine	1.80	5.8	6.5

[a] $pK_S = -2 \log L_{50\%}$, where $L_{50\%}$ is the concentration of ligand for 50% formation of the octahedral complex (hemochrome). The measurements were carried out at 20°C. in 2.5% aqueous CTAB (Lowe and Phillips, unpublished).

that the iron chelate of the less basic protoporphyrin has the greater affinity for pyridine. The decreased electron density of the ring nitrogens, due to the vinyl side chains of protoporphyrin, is presumably reflected in a decreased electron density on the metal ion, which in turn shows an increased affinity for the extra ligands. On the other hand the donor power of the ligand as such is not necessarily correlated with its affinity, although in a particular series of imidazoles studied by Cowgill and Clark (1952) this was so. As Table V shows, 4-aminopyridine is outstandingly the best electron donor, being some 10^7 times stronger than its 4-cyano analog insofar as proton coordination is concerned, yet its affinity for Fe(II) porphyrins is considerably lower. This would seem to indicate the importance of the electron-accepting capacity of the ligand. It seems likely that the stability of the metal–ligand bond will be dependent upon both the donor and acceptor capacity of the ligand and that maximal stability will be achieved when these two factors are properly balanced.

The affinities of a variety of substituted pyridines for Fe(II) mesoporphyrin dimethyl ester are shown in Table VI. The lower affinity of 2-substituted pyridine derivatives for Fe(II) mesoporphyrin seems likely to be due to a steric effect, similar behavior having been observed with 2- and 4,5-substituted imidazoles by Cowgill and Clark (1952). The surprising feature of these results is that 2,6-dimethylpyridine has a greater affinity than 2-methylpyridine itself.

TABLE VI
AFFINITY OF SUBSTITUTED PYRIDINES FOR FE(II) MESOPORPHYRIN

Ligand (pyridine substituent)	pK_a (H_2O) of ligand ring N	pK_S[a]
2-Cyano	—	<1
2-Methyl	6.1	<1
2-Amino	6.7	<1
2,6-Dimethyl	7.0	2.8
4-Amino	9.35	4.0
3-Amino	6.0	4.2
Nil(pyridine)	5.17	4.6
3-Cyano	1.35	5.1
3-Methyl	5.5	5.2
4-Methyl	6.1	5.5
4-Cyano	1.8	5.8

[a] See legend Table V.

F. d^5 SYSTEMS

d^5 Metalloporphyrins, for example the Mn(II) and Fe(III) chelates, readily add two extra ligands to form distorted octahedral-type complexes. The equilibrium between pyridine and Mn(II) mesoporphyrin in aqueous alkaline solution is such that the bispyridine complex is half-formed in 0.16 *M* pyridine (Taylor, 1940).

The Fe(III) chelates of protoporphyrin and other dicarboxylic acid–containing porphyrins show a strong tendency to polymerization in aqueous alkali. Such solutions tend to be polydispersed, but a predominantly dimeric structure is favored. This property, which seems to be characteristic of iron since it is shown also by Fe(II) but not by Co(II), Co(III), Mn(II), or Mn(III) porphyrin chelates, has impeded electrochemical and coordination studies of the iron porphyrins. It seems likely that polynuclear complexes involving Fe—O—Fe bridges are formed, though the importance of intermolecular forces between the large flat porphyrin rings cannot be ignored. This difficulty can be overcome either by increasing the charge of the porphyrin, for example by using a porphyrin bearing many carboxylic acid side chains (e.g., the tetracarboxylic coproporphyrins or the octacarboxylic uroporphyrins), or by adding a coordinating ligand such as pyridine or cyanide ion.

Fe(III) porphyrin chelates interact with a wide variety of ligands, including anionic ones such as hydroxide, halide, or cyanide ions, and neutral ones such as the aliphatic amines, aromatic nitrogen heterocycles, alcohols, water, etc. In general they tend to react with one anionic and one neutral ligand to give uncharged complexes, or with two neutral ligands to give positively charged complexes. They will, however, react with two cyanide ions to form anionic complexes. In contrast to the Fe(II) chelates they show no great preference for π-acceptor-type ligands. It is interesting to note that in aqueous solutions, much higher concentrations of the π-acceptor ligand pyridine are required to form the bispyridine Fe(III) than the bispyridine Fe(II) porphyrin complex (cf. Figs. 6 and 7), whereas with the weak

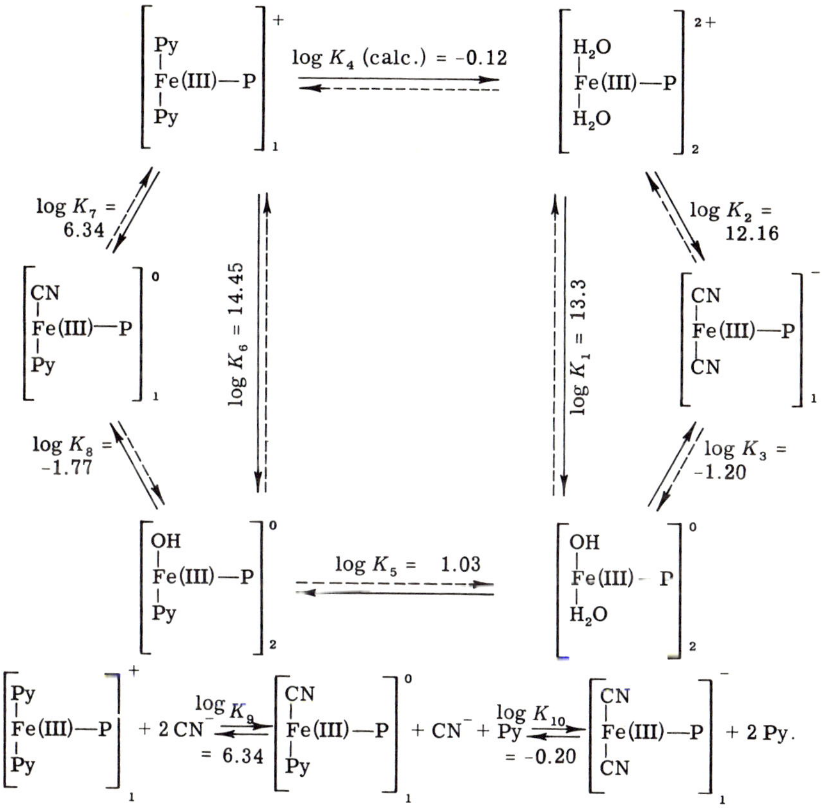

FIG. 6. Equilibria for reactions of Fe(III) protoporphyrin with further ligands. The log K values represent the formation constants at 30°C. for the reactions proceeding in the direction of the solid arrows. From Phillips (1960).

π-acceptor ligand imidazole the reverse is true. A possible explanation of these different affinities is given in Section III.

The extensive data of Shack and Clark have been re-examined by Phillips (1960), who has given the equilibrium constants shown in Fig. 6 for the reactions of Fe(III) protoporphyrin with hydroxide and cyanide ions and with pyridine in aqueous alkaline solution; similar constants for Fe(II) protoporphyrin are given in Fig. 7.

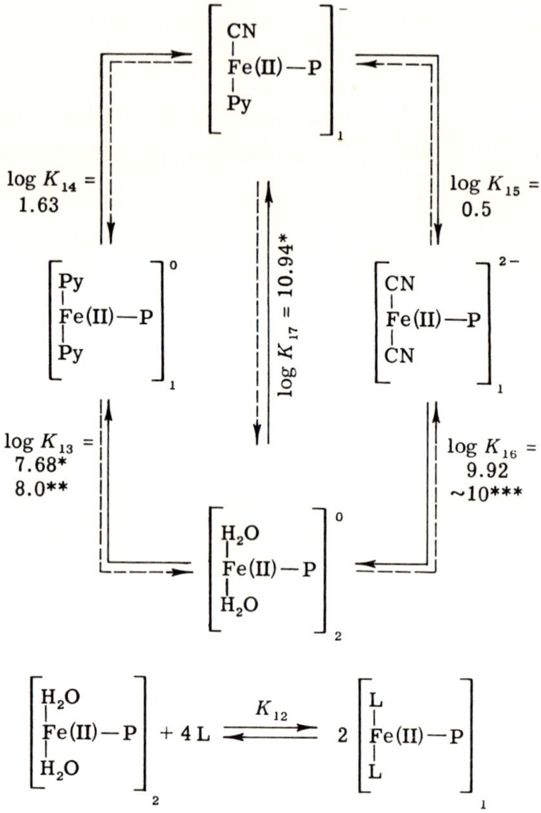

FIG. 7. Equilibria for reactions of Fe(II) protoporphyrin with further ligands. From Phillips (1960). See legend to Fig. 6. * Calculated from the other equilibrium values. ** Holden and Freeman (1929). *** Hill (1929).

G. d^4 SYSTEMS

The only metalloporphyrin of this type to have been studied is Mn(III) mesoporphyrin; 50% formation of its bispyridine complex requires 3 *M* pyridine, the affinity thus being much lower than that of the Mn(II) chelate, which requires 0.16 *M* pyridine only (Taylor, 1940). A similar preference for pyridine shown by Fe(II) mesoporphyrin as compared with the Fe(III) chelate is mentioned above.

H. d^3 Systems

No d^3 metalloporphyrins have been studied, although recently some related phthalocyanines have been described (Elvidge and Lever, 1961). These include the bishydroxy Mn(IV) and aquohydroxy Cr(III) phthalocyanine complexes which have the ability to form dianions, proton displacement occurring from the coordinated hydroxide ions as well as from the water molecule (cf. Section III,G).

I. d^2 and d^1 Systems

These have not been studied in the porphyrin series.

III. The Nature of the Metal-Ligand Bond

The nature of the metal-ligand bond in the metalloporphyrins is extremely variable, ranging from the stable and presumably covalent bonds of many transition metal porphyrin chelates to the readily dissociated bonds associated with their alkali metal compounds. There is an apparent correlation between bond-type, as far as this may be deduced from the magnetic properties, and the intensities of the α and β visible absorption bands. In the low spin type of complexes $\alpha/\beta > 1$, whereas in the high spin type $\alpha/\beta < 1$. The nature of the bond is usually dependent on whether or not extra ligands are coordinated to the metalloporphyrin, and in some instances on the nature of the ligands so coordinated. The results of measurements of magnetic susceptibilities of porphyrin chelates and their further complexes are collected in Table VII.

A. d^{10} Systems

Closed shell metal ions such as Zn(II) normally prefer a tetrahedral distribution of ligands involving sp^3 hybridization, but in the porphyrin chelates a square planar configuration is likely to be imposed. Thus the diamagnetic Zn(II) chelate of protoporphyrin (Falk and Nyholm, 1958) is regarded as a $4s\,4p^2\,4d$ hybrid involving the eight σ-bonding electrons of the porphyrin nitrogens. However, the addition of an extra ligand in a direction perpendicular to the porphyrin plane may allow the metal ion partly to escape from the sp^2d to the preferred sp^3 configuration by becoming bonded to the donor ligand and to three, not four, porphyrin nitrogen atoms. This could account for the high affinity of such d^{10} chelates for just one extra ligand molecule. The effect would be enhanced if, as seems likely, the d^{10} ion is somewhat out of the porphyrin plane. The spectroscopic shift to the red associated with perpendicular coordination would be con-

TABLE VII

MAGNETIC SUSCEPTIBILITIES OF METALLOPORPHYRINS

Square-planar chelates

System	Metal	Ligand	μ(B.M.)	Unpaired electrons	References[b]
d^9	Cu(II)	Protoporphyrin	1.86	1	1
d^9	Ag(II)	Protoporphyrin	1.87	1	1
d^8	Ni(II)	Protoporphyrin	0	0	1, 3
d^8	Ni(II)	Phthalocyanine	0	0	4
d^7	Co(II)	Protoporphyrin	2.78	1	1
d^7	Co(II)	Phthalocyanine	2.16	1	3
d^6	Fe(II)	Phthalocyanine	3.96	3[a]	4

Square-pyramidal complexes

System	Metal	Ligand	Extra ligand	μ(B.M.)	Unpaired electrons	References[b]
d^5	Fe(III)	Protoporphyrin	Cl^-	5.1, 5.80, 5.85	5	1, 2, 3
d^4	Mn(III)	Mesoporphyrin	Cl^-	2.6	2	2

Octahedral complexes

System	Metal	Ligand	Extra ligand	μ(B.M.)	Unpaired electrons	References[b]
d^6	Fe(II)	Protoporphyrin	$(OH_2)_2$	4.9	4	3
d^5	Fe(III)	Protoporphyrin	OH_2; OH^-	4.47, 5.1	5	2, 5
d^6	Fe(II)	Protoporphyrin	OH_2; CO	0	0	6
d^6	Fe(II)	Protoporphyrin	$(Pyridine)_2$	0	0	3
d^6	Fe(II)	Protoporphyrin	$(Nicotine)_2$	0	0	3
d^6	Fe(II)	Protoporphyrin	$(CN^-)_2$	0	0	3
d^6	Fe(II)	Protoporphyrin	CO; pyridine	0	0	6

[a] Calculations from variations of susceptibility with temperature indicate a thermal mixture containing 28% of the high spin form (4).

[b] *References:* (1) Falk and Nyholm (1958); (2) Haurowitz (1935); (3) Pauling and Coryell (1936); (4) Senff and Klemm (1939); (5) Coryell *et al.* (1937); (6) Wang *et al.* (1958).

sistent with a greater lability of electrons on the inner nitrogen atoms; moreover, such lability may explain the photochemical sensitivity of such metalloporphyrin complexes.

B. d^9 SYSTEMS

Cu(II) and Ag(II) form highly stable paramagnetic chelates (Haurowitz, 1935; Falk and Nyholm, 1958). The unpaired electron is presumably situated in the $d_{x^2-y^2}$-orbital, since electron spin resonance studies indicate that the odd electron is involved in the bonding of both Cu and Ag to the nitrogen atoms of the porphyrin (Roberts and Koski, 1960; Kneubühl *et al.*, 1960; Roberts *et al.*, 1961). The filled d_{z^2}-orbital would then explain the great reluctance of such chelates to add ligands in the perpendicular z direction.

C. d^8 SYSTEMS

Ni(II) forms extremely stable diamagnetic square-planar porphyrin chelates in which the $d_{x^2-y^2}$-orbital is presumably vacant and the d_{z^2}-orbital filled (Falk and Nyholm, 1958). It is possible, however, to disturb this arrangement, extra ligands in a fifth and sixth position splitting the electron pair in the d_{z^2}-orbital and forming a paramagnetic complex in which both the d_{z^2}- and $d_{x^2-y^2}$-antibonding orbitals are half-occupied. This coordination is favored by strongly electron-withdrawing (i.e., weakly basic) porphyrins and by strong field donor ligands (cf. Section II,C).

D. d^7 SYSTEMS

Co(II) forms highly stable square-planar chelates. Though one unpaired electron only is expected for a low-spin chelate of Co(II), magnetic susceptibilities greater than 2 B.M. are found in the square-planar Co(II) porphyrins and in Co(II) phthalocyanine (cf. Table VII), while values of about 1.8 B.M., nearer to the 1.73 B.M. expected for one unpaired electron, are found in the octahedral Co(II) porphyrin complexes (Havemann *et al.*, 1961).

It was suggested by Falk and Nyholm (1958) that in Co(II) protoporphyrin (2.78 B.M.) there is a large orbital contribution to the moment, and that the compound is in fact a low-spin chelate with one unpaired electron. The solvent dependence of the g-value of Co(II) phthalocyanine, measured by electron spin resonance (Gibson *et al.*, 1958) suggests strongly that the single unpaired electron occupies the d_{z^2} orbital.

Havemann *et al.* (1961) have studied Co(II) mesoporphyrin in water-pyridine mixtures, and have found that transition from the square-planar chelate to the octahedral complex may be measured by the decrease in susceptibility. Assuming that in the ground state Co(II) has six t_{2g} electrons and one $e_g(z^2)$ electron, they suggest that in the square-planar chelates the most important excited state involves raising a t_{2g} electron to the $e_g(z^2)$

level, the resulting "unsymmetrical" t_{2g} level being responsible for the orbital moment. In octahedral complexes, the energy level of d_{z^2} electrons is much higher than in square-planar compounds; the splitting cannot occur, and large orbital moments are not found.

E. d^6 Systems

Fe(II) porphyrins show great variability in behavior. Square-planar Fe(II) porphyrin chelates are very readily autoxidized, but some have been prepared under special conditions. Fischer *et al.* (1931) prepared crystalline Fe(II) porphyrins, and obtained powder spectra. Corwin and Erdman (1946) prepared crystalline Fe(II) mesoporphyrin; in dioxan solution it had absorption maxima at 585 and 540 mμ, suggesting a high spin complex (cf. Falk and Nyholm, 1958). The magnetic susceptibility of Fe(II) phthalocyanine was measured at +20 and −183°C. by Senff and Klemm (1939); the values found were consistent with a thermal mixture of about 30% of the high spin and 70% of the low spin form.

In the octahedral complexes the nature of both the ligand and the porphyrin is important. In aqueous and alcoholic solvents the Fe(II) complexes of the natural porphyrins are of the high spin type (four unpaired electrons) (Table VII). However, in the presence of nitrogen donors (e.g., pyridine, imidazole) or carbon donors (e.g., cyanide ions, alkyl isocyanides) they become diamagnetic; that is, the d_{xy}-, the d_{xz}- and the d_{yz}-orbitals are filled. The great affinity of Fe(II) porphyrins for π-acceptor ligands such as pyridine, 4-cyanopyridine, carbon monoxide, etc., can be regarded as a consequence of the vacant d_{z^2}-orbital; the metal and ligand are able to approach sufficiently close to each other for back π-bonding to occur between the filled d_{xz} and d_{yz} metal orbitals and the p_π-orbitals of the donor atom.

F. d^5 Systems

In Fe(III) porphyrins the iron atom carries, of course, one positive charge; even the simplest associated anions are probably coordinately bound (cf. Falk and Nyholm, quoted by Falk and Perrin, 1961); square-planar Fe(III) porphyrin chelates thus cannot exist. The simplest and best known compound of this type is ferriprotoporphyrin chloride ("chlorohemin"), and a number of measurements on the crystalline material reveal a typical high spin complex with five unpaired electrons (Table VII). The analogous Fe(III) phthalocyanine chloride on the other hand has been found by measurements of electron spin resonance (Griffith, 1958) to have three unpaired electrons, and it may, like Fe(II) phthalocyanine (see above), be a thermal mixture of low spin and high spin forms.

TABLE VIII
MAGNETIC SUSCEPTIBILITIES OF HEMOGLOBIN (FE PROTOPORPHYRIN-GLOBIN) DERIVATIVES

Hemoprotein	Sixth ligand	μ(B.M.)	Unpaired electrons	References[b]
Fe(II) (hemoglobin)	OH_2	4.47	5	1
	O_2	0	0	1
	CO	0	0	1
	CN^-	0	0	1
	PF_3	0	0	2
Fe(III) (hemiglobin)	OH_2	5.77, 5.65	5	1
	OH^-	4.47, 4.66	[a]	3, 4, 5
	F^-	5.92, 5.76	5	3, 4, 5
	$HCOO^-$	5.44	5	1
	CN^-	2.50	1	4, 5
	SH	2.26	1	6
	N_3^-	2.84	1	4
	NH_3	2.98, 2.93	1	7
	Imidazole	2.87	1	1

[a] Thermal mixture of high and low spin forms; (George *et al.*, 1961).
[b] *References:* (1) Coryell *et al.* (1937); (2) Wilkinson (1951); (3) Theorell and Ehrenberg (1951); (4) Scheler *et al.* (1957); (5) Keilin and Hartree (1951); (6) Keilin (1933); (7) Scheler *et al.* (1958).

In hemoglobin [Fe(II)] and hemiglobin [Fe(III)] the fifth coordination is probably through the —N= atom of an imidazole residue in the globin, and the sixth position is occupied by a water molecule. Table VIII shows the effect upon magnetic susceptibility of replacement of the water molecule by other ligands. In some Fe(III) hemoproteins there is evidence (George *et al.*, 1961) of a thermal mixture of a high and a low spin component, the composition of the mixture being dependent on the nature of the porphyrin and of the extra ligand groups present.

G. LOWER *d* SYSTEMS

A rather unusual type of bond has been proposed to account for the stability of dianions derived from the bishydroxy-Mn(IV) and aquohydroxy-Cr(III) phthalocyanine complexes, both of which are of the high spin type. It has been suggested (Elvidge and Lever, 1961) that the anions are stabilized through delocalization of their negative charge to the aromatic phthalocyanine nucleus. As they point out, this implies not only d_π–p double bonding between the metal and the extra ligand, but also conjugation between the aromatic plane and the z-axis perpendicular to it. Elvidge and Lever suggest that this perpendicular conjugation is possible only if d_{xz}- and/or d_{yz}-orbitals of the metal are vacant, so that it might be observed in octahedral complexes of titanium and vanadium besides Cr(III) (spin-free) and Mn(IV) (spin-free), but not in complexes of the next transition metals: Fe, Co, and Ni.

IV. Physicochemical Properties

A. Ionization Behavior of Porphyrins

If the carboxylic acid side chains which occur on many porphyrins are ignored, the porphin nucleus itself may be regarded as an ampholyte: the two iminonitrogen atoms (—N═; the "pyrrolenine" nitrogens) are capable of accepting protons, and the two pyrrole (>N—H) nitrogens are capable of either losing or accepting protons. Of the six possible ionic species, (X)–(XV), thus derivable from the neutral porphin (PH_2), the dianion (P^{--}), the monocation (PH_3^+), and the dication (PH_4^{++}) only have been observed.

Several conventions have been used in assigning pK values to the porphin nucleus (Conant *et al.*, 1934; Neuberger and Scott, 1952; Aronoff, 1958); the following scheme, introduced by Phillips (1960), will be used here:

$$pK_1 = pH - \log_{10} [P^{2-}]/[PH^-]$$
$$pK_2 = pH - \log_{10} [PH^-]/[PH_2]$$
$$pK_3 = pH - \log_{10} [PH_2]/[PH_3^+]$$
$$pK_4 = pH - \log_{10} [PH_3^+]/[PH_4^{++}]$$

Thus pK_1 and pK_2 refer to the acidic equilibria involved in the dissociation of protons from the pyrrole nitrogens, and pK_3 and pK_4 to the basic equilibria involved in the addition of protons to the pyrrolenine nitrogens.

The pyrrole nitrogen atoms of porphyrins behave as very weak acids; both pK_1 and pK_2 have been estimated in the case of etioporphyrin to be of the order of +16 (McEwen, 1936).

Measurements of the basicity of the pyrrolenine nitrogens of porphyrins are limited by solubility considerations, most porphyrins having isoelectric points at about pH 4, at which their solubility in water is of the order of 10^{-10} M only. This has led to the use, for determinations of basicity, of porphyrins with side chains which increase their solubility in water (Neu-

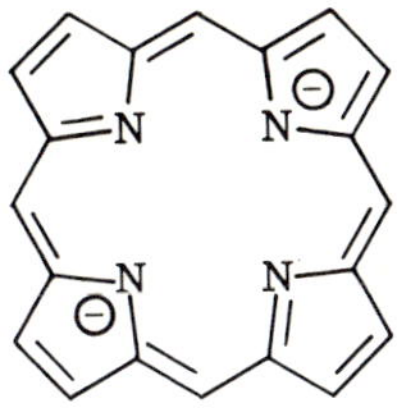

P^{--}

(X)

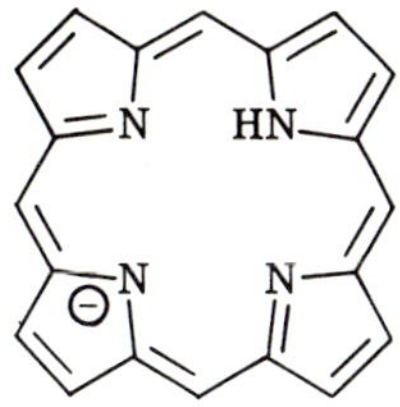

PH^{-}

(XI)

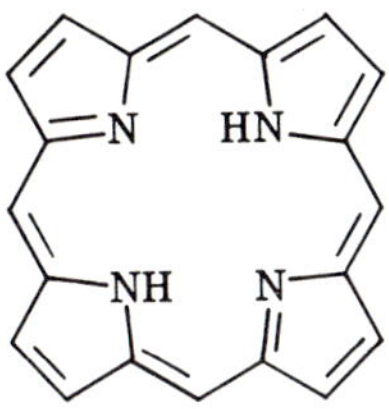

PH_2

Neutral molecule

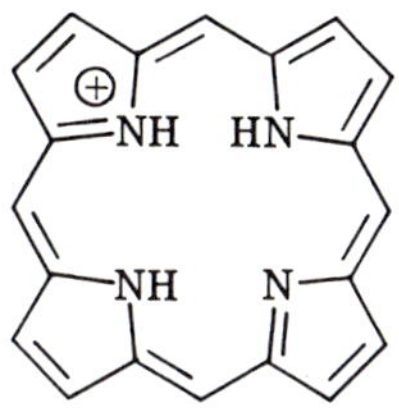

PH_3^{+}

(XII)

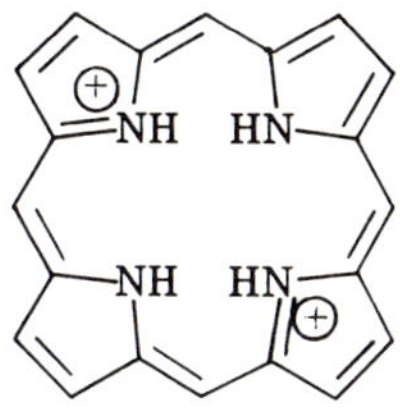

PH_4^{++}

(XIII)

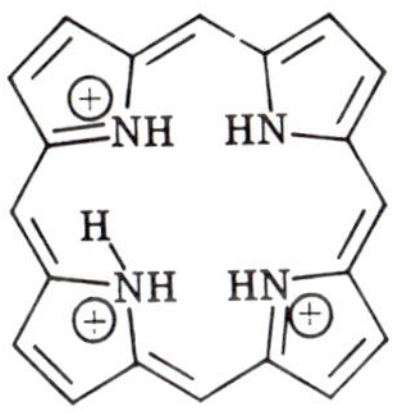

PH_5^{3+}

(XIV)

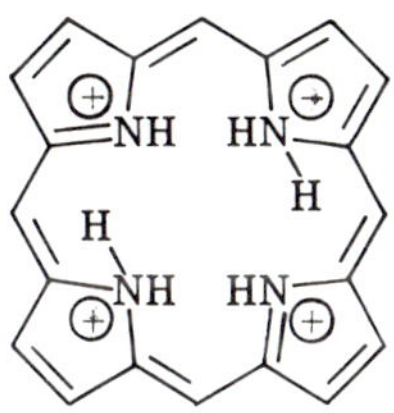

PH_6^{4+}

(XV)

berger and Scott, 1952), and of techniques such as potentiometric titration in glacial acetic acid (Conant *et al.*, 1934; Hall, 1930) and spectroscopic titration in nitrobenzene (Aronoff, 1958). The results of such studies are discussed in detail by Phillips (1960) who has also calculated the basicity of some porphyrins from phase-distribution data.

The finding (Phillips, 1958; Dempsey *et al.*, 1961) that porphyrins and their esters may be dispersed monomolecularly in aqueous detergent solutions has permitted the determination of their basic dissociation constants in that medium. Table XI summarizes pK_3 values determined in 2.5% (w/v) sodium dodecylsulfate for a series of 2,4-disubstituted deutero-

TABLE IX
THE BASICITY OF A SERIES OF SUBSTITUTED DEUTEROPORPHYRINS[a]

Porphyrin (methyl ester)	Substituent in position 2	Substituent in position 4	pK_3[a]	Absorption[b] band *I* (mμ)
Meso-	CH_2CH_3	CH_2CH_3	5.85	620
Deutero-	H	H	5.50	618
Copro-	$CH_2CH_2COOCH_3$	$CH_2CH_2COOCH_3$	5.50	620
Proto-	$CH{=}CH_2$	$CH{=}CH_2$	4.80	630
4-Formyldeutero-	H	CHO	3.80	640
2-Formyl-4-vinyldeutero-	CHO	$CH{=}CH_2$	3.75	644
2,4-Diacetyldeutero-	$COCH_3$	$COCH_3$	3.35	639
2,4-Diformyldeutero-	CHO	CHO	3.0	651

[a] Measured in 2.5% sodium dodecylsulfate at 25°C. (cf. Phillips, 1960; Dempsey *et al.*, 1961; Caughey and Phillips, unpublished).

[b] In dioxan. A similar relationship between λ and porphyrin p*K* is found in metalloporphyrins and their further complexes (cf. Falk and Perrin, 1961).

porphyrin esters. It will be noted that the greater the electron-attracting character of the substituent, that is, the greater the displacement of π-electrons to the periphery of the porphyrin nucleus, the less basic the central nitrogen atoms.

Most natural porphyrins and metalloporphyrins have propionic acid side chains, and the uroporphyrins have acetic acid side chains as well. For the determination of dissociation constants of the porphyrin nitrogen atoms it is of course convenient to remove the electrostatic interference of these groups by esterifying them. Nevertheless, in natural conditions, the side chain carboxyl groups dissociate simultaneously with, and modify the dissociation of, the ring nitrogen atoms; intrinsic p*K* values for the basic dissociation of several unesterified porphyrins in an aqueous environment have been calculated (Neuberger and Scott, 1952; Phillips, 1960, 1963).

B. Redox Behavior of Metalloporphyins

1. *Introduction*

Because of the importance of the oxidation-reduction properties of the natural compounds, e.g., in electron transport in the cytochromes, it is necessary to understand the factors affecting the potentials of simpler metalloporphyrins.

Complexes of porphyrins with iron, manganese, and cobalt undergo reversible oxidation-reduction. The reversibility is often rather sluggish in these complexes, their solubilities are low, and in the alkaline conditions which are necessary the iron complexes have a great tendency to polymerize (cf. Phillips, 1960). The coordination of further ligands, e.g., pyridine and other bases, improves all these properties to some extent and most studies of redox potential have been done with such complexes (hemochromes). The redox potentials of metalloporphyrins and of hemoproteins have been discussed in detail by Lemberg and Legge (1949) and Martell and Calvin (1952).

Redox potentials are expressed here* according to the equation:

$$E = E^\circ + \frac{RT}{F} \ln \frac{M^{3+}}{M^{++}} \tag{2}$$

i.e., the more stable the oxidized state, the more negative the value of E°.

2. *The Stabilization of the Oxidized State of Metal Ions by Chelation with Porphyrins*

Coordination by electronegative ligands tends generally to stabilize the oxidized state of an ion and so lower the redox potential of the complex as compared with the hydrated metal ion (Martell and Calvin, 1952, p. 57 ff.). This has a simple electrostatic explanation in the attraction between the negative dipole of the ligand and the extra positive charge of the oxidized metal ion, and the effect is exaggerated when the ligand has a negative charge, as in CN^-. Thus the redox potential of iron protoporphyrin, in the reaction

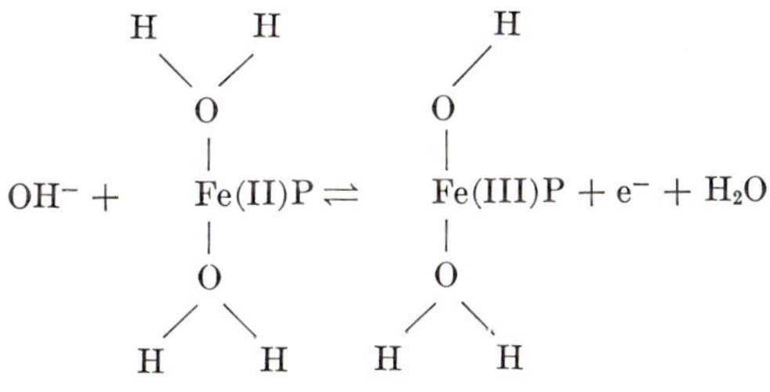

(3)

is approximately −150 mv. at pH 7, compared with the value of +771 mv. for the hydrated ferrous-ferric ion couple. The pH-dependence of this reaction and of other metalloporphyrin oxidation-reductions is discussed by Martell and Calvin (1952, p. 368 ff.).

* In this chapter, the signs of the E° values imply that the half cell reactions are written as reductions from left to right. These values are sometimes referred to as "reduction potentials." The above sign convention, which is opposite to the one adopted in Chapter 6, is used internationally in biochemical literature.

Other examples of this relative stabilization of the oxidized state are given in Table X, where the potentials of the bispyridine Mn, Fe, and Co

TABLE X
REDOX POTENTIALS OF HYDRATED AND CHELATED IONS[a]

	$E°$ (mv.)		
Metal	Hydrated ion	Metalloporphyrin bispyridine complex	$\Delta E°$ (mv.)
Mn	+1510	−387	1897
Fe	+771	−63	834
Co	+1814	−265	2079

[a] From Martell and Calvin, 1952.

porphyrins are compared with those of the respective hydrated ions. In addition to this effect, however, the redox potential is also influenced by the electronic configurations of the metal ions—aquo ions tending to exist in high spin forms whilst some metalloporphyrins exist in low spin forms. A strong-field π-bonding ligand such as pyridine has a particular affinity for d^6 [Fe(II) and Co(III)] metalloporphyrins forming diamagnetic octahedral complexes having the inert gas configuration. In the case of cobalt this electronic configuration effect reinforces the electrostatic effect in favoring the oxidized form whereas with iron the two effects oppose each other. Thus the $\Delta E°$ values in Table X are much greater for cobalt (2079 mv.) than for iron (834 mv.). The CN^- ion also forms low-spin octahedral iron porphyrin complexes but in this case the electrostatic effect of the anionic ligand further stabilizes the Fe(III) form compared with the neutral ligand, pyridine (cf. Table XII).

3. *The Effect of the Basicity of Nitrogen-Donor Ligands*

As has been found for a number of other types of chelate (Martell and Calvin, 1952; Perrin, 1959), among a closely related series of ligands, the greater the donor power (high pK_a) the more negative the potential. This effect may be shown with octahedral complexes of iron porphyrins (Table XI). There is a roughly linear dependence of E'_0 on pK_a, with a slope (Falk and Perrin, 1961) for $-\Delta E/\Delta pK$ of the order of 0.04 volt. The stability of these complexes is influenced by a variety of other factors, and the correlation between redox potential and ligand basicity is probably true only when other factors do not interfere (cf. Tables V and VI).

4. *The Effect of Substituents on the Porphyrin Nucleus*

As is to be expected, the lower the donor power of the chelating porphyrin nitrogen atoms (i.e., the stronger the electron-withdrawing sub-

TABLE XI
THE EFFECT OF BASICITY OF EXTRA LIGANDS ON THE POTENTIAL OF FE PROTOPORPHYRIN

Extra ligand	pK_a	E'_0, pH 7[a] (mv.)
Nicotine	3.15	200
Pyridine	5.2	158
α-Picoline	6.2	115
Histidine	6.0	>79
Pilocarpine	7.0	>52

[a] Calculated from data of Barron (1937) by Falk and Perrin (1961), for the reaction $Fe(II)B_2 + OH^- \rightleftharpoons Fe(III)B{\cdot}OH + B + e^-$.

stituents on the porphyrin nucleus) the more stabilized the reduced form, and the more positive the $E°$. This effect is illustrated in Table XII; its direction is the same whether the bispyridine, bis-α-picoline, or biscyano complexes are compared, and in all cases is correlated with shifts in the absorption to longer wavelength as basicity of the porphyrin decreases (cf. Table IX).

TABLE XII
INFLUENCE OF PORPHYRIN SIDE CHAINS ON REDOX POTENTIALS

Ferroporphyrin	Side chains at positions 2 and 4	pK_3[a]	E^0 (mv.) at pH 9.6, with the extra ligands: (Pyridine)$_2$[b]	(α-Picoline)$_2$[b]	$(CN^-)_2$[c]
Meso-	—C_2H_5	5.85	−63	—	−229
Copro-	—CH_2CH_2COOH	5.50	−36	—	−247
Etio-	All alkyl[d]	—	−29	—	—
Hemato-	—$CHOHCH_3$	—	+4	−99	−200
Proto-	—$CH{=}CH_2$	4.80	+15	−33	−183
Chlorocruoro-	—CHO (2); —$CH{=}CH_2$ (4)	3.75	—	−10	−113

[a] Of the methyl esters (cf. Table IX).
[b] For the reaction $Fe(II)B_2 + OH \rightleftharpoons Fe(III)B{\cdot}OH + B + e^-$ (cf. Falk and Perrin, 1961).
[c] From Martell and Calvin, 1952.
[d] See Table I.

C. SPECTRA

1. *Introduction*

The intense color of the porphyrins and their metal complexes, which has led to the use of the name "pyrrole pigments" and has stimulated and facilitated studies of them, is of course due to the highly conjugated nature of the pyrrole nucleus. This leads to the intense absorption band at about 400 mμ known as the "Soret" band which is present in all compounds in which the inner ring is conjugated, but not if this conjugation is broken, as it is, for example, in the porphyrinogens (hexahydroporphyrins) in which the bridge carbon atoms are saturated, or in the bile pigments, in which one bridge carbon has been lost. The Soret band is very intense, reaching molar extinction coefficients of the order 5×10^5. The strongest bands in the visible region, responsible for the color, are in the simpler porphyrins and their metal complexes approximately one twentieth to one tenth as intense as the Soret band.

2. *The Nature of Porphyrin Absorption Spectra*

The characteristic visible spectra of some simpler porphyrins and their metal complexes are illustrated in Figs. 8 and 9.

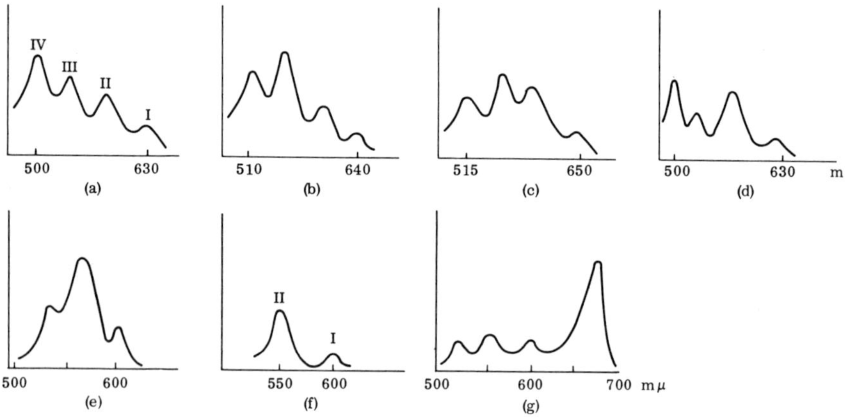

FIG. 8. Visible spectra of (a) etio-, (b) rhodo-, (c) oxorhodo- and (d) phyllo-type neutral porphyrins; (e) mono- and (f) dication spectra of an etio-type porphyrin; (g) neutral chlorin spectrum.

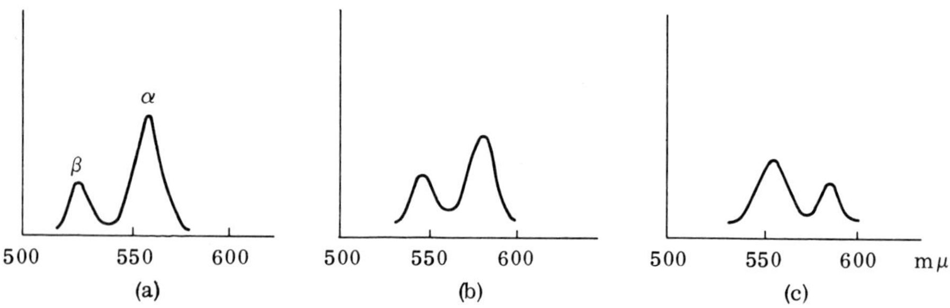

FIG. 9. Visible spectra of square-planar porphyrin chelates: (a) Ni(II); (b) Zn(II); (c) Cd(II).

The visible spectrum of most common porphyrins in neutral solvents consists of four relatively sharp bands, increasing stepwise in intensity from band I, in the region of 620 mμ, to band IV at about 500 mμ (Fig. 8a). Alkyl substituents symmetrically placed around the porphyrin nucleus (e.g., 1 to 8-octalkyl porphins) have little effect on the spectra, and the propionic acid side chains, common among natural porphyrins, have minimal effects. Electronegative substituents such as the vinyl group of protoporphyrin, the formyl group of chlorocruoroporphyrin and porphyrin *a* and of the chlorophylls *b*, and the acetyl group of bacteriochlorophyll have marked effects, causing shifts to longer wavelength (cf. Table IX) and, depending on their orientation around the nucleus, changes in the ratios of intensity of the visible bands as illustrated in Fig. 8b and c. An alkyl substituent on a methene bridge carbon leads to the characteristic spectrum illustrated in Fig. 8d. These effects have been discussed in detail elsewhere (Lemberg and Falk, 1951; Falk, 1963).

Reduction of a porphin to a chlorin (formula II) usually causes a color change from red to green, as in the chlorophylls; this is due to the presence in chlorin spectra of a predominant band I in the red (Fig. 8g). Band I occurs at longer wavelengths in a chlorin than in the corresponding porphyrin, and is much more intense, the ratio of the Soret band to band I being about 5 in the chlorin and about 20 in the porphyrin. In the cationic and anionic species of chlorins the spectra resemble those of the free bases, with displacement of the Soret and visible bands to shorter wavelength in the monocation and to longer wavelength in the dianion and dication (Phillips, 1958).

Porphyrin monocations usually have only three obvious visible bands (Fig. 8e), the relative intensities being altered by electronegative substituents. The Soret band also is present, and is displaced to shorter wavelength relative to the free base.

The porphyrin dianion and dication have very similar spectra, two predominant visible bands only being found (Fig. 8f). The Soret band is displaced in both instances to longer wavelength relative to the Soret band of the neutral molecule.

3. *The Nature of Metalloporphyrin Absorption Spectra*

The square-planar chelates with divalent metal ions have two predominant visible bands only, which are often termed the α (longest wavelength) and β bands (Fig. 9a, b, c) to avoid confusion with the bands (I to IV) of the metal-free compounds; the Soret band also is present. The ratio of intensities α/β varies (Fig. 9) from $\gg 1$ in the thermodynamically more stable chelates, e.g., Ni(II), to ~ 1, e.g., Zn(II), to <1, e.g., Cd(II), in the less stable chelates (cf. Section I,D).

Perpendicular coordination of one or two extra ligands to the porphyrin chelates causes changes in position, usually to longer wavelength, and decreases the ratio α/β of the two visible bands. Perpendicular complexes of iron porphyrins are by far the most important and the most studied, and coordination by two extra ligands, to give octahedral complexes, occurs with ligand molecules as diverse as water, pyridine, and other heterocyclic bases, amines, thiols, carbon monoxide, oxygen, and ions such as hydroxide and cyanide. The spectra of some of these complexes are illustrated in Fig. 10.

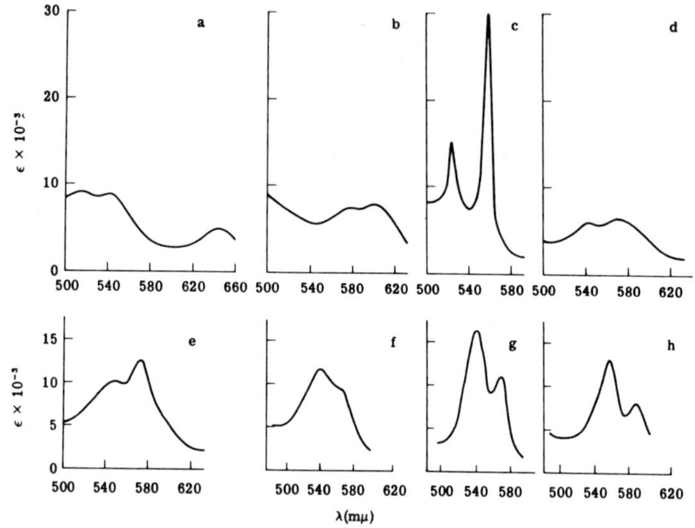

FIG. 10. Visible spectra of perpendicular complexes of iron protoporphyrin chelates. (a) Hemin [Fe(III) porphyrin chloride; measured in 2.5% CTAB]; (b) hematin [Fe(III) porphyrin hydroxide; measured in 2.5% CTAB]; (c) pyridine hemochrome [bispyridine Fe(II) porphyrin]; (d) dimeric hem [Fe(II) porphyrin in 0.1 *N* NaOH]; (e) bisaquohem [Fe(II) porphyrin in 2.5% CTAB, 0.1 *N* in respect to NaOH]; (f) CO—CN^- Fe(II) porphyrin (J. Keilin, 1949a); (g) bis-CN^- Fe(II) porphyrin (J. Keilin, 1949b); (h) bishydroxyl hem [Fe(II) porphyrin in 50% ethanol, 1.3 *N* in respect to NaOH; J. Keilin, 1949b].

The spectra of the ferric complexes are rather diffuse (Fig. 10a and b) containing a component [the charge-transfer band (Williams, 1956)] due to the interaction between electronic transitions associated with the porphyrin nucleus and the charge-transfer from the nucleus to the positively charged ferric ion. Brill and Williams (1961a) have found the charge-transfer band of Fe(III) hemoproteins and their derivatives to be useful in interpretations of their spin state.

The sharp, two-banded spectrum of bispyridine Fe(II) porphyrins

(Fig. 10c; the "pyridine hemochrome" spectrum) is used extensively for quantitative determinations, and in addition provides an important point of reference, being characteristic of low spin, octahedral Fe(II) porphyrin complexes. Bond type (cf. Section II) is sensitive, in the iron complexes, to the nature of the perpendicular ligands, and changes in bond type are reflected in the spectra. Thus the spectra of a Fe(II) porphyrin (heme) in aqueous alkali and in aqueous detergent (Fig. 10d and e) are displaced to much longer wavelength than the low spin pyridine hemochrome spectrum (Fig. 10c).

Mixed ligand complexes such as CO-pyridine, CO-water, and CN-water Fe(II) porphyrins have rather typical hemochrome spectra, but in the CO—CN and the CN—CN complexes the α and β bands are displaced to longer wavelength and their relative intensities reversed (Fig. 10f and g).

4. *Infrared Spectra*

The infrared spectra of a number of natural and synthetic porphyrins have been reported (Falk and Willis, 1951; Craven *et al.*, 1952; Thomas and Martell, 1958; Mason, 1958; Rimington *et al.*, 1958). The N—H stretching and bending frequencies have been determined by comparing deuterated and nondeuterated porphyrins; the N—H stretching frequency, at about 3300 cm.$^{-1}$, is virtually unaffected by changing from the solid state to carbon tetrachloride solution, suggesting that there is a high degree of N—H—N bonding within the porphyrin nucleus. The in-plane and out-of-plane N—H bending frequencies are at 980 and 720 cm.$^{-1}$. The spectra are consistent with a porphyrin model in which hydrogens are attached to opposite N atoms and are hydrogen-bonded to adjacent N atoms. The infrared spectra can be used to characterize porphyrin substituents and some position isomers (Falk and Willis, 1951; Wetherell and Hendrickson, 1959; Sidorov and Terenin, 1960).

It has been suggested (Thomas and Martell, 1958) that the position of a strong band in the region of 1000 cm.$^{-1}$ can be correlated with the thermodynamic stability of metal porphyrin complexes in a closely related series; in general, the higher the wave number (greater energy), the more stable the complex.

5. *Fluorescence Spectra*

Metal-free porphyrins have a very intense red fluorescence under ultraviolet light filtered by Wood's glass, allowing their photoelectric determination at concentrations as low as 10^{-10} *M* (Schwartz *et al.*, 1951). The fluorescence is most intense, as is absorption, in the porphyrin dication; it is not usually apparent in the solid state or in colloidal dispersions, and is readily quenched by some solutes.

As mentioned above (Section II) solutions of some metalloporphyrins [Zn(II), Cd(II), Mg(II), Sn(II)] do not fluoresce in pure, dry, noncoordinating solvents (e.g., carbon tetrachloride, cyclohexane), but do so if a little water or a polar ligand (e.g., alcohols, amines) is added, or if the solvent is polar (pyridine, dioxan, water, etc.). The more thermodynamically stable metalloporphyrins [Co(II), Ni(II), Cu(II), Ag(II), Fe(II), Fe(III), etc.] do not fluoresce but some of them phosphoresce (see below).

Fluorescence emission maxima for protoporphyrin (634 mμ), its monocation (612.5 mμ), its dication (606 mμ), and its Zn(II) chelate (589 mμ) have been measured in 2.5% sodium dodecylsulfate by Phillips (1963). Allison and Becker (1960) have measured the phosphorescence spectra of mesoporphyrin and its Zn(II), Cd(II), Ba(II), Co(II), Cu(II), Ni(II), and Pd(II) complexes, and the Mg(II) complex of etioporphyrin. The lifetime of the triplet state was shorter for Co(II) and Cu(II) than for the others mentioned, and Ag(II) mesoporphyrin had no phosphorescence emission.

Gouterman (1961) has discussed theoretical aspects of these emission spectra, and also of the recently discovered n–π triplet emission and n–π singlet absorption of porphyrins (Fernandez and Becker, 1959), as well as the triplet-triplet spectra of porphyrins (Linschitz and Sarkanen, 1958; Linschitz and Pekkarinen, 1960).

6. *N.M.R. Spectra*

Several reports of the N.M.R. spectra of porphyrins have appeared recently (Becker and Bradley, 1959; Ellis *et al.*, 1960; Becker *et al.*, 1961; Abraham, 1961; Abraham *et al.*, 1961; Caughey and Koski, 1962). The spectra have been found to be characterized by effects due to strong ring currents, as is to be expected for such large ring aromatic compounds. The wide difference in chemical shift for different protons, as well as the apparent lack of spin-spin interaction between groups at different positions on the porphyrin ring, indicate that the N.M.R. spectra of porphyrins can be more readily interpreted than might have been expected. The technique promises to provide a very useful and much-needed means of elucidating the detailed molecular structure of these compounds. It has been possible to make remarkably detailed proton assignments for many of the commonly encountered substituents of natural porphyrins and their derivatives.

7. *Theoretical Aspects of Absorption Spectra*

The recent analysis by Gouterman (1959, 1960, 1961) is summarized here. References to earlier treatments are given by Gouterman, who acknowledges in particular his debt to Platt's important work (1956).

Gouterman has based his interpretations on the metal p-orbitals; the justification for this is not strong, and it appears likely that consideration of the *d*-orbitals would have led him to much the same conclusions. In the discussion of bond type in these chelates (Section III), the arguments are based mainly on the metal *d*-orbitals.

The numbering and axes used are shown in Fig. 11a, and Gouterman takes as starting point the spectrum of the Zn(II) chelate of tetraphenylporphin (VIII). The visible bands (Fig. 12) are ascribed to π–π transitions

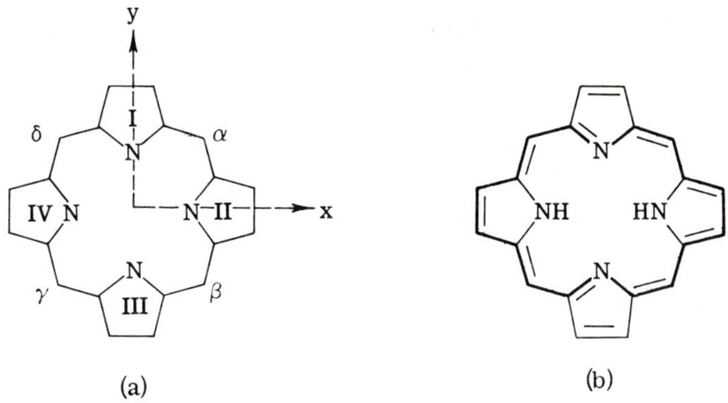

FIG. 11. (a) Porphin nucleus; (b) eighteen-membered ring (heavy bonds) stabilized by the central H atoms.

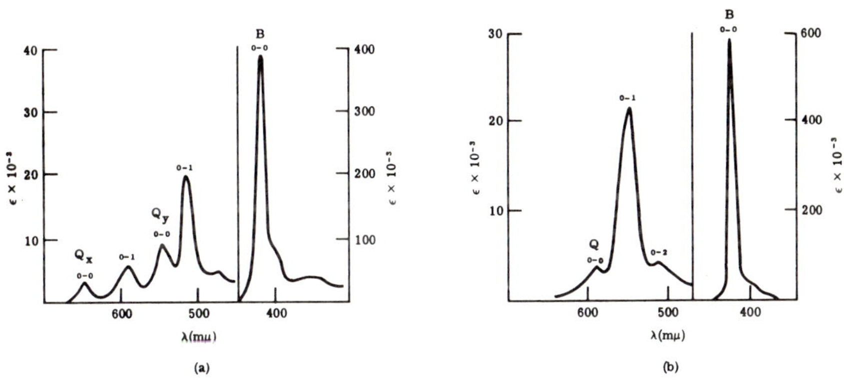

FIG. 12. Absorption spectra of (a) tetraphenylporphin and (b) its Zn(II) chelate. From Gouterman (1961); cf. Dorough *et al.* (1951).

and are associated with electronic displacement towards the periphery of the nucleus; for a metal porphin with D_{4h} symmetry (cf. Spooner and Teller, 1941) such transitions are of e_{1u} symmetry (cf. Longuet-Higgins *et al.*, 1950) consisting of two equivalent dipole transitions in the x and y directions. Gouterman uses the terms B and Q for the Soret and visible bands respectively, and also to define the pairs of excited states of the transitions.

It is known that the positions of bands I and III in the neutral porphyrin and the α band in metalloporphyrins vary greatly with the nature of the porphyrin substituents, while the positions of bands II, IV, and β are relatively insensitive. This, coupled with the constancy of energy separation between bands I and II, III and IV, and α and β, has led to the assignment of bands I, III, and α as 0–0 vibrational transitions, and bands II, IV, and β as 0–1 vibrational transitions (cf. Platt, 1956; Mason, 1958). Williams (1956) has shown, in addition, a direct relationship between the intensity of band III of certain porphyrins and the intensity of the α band of their copper complexes.

X-ray crystallographic analyses (cf. Section I,A) have shown that while Ni(II) phthalocyanine and Ni(II) etioporphyrin have square symmetry, metal-free phthalocyanine molecules are slightly distorted from square because of the opposite —NH groups (Fig. 11b). It is assumed that metal-free porphyrins are similarly distorted, so that the x- and y-axes are no longer equivalent in the neutral porphyrin molecule, and splitting of the Q_x and Q_y bands is thus not surprising. Gouterman uses B_x, B_y and Q_x, Q_y to identify the individual components of the transition pairs (Fig. 12).

The top filled and the lowest empty orbitals, obtained from simple molecular orbital calculations (Longuet-Higgins *et al.*, 1950), are shown in Fig. 13. The top filled orbitals, having symmetry a_{2u} and a_{1u}, are called b_1 and b_2 and the lowest empty orbitals, having e_g symmetry, are called c_1 and c_2. The orbital b_1 is calculated to have much higher energy than b_2, and the visible bands are identified with the transition $b_1 \rightarrow c_1$, c_2 and the Soret band with $b_2 \rightarrow c_1$, c_2.

This would require both bands to be of equal intensity, and to account for the great difference in observed intensity it is necessary to assume that the top filled orbitals, i.e., b_1 and b_2, are accidentally degenerate, making transitions to the visible bands forbidden.

According to Longuet-Higgins *et al.* (1950), in a metalloporphyrin the four porphyrin nitrogen atoms have four σ-orbitals pointing towards the metal, and contribute six electrons to them, the six remaining π-electrons being contributed to the ring system. It is assumed that the metal contributes two of its electrons to close the "octet" shell, and that the main effect of the metal on porphyrin spectra is due to the conjugation of its p-electrons with the π-electrons of the ring (Gouterman, 1959). This p-orbital can only interact with the a_{2u}-orbital (b_1), and Gouterman has shown that the less electronegative the metal, the greater this interaction. This leads to an increase in the energy of the a_{2u} relative to the e_g-orbitals, and thus to a shift of the visible bands to longer wavelength. Simultaneously, the

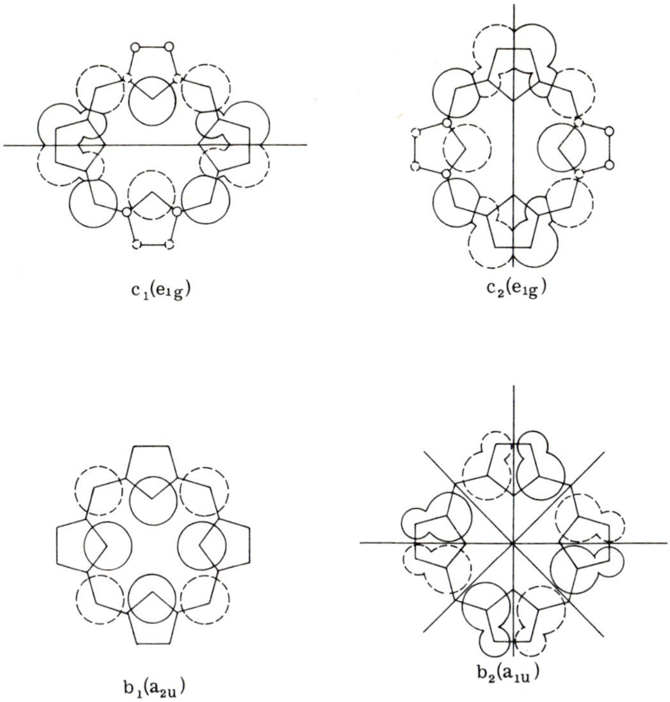

FIG. 13. Porphin molecular orbitals. The atomic orbital coefficients are proportional to the size of the circles; solid or dashed circles indicate sign. Symmetry nodes are drawn in heavy lines. From Gouterman (1961); cf. Longuet-Higgins *et al.* (1950).

b_1- and b_2-orbitals lose their degeneracy, so that there is an increase in the intensity of the visible absorption. Gouterman has demonstrated relationships between intensity and energy in the lowest visible band in metal tetraphenylporphins, and between electronegativity of some metals and the energy of visible bands in these complexes.

In the metal-free porphyrin, the eighteen-membered conjugated ring shown by the heavy lines in Fig. 11b is assumed to be stabilized by the opposite protons. There is evidence from fluorescence polarization experiments (Weigl, 1957) and from low temperature splitting of the Soret band of porphin (Rimington *et al.*, 1958) that the Q_x band is polarized perpendicularly to Q_y. From a cyclic polyene model (cf. Platt, 1956; Gouterman 1960), it follows also that if the H—H axis is in the x direction, this is the direction of the Q_x polarization since the Q_x band is that of lowest energy.

This idea is confirmed by a number of considerations of the effect of substituents on porphyrin spectra (Gouterman, 1960). Strongly electronegative substituents, as well as causing shifts to longer wavelength in porphyrins and all their complexes, cause marked effects upon the relative intensities of the visible bands of neutral porphyrins and their monocations (cf. Section IV,C,2). These effects are clearly related to considerations of the symmetry of the molecule. Platt (1956) developed a spectroscopic vector theory to explain these substituent effects, and Gouterman (1960) has introduced a modified theory based on a one-electron model.

As Gouterman points out, the present theoretical treatment is partly rigorous and partly empirical, and needs further development. In particular, the doubling of the visible band system needs rigorous explanation, as does the lack of splitting of the *B* (Soret) transition to an extent comparable to the splitting of the *Q* (visible) transitions, the observed values being 240 and 2960 $cm.^{-1}$ respectively (Rimington *et al.*, 1958). Another phenomenon which requires explanation is the lack of a strong splitting of the *Q* bands of a metal porphyrin even in the presence of very asymmetric substituents (Gouterman, 1960).

V. Hemoproteins: Some Aspects of Their Coordination Chemistry and Some Model Compounds

A. Introduction

It was suggested in 1932 by Conant that the heme iron of hemoglobin is bound by a nitrogen atom of a histidine residue in the globin; further studies (cf. Martell and Calvin, 1952; Lemberg and Legge, 1949) led to the suggestion that two histidine residues were coordinated, one in each perpendicular position, one strongly and the other weakly, the latter being displaced on reaction with O_2. From X-ray crystallographic analyses (cf. Kendrew *et al.*, 1960), it is now clear that in myoglobin the primary bond from the protein is indeed provided by a histidine residue. The X-ray evidence shows also that the sixth position in myoglobin is occupied not by a second loosely bound histidine residue but by a molecule of water, and a similar situation applies in hemoglobin (cf. Haurowitz, 1951; Keilin and Hartree, 1952). The chelated iron is in a poised state; on replacement of this molecule of water by O_2, the whole complex changes from the high spin to the diamagnetic, low spin state. That metal-to-O_2 π bonds are involved was suggested as early as 1936 by Pauling and Coryell, and the nature of the bonding has been discussed in detail more recently by Griffith (1956).

Wang *et al.* (1958) have suggested that perhaps hydrocarbon side chains of the protein, in the environment of the oxygen-reacting site on the heme iron, may tend to lower the dielectric constant of the medium, and that such an effect may contribute to the passivity of the Fe(II) of hemoglobin to oxidation, since the decomposition of Fe(II)HbO_2 to Fe(III)Hb and O_2^- or HO_2 should be slower in media of low dielectric constant.

That the nature of the porphyrin side chains is not very important for either hemoglobinlike or peroxidaselike activity, as long as the correct protein is used, has been shown clearly by the recombination experiments summarized in Tables XIII and XIV. Hemes with a wide variety of side

TABLE XIII
FE(II) PORPHYRINS GIVING "SYNTHETIC" HEMOGLOBINS WHEN COMBINED WITH NATIVE GLOBIN

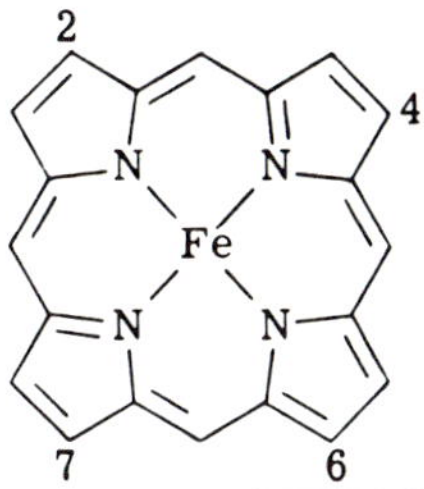

Substituent* at position

2	4	6	7
E	E	P	P
H	H	P	P
V	V	P	P
P	P	P	P
Pm	Pm	Pm	Pm
CHO	V	P	P
Ac	Ac	P	P
H	E	C	P

*For abbreviations, see Table 1. For references, see Lemberg & Legge (1949) and O'Hagan (1961)

chains, on combination with native globin, formed "synthetic" hemoglobins which combine reversibly with O_2. "Synthetic" peroxidases made by combining peroxidase protein (apoperoxidase) with meso- and hematoheme were indeed found by Paul (1959) to have greater peroxidase activity than when the natural prosthetic group, protoheme, was used. No peroxidase activity was found, however, when the Cu(II), Co(II), Mn(II), or Ni(II) chelates of protoporphyrin were added to apoperoxidase (Theorell, 1945).

The "synthetic" models of course differed in respect to some of the finer properties of the natural hemoproteins—e.g., the Bohr effect and the

TABLE XIV
FE(II) PORPHYRINS GIVING "SYNTHETIC" PEROXIDASES WHEN COMBINED WITH NATIVE APOPEROXIDASE

Substituent[a] at position				Relative peroxidative activity[b]
2	4	6	7	
H	H	P	P	50
E	E	P	P	132
B	B	P	P	130
V	V	P	P	100
Ac	Ac	P	P	2

[a] For abbreviations, see Table I.
[b] Paul, 1959.

shape of their oxygen dissociation curves in the case of the hemoglobins (cf. Martell and Calvin, 1952). It is, in fact, well known that the structure of the whole hemoprotein macromolecule is important in such respects (cf. Kaziro and Tsushima, 1961).

Whether a Fe(II) hemoprotein combines reversibly with O_2 without oxidation to the Fe(III) state (hemoglobins and myoglobins), whether it combines with H_2O_2 and catalyzes its decomposition (catalases), or whether it combines with H_2O_2 which it activates in the oxidation of substrate molecules (peroxidases) is thus determined mainly by the nature of the specific protein involved. This is the more evident when it is realized that Fe(II) or Fe(III) chelates of the same tetrapyrrole, i.e., protoporphyrin, form the prosthetic groups of all the hemoglobins, myoglobins, catalases, and peroxidases.

B. HEMOGLOBINS

Hemoglobins, then, are essentially square-planar Fe(II) protoporphyrin chelates, further coordinated with one strong field and one weak field ligand. Mixed complexes of this type have not yet been obtained in solution, because once one strong field ligand has become coordinated, the affinity for a second becomes great (cf. Fig. 7 and also Nakahara and Wang, 1958), and water is displaced from the transient mixed complex. Nature overcomes this difficulty by the use of macromolecules—the proteins—which maintain the potentially unstable mixed (protein-water) complexes in an activated state by steric restrictions. A model in which this situation has been imitated has been described by Wang (1958, 1961). Protoheme methyl ester was coordinated to 1-(2-phenylethyl)-imidazole held in a polystyrene film. The reactive sixth position of the heme was protected by CO, which was removed by evacuation after formation of the film. Reversible combination with O_2 could then be demonstrated, even in the presence of water; the O_2 complex had an oxyhemoglobinlike spectrum. The model (Fig. 14)

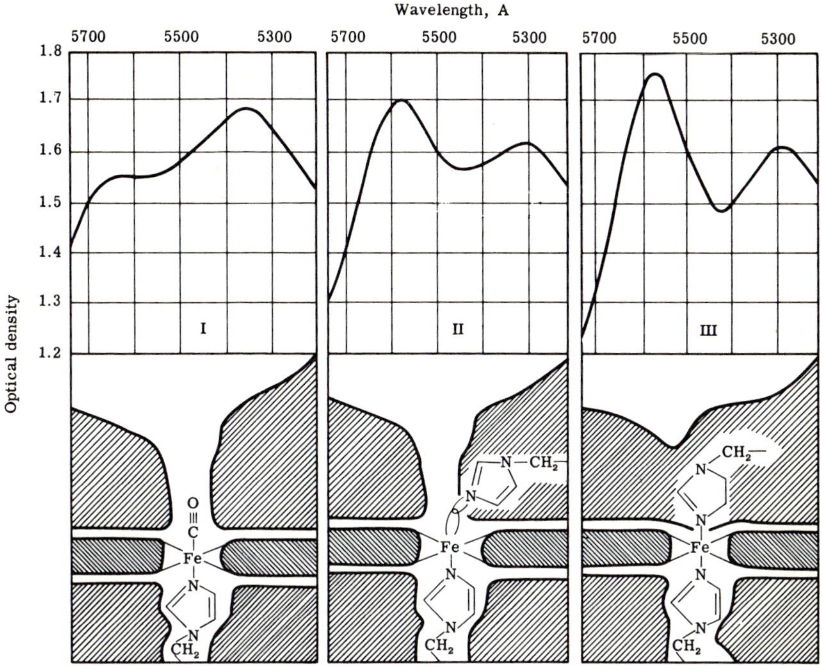

FIG. 14. A hemoglobin model. Spectra and structural diagrams of a high polystyrene content film at various stages of the experiment. Curve I, spectrum of the film saturated with carbon monoxide; Curve II, spectrum of the active form of the film; Curve III, spectrum of the same film after thermal denaturation. The schematic diagram under each curve represents the structure of the active center at the corresponding stage of the experiment. From Wang (1961).

could be "denatured" by heat, and "renatured" by heating with CO which was then removed. When hemin, with free carboxylic acid groups, was used instead of its ester, the film was inactive, possibly because the ester groups are required to anchor the heme molecules in the correct orientation to the lipophilic polystyrene.

Corwin and Erdman had found in 1946 that in anhydrous solvents hemes and some of their perpendicular complexes are passive to oxidation by O_2, and Corwin and Reyes (1956) and Corwin and Bruck (1958) have described reversible oxygenation in the solid state of crystalline bisimidazole and bispyridine complexes of proto- and mesohemes. Elvidge and Lever (1959) have found that Mn(II) phthalocyanine in pyridine solution is reversibly oxygenated, the active species probably being the square-pyramidal monopyridine complex (cf. Orgel, 1961). A variety of O_2-carrying complexes of Fe(II), Co(II), and even Re, with chelators other than porphyrins, have been described (cf. Larkworthy and Nyholm, 1959).

C. Catalases and Peroxidases

Models based on hemes have not yet been evolved for catalases or peroxidases, though all heme compounds have, in fact, some catalase and peroxidase activity (cf. Lemberg and Legge, 1949, p. 402).

Catalase and peroxidase appear to exist in tissues mainly in the Fe(III) high spin state (cf. Brill and Williams, 1961a,b). Recent evidence (Nakatani, 1960, 1961; cf. Little and Neilands, 1960) points strongly to binding of the heme iron in catalase by one or even two histidine residues of the catalase protein. This is inconsistent with the fact that in catalase the ferric state is much more highly stabilized than in peroxidase, in which the protein-to-iron bond appears to be through a protein carboxyl group (Theorell and Paul, 1944). A carboxyl ligand should, of course, stabilize the oxidized state much more than should a histidine nitrogen. On the other hand, imidazole has a high affinity for Fe(III) porphyrins, probably due to charge delocalization (cf. Phillips, 1960), and being an ampholyte may even bond to the Fe(III) of catalase through a negatively charged nitrogen atom.

An interesting aspect of these compounds is the change in properties which occurs on addition of certain extra ligands. With Fe(III) hemoglobins and myoglobins, strong field ligands (e.g., N_3^- or CN^-) are required to cause spin-pairing, with a change from high spin to low spin type (cf. Table VIII). With catalase, neither N_3^- or CN^- will cause this change (azide complex 5.86 B.M., cyanide complex 4.02 B.M.); the pairing energy of the Fe(III) ion is clearly greater in catalase than in hemoglobins or myoglobins. In the case of Fe(III) peroxidase (5.48 B.M.) in which the existing evidence indicates a —COO^--to-Fe bond in the fifth position, the replacement of water by OH^- in the sixth position converts the complex to the low spin type (2.66 B.M.) (cf. George *et al.*, 1961; Brill and Williams, 1961a). This seems surprising, since OH^- is usually found to have a weaker ligand field than water (Basolo and Pearson, 1958). Whatever the mechanism of this effect of the OH^- ion, presumably a similar mechanism is responsible for the existence of Fe(III) hemoglobin hydroxide partly in the low spin form while the aquo complex is clearly high spin (cf. Section III,F and Table VIII).

D. Cytochromes

In cytochromes of the *b* and *c* types which are concerned in electron-transport, it seems that two strong field ligand atoms of the protein are attached to the heme iron, one on each side. Mammalian cytochrome *c* is a low spin complex (Theorell, 1941), and from its spectroscopic properties cytochrome *b* also appears to be low spin. It is clear that one of the protein-iron bonds in cytochrome *c* is from a histidine residue, and the second may be from a lysine amino group in the same protein molecule (Margoliash *et al.*, 1959). The biological activity of these cytochromes involves one-electron oxidations and reductions in the "electron transport chain"; the iron is not required to react with further small molecules, and indeed cannot readily do so. Cytochrome a_3 (cytochrome oxidase), however, is not only reduced by an electron passed to it along the electron transport chain, but combines with molecular oxygen which removes the electron, bringing the cytochrome back to the ferric state. The heme group of cytochrome a_3 must, therefore, have its sixth position available for reaction with O_2. The nature of the protein-iron bond in the fifth position is not known.

An interesting cytochrome oxidase model has been described by Wang and Brinigar (1960). Fe(III) protoporphyrin was mixed with the "connecting ligand" 4,4′-dipyridyl, the hemichrome so formed reduced with $Na_2S_2O_4$, and mixed with buffer containing poly-L-lysine. A solid polymer separated, in which the heme groups were linked through dipyridyl groups to give a linear polymer which was presumably stabilized through interaction of the dissociated propionic acid side chains of the heme with cationic groups of the polylysine (Fig. 15). The finely ground polymer, suspended in buffer, was found to catalyze the oxidation of Fe(II) cytochrome *c*; it appeared "that the O_2 bound at each terminal heme group was reduced directly to water through a 4-electron transfer mechanism."

Among the heme prosthetic groups of the cytochromes there are side chain differences; in cytochrome *c* all the heme side chains are saturated, cytochromes *b* have protoheme with its two vinyl groups, and the heme of cytochromes *a* and a_3 has a vinyl and a formyl side chain. The spectroscopic and redox properties of the mammalian cytochromes *a*, *b*, and *c* do not follow the order expected from the physicochemical correlations described earlier in this chapter. When the proteins are replaced by pyridine molecules, however, the resulting hemochromes behave like other model compounds. The proteins thus influence the physical properties of these hemes profoundly.

Similar influences of the protein are seen when the physical properties are compared of those cytochromes *b* from different species which have been shown to have the same (protoheme) prosthetic group. Wide variation in spectra and in redox potential is found. A similar situation exists among cytochromes *c* (cf. Morton, 1958). These effects of the proteins have been discussed by Perrin and Falk (1961), and Falk (1961) found that the range of spectroscopic properties of these cytochromes could be obtained in some model hemochromes.

Many intriguing questions remain for the theoretical chemist and the

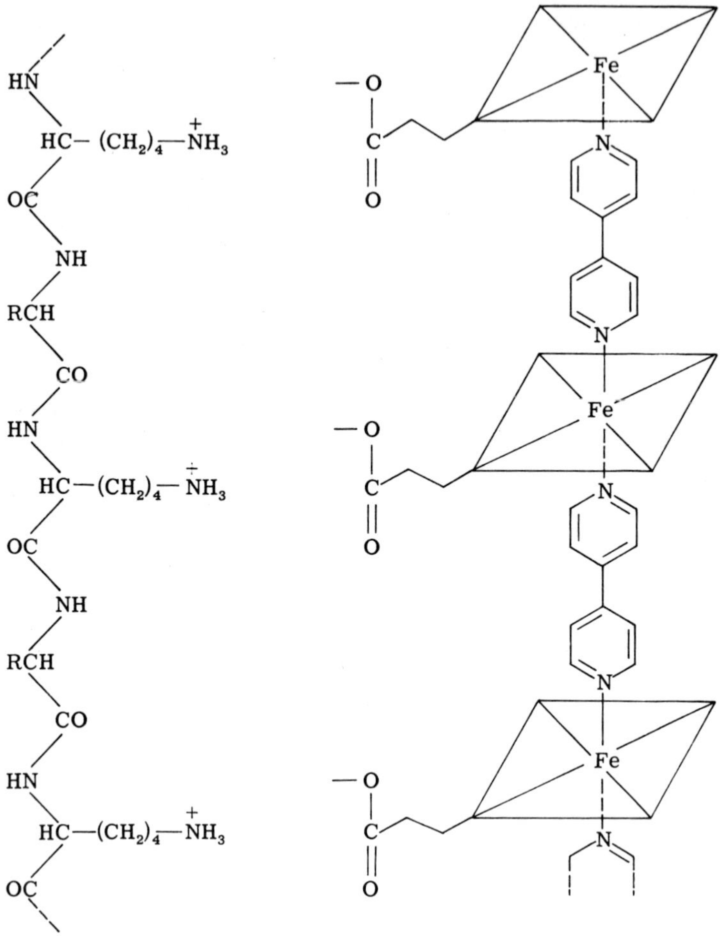

FIG. 15. Diagrammatic representation of Wang and Brinigar's (1960) cytochrome oxidase model. Fe(III) protoporphyrin (protohemin) molecules are polymerized through the "connecting ligand" 4,4′-dipyridyl, and the polymer is stabilized by charge interaction between the carboxylic side chains of the hemin and basic groups of polylysine.

coordination chemist. Thus, in spite of a number of theories, no really satisfactory explanation exists for the passivity of Fe(II) hemoglobin to oxidation when it is oxygenated; for the mechanism of the dissociation by light of the CO complex of cytochrome a_3; for the marked variations in spectrum and redox potential of the same heme when linked with different species-specific proteins as in the cytochromes *b* or *c*, and for many other interesting phenomena. It is perhaps not unjust to claim that the stimulus of such questions has led to many developments in our experimental and intellectual techniques; it is clear that research in this field is in a stage of great acceleration and promises many exciting advances.

References

Abraham, R. J. (1961). *Mol. Phys.* **4,** 145.

Abraham, R. J., Jackson, A. H., and Kenner, G. W. (1961). *J. Chem. Soc.* p. 3468.

Allison, J. D., and Becker, R. S. (1960). *J. Chem. Phys.* **32,** 1410.

Aronoff, S. (1958). *J. Phys. Chem.* **62,** 428.

Barker, H. A., Smyth, R. D., Weissbach, H., Munch-Petersen, A., Toohey, J. I., Rodd, J. N., Volcani, B. E., and Wilson, R. M. (1960). *J. Biol. Chem.* **235,** 181.

Barron, E. S. G. (1937). *J. Biol. Chem.* **121,** 285.

Basolo, F., and Pearson, R. G. (1958). "Mechanisms of Inorganic Reactions." Wiley, New York.

Becker, E. D., and Bradley, R. B. (1959). *J. Chem. Phys.* **31,** 1413.

Becker, E. D., Bradley, R. B., and Watson, C. J. (1961). *J. Am. Chem. Soc.* **83,** 3743.

Bogorad, L. (1960). *In* "Comparative Biochemistry of Photoreactive Systems" (M. B. Allen, ed.), p. 227. Academic Press, New York.

Borg, D. C., and Cotzias, G. C. (1958). *Nature* **182,** 1677.

Brill, A. S., and Williams, R. J. P. (1961a). *Biochem. J.* **78,** 246.

Brill, A. S., and Williams, R. J. P. (1961b). *Biochem. J.* **78,** 253.

Caughey, W. S., and Koski, W. S. (1962). *Biochemistry* **1,** 923.

Caughey, W. S., Deal, R. M., McLees, B. D., and Alben, J. O. (1962). *J. Am. Chem. Soc.* **84,** 1735.

Clezy, P. S., and Barrett, J. (1961). *Biochem. J.* **78,** 798.

Conant, J. B. (1932). *Harvey Lectures* **28,** 159.

Conant, J. B., Chow, B. F., and Dietz, E. M. (1934). *J. Am. Chem. Soc.* **56,** 2185.

Corwin, A. H., and Bruck, S. D. (1958). *J. Am. Chem. Soc.* **80,** 4736.

Corwin, A. H., and Erdman, J. G. (1946). *J. Am. Chem. Soc.* **68,** 2473.

Corwin, A. H., and Reyes, Z. (1956). *J. Am. Chem. Soc.* **78,** 2437.

Coryell, C. D. (1939). *J. Phys. Chem.* **43,** 841.

Coryell, C. D., and Stitt, F. (1940). *J. Am. Chem. Soc.* **62,** 2942.

Coryell, C. D., Stitt, F., and Pauling, L. (1937). *J. Am. Chem. Soc.* **59,** 633.

Cowgill, R. W., and Clark, W. M. (1952). *J. Biol. Chem.* **198,** 33.

Craig, D. P., Maccoll, A., Nyholm, R. S., Orgel, L. E., and Sutton, L. E. (1954). *J. Chem. Soc.* p. 332.

Craven, C. W., Reissman, K. R., and Chinn, H. I. (1952). *Anal. Chem.* **24,** 1214.

Crute, M. B. (1959). *Acta Cryst.* **12,** 24.

Dempsey, B., Lowe, M. B., and Phillips, J. N. (1961). *In* "Haematin Enzymes" (J. E. Falk, R. Lemberg, and R. K. Morton, eds.), p. 29. Pergamon Press, London.

Dorough, G. D., Miller, J. R., and Huennekens, F. M. (1951). *J. Am. Chem. Soc.* **73,** 4315.

Drabkin, D. L. (1942). *Ann. Rev. Biochem.* **11,** 531.

Eggerer, H., Stadtman, E. R., Overath, P., and Lynen, F. (1960). *Biochem. Z.* **333,** 1.

Ellis, J., Jackson, A. H., Kenner, G. W., and Lee, J. (1960). *Tetrahedron Letters* **2,** 23.

Elvidge, J. A., and Lever, A. B. P. (1959). *Proc. Chem. Soc.* p. 195.

Elvidge, J. A., and Lever, A. B. P. (1961). *J. Chem. Soc.* p. 1257.

Falk, J. E. (1961). *In* "Haematin Enzymes" (J. E. Falk, R. Lemberg, and R. K. Morton, eds.), p. 74. Pergamon Press, London.

Falk, J. E. (1963). *In* "Comprehensive Biochemistry" (M. Florkin and E. Stotz, eds.) Vol. 9, p. 3. Elsevier, Amsterdam.

Falk, J. E., and Nyholm, R. S. (1958). *In* "Current Trends in Heterocyclic Chemistry" (A. Albert, G. M. Badger, and C. W. Shoppee, eds.) Butterworths, London.
Falk, J. E., and Perrin, D. D. (1961). *In* "Haematin Enzymes" (J. E. Falk, R. Lemberg, and R. K. Morton, eds.), p. 56. Pergamon Press, London.
Falk, J. E., and Willis, J. B. (1951). *Australian J. Sci. Research* **A4,** 579.
Falk, J. E., Lemberg, R., and Morton, R. K. (1961). "Haematin Enzymes". Pergamon Press, London.
Fernandez, J., and Becker, R. (1959). *J. Chem. Phys.* **31,** 467.
Fischer, H., and Orth, H. (1937). "Die Chemie des Pyrrols," Vol. 2, Part I. Akademische Verlagsges., Leipzig.
Fischer, H., and Stern, A. (1940). "Die Chemie des Pyrrols," Vol. 2, Part 2. Akademische Verlagsges., Leipzig.
Fischer, H., Treibs, A., and Zeile, K. (1931). *Z. physiol. Chem. Hoppe-Seyler's* **195,** 1.
Fleischer, E. B. (1963a). *J. Amer. Chem. Soc.,* **85,** 146.
Fleischer, E. B. (1963b). *J. Amer. Chem. Soc.,* **85,** 1353.
George, P., Beetlestone, J., and Griffith, J. S. (1961). *In* "Haematin Enzymes" (J. E. Falk, R. Lemberg, and R. K. Morton, eds.), p. 105. Pergamon Press, London.
Gibson, J. F., Ingram, D. J., and Schonland, D. (1958). *Discussions Faraday Soc.,* No. 26, p. 72.
Goldberg, A., Aschenbrucker, H., Cartwright, G. E., and Wintrobe, M. M. (1956). *Blood* **11,** 821.
Gouterman, M. (1959). *J. Chem. Phys.* **30,** 1139.
Gouterman, M. (1960). *J. Chem. Phys.* **33,** 1523.
Gouterman, M. (1961). *J. Mol. Spectroscopy* **6,** 138.
Granick, S. (1961). *J. Biol. Chem.* **236,** 1168.
Granick, S., and Gilder, H. (1946). *J. Gen. Physiol.* **30,** 1.
Granick, S., and Mauzerall, D. (1961). *In* "Metabolic Pathways" (D. M. Greenberg, ed.), p. 525. Academic Press, New York.
Griffith, J. S. (1956). *Proc. Roy. Soc.* **A235,** 23.
Griffith, J. S. (1958). *Discussions Faraday Soc.* **No. 26,** 81.
Hall, N. F. (1930). *J. Am. Chem. Soc.* **52,** 5115.
Harned, H. S., and Owen, B. E. (1943). "The Physical Chemistry of Electrolyte Solutions," p. 547. Reinhold, New York.
Haurowitz, F. (1935). *Ber.* **68,** 1795.
Haurowitz, F. (1951). *J. Biol. Chem.* **193,** 443.
Havemann, R., Haberditzl, W., and Mader, K. H. (1961). *Z. phys. Chem.* **218,** 71.
Hill, R. (1929). *Proc. Roy. Soc.* **B105,** 112.
Hodgkin, D. C., Kamper, J., Lindsey, J., Mackay, M., Pickworth, J., Robertson, J. H., Shoemaker, C. B., White, J. G., Prosen, R. J., and Trueblood, K. N. (1957). *Proc. Roy. Soc.* **A242,** 228.
Holden, H. F., and Freeman, M. (1929). *Australian J. Exptl. Biol. Med. Sci.* **6,** 79.
Kaziro, K., and Tsushima, K. (1961). *In* "Haematin Enzymes" (J. E. Falk, R. Lemberg, and R. K. Morton, eds.), p. 80. Pergamon Press, London.
Keilin, D. (1933). *Proc. Roy. Soc.* **B113,** 393.
Keilin, D., and Hartree, E. F. (1951). *Biochem. J.* **49,** 88.
Keilin, D., and Hartree, E. F. (1952). *Nature* **170,** 161.
Keilin, J. (1949a). *Biochem. J.* **45,** 440.
Keilin, J. (1949b). *Biochem. J.* **45,** 448.
Kendrew, J. C. (1963). *Science* **139,** 1259.
Kendrew, J. C., Dickerson, R. E., Strandberg, B. E., Hart, R. G., Davies, D. R., Phillips, D. C., and Shore, V. C. (1960). *Nature* **185,** 4711.

Klemm, W. (1935). *Angew. Chem.* **48,** 617.
Kneubühl, F. K., Koski, W. S., and Caughey, W. S. (1960). *J. Am. Chem. Soc.* **83,** 1607.
Krueger, R. C., Melnick, I., and Klein, J. R. (1956). *Arch. Biochem. Biophys.* **64,** 302.
Labbe, R. F., and Hubbard, N. (1961). *Biochim. et Biophys. Acta* **52,** 130.
Larkworthy, L. F., and Nyholm, R. S. (1959). *Nature* **183,** 1377.
Lemberg, R. (1961). *In* "Haematin Enzymes" (J. E. Falk, R. Lemberg, and R. K. Morton, eds.), p. 216. Pergamon Press, London.
Lemberg, R., and Falk, J. E. (1951). *Biochem. J.* **49,** 674.
Lemberg, R., and Legge, J. W. (1949). "Haematin Compounds and Bile Pigments." Wiley (Interscience), New York.
Linschitz, H., and Pekkarinen, L. (1960). *J. Am. Chem. Soc.* **82,** 2411.
Linschitz, H., and Sarkanen, K. (1958). *J. Am. Chem. Soc.* **80,** 4826.
Little, H. N., and Neilands, J. B. (1960). *Nature* **188,** 913.
Livingstone, R., and Weil, S. (1952). *Nature* **170,** 750.
Livingstone, R., Watson, W. F., and McArdle, J. (1949). *J. Am. Chem. Soc.* **71,** 1542.
Lochhead, A. C., and Goldberg, A. (1961). *Biochem. J.* **78,** 146.
Longuet-Higgins, H. C., Rector, C. W., and Platt, J. R. (1950). *J. Chem. Phys.* **18,** 1174.
Lowe, M. B., and Phillips, J. N. (1961). *Nature* **190,** 262.
Lowe, M. B., and Phillips, J. N. (1962). *Nature* **194,** 1058.
McConnell, R. J., Overell, B. G., Petrow, V., and Sturgeon, B. (1953). *J. Pharm. and Pharmacol.* **5,** 179.
McEwen, W. K. (1936). *J. Am. Chem. Soc.* **58,** 1124.
Margoliash, E. (1961). *Ann. Rev. Biochem.* **30,** 549.
Margoliash, E., Frohwirt, N., and Wiener, E. (1959). *Biochem. J.* **71,** 559.
Martell, A. E., and Calvin, M. (1952). "The Chemistry of the Metal Chelate Compounds." Prentice-Hall, Englewood Cliffs, New Jersey.
Mason, S. F. (1958). *J. Chem. Soc.* p. 976.
Miller, J. R., and Dorough, G. D. (1952). *J. Am. Chem. Soc.* **74,** 3977.
Minakami, S. (1958). *J. Biochem. (Tokyo)* **45,** 833.
Minakami, S., Yoneyama, Y., and Yoshikawa, H. (1958). *Biochim. et Biophys. Acta* **28,** 447.
Morton, R. K. (1958). *Revs. Pure and Appl. Chem. (Australia)* **8,** 161.
Nakahara, A., and Wang, J. H. (1958). *J. Am. Chem. Soc.* **80,** 6526.
Nakatani, M. (1960). *J. Biochem. (Tokyo)* **48,** 476.
Nakatani, M. (1961). *J. Biochem. (Tokyo)* **49,** 98.
Neilands, J. B., and Tuppy, H. (1960). *Biochim. et Biophys. Acta* **38,** 351.
Neuberger, A., and Scott, J. J. (1952). *Proc. Roy. Soc.* **A213,** 307.
Nevé, R. A. (1961). *In* "Haematin Enzymes" (J. E. Falk, R. Lemberg, and R. K. Morton, eds.), p. 207. Pergamon Press, London.
Nishida, G., and Labbe, R. F. (1959). *Biochim. et Biophys. Acta* **31,** 519.
O'Hagan, J. E. (1961). *In* "Haematin Enzymes" (J. E. Falk, R. Lemberg, and R. K. Morton, eds.), p. 173. Pergamon Press, London.
Orgel, L. E. (1961). *In* "Haematin Enzymes" (J. E. Falk, R. Lemberg, and R. K. Morton, eds.), p. 104. Pergamon Press, London.
Orlando, J. A. (1958). Ph.D. Thesis, University of California, Berkeley.
Oyama, H., Sugita, Y., Yoneyama, Y., and Yoshikawa, H. (1961). *Biochim. et Biophys. Acta* **47,** 413.
Paul, K.-G. (1959). *Acta Chem. Scand.* **13,** 1239.
Paul, M. A., and Long, F. A. (1957). *Chem. Revs.* **57,** 1.
Pauling, L., and Coryell, C. D. (1936). *Proc. Natl. Acad. Sci. N. Y.* **22,** 159.

Perrin, D. D. (1959). *J. Chem. Soc.* p. 290.
Phillips, J. N. (1958). *In* "Current Trends in Heterocyclic Chemistry" (A. Albert, G. M. Badger, and C. W. Shoppee, eds.) p. 30. Butterworths, London.
Phillips, J. N. (1960). *Revs. Pure & Appl. Chem.* (*Australia*) **10,** 35.
Phillips, J. N. (1963). *In* "Comprehensive Biochemistry" (M. Florkin and E. Stotz, eds.) Vol. 9, p. 34. Elsevier, Amsterdam.
Plant, G. W. E. (1961). *Ann. Rev. Biochem.* **30,** 409.
Platt, J. R. (1956). *In* "Radiation Biology," Vol. 3, p. 101. McGraw-Hill, New York.
Porra, R. J., and Falk, J. E. (1961). *Biochem. Biophys. Research Communs.* **5,** 179.
Porra, R. J., and Falk, J. E. (1963). *Biochem. J.* **90,** 69.
Porra, R. J., and Jones, O. T. G. (1963). *Biochem. J.* **87,** 186.
Rimington, C. (1958). *Revs. Pure and Appl. Chem.* (*Australia*) **8,** 129.
Rimington, C., Mason, S. F., and Kennard, O. (1958). *Spectrochim. Acta* **12,** 65.
Roberts, E. M., and Koski, W. S. (1960). *J. Am. Chem. Soc.* **82,** 3006.
Roberts, E. M., Koski, W. S., and Caughey, W. S. (1961). *J. Chem. Phys.* **33,** 591.
Robertson, J. M. (1936). *J. Chem. Soc.* p. 1195.
Scheler, W., Schoffa, G., and Jung, F. (1957). *Biochem. Z.* **329,** 232.
Scheler, W., Schoffa, G., and Jung, F. (1958). *Biochem. Z.* **330,** 538.
Schwartz, S., Zieve, L., and Watson, C. J. (1951). *J. Lab. Clin. Med.* **37,** 843.
Selwood, P. W. (1956). "Magnetochemistry," 2nd ed. Wiley (Interscience), New York.
Senff, H., and Klemm, W. (1939). *J. prakt. Chem.* **154,** 73.
Shemin, D. (1956). *In* "Currents in Biochemical Research" (D. E. Green, ed.), p. 518. Interscience, New York.
Sidorov, A. N., and Terenin, A. N. (1960). *Optics and Spectroscopy* (*USSR*) *English Transl.* **8,** 254.
Spooner, H., and Teller, E. (1941). *Revs. Modern Phys.* **13,** 75.
Tait, G. H., and Gibson, K. D. (1961). *Biochim. et Biophys. Acta* **52,** 614.
Taylor, J. F. (1940). *J. Biol. Chem.* **135,** 569.
Theorell, H. (1941). *J. Am. Chem. Soc.* **63,** 1820.
Theorell, H. (1945). *Nature* **156,** 474.
Theorell, H., and Ehrenberg, A. (1951). *Acta Chem. Scand.* **5,** 823.
Theorell, H., and Paul, K.-G. (1944). *Arch. Biochem.* **A18,** No. 12.
Thomas, D. W., and Martell, A. E. (1958). *Arch. Biochem. Biophys.* **76,** 286.
Wang, J. H. (1958). *J. Am. Chem. Soc.* **80,** 3168.
Wang, J. H. (1961). *In* "Haematin Enzymes" (J. E. Falk, R. Lemberg, and R. K. Morton, eds.), p. 98. Pergamon Press, London.
Wang, J. H., and Brinigar, W. S. (1960). *Proc. Natl. Acad. Sci. U. S.* **46,** 958.
Wang, J. H., Nakahara, A., and Fleischer, E. B. (1958). *J. Am. Chem. Soc.* **80,** 1109.
Weigl, J. W. (1957). *J. Mol. Spectroscopy* **1,** 133.
Wetherell, H. R., Hendrickson, M. J., and McIntyre, A. R. (1959). *J. Am. Chem. Soc.* **81,** 4517.
Wilkinson, G. (1951). *Nature* **168,** 514.
Williams, R. J. P. (1956). *Chem. Revs.* **56,** 299.
Willstätter, R., and Stoll, A. (1913). "Untersuchungen über Chlorophyll." Springer, Berlin.
Wolstenholme, G. E. W., and Millar, E. C. P. (1955). "The Biosynthesis of Porphyrins and Porphyrin Metabolism." Churchill, London.

Author Index

Numbers in italics indicate the page on which the reference is listed.

F

G

H

I

J

M

T

U

V

W

Subject Index

Single-letter prefixes like *o*- and *m*- are ignored in arranging compounds in alphabetical order.

Metal chelates are listed under the appropriate chelating agent.

Where two chelating agents are involved (mixed metal chelates), the compound is listed under the chelating agent which is first according to the rules of nomenclature (see page 4); e.g., acetylacetonatobis-2,2′-bipyridine ruthenium chloride is listed under acetylacetone.

Metal chelates in which a unidentate ligand is present are sometimes listed under the unidentate ligand, e.g., oxymetalloporphyrins. This convention is usually restricted to classes of compounds like the metalloporphyrins. Individual compounds like chloroaquobisethylenediamine cobalt(III) chloride are listed under the appropriate chelating agent, in this case, ethylenediamine.

Many of the more complicated multidentate chelating agents derived from two or more simpler chelating agents are not listed under their systematic names but under the names of either or both of the simpler constituents, e.g., the tridentate from salicylaldehyde and (*o*-aminophenol) is listed under salicylaldehyde.

A

C

D

E

H

I

K

L

M